KT-199-560

APPLIED MODELING IN CATCHMENT HYDROLOGY

Edited by

Vijay P. Singh

WATER RESOURCES PUBLICATIONS

UNIVERSITY LIBRARY
- 5 ''' 1983
LANCASTER

For information and correspondence:
WATER RESOURCES PUBLICATIONS
P.O. Box 2841
Littleton, Colorado 80161, U.S.A.

APPLIED MODELING IN CATCHMENT HYDROLOGY

Proceedings of the International Symposium on Rainfall-Runoff Modeling held May 18-21, 1981 at Mississippi State University, Mississippi State, Mississippi, U.S.A.

Edited by
Vijay P. Singh

Department of Civil Engineering
Louisiana State University
Baton Rouge, Louisiana 70803, U.S.A.

ISBN-0-918334-43-8

U.S. Library of Congress Catalog Card Number 81-71292

Copyright © 1982 by Water Resources Publications. All rights reserved. Printed in the United States of America. The text of this publication may not be reproduced, stored in a retrieval system, or transmitted, in any form or by any means, without a written permission from the Publisher.

This publication is printed and bound by BookCrafters, Inc., Chelsea, Michigan, U.S.A.

82 009994

PREFACE

In the last three decades there has been a proliferation of research on rainfall-runoff modeling. As a result there exists an abundance of literature in this area. As we enter into a new decade with new possibilities and challenges, it appears appropriate to pause and determine where we are, where we are going, where we ought to be going, and what the most outstanding problems are that ought to be addressed on a priority basis for rapid progress of hydrology. To address these issues in a scientific forum is what constituted essentially the rationale for organizing the International Symposium on Rainfall-Runoff Modeling which was held May 18-21, 1981 at Mississippi State University, Mississippi State, Mississippi.

The objectives of this Symposium were therefore (1) to assess the state of the art of rainfall-runoff modeling, (2) to demonstrate the applicability of current models, (3) to determine directions for future research, (4) to assemble unreported research, (5) to establish complementary elements of seemingly different approaches, and (6) to augment interdisciplinary interaction.

We received an overwhelming response to our call for papers. It was indeed a difficult task to select among the many excellent papers that were submitted, and we regret that we could not include all of them in the Symposium program. The sole criterion for selection of a paper was its merit in relation to the Symposium objectives. The subject matter of the Symposium was divided into 26 major topics encompassing virtually the entire spectrum of rainfall-runoff cycle. Each topic entailed an invited state-of-the-art paper and a number of contributed papers. These contributions blended naturally to evolve a synthesized body of knowledge on that topic. Extended abstracts of all the invited and contributed papers were assembled in a pre-Symposium proceedings volume. Each registered Symposium participant was given this volume. This helped stimulate discussion and exchange of ideas during the Symposium.

The papers presented at the Symposium were refereed in a manner similar to that employed for publishing a journal article. As a result, nearly 40 percent of the papers did not pass the review and were therefore eliminated from inclusion in the final procedings. The accepted papers were divided in four parts. The papers contained in this book, APPLIED MODELING IN CATCHMENT HYDROLOGY, represent one part of the Symposium contributions. The other parts are embodied in three separate books, RAINFALL-RUNOFF RELATIONSHIP, MODELING COMPONENTS OF HYDROLOGIC CYCLE and STATISTICAL ANALYSIS OF RAINFALL AND RUNOFF, which are being published simultaneously. Arrangement of papers in these books under four different titles was a natural consequence of the diversity of techical material discussed in these papers. These books can be treated almost independently, although some overlap does exist between them.

This book contains six sections. Each section starts normally with an invited state-of-the-art paper followed by contributed papers. Beginning with development of various catchment models the papers go on to discuss evaluation of catchment models, application of catchment models, application in urban environment, application in forest environment and remote sensing applications.

The book will be of interest to researchers as well as those engaged in practice of Civil Engineering, Agricultural Engineering, Hydrology, Water Resources, Earth Resources, Forestry, and Environmental Sciences. The graduate students as well as those wishing to conduct research in rainfall-runoff modeling will find this book to be of particular significance.

I wish to take this opportunity to express my sincere appreciation to all the members of the Organizing Committee and the Mississippi State University administration for their generous and timely help in the organization of the symposium. A lack of space does not allow me to list all of them here, but I would like to single out Dr. Victor L. Zitta who chaired the local arrangements pertaining to the Symposium. Numerous other people contributed to the Symposium in one way or another. The authors, including the invited speakers, contributed to the Symposium technically and made it what it was. The session chairmen and co-chairmen administered the sessions in a positive and professional manner. The referees took time out from their busy schedules and reviewed the papers. I owe my sincere gratitude to all these individuals.

If the success of a Symposium is measured in terms of the quality of participants and presentations then most people would agree that this Symposium was a resounding success. A very large number of inter-nationally well-known people, who have long been recognized for their contributions and have long been at the forefront of hydrologic research, came to participate in the Symposium. More than 25 countries, covering the five continents and most of the countries of the world active in hydrologic research, were represented. It is hoped that many long and productive friendships will develop as a result of this Symposium.

Vijay P. Singh
Symposium Director

iv

TABLE OF CONTENTS

SECTION 3   APPLICATION OF CATCHMENT MODELS

SECTION 4   APPLICATION IN URBAN ENVIRONMENT

ACKNOWLEDGEMENTS

The International Symposium on Rainfall-Runoff Modeling was sponsored and co-sponsored by a number of organizations. The sponsors supported the Symposium financially without which it might not have come to fruition. Their financial support is gratefully acknowledged. The co-sponsors extended their help in announcing the Symposium through their journals, transactions, newletters or magazines. This publicity helped increase participation in the Symposium, and is sincerely appreciated. The following is the list of Symposium sponsors and co-sponsors.

SYMPOSIUM SPONSORS

National Science Foundation

U.S. Department of Agriculture, Science and Education Administration

U.S. Army Research Office

U.S. Department of the Interior, Office of Water Research and Technology

United Nations Educational Scientific and Cultural Organization

Mississippi State University, Agricultural and Forestry Experiment Station, Department of Civil Engineering, Office of Graduate Studies and Research, Water Resources Research Institute

Mississippi-Alabama Sea Grant Consortium

SYMPOSIUM CO-SPONSORS

American Geophysical Union

American Society of Civil Engineers, Hydraulics Division and Irrigation and Drainage Division

Institute of Hydrology, Great Britain

International Association of Hydrological Sciences

International Union of Forestry Research Organizations

International Water Resources Association

Mississippi State Highway Department

Regional Committee for Water Resources in Central America

Tennessee-Tombigbee Water Development Authority

U.S. Army Corps of Engineers, Vicksburg District

U.S. Department of Agriculture, Forest Service

U.S. Geological Survey, Water Resources Division, Gulf Coast Hydroscience Center; and Water Resources Division, Jackson, Mississippi

Universidad de San Carlos de Guatemala

Section 1
MODELING OF CATCHMENT HYDROLOGY

A FAST-TRANSIENT, TWO-DIMENSIONAL, DISCRETE-ELEMENT RAINFALL-RUNOFF MODEL FOR CHANNELIZED, COMPOSITE SUBSURFACE-SURFACE FLOWS IN VALLEYS WITH STEEP TERRAIN

Arsev H. Eraslan, Ph.D.
Professor, Department of Engineering Science and Mechanics
The University of Tennessee, Knoxville, Tennessee 37916 U.S.A.

Ismail H. Erhan
Senior Environmental Scientist, Sciences Division
Henningson Durham and Richardson, Suite 400, 9051 Executive Park Drive
Knoxville, Tennessee 37923 U.S.A.

Wen L. Lin, Ph.D.
Deputy Chief Scientist, Environmental Research Application Systems
ERAS, Inc., 1826 United American Plaza
Knoxville, Tennessee 37927 U.S.A.

ABSTRACT

A fast-transient, two-dimensional discrete-element mathematical model is presented for simulating channelized, composite subsurface-surface rainfall-runoff flow conditions in valleys with steep terrain. The state-of-the-art, direct computational model was developed for the assessment of the potential hydrological impact of the construction and deployment of the proposed M-X missile system in valleys in Nevada and Utah which constitute a major part of the Great Basin physiographic region. The development of the mathematical model incorporated all the special characteristics of rainfall-runoff flow, including dry-ground conditions and three distinct flow regimes, consisting of (1) initial subsurface groundwater flow, (2) subsequent first-level composite flow, as the combination of subsurface groundwater flow and surface-water flow in water courses with directional channels, and (3) final second-level composite flow, as the combination of subsurface groundwater flow, surface-water flow in water courses with directional channels and surface-layer flow above the ground. The results of application of the mathematical model and the preliminary development-program version of its associated RAINER computer code indicated that the construction of an M-X cluster group could alter the natural rainfall-runoff flow conditions; and consequently, it could cause short-term (and possibly long-term) impact on the geohydrological conditions in the valley and on the recharge characteristics of the groundwater in the aquifer systems.

INTRODUCTION

The geohydrological conditions in the valleys of the Great Basin physiographic region in Nevada and Utah control the availability of water (1) to the local vegetation cover, (2) to the playas, and (3) to the individual drainage systems that recharge the groundwater in shallow aquifers (perched) and in deep aquifers (in carbonate rock formations) which cumulatively form the regional groundwater storage system. Hence, the alteration of the geohydrological conditions in a valley could lead to the alteration of one or more ecologically important aspects of water availability in the region.

Recharge conditions of the groundwater in most of valleys in the Great Basin physiographic region depend on the limited amount of precipitation in the forms of (1) snowmelt runoff during winter and spring, and (2) rainfall runoff during spring and summer. The precipitation conditions in spring and summer are generally in the form of localized thundershowers which supply large amounts of water for short periods of time. The contribution of thundershowers to the recharge conditions of the groundwater can be substantial, particularly in a hydrographicaly closed (or only surficially closed) valley which collects water from precipitation on mountains.

Rainfall-runoff flows from precipitation on the valley surface and on the mountains do not infiltrate directly and uniformly into the stored groundwater in the deep aquifer systems. In a typical valley, the existence of transverse faults, clay layers and other geological formations generally inhibit the uniform infiltration of groundwater into the deep aquifer, except in regions where the geological structure of the valley floor has the necessary cracks, faults and solution openings (in the carbonate rocks). In a large number of valleys, particularly in hydrographically closed (or only surficially closed) valleys, the areas where the surface and subsurface flows can infiltrate into the groundwater in the deep aquifer systems are generally located at the topographically lowest areas.

The loss by evaporation is generally negligible in the groundwater stored in the deep aquifer system of a valley. However, the groundwater collected from the precipitation in the shallow subsurface layer can loose substantial amounts of water by direct evaporation or by transpiration by plants (phreatophytes) in the low humidity and high temperature climate of a typical valley of the Great Basin physiological region. The recharge of the groundwater in the deep aqufier system of a valley is always adversely affected by the evaporation losses which occur through the ground surface.

Rainfall-runoff flow conditions are generally controlled by the local channelization characteristics of the ground surface which can include the effects of large, relatively deep, single water courses and also the effects of combined, highly porous fill in the valley floor. The effect of the slope of the ground surface, the slopes of the water courses, and the spatial variation of the slope of the impermeable layers in the valley fill also control rainfall-runoff flow conditions.

Rainfall-runoff flow conditions are critically important during relatively short periods of heavy rainfall (thundershowers) which could occur in different regions of a valley at different times. Since spatially uniform, relatively moderate rainfall conditions over the entire surface area do not occur for relatively long periods of time, rainfall-runoff flow conditions do not exhibit gradually developing characteristics in the valley.

Considering the large number of geological and geomorphological characteristics that control the rainfall-runoff flow conditions, it becomes evident that the alteration of one or more of these characteristics could lead to substantial alterations of the geohydrological conditions in a valley which in turn could lead to alterations of the recharge characteristics of the aquifer systems in the region.

Overview of the Presently Available Mathematical Models

At present the analysis of the rainfall-runoff flow conditions are generally based on the applications of the computer programs associated with surface-water flow models which were developed for the simulations

4

of surface-water hydrographs in river channels and flood plains.

The most commonly employed computer simulation models are the various versions of the Flood Hydrograph Package (HEC-1 ,1973, 1976, 1978) which was developed by U.S. Army Corps of Engineers. This mathematical model has the capability to simulate flood hydrographs that result from rainfall runoff conditions in a complex network of multi-channel river systems with multi-basin drainage areas.

A similar mathematical model for networks of multi-channel river systems (Burnash, et al, 1973) was developed by the Joint Federal-State Flood Forecasting Center of California. This model was extensively used for predicting regional flooding conditions on a real-time basis.

The state of the art in mathematical modeling of surface-water flows have advanced significantly in recent years. There exists numerous transient, multi-dimensional hydrodynamic models which were developed and applied for the simulation of the natural and plant-induced flow conditons in lakes, rivers, reservoirs, estuaries and coastal zones (e.g., Eraslan, 1973, 1981a, 1981b, 1981c, Lin and Eraslan, 1981, also Sundermann and Holz, 1980, and Johnson, 1981)

The mathematical models which were developed for the analysis of surface-water flow problems cannot be applied directly to the simulation of rainfall-runoff flow conditions in valleys with steep terrain. These models do not include any provision for incorporating local dry-bed conditions which always occur in valleys with steep terrain under fast-transient, intermittent rainfall conditions. Furthermore, the surface-water flow models cannot include the effects of the subsurface ground-water flow conditions which always occur in valleys with relatively porous valley fill and with large numbers of small channelized water courses. Another limitation of the surface-water flow models is their inability to simulate fast-transient downhill flow conditions which always occur in valleys with steep terrain during short periods of heavy rainfall on different parts of the valley surface and on the mountains which enclose the valley.

A historically significant study which attempted the coupling of surface-water flow with groundwater flow was presented by Pinder and Sauer (1971). The mathematical development of this model considered one-dimensional surface-water flow conditions in a straight rectangular open channel which was enclosed on the sides by a homogeneous, isotro-pic, unconfined aquifer. The governing mathematical systems for one-dimensional open-channel flow and two-dimensional groundwater flow were solved simultaneously to study the effects of bank storage on the flood hydrograph for a hypothetical flood plain. The results of the numerical solutions for the hypothetical case study indicated that the effect of the groundwater flow through the porous aquifer could be considerable in controlling the surface-water flow conditions in the stream channel.

A detailed mathematical model which includes the one-dimensional surface-water flow with groundwater flow was presented by Freeze (1972). The mathematical formulation of the model considers transient, three-di-mensional, unsaturated-saturated subsurface flow and gradually varying, one-dimensional open-channel flow for studying the contributions of the base flow to perennial flow in streams. The applications of the model to a hypothetical drainage basin indicated that the duration and rate of rainfall, hydrogeological configuration, saturated permeabilities and unsaturated characteristics of the ground considerably influenced the surface-water flow conditions in streams.

Detailed site-specific studies concerned with the applications of mathematical models to coupled surface-water flow and groundwater flow problems were completed by Cooley and Westphal (1974). These studies were mainly concerned with the comparisons of the predictive capabi-lities of different models with different levels of complexity in

5

application to the simulation of the same hydrological conditions in selected reaches of the Humbolt River in northern Nevada. The results of the study concluded that if appropriate geophysical and geohydrological data could be obtained as input to the models, the results of the different models were quite similar in predicting the general surface-water and groundwater flow conditions in the region.

A recent mathematical model for coupled transient, one-dimensional open-channel flow and transient two-dimensional groundwater flow in an unconfined aquifer was developed by Cunningham (1977) based on the application of the finite-element method (FEM). The model was applied to a site-specific problem concerned with the subsurface and surface flow conditions in the region of the Truckee River in northern Nevada both for calibrating the model and also for investigating the predictive capabilities of its computer simulation results. The model was also applied to the simulation of the geohydrological conditions in a hypothetical region which consisted of a stream bounded on both sides by a large unconfined aquifer with closed enclosure boundaries.

All mathematical models for simulating coupled surface-water flow and groundwater flow conditions (Pinder and Sauer, 1971, Freeze, 1972 and Cunningham, 1977) consider distinctly separated hydrological regions where the mathematical modeling of the flow conditions are formulated differently. The mathematical development of any of the existing, advanced models considers both the conservation of mass principle and the momentum principle in formulating the governing system of computational equations for the hydrodynamic conditions in the open-channel, which constitutes the surface-water hydrological region of the system. However, the mathematical development of the groundwater flow conditions in the aquifer considers only the mass conservation principle, supplemented by the subsidiary law of Dupuit-Forchheimer assumption which represents the generalized form of the Darcy's Law (Bear, 1972). The solutions for the surface-water flow conditions are coupled with the solutions for the groundwater flow conditions only at the common boundary of the distinctly separated hydrological regions.

Since the governing mathematical system for transient, one-dimensional free-surface flow conditions does not apply directly to the groundwater flow conditions in the aquifer, and similarly, since the governing mathematical system for transient multi-dimensional groundwater flow conditions cannot apply to the surface-flow conditions in the open channel, two distinctly separated hydrological regions can only have two distinctly different flow regimes at all times, according to the mathematical formulations of all presently available mathematical models for the simulation of rainfall-runoff flow conditions.

Special Unique Features of the Present Rainfall-Runoff Model

Considering the general rainfall-runoff flow conditions in a typical valley with steep terrain, all parts of the hydrological region of the valley can have three different regimes of rainfall-runoff flow, consisting of (1) initial subsurface groundwater flow, (2) subsequent first-level composite flow, as the combination of subsurface groundwater flow and surface-water flow in water courses with directional channels, and (3) final second-level composite flow, as the combination of subsurface groundwater flow, surface-water flow in water courses with directional channels and surface-layer flow above the ground. Furthermore, all parts of the geohydrological region of the valley can have the same four flow regimes in reverse sequence after the termination of the rainfall due to the downhill flow of the residual water from higher elevations (Fig. 1).

Therefore, the geohydrological region of a valley with steep terrain cannot be separated into two distinctly different parts, corres-

6

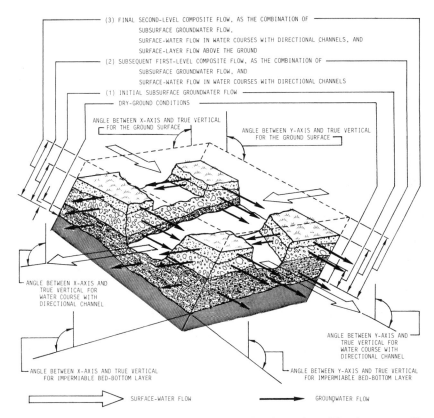

(3) FINAL SECOND-LEVEL COMPOSITE FLOW, AS THE COMBINATION OF
    SUBSURFACE GROUNDWATER FLOW,
    SURFACE-WATER FLOW IN WATER COURSES WITH DIRECTIONAL CHANNELS, AND
    SURFACE-LAYER FLOW ABOVE THE GROUND
(2) SUBSEQUENT FIRST-LEVEL COMPOSITE FLOW, AS THE COMBINATION OF
    SUBSURFACE GROUNDWATER FLOW, AND
    SURFACE-WATER FLOW IN WATER COURSES WITH DIRECTIONAL CHANNELS
(1) INITIAL SUBSURFACE GROUNDWATER FLOW
    DRY-GROUND CONDITIONS

ANGLE BETWEEN X-AXIS AND TRUE VERTICAL
FOR THE GROUND SURFACE

ANGLE BETWEEN Y-AXIS AND TRUE VERTICAL
FOR THE GROUND SURFACE

ANGLE BETWEEN X-AXIS AND
TRUE VERTICAL FOR
WATER COURSE WITH
DIRECTIONAL CHANNEL

ANGLE BETWEEN Y-AXIS AND
TRUE VERTICAL FOR
WATER COURSE WITH
DIRECTIONAL CHANNEL

ANGLE BETWEEN X-AXIS AND TRUE VERTICAL
FOR IMPERMIABLE BED-BOTTOM LAYER

ANGLE BETWEEN Y-AXIS AND TRUE VERTICAL FOR
IMPERMIABLE BED-BOTTOM LAYER

⇨ SURFACE-WATER FLOW   → GROUNDWATER FLOW

Fig. 1. Schematic representation of the channelized, composite subsurface-surface flow regimes of rainfall-runoff for inclined terrain.

ponding to (1) surface-water flow region at all times, and (2) ground-water flow region at all times. Hence, the rainfall-runoff flow conditions in a valley with steep terrain can not be realistically simulated by the presently available mathematical models for coupled surface-water flow and groundwater flow conditions, since all these models are based on the fundamental assumption that two distinctly different flow regimes always exit in their corresponding two distinctly separated geohydrological regions.

The mathematical formulation of the present fast-transient, two-dimensional discrete-element rainfall-runoff model (Eraslan, Erhan and Lin, 1981) is completely different from all existing mathematical models which are being utilized for the analysis of rainfall-runoff conditions. In contrast to all existing mathematical models, the present rainfall-runoff model can consider local, temporally varying flow conditions which can continuously go through the necessary transitions between dry ground conditions and three different flow regimes of rainfall-runoff flow, consisting of (1) initial subsurface groundwater flow, (2) subsequent first-level composite flow, as the combination of subsurface groundwater flow and surface-water flow in water courses with directional channels, and (3) final second-level composite flow, as the combination of subsurface groundwater flow, surface-water flow in water courses with directional channels and surface-layer flow above the ground (Fig. 1).

7

SUMMARY OF THE MATHEMATICAL DEVELOPMENT OF THE MODEL

The development of the fast-transient two-dimensional rainfall-runoff model is based on the application of the discrete-element method (DEM) which represents an advanced direct computer simulation approach. The discrete-element method (DEM) is conceptually similar to earlier direct numerical simulation methods developed at Los Alamos Scientific Laboratory by Evans and Harlow (1967) for the particle-in-cell (PIC) method, by Harlow and Welch (1965, 1966) and Welch et al, (1966) for the marker-and-cell (MAC) method, and particularly by Gentry, Martin and Daly (1966) for the Fluid-in-Cell (FLIC) method.

The development of the general discrete-element method (DEM) is already presented in various studies (Eraslan, 1973, 1974a, 1974b, 1975a, 1975b, 1976, 1981a, 1981b, 1981c, Lin, 1981, Lin and Eraslan, 1981). The method is called the "discrete-element method" because the development of the mathematical system of computational equations for the solution of the flow variables is based on the application of the integral forms of the physical principles directly to "discrete-elements" with finite volumes and finite enclosure surfaces.

In the application of the discrete-element method to the modeling of rainfall-runoff flow conditions in a valley with steep terrain, the surface area of the geohydrological region in the (x,y) plane is divided into two-dimensional elements, according to a special geometrical discretization technique. The geometrical discretization technique is based on a specified nonuniform (or uniform) rectangular grid system which automatically allows for different spatial resolution scales in different subregions. The two-dimensional elements can be obtained by utilizing the automated discrete-element-generator computer code PREPR2 (Holdeman, 1977) or its modified and improved version PREPAR (Lin and Erhan, 1978).

Considering the spatially varying slope conditions of the terrain, each discrete-element is oriented according to three generalized orthogonal coordinate axes, such that (1) x-axis and y-axis are parallel to the surface plane of the terrain (or approximately parallel to the surface plane of the impermeable bed-bottom layer below the ground) and (2) z-axis is normal to the surface plane of the terrain. Hence, the z-axis of the element is not parallel to the true vertical direction of the geophysical surface plane (approximately spherical) of the earth (Eraslan, Erhan and Lin, 1981); and therefore, x-axis and y-axis do not form a horizontal plane parallel to the geophysical surface plane of the earth (Fig. 1).

Formulation of the Discrete-Element Equations

The formulation of the fast-transient, two-dimensional discrete-element rainfall-runoff model is based on the applications of the integral forms of two physical principles to the discrete elements which can contain water in the flow region. The detailed derivation of the discretized computational equations of the model is presented in a report by Eraslan, Erhan and Lin (1981).

The general integral forms of the two physical principles for water contained in a control volumes v_{cv} enclosed by the control surface area A_{cv} can be stated as

1. mass conservation principle:

$$\frac{\partial}{\partial t}\iiint_{V_{cv}} \rho \, dv + \iint_{A_{cv}} \rho \vec{V} \cdot \hat{n} \, dA = 0 . \qquad (1)$$

8

2. momentum principle (second law of motion):

$$\frac{\partial}{\partial t}\iiint_{V_{cv}}(\rho\vec{V})\,dv + \iint_{A_{cv}}(\rho\vec{V})\vec{V}\cdot\hat{n}\,dA = \iiint_{V_{cv}}\vec{f}_b\,dv + \iint_{A_{cv}}\vec{f}_s\,dA \ . \qquad (2)$$

The variables used in the statements of the integral forms of the two physical principles Eqs. (1) and (2) represent

t = time [s] ,

V_{cv} = control volume [m^3],

A_{cv} - control surface area [m^2],

ρ = mass density of water [kg/m^3],

$\vec{V}$ = velocity (vector) of the flow of water [m/s],

$\hat{n}$ - unit normal vector (outward positive) on the control surface area,

$\vec{f}_b$ = body force (vector) acting on per unit volume of water [N/m^3],

$\vec{f}_s$ - surface force (vector) acting on per unit area in water [N/m^2].

The general integral forms of the two physical principles [Eqs. (1) and (2)] can be simplified by considering the fundermental assumptions of the mathematical model.

The water is assumed as incompressible on the ground surface and in the porous media in the subsurface. The flow is predominantly in the curvilinear surface plane (x,y) of the flow region.

The inertia effects can be neglected in the momentum principle, since both the subsurface flow and the surface flow in the water courses and above ground will have low characteristic Froude number values for rainfall-runoff flow conditions.

The simplified integral forms of two physical principles Eqs. (1) and (2) are formulated according to the staggered-element form of the general discrete-element method. This formulation is conceptually similar to the conventional staggered mesh systems used in finite-difference formulations of fluid flow problems (Harlow and Welch, 1965, Leendertse, 1973, Spraggs and Street, 1975, also Roache, 1972).

The simplified integral form of the conservation of mass principle Eq. (1) is formulated for the central element (i,j). The simplified integral form of the x-component of the momentum principle Eq. (2) is formulated for two staggered elements along x-direction as x-upstream element (i-½,j) and x-downstream element (i+½,j). The simplified integral form of the y-component of the momentum principle Eq. (2) is formulated for two staggered elements along y-direction as y-upstream element (i,j-½) and y-downstream element (i,j+½).

Hence, the five scalar equations representing the simplified integral forms of two physical principles Eqs. (1) and (2) for the central element and four staggered elements become
mass conservation principle for central element (i,j):

$$\frac{\partial}{\partial t}\iiint_{\overline{V}_{i,j}}dv + \iint_{\overline{A}_{i,j}}\vec{V}\cdot\hat{n}\,dA = 0 \ , \qquad (3)$$

x-component of momentum principle for x-upstream element (i-½,j) and x-downstream element (i+½,j):

$$\frac{\partial}{\partial t}\iiint_{\overline{V}_{i\pm\frac{1}{2},j}} V_x \, dv = \frac{1}{\rho}\iiint_{\overline{V}_{i\pm\frac{1}{2},j}} f_{bx} \, dv + \frac{1}{\rho}\iint_{\overline{A}_{i\pm\frac{1}{2},j}} f_{sx} \, dA \,, \tag{4}$$

y-component of the momentum principle for y-upstream element $(i, j-\frac{1}{2})$ and y-downstream element $(i, j+\frac{1}{2})$:

$$\frac{\partial}{\partial t}\iiint_{\overline{V}_{i,j\pm\frac{1}{2}}} V_y \, dv = \frac{1}{\rho}\iiint_{\overline{V}_{i,j\pm\frac{1}{2}}} f_{by} \, dv + \frac{1}{\rho}\iint_{\overline{A}_{i,j\pm\frac{1}{2}}} f_{sy} \, dA \,. \tag{5}$$

The additional variables used in five scalar equations representing the simplified integral forms of two physical principles [Eqs. (3) through (5)] are defined, as

$\overline{V}_{i,j}$ = available control volume (void space) in element (i, j) $[m^3]$,

$\overline{A}_{i,j}$ = available control surface flow area (porous area) of element (i, j) $[m^2]$,

$\overline{V}_{i\pm\frac{1}{2},j}$ = available control volume (void space) in x-upstream element $(i-\frac{1}{2}, j)$ or x-downstream element $(i+\frac{1}{2}, j)$ $[m^3]$,

$\overline{A}_{i\pm\frac{1}{2},j}$ = available control surface flow area (porous area) of x-upstream element $(i-\frac{1}{2}, j)$ or x-downstream element $(i+\frac{1}{2}, j)$ $[m^2]$,

$\overline{V}_{i,j\pm\frac{1}{2}}$ = available control volume (void space) in y-upstream element $(i, j-\frac{1}{2})$ or y-downstream element $(i, j+\frac{1}{2})$ $[m^3]$,

$\overline{A}_{i,j\pm\frac{1}{2}}$ = available control surface flow area (porous area) of y-upstream element $(i, j-\frac{1}{2})$ or y-downstrean element $(i, j+\frac{1}{2})$ $[m^2]$,

V_x = x-component of velocity of the flow of water $[m/s]$,

V_y = y-component of velocity of the flow of water $[m/s]$,

f_{bx} = x-component of body force acting on per unit volume of water $[N/m^3]$,

f_{by} = y-component of body force acting on per unit volume of water $[N/m^3]$,

f_{sx} = x-component of surface force acting on per unit area in water $[N/m^2]$,

f_{sy} = y-component of surface force acting on per unit area in water $[N/m^2]$.

Considering the simplified integral form of mass conservation principle for central element (i, j), the time-derivative of the control volume integral on the left side of Eq. (3) can be evaluated in discretized form, as

$$\frac{\partial}{\partial t}\iiint_{\overline{V}_{i,j}} dv = \frac{\partial}{\partial t}(\Delta x_{i,j}\Delta y_{i,j}\overline{\alpha}_{z;i,j}\overline{D}_{i,j})$$

$$= \Delta x_{i,j}\Delta y_{i,j}\overline{\alpha}_{z;i,j}(\frac{\partial \overline{D}}{\partial t}i,j + \frac{\overline{D}_{i,j}}{\overline{\alpha}_{z;i,j}}\frac{\partial\overline{\alpha}}{\partial t}z;i,j) \,, \tag{6}$$

10

where

$\Delta x_{i,j}$ = specified length along x-axis of central element (i,j) $\left[m \right]$,

$\Delta y_{i,j}$ = specified length along y-axis of central element (i,j) $\left[m \right]$,

$\overline{D}_{i,j}$ - apparent water depth in central element (i,j) $\left[m \right]$,

$\alpha_{z;i,j}$ - areal flow porosity of the plane normal to z-axis in central element (i,j).

The areal porosity of the plane normal to z-axis varies with water depth in the element, since the formulation includes the transition of the flow conditions between three regimes of rainfall-runoff flow, consisting of (1) initial subsurface groundwater flow, (2) subsequent first-level composite flow, as the combination of subsurface groundwater flow and surface-water flow in water courses with directional channels, and (3) final second-level composite flow, as the combination of subsurface groundwater flow, surface-water flow in water courses with directional channels and surface-layer flow above the ground.

Hence, the term containing the time-derivative of the areal porosity in Eq. (6) can be reformulated, as

$$\frac{\overline{D}_{i,j}}{\overline{\alpha}_{z;i,j}} \frac{\partial \overline{\alpha}}{\partial t}_{z;i,j} = \frac{\overline{D}_{i,j}}{\overline{\alpha}_{z;i,j}} (\frac{\partial \overline{\alpha}}{\partial D} z)_{i,j} \frac{\partial \overline{D}}{\partial t}_{i,j} = \overline{\gamma}_{z;i,j} \frac{\partial \overline{D}}{\partial t}_{i,j} , \quad (7)$$

where, the definition

$$\overline{\gamma}_{z;i,j} = \frac{\overline{D}_{i,j}}{\overline{\alpha}_{z;i,j}} (\frac{\partial \overline{\alpha}}{\partial D} z)_{i,j} , \quad (8)$$

represents

$\overline{\gamma}_{z;i,j}$ - modified storativity coefficient associated with the variation of water depth along z-axis in central element (i,j).

Considering a constant reference depth in each element, representing the depth of the available space for the flow of water above the impermeable bottom layer of the flow bed as

$D_{bed;i,j}$ - specified reference depth of the flow bed in central element (i,j) above the impermeable bottom layer $\left[m \right]$,

the actual temporally varying apparent depth of water can be formulated, as

$$\overline{D}_{i,j} = D_{bed;i,j} + \overline{H}_{i,j}, \quad (9)$$

where

$\overline{H}_{i,j}$ - apparent water-surface elevation in the central element (i,j) relative to the surface at the top of the specified depth of the flow bed in the central element (i,j) $\left[m \right]$.

Hence, substituting Eq. (8) and Eq. (9) in Eq. (6), which represents the first part of the left side of the simplified integral form of mass conservation principle for the central element (i,j)

11

Eq. (3) , becomes

$$\frac{\partial}{\partial t} \iiint_{\overline{V}_{i,j}} dv = \Delta x_{i,j} \Delta y_{i,j} \overline{\alpha}_{z;i,j} (1 + \overline{\gamma}_{z;i,j}) \frac{\partial \overline{H}}{\partial t}_{i,j} \cdot \tag{10}$$

The control-surface integral on the left side of the simplified integral form of mass conservation principle Eq. (3) for central element (i,j) can be formulated in discretized form by considering the mass transport of water across all possible surface areas associated with central element (i,j), as:

$$\iint_{\overline{A}_{i,j}} \vec{V} \cdot \hat{n} \, dA = \tilde{G}_{x;i+\frac{1}{2},j} - \tilde{G}_{x;i-\frac{1}{2},j} + \tilde{G}_{y;i,j+\frac{1}{2}} - \tilde{G}_{y;i,j-\frac{1}{2}}$$

$$- \Delta x_{i,j} \Delta y_{i,j} g_{rnf;i,j} - \delta_{drn,m;i,j} (A_{drn,m} g_{rnf,m} + G_{drn,m})$$

$$- \Delta x_{i,j} \Delta y_{i,j} (\overline{g}_{inf;i,j} - \overline{g}_{evp;i,j} - \overline{g}_{ptp;i,j}) - \delta_{pmp,n;i,j} G_{pmp,n} , \tag{11}$$

where the variables associated with water flow represent

$\tilde{G}_{x;i\pm\frac{1}{2},j}$ = flow rate (marked) along x-axis at x-upstream enclosure surface at $(i-\frac{1}{2},j)$ or at x-downstream enclosure surface at $(i+\frac{1}{2},j)$ of central element (i,j), or
= flow rate (marked) along x-axis in x-upstream element $(i-\frac{1}{2},j)$ or x-downstream element $(i+\frac{1}{2},j)$ [m^3/s],

$\tilde{G}_{y;i,j\pm\frac{1}{2}}$ = flow rate (marked) along y-axis at y-upstream enclosure surface at $(i,j-\frac{1}{2})$ or at y-downstream enclosure surface at $(i,j+\frac{1}{2})$ of central element (i,j), or
= flow rate (marked) along y-axis in y-upstream element $(i,j-\frac{1}{2})$ or y-downstream element $(i,j+\frac{1}{2})$ [m^3/s],

$g_{rnf;i,j}$ = rainfall rate per unit area on the surface of central element (i,j) [m/s],

$\delta_{drn,m;i,j}$ = Kronecker delta for indicating the location of the inflowing (or outflowing) boundary drainage flow m,
0 : boundary drainage flow m is not in central element (i,j),
1 : boundary drainage flow m is in central element (i,j).

$A_{drn,m}$ - specified rainfall-collecting drainage area associated with the boundary-drainage flow m [m^2],

$g_{rnf,m}$ - specified rainfall rate per unit area on the surface of rainfall-collecting drainage area associated with the boundary drainage flow m [m/s],

$G_{drn,m}$ - specified drainage flow rate associated with the boundary-drainage flow m or associated with inflowing (or outflowing) stream or river m [m^3/s],

$\bar{g}_{inf;i,j}$ — specified (or modeled) infiltration flow rate per unit area at the bed-bottom surface of central element (i,j) $[m/s]$,

$\bar{g}_{evp;i,j}$ — specified (or modeled) evaporation flow rate per unit area at the ground surface (water surface) of central element (i,j) $[m/s]$,

$\bar{g}_{ptp;i,j}$ — specified (or modeled) transpoevaporation flow rate per unit area at the ground surface area with vegetation cover in central element (i,j) $[m/s]$,

$\delta_{pmp,n;i,j}$ — Kronecker delta for indicating the location of pump m,
  0 : pump m is not in central element (i,j),
  1 : pump m is in central element (i,j),

$G_{pmp,n}$ — specified pumping flow rate of pump $m\,[m^3/s]$.

The flow rate (marked) along x-axis at x-upstream enclosure surface at $(i-\frac{1}{2},j)$ or at x-downstream enclosure surface at $(i+\frac{1}{2},j)$ of central element (i,j), which also represents the flow rate (marked) along x-axis in x-upstream element $(i-\frac{1}{2},j)$ or x-downstream element $(i+\frac{1}{2},j)$ is defined, as

$$\tilde{G}_{x;i\pm\frac{1}{2},j} = \iint_{\overline{A}_{i\pm\frac{1}{2},j}} \vec{V}\cdot\hat{n}\,dA = \overline{\Delta y}_{i\pm\frac{1}{2},j}\,\overline{D}_{i\pm\frac{1}{2},j}\,\overline{\alpha}_{x;i\pm\frac{1}{2},j}\,\overline{V}_{x;i\pm\frac{1}{2},j}\,. \qquad (12)$$

Similarly, the flow rate (marked) along y-axis at y-upstream enclosure surface at $(i,j-\frac{1}{2})$ or at y-downstream enclosure surface at $(i,j+\frac{1}{2})$ of central element (i,j) which also represents the flow rate (marked) along y-axis in y-upstream element $(i,j-\frac{1}{2})$ or y-downstream element $(i,j+\frac{1}{2})$ is defined, as

$$\tilde{G}_{y;i,j\pm\frac{1}{2}} = \iint_{\overline{A}_{i,j\pm\frac{1}{2}}} \vec{V}\cdot\hat{n}\,dA = \overline{\Delta x}_{i,j\pm\frac{1}{2}}\,\overline{D}_{i,j\pm\frac{1}{2}}\,\overline{\alpha}_{y;i,j\pm\frac{1}{2}}\,\overline{V}_{y;i,j\pm\frac{1}{2}}\,. \qquad (13)$$

The additional variables associated with the flow rates (marked) in Eqs. (12) and (13) represent

$\overline{\Delta x}_{i,j\pm\frac{1}{2}}$ = length along x-axis of y-upstream enclosure surface at $(i,j-\frac{1}{2})$ or y-downstream enclosure surface at $(i,j+\frac{1}{2})$ of central element (i,j), or
length along x-axis of y-upstream element $(i,j-\frac{1}{2})$ or y-downstream element $(i,j+\frac{1}{2})$ $[m]$,

$\overline{\Delta y}_{i\pm\frac{1}{2},j}$ = length along y-axis of x-upstream enclosure surface at $(i-\frac{1}{2},j)$ or x-downstream enclosure surface at $(i+\frac{1}{2},j)$ of central element (i,j), or
= length along y-axis of x-upstream element $(i-\frac{1}{2},j)$ or x-downstream element $(i+\frac{1}{2},j)$ $[m]$,

$\overline{D}_{i\pm\frac{1}{2},j}$ = apparent water depth at x-upstream enclosure surface at $(i-\frac{1}{2},j)$ or x-downstream enclosure surface at $(i+\frac{1}{2},j)$ of central element (i,j), or
= apparent water depth in x-upstream element $(i-\frac{1}{2},j)$ or x-downstream element $(i+\frac{1}{2},j)$ $[m]$,

$\overline{D}_{i,j\pm\frac{1}{2}}$ — apparent water depth at y-upstream enclosure surface at

$(i, j-\frac{1}{2})$ or y-downstream enclosure surface at $(i, j+\frac{1}{2})$ of central element (i, j), or
- apparent water depth in y-upstream element $(i, j-\frac{1}{2})$ or y-downstream element $(i, j+\frac{1}{2})$ [m],

$\overline{\alpha}_{x;i\pm\frac{1}{2},j}$ - areal flow porosity of the plane normal to x-axis at x-upstream enclosure surface at $(i-\frac{1}{2}, j)$ or x-downstream enclosure surface at $(i+\frac{1}{2}, j)$ of central element (i, j), or
- areal porosity of the plane normal to x-axis in x-upstream element $(i-\frac{1}{2}, j)$ or x-downstream element $(i+\frac{1}{2}, j)$,

$\overline{\alpha}_{y;i,j\pm\frac{1}{2}}$ - areal flow porosity of the plane normal to y-axis at y-upstream enclosure surface at $(i, j-\frac{1}{2})$ or y-downstream enclosure surface at $(i, j+\frac{1}{2})$ of central element (i, j), or
- areal porosity of the plane normal to y-axis in y-upstream element $(i, j-\frac{1}{2})$ or y-downstream element $(i, j+\frac{1}{2})$,

$\widetilde{V}_{x;i\pm\frac{1}{2},j}$ - apparent x-component of velocity (marked) at x-upstream enclosure surface at $(i-\frac{1}{2}, j)$ or x-downstream enclosure surface at $(i+\frac{1}{2}, j)$ of central element (i, j), or
- apparent x-component of velocity (marked) in x-upstream element $(i-\frac{1}{2}, j)$ or x-downstream element $(i+\frac{1}{2}, j)$ [m/s],

$\widetilde{V}_{y;i,j\pm\frac{1}{2}}$ = apparent y-component of velocity (marked) at y-upstream enclosure surface at $(i, j-\frac{1}{2})$ or y-downstream enclosure surface at $(i, j+\frac{1}{2})$ of central element (i, j), or
apparent y-component of velocity (marked) in y-upstream element $(i, j-\frac{1}{2})$ or y-downstream element $(i, j+\frac{1}{2})$ [m/s].

Hence, substituting Eq. (10) and Eq. (11) in Eq. (3), the resulting computational discrete-element form of mass conservation principle for central element (i, j) can be rearranged to obtain the discrete-element equation for apparent water-surface elevation $\overline{H}_{i,j}$ in central element (i, j), as

$$\frac{\partial \overline{H}}{\partial t}_{i,j} = \frac{1}{\Delta x_{i,j} \Delta y_{i,j} \overline{\alpha}_{z;i,j}(1 + \gamma_{z;i,j})} (\widetilde{G}_{x;i-\frac{1}{2},j} - \widetilde{G}_{x;i+\frac{1}{2},j} + \widetilde{G}_{y;i,j-\frac{1}{2}} - \widetilde{G}_{y;i,j+\frac{1}{2}})$$

$$+ \frac{1}{\overline{\alpha}_{z;i,j}(1 + \overline{\gamma}_{z;i,j})} g_{rnf;i,j}$$

$$+ \frac{\delta_{drn,m;i,j}}{\Delta x_{i,j} \Delta_{i,j} \overline{\alpha}_{z;i,j}(1 + \overline{\gamma}_{z;i,j})} (A_{drn,m} g_{rnf,m} + G_{drn,m})$$

$$+ \frac{1}{\overline{\alpha}_{z;i,j}(1 + \overline{\gamma}_{z;i,j})} (\overline{g}_{inf;i,j} - \overline{g}_{evp;i,j} - \overline{g}_{ptp;i,j})$$

$$+ \frac{\delta_{pmp,n;i,j}}{\Delta x_{i,j} \Delta y_{i,j} \overline{\alpha}_{z;i,j}(1 + \overline{\gamma}_{z;i,j})} G_{pmp,n;i,j} \quad . \tag{14}$$

14

The discrete-element equation for apparent water surface elevation $H_{i,j}$ in central element (i,j) [Eq. (14)] represents the first set of semi-discretized (temporally-continuous, spatially-discretized) computational system of first-order differential equations for the solution of the composite subsurface-surface flow conditions.

Considering the simplified integral form of x-component of momentum principle for x-upstream element $(i-\frac{1}{2},j)$ or x-downstream element $(i+\frac{1}{2},j)$, the time-derivative of the control-volume integral on the left side of [Eq. (4)] can be evaluated in discretized form, as

$$\frac{\partial}{\partial t}\iiint_{\overline{V}_{i\pm\frac{1}{2},j}} V_x \, dv = \frac{\partial}{\partial t}(\overline{\Delta x}_{i\pm\frac{1}{2},j}\,\overline{\Delta y}_{i\pm\frac{1}{2},j}\,\overline{D}_{i\pm\frac{1}{2},j}\,\overline{\alpha}_{x;i\pm\frac{1}{2},j}\,\widetilde{V}_{x;i\pm\frac{1}{2},j}$$

$$= \overline{\Delta x}_{i\pm\frac{1}{2},j}\frac{\partial\widetilde{G}}{\partial t}_{x;i\pm\frac{1}{2},j}\,. \tag{15}$$

The x-component of body force acting on the water consists of the gravity force associated with the inclination of x-axis (relative to the normal to the geophysical surface plane), according to the local slope of the terrain parallel to the impermeable bed bottom surface, since the Coriolis acceleration (force) effect can be neglected for groundwater and surface-water flow conditions in meso-scale geohydrological regions. Hence, the control volume integral on the right side of Eq. (4), associated with x-component of body force in x-upstream element $(i-\frac{1}{2},j)$ or x-downstream element $(i+\frac{1}{2},j)$, can be evaluated in discretized form, as

$$\frac{1}{\rho}\iiint_{\overline{V}_{i\pm\frac{1}{2},j}} f_{bx} \, dv = -\overline{\Delta x}_{i\pm\frac{1}{2},j}\,\overline{\Delta y}_{i\pm\frac{1}{2},j}\,\overline{D}_{i\pm\frac{1}{2},j}\,\overline{\beta}_{x;i\pm\frac{1}{2},j}\,g_a\cos\overline{\theta}_{n,x;i\pm\frac{1}{2},j}\,, \tag{16}$$

where

$\overline{\Delta x}_{i\pm\frac{1}{2},j}$ = length along x-axis of x-upstream element $(i-\frac{1}{2},j)$ or x-downstream element $(i+\frac{1}{2},j)$ [m],

$\overline{\beta}_{x;i\pm\frac{1}{2},j}$ = areal gravity porosity of the plane normal to x-axis in x-upstream element $(i-\frac{1}{2},j)$ or x-downstream element $(i+\frac{1}{2},j)$, or

- areal gravity porosity of the plane normal to x-axis at x-upstream enclosure surface at $(i-\frac{1}{2},j)$ or x-downstream enclosure surface at $(i+\frac{1}{2},j)$ of center element (i,j),

g_a = gravitational acceleration constant [9.81 m/s^2],

$\overline{\theta}_{n,x;i\pm\frac{1}{2},j}$ = angle between x-axis and the true vertical direction normal to the geophysical surface plane in x-upstream element $(i-\frac{1}{2},j)$ or x-downstream element $(i+\frac{1}{2},j)$, or

- angle between x-axis and the true vertical direction normal to the geophysical surface plane at x-upstream enclosure surface at $(i-\frac{1}{2},j)$ or x-downstream enclosure surface at $(i+\frac{1}{2},j)$ of central element (i,j).

The x-component of surface force acting on the water consists of the modified hydrostatic pressure force along x-axis and the viscous drag force for general composite subsurface-surface flow conditions. Hence, the control-surface integral on the right side of Eq. (4), associated with the x-component of the surface force in x-upstream element $(i-\frac{1}{2},j)$ or x-downstream element $(i+\frac{1}{2},j)$, can be evaluated in discretized form, as

$$\frac{1}{\rho} \iint_{\overline{A}_{i\pm\frac{1}{2},j}} f_{3x}dv = \pm \overline{\Delta y}_{i\pm\frac{1}{2},j} \overline{D}_{i\pm\frac{1}{2},j} \overline{\kappa}_{i\pm\frac{1}{2},j} \ g_a \cos \overline{\theta}_{n,z;i\pm\frac{1}{2},j} (\overline{H}_{i,j} - \overline{H}_{i\pm1,j})$$

$$+ \overline{\Delta x}_{i\pm\frac{1}{2},j} \overline{f}^*_{grd,x;i\pm\frac{1}{2},j} \ , \qquad\qquad (17)$$

where

$\overline{\kappa}_{x;i\pm\frac{1}{2},j}$ - areal pressure porosity of the plane normal to x-axis in x-upstream element $(i-\frac{1}{2},j)$ or x-downstream element $(i+\frac{1}{2},j)$, or
- areal pressure porosity of the plane normal to x-axis at x-upstream enclosure surface at $(i-\frac{1}{2},j)$ or x-downstream enclosure surface at $(i+\frac{1}{2},j)$ of central element (i,j),

$\overline{\theta}_{n,z;i\pm\frac{1}{2},j}$ - angle between z-axis and the true vertical direction normal to the geophysical surface plane in x-upstream element $(i-\frac{1}{2},j)$ or x-downstream element $(i+\frac{1}{2},j)$, or
- angle between z-axis and the true vertical direction normal to the geophysical surface plane at x-upstream enclosure surface at $(i-\frac{1}{2},j)$ or x-downstream enclosure surface at $(i+\frac{1}{2},j)$ of central element (i,j),

$\overline{H}_{i\pm1,j}$ - apparent water surface elevation in x-upstream adjacent element $(i-1,j)$ or x-downstream adjacent element $(i+1,j)$ [m],

$\overline{f}^*_{grd,x;i\pm\frac{1}{2},j}$ - x-component of modified ground drag force (viscous effects) per unit length along x-axis for general composite subsurface-surface flow conditions in x-upstream element $(i-\frac{1}{2},j)$ or x-downstream element $(i+\frac{1}{2},j)$ $[m^3/s^2$ (modified flow rate consistant units)].

It is important to note that the flow region above the impermeable bed-bottom layer for composite subsurface-surface flow conditions can have particular arrangements of void spaces, associated with the particular packing conditions of the solid material, which can affect differently the flow, gravity force and hydrostatic pressure force conditions. Therefore, the general discretized formulation of the fast-transient, discrete-element rainfall-runoff model includes three different areal porosity definitions, as areal flow porosity α_x, areal gravity porosity β_x and areal pressure porosity κ_x, for the plane normal to x-axis in the central and staggered elements [Eqs. (12), (16) and (17)]. The formulation of the model can be readily simplified by considering the restricted form, with all three areal porosities as equal to each other for the plane normal to x-axis, as

$$\kappa_x = \beta_x = \alpha_x \quad \text{(for all elements).} \qquad\qquad (18)$$

Hence, substituting Eq. (15), Eq. (16), and Eq. (17) in Eq. (4), the resulting computational discrete-element form of x-component of momentum principle for x-upstream element $(i-\frac{1}{2},j)$ or x-downstream element $(i+\frac{1}{2},j)$ can be rearranged to obtain the discrete-element equation for x-component of flow rate (marked) $\overline{G}_{x;i\pm\frac{1}{2},j}$ in x-upstream element $(i-\frac{1}{2},j)$ or x-downstream element $(i+\frac{1}{2},j)$, as

16

$$\frac{\partial \tilde{G}}{\partial t}x_{;i\pm\frac{1}{2},j} = -\overline{\Delta y}_{i\pm\frac{1}{2},j}\,\overline{D}_{i\pm\frac{1}{2},j}\ \overline{\beta}_{x;i\pm\frac{1}{2},j}\,g_\alpha \cos \overline{\theta}_{n,x;i\pm\frac{1}{2},j}$$

$$\pm\frac{\overline{\Delta y}_{i\pm\frac{1}{2},j}\,\overline{D}_{i\pm\frac{1}{2},j}\,\overline{\kappa}_{x;i\pm\frac{1}{2},j}}{\overline{\Delta x}_{i\pm\frac{1}{2},j}}\ g_\alpha \cos \overline{\theta}_{n,z;i\pm\frac{1}{2},j}(\overline{H}_{i,j} - \overline{H}_{i\pm1,j})$$

$$+\overline{f}^{*}_{grd,x;i\pm\frac{1}{2},j}\ . \tag{19}$$

The discrete-element equation for x-component of flow rate (marked) $G_x;i\frac{1}{2},j$ in x-upstream element $(i-\frac{1}{2},j)$ or x-downstream element $(i+\frac{1}{2},j)$ [Eq. (19)] represents the second set of semi-discretized (temporally-continuous, spatially-discretized) computational system of first-order differential equations for the solution of the composite subsurface-surface flow conditions.

Similarly, evaluating in discretized form the control-volume integrals and control surface integrals of the simplified integral form of y-component of momentum principle [Eq. (5)], the resulting computational discrete-element form of y-component of momentum principle for y-upstream element $(i,j-\frac{1}{2})$ or y-downstream element $(i,j+\frac{1}{2})$ can be rearranged to obtain the discrete-element equation for y-component of flow rate (marked) $\tilde{G}_{y;i,j\pm\frac{1}{2}}$ in y-upstream element $(i,j-\frac{1}{2})$ or y-downstream element $(i,j+\frac{1}{2})$, as

$$\frac{\partial \tilde{G}}{\partial t}y_{;i,j\pm\frac{1}{2}} = -\overline{\Delta x}_{i,j\pm\frac{1}{2}}\,\overline{D}_{i,j\pm\frac{1}{2}}\ \overline{\beta}_{y;i,j\pm\frac{1}{2}}\,g_\alpha \cos \overline{\theta}_{n,y;i,j\pm\frac{1}{2}}$$

$$\pm\frac{\overline{\Delta x}_{i,j\pm\frac{1}{2}}\,\overline{D}_{i,j\pm\frac{1}{2}}\,\overline{\kappa}_{y;i,j\pm\frac{1}{2}}}{\overline{\Delta y}_{i,j\pm\frac{1}{2}}}\ g_\alpha \cos \overline{\theta}_{n,z;i,j\pm\frac{1}{2}}(\overline{H}_{i,j} - \overline{H}_{i\pm1,j})$$

$$+\overline{f}^{*}_{grd,y;i,j\pm\frac{1}{2}}\ . \tag{20}$$

where,

$\overline{\beta}y_{i,j\pm\frac{1}{2}}$ = areal gravity porosity of the plane normal to y-axis in y-upstream element $(i,j-\frac{1}{2})$ or y-downstream element $(i,j+\frac{1}{2})$, or
= areal gravity porosity of the plane normal to y-axis at y-upstream enclosure surface at $(i,j-\frac{1}{2})$ or y-downstream enclosure surface at $(i,j+\frac{1}{2})$ of center element (i,j),

$\overline{\theta}_{n,y;i,j\pm\frac{1}{2}}$ - angle between y-axis and the true vertical direction normal to the geophysical surface plane in y-upstream element $(i,j-\frac{1}{2})$ or y-downstream element $(i,j+\frac{1}{2})$, or
= angle between y-axis and the true vertical direction normal to the geophysical surface plane at y-upstream enclosure surface at $(i,j-\frac{1}{2})$ or y-downstream enclosure surface at $(i,j+\frac{1}{2})$ of central element (i,j).

$\overline{\kappa}_{y;i,j\pm\frac{1}{2}}$ - areal pressure porosity of the plane normal to y-axis in y-upstream element $(i,j-\frac{1}{2})$ or y-downstream element $(i,j+\frac{1}{2})$, or
= areal pressure porosity of the plane normal to y-axis at y-upstream enclosure surface at $(i,j-\frac{1}{2})$ or y-downstream

enclosure surface at $(i,j+\frac{1}{2})$ of central element (i,j),

$\overline{\Delta y}_{i,j\pm\frac{1}{2}}$ = length along y-axis of y-upstream element $(i,j-\frac{1}{2})$ or y-downstream $(i,j+\frac{1}{2})$ $[m]$,

$\overline{\theta}_{n,z;i,j\pm\frac{1}{2}}$ = angle between z-axis and the true vertical direction normal to the geophysical surface plane in y-upstream element $(i,j-\frac{1}{2})$ or y-downstream element $(i,j+\frac{1}{2})$, or
- angle between z-axis and the true vertical direction normal to the geophysical surface plane at y-upstream enclosure surface at $(i,j-\frac{1}{2})$ or y-downstream enclosure surface at $(i,j+\frac{1}{2})$ of central element (i,j),

$\overline{H}_{i,j\pm 1}$ = apparent water surface elevation in y-upstream adjacent element $(i,j-1)$ or y-downstream adjacent element $(i,j+1)$ $[m]$,

$\overline{f}^*_{grd,y;i,j\pm\frac{1}{2}}$ = y-component of modified ground drag force (viscous effects) per unit length along y-axis for general composite subsurface-surface flow conditions in y-upstream element $(i,j-\frac{1}{2})$ or y-downstream element $(i,j+\frac{1}{2})$ $[m^3/s^2$ (modified flow rate consistant units)$]$.

The discrete-element equation for y-component of flow rate (marked) $\tilde{G}_{y;i,j\pm\frac{1}{2}}$ in y-upstream element $(i,j-\frac{1}{2})$ or y-downstream element $(i,j+\frac{1}{2})$ [Eq. (20)] represents the third and final set of semi-discretized (temporally-continuous, spatially-discretized) computational system of first-order differential equations for the solution of the composite subsurface-surface flow conditions.

The computational system, consisting of (1) discrete-element equation for apparent water-surface elevation $\overline{H}_{i,j}$ [Eq. (14)], (2) discrete-element equation for x-component of flow rate (marked) $\tilde{G}_{x;i\pm\frac{1}{2},j}$ [Eq. (19)], and (3) discrete-element equation for y-component of flow rate (marked) $\tilde{G}_{y;i,j\pm\frac{1}{2}}$ Eq. (20) , represents 3N number of first-order differential equations for 3N number of unknown principal flow variables, consisting of (a) N number of apparent water surface elevations $\overline{H}_{i,j}$ in central elements (i,j), (b) N number of x-components of flow rate (marked) $\tilde{G}_{x;i\pm\frac{1}{2},j}$ in x-upstream elements $(i-\frac{1}{2},j)$ or x-downstream elements $(i+\frac{1}{2},j)$, and (3) N number of y-components of flow rate (marked) $\tilde{G}_{y;i,j\pm\frac{1}{2}}$ in y-upstream elements $(i,j-\frac{1}{2})$ or y-downstream elements $(i,j+\frac{1}{2})$. However, the computational system is not mathematically consistent, since 3N number of discrete-element equations [Eqs. (14), (19) and (20)] also contain auxilary flow variables and parameters in addition to 3N number of principal flow variables.

Closure of the Computational System of Discrete-Element Equations

The closure of the computational system of discrete-element equations [Eqs. (14), (19) and (20)] of the composite subsurface-surface rainfall-runoff model utilizes various special discretization features of the staggered discrete-element method (Eraslan, et al, 1981 and Eraslan, Erhan and Lin, 1981) and special formulations of the components of modified ground drag force per unit length based on subsidiary models which can consider the composite flow of water as groundwater through porous media in the ground and also as surface-water flow in water courses and above the ground.

Formulation of the computational algorithm of variable-size staggered discrete-element method

The geometrical properties of variable-size central discrete elements (i,j) which span the geohydrological region can be determined by the automated discrete-element-generator computer program PREPR2 (Holdeman, 1977) or by its modified and improved version PREPAR (Lin and Erhan, 1978). The geometrical discretization technique of the general discrete-element method uses variable size rectangular cells for the interior central elements and truncated rectangular cells for the boundary central elements (Eraslan 1975a, 1975b, 1981b, 1981c). Therefore, the dimensions of the staggered elements are determined based on special considerations which guarantee the validity of the linear interpolation of the flow variables (Eraslan, Erhan and Lin 1981).

The length along x-axis of x-upstream element $(i-\frac{1}{2},j)$ or x-downstream element $(i+\frac{1}{2},j)$ is formulated as the minimum length along x-axis of the two adjacent central elements $(i+1,j)$ or (i,j), as

$$\overline{\Delta x}_{i\pm\frac{1}{2},j} = \min(\Delta x_{i\pm1,j}, \Delta x_{i,j}) . \tag{21}$$

Similarly, the length along y-axis of y-upstream element $(i,j-\frac{1}{2})$ or y-downstream element $(i,j+\frac{1}{2})$ is formulated as the minimum length along y-axis of the two adjacent central elements $(i,j+1)$ or (i,j), as

$$\overline{\Delta y}_{i,j\pm\frac{1}{2}} = \min(\Delta y_{i,j\pm1}, \Delta y_{i,j}) . \tag{22}$$

The length along y-axis of an x-upstream element $(i-\frac{1}{2},j)$ or x-downstream element $(i+\frac{1}{2},j)$ for an adjacent interior central element is determined differently from the one for an adjacent boundary central element.

For an interior central element $(i+1,j)$ and an interior central element (i,j):

$$\overline{\Delta y}_{i\pm\frac{1}{2},j} = \Delta y_{i,j} = \Delta y_{i\pm1,j} . \tag{23a}$$

For a boundary central element $(i+1,j)$ and/or for a boundary central element (i,j):

$$\overline{\Delta y}_{i\pm\frac{1}{2},j} = \text{specified input data.} \tag{23b}$$

Similarly the length along x-axis of a y-upstream element $(i,j-\frac{1}{2})$ or y-downstream element $(i,j+\frac{1}{2})$ for an adjacent interior central element is determined differently from the one for an adjacent boundary central element.

For an interior central element $(i,j+1)$ and an interior central element (i,j):

$$\overline{\Delta x}_{i,j\pm\frac{1}{2}} = \Delta x_{i,j} = \Delta x_{i,j\pm1} . \tag{24a}$$

For a boundary central element $(i,j+1)$ and/or a boundary central element (i,j),

$$\overline{\Delta x}_{i,j\pm\frac{1}{2}} - \text{specified input data.} \qquad (24b)$$

The specified input data for the lengths of the staggered boundary elements are obtained directly from the output of the automated discrete-element-generator computer program PREPR2, (Holdeman, 1976) or from its modified and improved version PREPAR (Lin and Erhan, 1978). Considering that a flow area is not a part of the impermeable enclosure boundary of the flow region, it must always be common to two adjacent elements, x-downstream enclosure surface at $(i+\frac{1}{2},j)$ of central element (i,j), which corresponds to x-upstream enclosure surface at $(i+1)-\frac{1}{2},j$ of central element $(i+1,j)$, and y-downstream enclosure surface at $(i,j+\frac{1}{2})$ of central element (i,j), which corresponds to y-upstream enclosure surface at $i,(j+1)-\frac{1}{2}$ of central element $(i,j+1)$.

The computational algorithm of the staggered discrete-element method automatically eliminates duplicative calculations at common locations, by evaluating the flow variables only at the downstream enclosure surfaces of central elements, consisting of (1) x-downstream enclosure surface at $(i+\frac{1}{2},j)$ of central element (i,j) which corresponds to the location of x-downstream element $(i+\frac{1}{2},j)$ and (2) y-downstream enclosure surface at $(i,j+\frac{1}{2})$ of central element (i,j) which corresponds to the location of y-downstream element $(i,j+\frac{1}{2})$ (Eraslan, Erhan and Lin, 1981).

Formulation of the components of modified ground drag force per unit length for composite subsurface-surface flow conditions

The development of the fast-transient, discrete-element rainfall-runoff model considers three regimes of rainfall-runoff flow, consisting of (1) initial subsurface groundwater flow, (2) subsequent first-level composite flow, as the combination of subsurface groundwater flow and surface-water flow in water courses with directional channels, and (3) final second-level composite flow, as the combination of subsurface groundwater flow, surface-water flow in water courses with directional channels and surface-layer flow above the ground. Therefore, the rainfall-runoff model requires a plausible subsidiary model for the components of modified ground drag force per unit length which can approximately represent both groundwater flow conditions and surface-water flow conditions.

The development of a subsidiary model for the modified ground drag force per unit length for groundwater flow through porous media is difficult since it requires considerations of detailed micro-scale packing conditions of the solid material which determine the connectivity conditions of the micro-scale flow conduits in the ground (Scheidegger, 1953, Irmay, 1958, also Bear, 1972). Furthermore, the subsidiary model must include the capability to determine approximately the components of modified ground drag force per unit length for surface-water flow conditions in channelized water courses and also for surface-layer flow conditions above the ground.

The proposed subsidiary model for the components of modified ground drag force per unit length is based on the assumption of nonisotopic, directionally aligned, parallel conduits along x-axis and along y-axis which can approximately incorporate the viscous drag effects in the ground and above the ground for all three flow regimes of the composite subsurface-surface flow conditions (Eraslan, Erhan and Lin, 1981).

The x-component of modified ground drag force per unit length along x-axis is formulated as the combined effect of viscous drag forces in all parallel, identical, individual flow conduits, which allow the water to flow along x-axis, as

$$\bar{f}^*_{grd,x} = N_x \bar{f}^*_{cnd,x},$$ (25a)

where

N_x = number of parallel flow conduits along x-axis,

$\bar{f}^*_{cnd,x}$ = viscous drag force per unit length along x-axis in a flow conduit along x-axis [m^3/s (modified flow rate consistent units)].

Similarly, the y-component of modified ground drag force per unit length along y-axis is formulated as the combined effect of viscous drag forces in all parallel, identical, individual flow conduits, which allow the water to flow along y-axis, as

$$\bar{f}^*_{grd,y} = N_y \bar{f}^*_{cnd,y},$$ (25b)

where

N_y = number of parallel flow conduits along y-axis,

$\bar{f}^*_{cnd,y}$ = viscous drag force per unit length along y-axis in a flow conduit along y-axis [m^3/s (modified flow rate consistent units)].

The development of the viscous drag forces in the individual flow conduits is based on the modified formulation of Blasius power-law velocity distribution which is applicable to flow conditions in closed ducts and pipes (Goldstein, 1938) and also in open channels (Eraslan, 1973, 1981a, Eraslan, Kim and Harris, 1977, and Eraslan, Erhan and Lin, 1980).

Hence, viscous drag force per unit length along x-axis in a flow conduit along x-axis is formulated, as

$$\bar{f}^*_{cnd,x} = \nu \left[C^*_{lam,x} + C^*_{trb,x} \left(\frac{n+1}{n} \right)^{\frac{2}{n+1}} \frac{2n}{n+1} C^*_{fr,cnd,x} Re_x^{\frac{n-1}{n+1}} \right] V_x,$$ (26a)

where

ν = kinematic viscosity of water [approximately $10^{-6}\, m^2/s$],

$C^*_{lam,x}$ = specified coefficient associated with laminar drag force along x-axis in a flow conduit along x-axis,

$C^*_{trb,x}$ = specified coefficient associated with turbulent drag force along x-axis in a flow conduit along x-axis,

n = specified power in the formulation of Blasius power-law velocity distribution (generally assumed as 8, or between 6 and 12),

21

$C^*_{fr,cnd,x}$ — specified friction coefficient associated with the formulation of Blasius power-law drag for a flow conduit along x-axis,

Re_x — Reynolds number for flow conduit along x-axis.

Similarly, viscous drag force per unit length along y-axis in a flow conduit along y-axis is formulated, as

$$\bar{f}^*_{cnd,y} = \nu \left[C^*_{lam,y} + C^*_{trb,y} \left(\frac{n+1}{n} \right)^{\frac{2}{n+1}} C^*_{fr,cnd,y} Re_y^{\frac{2n}{n+1}} \right]^{\frac{n-1}{n+1}} \tilde{V}_y , \qquad (26b)$$

where

$C^*_{lam,y}$ — specified coefficient associated with laminar drag force along y-axis in a flow conduit along y-axis,

$C^*_{trb,y}$ — specified coefficient associated with the turbulent drag along y-axis in a flow conduit along y-axis,

$C^*_{fr,cnd,y}$ — specified friction coefficient associated with the formulation of Blasius power-law drag for a flow conduit along y-axis,

Re_y — Reynolds number for flow conduit along y-axis.

Reynolds numbers for the two sets of flow conduits are defined, as

$$Re_x = \frac{d_x \tilde{V}_x}{\nu} , \qquad Re_y = \frac{d_y \tilde{V}_y}{\nu} , \qquad (27)$$

where

d_x = specified characteristic length of the cross-sectional area of a flow conduit along x-axis [m],

d_y = specified characteristic length of the cross-sectional area of a flow conduit along y-axis [m].

In the development of the components of modified ground drag force per unit length, the subscripts which identify x-downstream element $(i+\frac{1}{2}, j)$ and y-downstream element $(i, j+\frac{1}{2})$ are intentionally neglected for reasons of brevity in the formulations.

It is important to note that special case with the number of flow conduits Nx = 1 (and/or Ny = 1), the x-component (and/or y-component) of modified ground drag force per unit length [Eqs. (25a) and (25b)] represents the viscous drag force per unit length in a single flow conduit along x-axis (and/or y-axis) [Eqs. (26a) and (26b)]. Hence, in the limiting case for the minimum number of flow conduits, the general formulation of the components of modified ground drag force per unit length automatically represents a plausible subsidiary model for modified ground drag force per unit length for (1) surface-water flow conditions in a water course, or (2) surface-layer flow conditions above the ground.

The number of flow conduits along x-axis (and/or the number of flow conduits along y-axis) in a central element (or in a staggered element) are determined based on (1) the areal porosity of the plane normal to x-axis (and/or areal porosity of the plane normal to y-axis), and (2) the characteristic length of the cross-sectional area of the flow conduits along x-axis (and/or the characteristic length of the flow conduits along y-axis), as

$$N_x = \frac{\overline{\Delta y} \, \overline{D} \, \overline{\alpha}_x}{C_{N,x} \, d_x^2}, \qquad N_y = \frac{\overline{\Delta x} \, \overline{D} \, \overline{\alpha}_y}{C_{N,y} \, d_y^2}, \qquad (28)$$

where

$C_{N,x}$ = specified shape-factor coefficient associated with the cross-sectional area of a flow conduit along x-axis,

$C_{N,y}$ = specified shape-factor coefficient associated with the cross-sectional area of a flow conduit along y-axis.

For small values of the characteristic length of the cross-sectional area of a flow conduit, which could be comparable to grain size of the packed solid material in the ground, the number of flow conduits would be large. Therefore, according to the general formulation [Eqs. (25a) through (26b)], the component of modified ground drag force per unit length becomes the combined effect of a large number of viscous drag forces per unit length in a large number of parallel, identical conduits with small cross-sectional areas. Hence, in the limiting case of a maximum number of flow conduits, the general formulation of the components of modified ground drag force per unit length automatically represents a plausible subsidiary model for modified ground drag force per unit length for groundwater flow conditions through porous media.

The formulations of the subsidiary model indirectly accomplishes the elimination of the auxilary flow variables, x-component of modified ground drag force per unit length along x-axis $\overline{f}^*_{grd,x}$ and y-component of modified ground drag force per unit length along y-axis $f^*_{grd,y}$ [Eqs. (25a) through (28)], in terms of the principle flow variables, consisting of (1) apparent water-surface elevation H (associated with apparent water depth D), (2) flow rate (marked) along x-axis $\tilde{G}_x$ associated with apparent x-component of velocity (marked) $\tilde{V}_x$, (3) flow rate (marked) along y-axis $\tilde{G}_y$ associated with apparent y-component of velocity (marked) $\tilde{V}_y$.

Formulation of the variations of geometrical and geophysical properties of the ground with apparent water depth in central elements

All geometrical and geophysical properties of the ground, which control the composite subsurface-surface flow conditions associated with the rainfall-runoff in valleys with steep terrain, are specified in central elements (i,j) as variable conditions which depend on apparent depth of water above the impermeable bed-bottom layer. The necessary information about the variations of the geometrical and geophysical properties of the ground with apparent depth of water in central elements (i,j) can be obtained from two sources.

First, the geophysical properties of the ground, which specify the general characteristics of the valley fill below the ground surface, can be determined from the available information about the geological and geomorphological characteristics of the valley. This information must include the approximate depth to the impermeable bed-bottom layer, which usually corresponds to the basement rock or a densely packed clay layer. In parts of the geohydrological region where a reasonabley impermeable layer can not be found, the depth to the water table, associated with the standing deep groundwater level, can be used to represent the

23

impermeable bed-bottom layer for subsurface flow conditions.

The information about the nonhomogenous and nonisotropic porosity characteristics of the different layers of valley fill above the impermeable bed-bottom layer is used to specify the variations of porosity conditions with apparent water depth in the elements for subsurface flow conditions (Eraslan, Erhan and Lin, 1981). This information must also include the effects of the directional small channels and water courses on the nonhomogeneous and nonisotropic porosity characteristics of the different layers.

The available information about the orientation and inclination (relative to the geophysical plane) of the basement rock or a densely packed clay layer is used to determine approximately the angles between the three orthogonal axes (x,y,z) of the discretized formulation and the true vertical direction normal to the geophysical surface plane.

Second, the geometrical properties of the directional channels of large water courses can be obtained directly from the available topographic information about the ground elevation conditions in the valley.

The general slope conditions of the ground surface can be obtained from the elevation contours on 1:24,000 scale maps of the valley.

The number of large water courses and their sizes, shapes, directions and slopes can also be obtained from the elevation contours on 1:24,000 scale maps of the valley.

The variations of the geometrical and geophysical properties of the ground with apparent depth of water in central element (i,j) are formulated based on specified functions (linear and/or second order) of apparent water depth in different specified layers, between specified values of apparent water depth, in central elements (i,j). Hence, any geometrical or geophysical property $\overline{\Psi}_{i,j}$ of the ground in central element (i,j) between layer 1 and layer $l+1$ is formulated, as

$$\overline{\Psi}_{i,j}(\overline{D}_{i,j}) = \Psi_{l;i,j}(\overline{D}_{i,j}) \quad \text{for } D_{l;i,j} \leq \overline{D}_{i,j} \leq D_{l+1;i,j}, \qquad (29)$$

where

$\overline{\Psi}_{i,j}$ — average values of a geometrical or geophysical property of the ground for subsurface-surface flow conditions in central element (i,j),

$\Psi_{l;i,j}(\overline{D}_{i,j})$ — specified functional value, based on the value of apparent water depth, of a geometrical or geophysical property of the ground in layer 1 for subsurface-surface flow conditions in central element (i,j),

$D_{l;i,j}$ — specified data for the depth of the bottom surface of the specified layer 1 in central element (i,j),

$D_{l+1;i,j}$ — specified data for the depth of the top surface of the specified layer 1 in central element (i,j).

The general formulation of the specified variation of any geometrical or geophysical property $\overline{\Psi}_{i,j}$ of the ground with the apparent water depth in central element (i,j) accomplishes the elimination of all the auxilary flow variables which represent the geometrical and geophysical properties of the ground, consisting of (1) areal flow porosities of the planes normal to x-axis and y-axis, $\overline{\alpha}_x$ and $\overline{\alpha}_y$, respectively, (2) modified storativity coefficient associated with the variation of water depth along z-axis, $\overline{\gamma}_z$, (3) areal gravity porosities of the planes normal to x-axis and y-axis, $\overline{\beta}_x$ and $\overline{\beta}_y$, respectively, (4)

24

areal pressure porosities of the planes normal to x-axis and y-axis, κ_x and κ_y, respectively, (5) the angles between x-axis, y-axis and z-axis and the true vertical direction normal to the geophysical surface plane, $\theta_{n,x}$, $\theta_{n,y}$ and $\theta_{n,z}$, respectively, (6) specified coefficients associated with laminar drag forces along x-axis and y-axis in flow conduits along x-axis and y-axis, $C^*_{lam,x}$ and $C^*_{lam,y}$, respectively, (7) specified coefficients associated with turbulent drag forces along x-axis and y-axis in flow conduits along x-axis and y-axis, $C^*_{trb,x}$ and $C^*_{trb,y}$, respectively, (8) friction coefficients associated with the formulations of Blasius power-law drag for turbulent drag forces along x-axis and y-axis in flow conduits along x-axis and y-axis, $C^*_{fr,cnd,x}$ and $C^*_{fr,cnd,y}$, respectively, (9) specified characteristic lengths of the cross-sectional areas of the flow conduits along x-axis and y-axis, dx and dy, respectively, and (10) specified shape-factor coefficients associated with the cross-sectional areas of the flow conduits along x-axis and y-axis, $C_{N,x}$ and $C_{N,y}$, respectively, in terms of the principle flow variable, apparent water surface elevation H (associated with apparent water depth D) in central element (i,j).

Linear interpolation of the geometrical and geophysical properties of the ground in staggered elements

The geometrical and geophysical properties of the ground, associated with the composite subsurface-surface flow conditions in valleys with steep terrain, are specified as conditions in central elements (i,j). Therefore, according to the special computational algorithm of the staggered discrete-element method, it is necessary to interpolate the values of the geometrical and geophysical properties at x-downstream enclosure surface at (i+½,j) of central element (i,j), which corresponds to the location of x-downstream element at (i+½,j), and at y-downstream enclosure surface at (i,j+½), which corresponds to the location of y-downstream element (i,j+½).

According to the geometrical discretization technique of the variable-size, staggered discrete-element method, the length along x-axis of x-downstream element (i+½,j) is formulated as the minimum length along x-axis of the two adjacent central elements (i+1,j) or (i,j) Eq. (21) . Therefore, linear interpolation of any geometrical or geophysical property of the ground in x-downstream element (i+½,j) is based on one of two formulations, as

for $\Delta x_{i,j} < \Delta x_{i+1,j}$:

$$\overline{\Psi}_{i+\frac{1}{2},j} = \overline{\Psi}_{i,j} + \frac{2\,\Delta x_{i,j}}{(\Delta x_{i,j} + \Delta x_{i+1,j})}\ (\overline{\Psi}_{i+1,j} - \overline{\Psi}_{i,j})\ , \tag{30a}$$

or for $\Delta x_{i,j} \geq \Delta x_{i+1,j}$:

$$\overline{\Psi}_{i+\frac{1}{2},j} = \overline{\Psi}_{i+1,j} - \frac{2\,\Delta x_{i+1,j}}{(\Delta x_{i,j} + \Delta x_{i+1,j})}\ (\overline{\Psi}_{i+1,j} - \overline{\Psi}_{i,j})\ . \tag{30b}$$

Similarly, the length along y-axis of y-downstream element (i,j+½) is formulated as the minimum length along y-axis of the two adjacent central elements (i,j+1) and (i,j) [Eq. (22)]. Therefore, linear interpolation of any geometrical or geophysical property of the ground in y-downstream element (i,j+½) is based on one of two formulations, as for $\Delta y_{i,j} < \Delta y_{i,j+1}$:

$$\overline{\Psi}_{i,j+\frac{1}{2}} = \overline{\Psi}_{i,j} + \frac{2 \Delta y_{i,j}}{(\Delta y_{i,j} + \Delta y_{i,j+1})} (\overline{\Psi}_{i,j+1} - \overline{\Psi}_{i,j}), \tag{31a}$$

or for $\Delta y_{i,j} \geq \Delta y_{i,j+1}$:

$$\overline{\Psi}_{i,j+\frac{1}{2}} = \overline{\Psi}_{i,j+1} - \frac{2 \Delta y_{i,j+1}}{(\Delta y_{i,j} + \Delta y_{i,j+1})} (\overline{\Psi}_{i,j+1} - \overline{\Psi}_{i,j}). \tag{31b}$$

The general representation of the linear interpolation formulas for any geometrical or geophysical property of the ground Eqs. (30a) through (31b) accomplish the elimination of all the auxilary variables which represent the values of the geometrical and geophysical properties of the ground in x-downstream element $(i+\frac{1}{2},j)$ and in y-downstream element $(i,j+\frac{1}{2})$ in terms of the values of the geometrical and geophysical properties of the ground in three adjacent element (i,j), $(i+1,j)$ and $(i,j+1)$.

Marked-upwind-differencing technique for the components of flow rate in staggered elements

A mathematical model for simulating rainfall-runoff conditions in a valley with steep terrain must include the capability to consider local dry-ground conditions, since during the time periods before and after a rainfall event, most parts of the valley could effectively be without sufficient water to sustain subsurface and/or surface flow conditions. The computational algorithm of the fast-transient, discrete-element, composite subsurface-surface rainfall-runoff model utilizes a marked-upwind-differencing technique for considering the dry-ground conditions based on apparent water depth $D^*_{min;i,j}$ in central elements (i,j).

In each central element (i,j), a minimum water depth is considered, as

$D^*_{min;i,j}$ = specified minimum water depth below which subsurface- and/or surface-flow conditions cannot be sustained in element (i,j).

The minimum water depth $D^*_{min;i,j}$ can be specified as zero without any computational difficulty. However, it is generally more realistic to assume a reasonably small value greater than zero for the minimum depth $\overline{D}_{i,j}$ in element (i,j), since subsurface groundwater flow conditions and/or surface-flow conditions in channelized water courses usually terminate below a finite value of apparent water depth in an actual valley with steep terrain.

Based on consideration of physical conditions, the computational algorithm of the fast-transient, discrete-element rainfall-runoff model includes a special feature which guarantees that no water can flow out of a dry-ground element. The computational algorithm guarantees that water cannot flow out of a dry-ground element by implementing a marked-upwind-differencing technique in calculating the flow rates in staggered elements $(i+\frac{1}{2},j)$ and $(i,j+\frac{1}{2})$, which correspond to the locations of the upstream and downstream enclosure surfaces of central element (i,j) (Eraslan, Erhan and Lin, 1981).

According to the marked-upwind-differencing technique, the computational algorithm considers two comparison steps in calculating the flow rate along x-axis in x-upstream enclosure surface at $(i+\frac{1}{2},j)$ of central element (i,j), which corresponds to flow rate along x-axis in x-upstream element $(i+\frac{1}{2},j)$, as

for $\overline{G}_{x;i+\frac{1}{2},j} \geq 0$ (calculated outflow rate):

if $\overline{D}_{i,j} \leq D^*_{min;i,j}$,

$$\widetilde{G}_{x;i+\frac{1}{2},j} = 0 \text{ (flow rate marked as zero)}, \tag{32a}$$

$$\frac{\partial \widetilde{G}}{\partial t}x;i+\frac{1}{2},j = 0 \text{ (time-derivative of flow rate marked as zero)}, \tag{32b}$$

or for $\overline{G}_{x;i+\frac{1}{2},j} < 0$ (calculated inflow rate):

if $\overline{D}_{i+1,j} \leq D^*_{min;i+1,j}$,

$$\widetilde{G}_{x;i+\frac{1}{2},j} = 0 \text{ (flow rate marked as zero)}, \tag{32c}$$

$$\frac{\partial \widetilde{G}}{\partial t}x;i+\frac{1}{2},j = 0 \text{ (time-derivative of flow rate marked as zero)}. \tag{32d}$$

Similarly, the computational algorithm considers two comparison steps in calculating the flow rate along y-axis in y-upstream enclosure surface at $(i,j+\frac{1}{2})$ of central element (i,j), which corresponds to flow rate along y-axis in y-upstream element $(i,j+\frac{1}{2})$, as

for $\overline{G}_{y;i,j+\frac{1}{2}} \geq 0$ (calculated outflow rate):

if $\overline{D}_{i,j} \leq D^*_{min;i,j}$,

$$\widetilde{G}_{y;i,j+\frac{1}{2}} = 0 \text{ (flow rate marked as zero)}, \tag{33a}$$

$$\frac{\partial \widetilde{G}}{\partial t}y;i,j+\frac{1}{2} = 0 \text{ (time-derivative of flow rate marked as zero)}, \tag{33b}$$

or for $\overline{G}_{y;i,j+\frac{1}{2}} < 0$ (calculated inflow rate):

if $\overline{D}_{i,j+1} \leq D^*_{min;i,j+1}$,

$$\widetilde{G}_{y;i,j+\frac{1}{2}} = 0 \text{ (flow rate marked as zero)}, \tag{33c}$$

$$\frac{\partial \widetilde{G}}{\partial t}y;i,j+\frac{1}{2} = 0 \text{ (time-derivative of flow rate marked as zero)}. \tag{33d}$$

The marked-upwind-differencing technique of the computational algorithm guarantees that the mass conservation principle for water is satisfied identically in every central element (i,j) and staggered element $(i+\frac{1}{2},j)$ or $(i,j+\frac{1}{2})$. Furthermore, and more importantly, it also guarantees that the mass conservation principle for water is satisfied identically for the entire flow region.

Since the computational algorithm can readily consider dry-ground conditions in the elements, the present fast-transient, two-dimensional, discrete-element rainfall-runoff model is superior to all other existing

27

mathematical models for simulating rainfall-runoff flow conditions in actual valleys with steep terrain.

Numerical solution algorithm and structure of RAINER computer code

The governing computational system of discrete-element equations, consisting of (1) discrete-element equation for apparent water surface elevation $\overline{H}_{i,j}$ in element (i,j) [Eq.(14)], (2) discrete-element equation for x-component of flow rate (marked) $\widetilde{G}_{x;i+\frac{1}{2},j}$ in x-upstream element (i-½,j) or x-downstream element (i+½,j) [Eq. (19)], and (3) discrete-element equation for y-component of flow rate (marked) $\widetilde{G}_{y;i,j\pm\frac{1}{2}}$ in y-upstream element (i,j-½) or y-downstream element (i,j+½) [Eq. (20)], represent a moderately coupled, generally nonautonomous system of first-order, nonlinear (quasi-linear), ordinary differential equations, as

$$\frac{d\overline{H}}{dt}_{i,j} = \overline{F}_{H;i,j} = \left[\overline{F}_H(t,\overline{H},\widetilde{G}_x,\widetilde{G}_y)\right]_{i,j}, \tag{34a}$$

$$\frac{d\widetilde{G}}{dt}_{x;i\pm\frac{1}{2},j} = \overline{F}_{G,x;i\pm\frac{1}{2},j} = \left[\overline{F}_{G,x}(t,\overline{H},\widetilde{G}_x)\right]_{i\pm\frac{1}{2},j}, \tag{34b}$$

$$\frac{d\widetilde{G}}{dt}_{y;i,j\pm\frac{1}{2}} - \overline{F}_{G,y;i,j\pm\frac{1}{2}} - \left[\overline{F}_{G,y}(t,\overline{H},\widetilde{G}_y)\right]_{i,j\pm\frac{1}{2}}. \tag{34c}$$

The computational system of first-order ordinary differential equations are numerically integrated according to a two-time-step, three-time-level (mid-point-time-splitting) technique, as

$$\overline{H}_{i,j}^{n+\frac{1}{2}} = \overline{H}_{i,j}^{n} + \frac{\Delta t}{2}\overline{F}_{H;i,j}^{n}, \tag{35a}$$

$$\overline{F}_{H;i,j}^{n} = \left[\overline{F}_H(t^n,\overline{H}^n,\widetilde{G}_x^n,\widetilde{G}_y^n)\right]_{i,j}, \tag{35b}$$

$$\widetilde{G}_{x;i\pm\frac{1}{2},j}^{n+\frac{1}{2}} - \widetilde{G}_{x;i\pm\frac{1}{2},j}^{n} + \frac{\Delta t}{2}\overline{F}_{G,x;i\pm\frac{1}{2},j}^{n}, \tag{36a}$$

$$\overline{F}_{G,x;i\pm\frac{1}{2},j}^{n} = \left[\overline{F}_{G,x}(t^n,\overline{H}^n,\widetilde{G}_x^n)\right]_{i\pm\frac{1}{2},j}, \tag{36b}$$

$$\widetilde{G}_{y;i,j\pm\frac{1}{2}}^{n+\frac{1}{2}} - \widetilde{G}_{y;i,j\pm\frac{1}{2}}^{n} + \frac{\Delta t}{2}\overline{F}_{G,y;i,j\pm\frac{1}{2}}^{n}, \tag{37a}$$

$$\overline{F}_{G,y;i,j\pm\frac{1}{2}}^{n} = \left[\overline{F}_{G,y}(t^n,\overline{H}^n,\widetilde{G}_y^n)\right]_{i,j\pm\frac{1}{2}}, \tag{37b}$$

$$\overline{H}_{i,j}^{n+1} = \overline{H}_{i,j}^{n} + \Delta t\,\overline{F}_{H;i,j}^{n+\frac{1}{2}}, \tag{38a}$$

$$\overline{F}_{H;i,j}^{n+\frac{1}{2}} - \left[\overline{F}_H(t^{n+\frac{1}{2}},\overline{H}^{n+\frac{1}{2}},\widetilde{G}_x^{n+\frac{1}{2}},\widetilde{G}_y^{n+\frac{1}{2}})\right]_{i,j}, \tag{38b}$$

$$\widetilde{G}_{x;i\pm\frac{1}{2},j}^{n+1} = \widetilde{G}_{x;i\pm\frac{1}{2},j}^{n} + \Delta t\,\overline{F}_{G,x;i\pm\frac{1}{2},j}^{n+\frac{1}{2}}, \tag{39a}$$

$$\overline{F}_{G,x;i\pm\frac{1}{2},j}^{n+\frac{1}{2}} - \left[\overline{F}_{G,x}(t^{n+\frac{1}{2}},\overline{H}^{n+\frac{1}{2}},\widetilde{G}_x^{n+\frac{1}{2}})\right]_{i\pm\frac{1}{2},j}, \tag{39b}$$

$$\widetilde{G}_{y;i,j\pm\frac{1}{2}}^{n+1} = \widetilde{G}_{y;i,j\pm\frac{1}{2}}^{n} + \Delta t \ \overline{F}_{G,y;i,j\pm\frac{1}{2}}^{n+\frac{1}{2}} , \qquad (40a)$$

$$\overline{F}_{G,y;i,j\pm\frac{1}{2}}^{n+\frac{1}{2}} = \left[\overline{F}_{G,y} (t^{n+\frac{1}{2}}, \overline{H}^{n+\frac{1}{2}}, \widetilde{G}_y^{n+\frac{1}{2}}) \right]_{i,j\pm\frac{1}{2}} . \qquad (40b)$$

The stability of the numerical solutions is based on the generalized Courant-Friedrichs-Lewy (CFL) criterion (Courant, Friedrichs and Lewy, 1938, 1965) for free-surface flow conditions (Eraslan, 1981a, 1981b, and Eraslan, Erhan and Lin, 1981). A sufficient condition for the maximum allowable time step which conservatively guarantees the stability of the numerical solutions can be formulated, as

$$\Delta t \leq \frac{\min (\Delta x_{i,j}, \Delta y_{i,j})}{\left[g_a \max (\overline{D}_{i,j}) \right]^{\frac{1}{2}} + \max (\widetilde{V}_{x;i\pm\frac{1}{2},j}, \widetilde{V}_{y;i,j\pm\frac{1}{2}})} . \qquad (41)$$

It is important to note that the maximum allowable time step Δt cannot be accurately estimated a priori for numerical integration of the computational system of ordinary differential equations, since the apparent water depth and the apparent velocity components Vx and Vy can increase rapidly for surface-water flow conditions over steep terrain during the numerical integration. Therefore, the numerical solution requires systematic considerations of the stability criterion Eq. (41) at selected intervals during the numerical integration process.

Preliminary, development-program version of the associated RAINER computer code of the fast-transient, two-dimensional, discrete-element rainfall-runoff model is a compartmentalized computer program which consists of MAIN and the following eleven subroutines:
1. REGION for calculating discretized geometrical properties of the central elements,
2. SOLVER for numerical integration of computational system of discrete-element equations,
3. GROUND for calculating geometrical and geophysical properties of the ground in the central elements,
4. SURFER for calculating time-rate-of-change of apparent water surface elevation Eqs. (14), (34a), (35a), (35b), (38a) and (38b) ,
5. FORCER for calculating time-rate-of-change of components of flow rate (marked) Eqs. (19), (20), (34b), (34c), (36b) through (37b), and (39a) through (40b) ,
6. PUMPER for calculating contribution of pumping conditions,
7. RAININ for calculating contribution of rainfall conditions,
8. BOUNDS for calculating boundary conditions along assumed enclosure boundary of the hydrological region,
9. OUTPUT for outputing the numerical solution results in different formats associated with input data for different computer graphics programs and for restarting the computer simulations,
10. INITAL for starting the computer simulations based on specified input data for initial conditions,
11. TRENCH for calculating the altered geometrical and geophysical properties of the ground as a result of construction.

The development-program version of the RAINER computer code was specifically designed in a compartmentalized form to enable the user to modify its subroutines, based on updated information about the geological and geophysical characteristics of the rainfall-runoff flow conditions in the hydrological region, without requiring major modifications in the fundemental computational structure of its general

29

numerical solution algorithm.

PRELIMINARY APPLICATION OF THE MATHEMATICAL MODEL (RAINER COMPUTER CODE) TO THE SIMULATION OF RAINFALL-RUNOFF FLOW CONDITIONS IN DRY LAKE VALLEY

Dry Lake Valley represents a typical hydrological region designated for the deployment of the proposed M-X missile system. The geotechnically suitable area designated for construction can include up to eight hypothetical clusters of the M-X missile system in the valey.

The surface area of Dry Lake Valley, which extends approximately 56 km along the north-south direction and approximately 22 km along the east-west direction, is shown in Fig. 2 in the standard computer graphics format according to the general-usage computer graphics package PLOTER which was specifically developed for the applications of the general discrete- element method (Lin, Erhan and Eraslan, 1981).

The hydrological region of Dry Lake Valley for the simulation of the rainfall-runoff flow conditions was assumed to be the area within the 300 m topographic-elevation contour above the central area of maximum depression which corresponds to the location of the playa. The land area of the valley outside of the 300 m topographic-elevation contour predominantly consists of the mountains which enclose the valley. The slope conditions of the ground surface in the assumed geohydrological region generally vary between zero and 1.5 m/km along the north-south direction and between zero and 60 m/km along the east-west direction. The selection of Dry Lake Valley for the preliminary analysis of the potential impact of the construction of a M-X cluster group on the rainfall-runoff flow conditions in the valley was based on the considerations of the special geohydrological features of the valley.

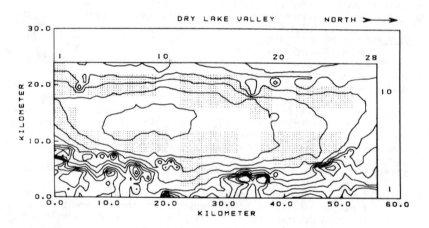

Fig. 2. Computer graphics representation of the plan view of Dry Lake Valley. Ground-surface (topographic) elevation contours (50 m gradation). Shaded areas represent designated ground for construction of a M-X cluster group.

Dry Lake Valley is a typical surficially closed valley where the groundwater in the aquifer systems can only be recharged by the drainage from the precipitation on the valley surface and largely from the precipitation on the mountains enclosing the valley. The drainage conditions are predominantly toward the playa located at the center of

30

the valley. The playa and its surroundings represent the area of maximum depression in the valley which serves as the major collection area for all the drainage flows. The ground surface of the valley floor includes both major and minor water courses which control the subsurface and surface drainage conditions in the valley. Infiltration through the impermeable layers of the valley floor is limited to cracks, faults, and solution holes (in the carbonate rocks) located near the mountains, enclosing the valley, and in the central area which contains the playa. The depth to the groundwater below the playa in Dry Lake Valley limits the evapotranspiration losses from the groundwater reservoir to a minimum.

The groundwater in the deep aquifer system of Dry Lake Valley flows only into Delamar Valley which is to the south/southwest of Dry Lake Valley. Dry Lake Valley is closed to all other groundwater flows due to the formation of the bedrock below the mountains enclosing the valley. Hence, all the water collected from precipitation on the valley surface and on the mountains enclosing the valley drains toward the playa which is located in the area of the maximum depression in the valley, where it recharges the groundwater aquifer systems in the geohydrological region.

The distribution of the types of vegetation in the water courses and in the intermittent streams of the valley has the potential to cause substantial loss of water through transpoevaporation by special plants (phreatophytes) which can readily utilize the water in areas where the soil is saturated.

Hence, the construction of a large M-X cluster group on the designated geotechnically suitable areas in Dry Lake Valley can conceivably alter the rainfall-runoff flow conditions; and consequently it can reduce the rate of collection of the water from the drainage areas in the mountains enclosing the valley. Ample distribution of special plants (phreatophytes), could cause rapid evaporation of the water that could accumulate in the water courses and in the subsurface layers which could be blocked by the construction of the M-X clusters. The combined effect of the reduction of rainfall-runoff flow toward the playa and the increased rate of transpoevaporation, caused by the redistribution of the accumulated water on the surface and in the saturated zone in the subsurface layer, could lead to increased rates of evaporation from the surface area of the valley; and it can ultimately lead to the reduction in recharge conditions of the groundwater aquifer systems in the geohydrological region.

The simulation studies of the rainfall-runoff flow conditions in Dry Lake Valley required various sets of input data to the RAINER computer code about the geometrical and geophysical characteristics of the valley. Considering extremely limited quantity and quality of the available field-measured data, the computer simulation results obtained by implementing RAINER computer code can only be classified as preliminary results which can be used for identifying potential problems associated with the geohydrological impact of the construction of a M-X cluster group on designated geotechnically suitable areas in Dry Lake Valley.

The application of RAINER computer code was based on approximate but realistic sets of input data. The slope conditions of the valley floor were determined from the topographic information available on the 1:24,000 scale maps of Dry Lake Valley. The channelization conditions associated with the number and size of the watercources and intermittent streams on the valley floor were also determined from the topographic information available on the 1:24,000 scale maps of Dry Lake Valley.

The depth of the flow bed, representing the subsurface- and surface-flow layer, above the impermeable bed-bottom layer in the valley floor, was estimated by considering the available geotechnical

31

information about Dry Lake Valley. The maximum depth of the porous ground for the composite subsurface-surface flow conditions was assumed to be less than 20 meters. Since, the actual depth to the stationary water table in Dry Lake Valley is more than 30 meters, the computer simulation results for the altered rainfall-runoff flow conditions after the construction of the M-X cluster group on designated geotechnically suitable areas, represents a conservative analysis of the impact on the geohydrological conditions in the valley.

The drainage areas in the mountains enclosing the valley were determined from the topographic information avaiable on the 1:24,000 scale maps of Dry Lake Valley.

The necessary input data for the geophysical conditions of the porous ground was approximately determined based on the available information from field-measurements and also from the literature. The limited available data about the permeability characteristics of the valley fill was converted into appropriate forms as required by the input data sets for specifying the geophysical conditions in central elements (i,j) which represent the two-dimensional variations of the geophysical properties over the geohydrological region of the valley (Eraslan, Erhan and Lin, 1981). The values for all areal porosities on planes normal to north-south direction were estimated as varying between 0.12 and 0.40. The values for all areal porosities on planes normal to east-west direction were estimated as varying between 0.20 and 0.40. The values for the characteristic length (including the effect of channelized water courses) of the cross-sec- tional area of the flow conduits along the north-south direction were estimated as varying between 0.01 m and 1.0 m. The values for the characteristic length (including the effect of channelized water courses) of the cross-sectional areas of the conduits along the east-west direction were estimated as varying between 0.5 m and 1.0 m. The values for the friction coefficient associated with the formulation of Blasius power-law drag forces in the flow conduits along the north-south direction and along east-west direction were estimated as 0.02. The values for the shape-factor coefficients associated with the cross-sectional areas of the flow conduits along the north-south direction and along the east-west direction were assumed as 1.0.

The duration of the rainfall during the rainstorm was assumed as 1 h. The rainfall rate was assumed as 2.0 cm/h approximately. The rainfall was assumed to be spatially uniform over the entire surface area of Dry Lake Valley. Although the assumed duration and rate of rainfall can be considered as representative of typical thundershower conditions in the climatological region, the assumed spatially uniform distribution over the entire valley surface is definitely not representative of the typical rainstorm conditions over Dry Lake Valley. The spatially uniform rainfall condition was considered in the preli- minary computer simulations of the rainfall-runoff flow conditions, since the study was only concerned with a hypothetical case for the assessment of the potential geohydrological impact of the construction of a M-X cluster group on designated geotechnically suitable areas in Dry Lake Valley.

The results of the computer simulations were considered in two forms as: (1) effective-velocity vectors for rainfall-runoff flow con- ditions and (2) constant-depth contours for accumulated water.

Since values of the components of apparent velocity cannot be defined for dry-ground conditions, another representative set as components of effective velocity were considered in central element (i,j), as

$$\overline{V}_{eff,x;i,j} = \frac{\overline{G}_{x,i,j}}{\Delta y_{i,j} D_{bed;i,j}} \quad , \quad \overline{V}_{eff,y;i,j} = \frac{\overline{G}_{y,i,j}}{\Delta x_{i,j} D_{bed;i,j}} \quad , \qquad (42)$$

in terms of linearly-averaged components of flow rate $\overline{G}_{x,i,j}$ and $\overline{G}_{y,i,j}$ and specified, temporally-constant reference depth of flow bed $D_{bed;i,j}$ in central element (i,j).

The results of the computer simulations were presented in computer graphics format by utilizing the standard general-usage computer graphics package PLOTER of the discrete-element method (Lin, Erhan and Eraslan, 1981).

Computer Simulation Results for Natural Rainfall-Runoff Flow Conditions in Dry Lake Valley

RAINER computer code was applied to the simulation of natural rainfall-runoff flow conditions in Dry Lake Valley for a typical rainstorm scenario, with a rainfall duration of 1 h and with a rainfall rate of 2 cm/h, on the entire surface area of the valley and on the drainage areas in the mountains. The computer simulation results for natural rainfall-runoff flow conditions were obtained for a four-hour real-time period after the start of the rainstorm, which corresponded to a three-hour real-time period after the end of the rainstorm, in the hydrological region.

The effective-velocity vectors and constant-depth contours for accumulated water (subsurface and surface), representing the natural rainfall-runoff flow conditions in Dry Lake Valley at one hour after the start of the rainstorm (at the end of the rainstorm) are presented in Figs. 3a and 3b, respectively.

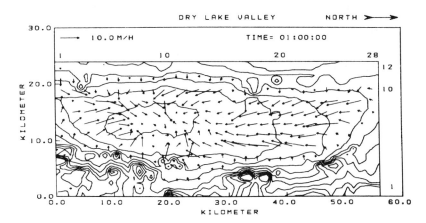

Fig. 3a. Computer simulation results for effective-velocity vectors, representing natural rainfall-runoff flow conditions in Dry Lake Valley at 1 h after the start of the rainstorm (at the end of the rainstorm).

The computer simulation results indicate regionally developed rainfall-runoff flow conditions, with effective-velocity vectors depending on the local slope and channelization characteristics of the specified water courses in the valley floor (Fig. 3a). The results also indicate that the inflow conditions from the drainage areas along the assumed enclosure boundaries of the geohydrological region of Dry Lake Valley are the dominating factors which control the accumulation of water in areas at higher elevations, near the mountains (see Fig. 3b), where the local depth of accumulated water layer increases to approximately 45 cm. Furthermore, as expected, the depth of accumulated

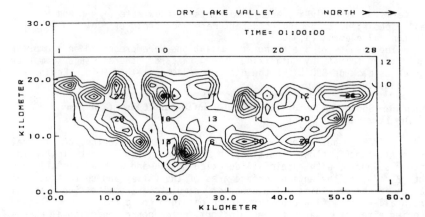

Fig. 3b. Computer simulation results for constant-depth contours for accumulated water (5 cm gradation), representing natural rainfall-runoff flow conditions in Dry Lake Valley at 1 h after the start of the rainstorm (at the end of the rainstorm).

water layer in the areas which include the playa, corresponding to the location of the area of maximum depression in the valley, also starts increasing (see Fig. 3b), with maximum local depth of water layer reaching approximately 19 cm.

The effective-velocity vectors and constant-depth contours for accumulated water (subsurface and surface), representing the natural rainfall-runoff flow conditions in Dry Lake Valley at two hours after the start of the rainstorm (at one hour after the end of the rainstorm) are presented in Figs. 4a and 4b, respectively.

At two hours after the start of the rainstorm, which corresponds to one hour after the end of the rainstorm, there is no inflow from the

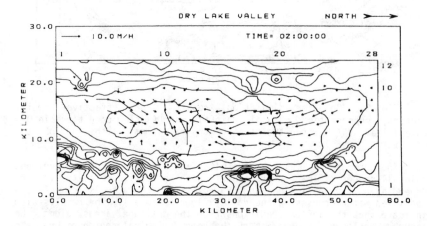

Fig. 4a. Computer simulation results for effective-velocity vectors, representing natural rainfall-runoff flow conditions in Dry Lake Valley at 2 h after the start of the rainstorm (at 1 h after the end of the rainstorm).

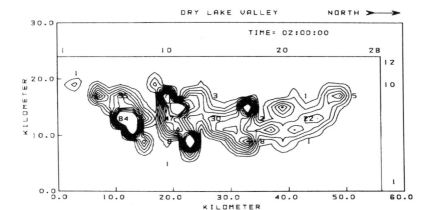

Fig. 4b. Computer simulation results for constant-depth contours for accumulated water (5 cm gradation) representing natural rainfall-runoff flow conditions in Dry Lake Valley at 2 h after start of rainstorm (at 1 h after the end of the rainstorm).

drainage areas along the assumed enclosure boundaries of the geohydro-logical region of Dry Lake Valley. Therefore, as expected, the computer simulation results indicate that depth of the accumulated water in areas at higher elevation, near the mountains, starts decreasing rapidly (Fig. 4b). The rainfall-runoff flow conditions continue mainly toward the area of maximum depression, depending on the special channelization characteristics of the major water courses in the valley floor (Fig. 4a). Furthermore, also as expected the depth of accumulated water layer in areas which include the playa, corresponding to the location of maximum depression in the valley, increases rapidly, with maximum local depth of accumulated water exceeding 80 cm in the vicinity of the playa.

The effective-velocity vectors and constant-depth contours for accumulated water (subsurface and surface), representing natural rain-fall-runoff flow conditions in Dry Lake Valley at four hours after the start of the rainstorm (at three hours after the end of the rainstorm) are presented in Fig. 5a and 5b, respectively.

The computer simulation results indicate that at four hours after the start of the rainstorm (at three-hours after the end of the rainstorm) the continuing rainfall-runoff flow conditions are limited to the major water course which extends from the northern end of toward the playa at the approximate location of the area of maximum depression in the valley (Fig. 5a). This result can be readily justified based on the consideration of the physical conditions in the valley.   The rainfall-runoff flow from areas with higher elevation move rapidly from the east and west enclosure boundaries toward the center of the valley, since the slopes along the east-west direction are much more pronounced than the slopes along the north-south direction. Therefore, the water collects rapidly in the main channelized water course which extends along the north-south direction toward the area of maximum depression which includes the playa in the valley.

The computer simulation results for the depth of accumulated water (subsurface and surface) at four hours after the start of the rainstorm (at three hours after the end of the rainstorm) indicate that the major portion of the water from the precipitation on the valley surface and on the mountains enclosing the valley, during the one-hour duration of the

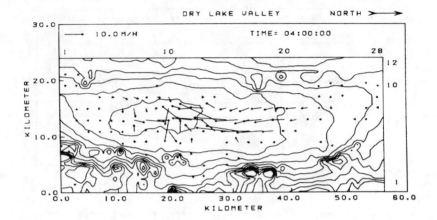

Fig. 5a. Computer simulation results for effective-velocity vectors, representing natural rainfall-runoff flow conditions in Dry Lake Valley at 4 h after the start of the rainstorm (at 3 h after the end of the rainstorm).

rainstorm, would already be collected in the area of maximum depression which includes the playa in the valley (Fig. 5b). The results also indicate, as expected, that the areas at higher elevations would return to local dry-ground conditions three hours after the end of the rainstorm.

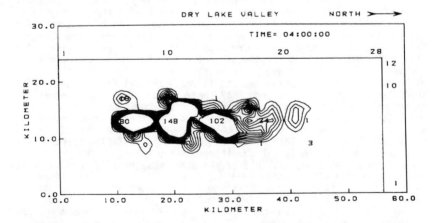

Fig. 5b. Computer simulation results for constant-depth contours for accumulated water (5 cm gradation), representing natural rainfall-runoff flow conditions in Dry Lake Valley at 4 h after the start of the rainstorm (at 3 h after the end of the rainstorm).

The results of the computer simulations conclude that three hours after the end of a rainstorm, with the one-hour duration and 2 cm/h rainfall rate on the entire land area of the valley and on the drainage areas in the mountains, the rainfall-runoff flow conditions in Dry Lake Valley could result in accumulation of approximately 1.5 m of water, with a maximum depth of 1.7 m, in an area of approximately 100 sq.km, in the middle of the valley corresponding to the area maximum depression which includes the playa (Fig. 5b).

Computer Simulation Results for Altered Rainfall-Runoff Flow Conditions with a Hypothetical (or Planned) M-X Cluster Group in Dry Lake Valley

The computer simulation results for altered rainfall-runoff flow conditions after the construction of a hypothetical (or planned) M-X cluster group on designated geotechnically suitable areas in Dry Lake Valley were also obtained for the same typical rainstorm scenario, with a rainfall duration of 1 h and with a rainfall rate of 2 cm/h on the entire land area of the valley and on the drainage areas in the mountains. The computer simulation results for altered rainfall-runoff flow conditions were again obtained for a four-hour real time period after the start of the rainstorm, which corresponded to a three-hour real-time period after the end of the rainstorm in the hydrological region.

Since detailed information about the planned construction of the M-X cluster group was not available, the altered values for the geophysical properties of the ground were based on various plausible assumptions. It was assumed that in the geotechnically suitable areas designated for the construction of the M-X cluster group, the solid material in the ground would be packed denser after the construction. Hence, after the construction, all values for the directional areal porosities would be lower than the presently existing values for unaltered natural state. Furthermore, it was assumed that the operation of heavy equipment and the construction of the roads and shelters could block some of the major and minor water courses in the geotechnically suitable areas designated for the construction of the M-X cluster group.

The altered ground conditions in Dry Lake Valley were approximated based on considerations of (1) reduced values for directional areal porosities, and (2) reduced values for directional characteristics lengths of the cross-sectional areas of the flow conduits (including the effect of channelized water courses) in the geotechnically suitable areas designed for the construction of the M-X cluster group (Fig. 1). The altered ground conditions were approximated by considering altered geophysical conditions in a depth of 4 m from the ground surface only in the designated geotechnically suitable areas. The values for all areal porosities on planes normal to the north-south direction were lowered, with reduced values varying between 0.1 and 0.2 in a depth of 4 m from the ground surface in the designated geotechnically suitable areas. Similarly, the values for all areal porosities along the east-west direction were lowered, with reduced values as varying between 0.1 and 0.2 in a depth of 4 m from the ground surface in the geotechnically suitable areas designated for the construction of the M-X missile cluster group. The values for the characteristic length (including the effect of channelized water courses) of the cross-sectional area of the flow conduits along the north-south direction were reduced to 0.01 m in a depth of 4 m from the ground surface in the designated geotechnically suitable areas. Similarly, the values for the characteristic length (including the effect of channelized water courses) of the cross-sectional area of the flow conduits along east-west direction were reduced to 0.01 m in a depth of 4 m from the ground surface in the designated geotechnically suitable areas.

The average composite values for the directional areal porosities and the directional characteristic lengths of the cross-sectional areas of the flow conduits (including the effect of channelized water courses) in the elements which contain the geotechnically suitable areas designated for the construction of the M-X cluster group were calculated based on the modification of the values corresponding to unaltered natural state, according to the computational algorithm of TRENCH subroutine of RAINER computer code (Eraslan, Erhan and Lin, 1981).

The effective-velocity vectors and constant-depth contours for accumulated water (subsurface and surface), representing the altered rainfall-runoff flow conditions in Dry Lake Valley, with the M-X cluster group on designated geotechnically suitable areas in Dry Lake Valley, at one hour after the start of the rainstorm (at the end of the rainstorm) are presented in Figs. 6a and 6b, respectively.

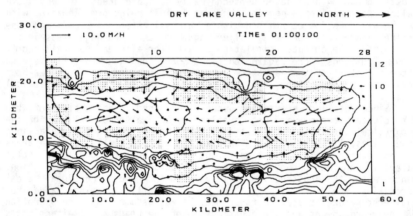

Fig. 6a. Computer simulation results for effective-velocity vectors, representing the altered rainfall-runoff flow conditions in Dry Lake Valley with the M-X cluster group at 1 h after the start of the rainstorm (at the end of the rainstorm).

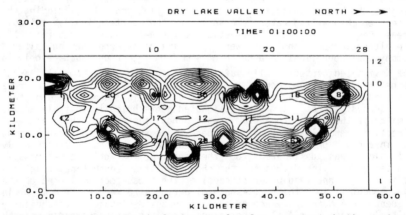

Fig. 6b. Computer simulation results for constant-depth contours for accumulated water (5 cm gradation), representing the altered rainfall-runoff flow conditions in Dry Lake Valley with the M-X cluster group at 1 h after the start of the rainstorm (at the end of the rainstorm).

The computer simulation results indicate regionally developed rainfall-runoff flow conditions except in the geotechnically suitable areas designated for the construction of the M-X cluster group, with the effective-velocity vectors depending on the local slope and channelization characteristics of the specified water courses in the valley floor (Fig. 6a). The results indicate that the inflow conditions from the drainage areas along the assumed enclosure boundaries of the geohydrological region of Dry Lake Valley are the dominating factors which control the accumulation of water in areas at higher elevation, near the mountains (see Fig. 6b), where the local depth of the accumulated water layer increases to more than 100 cm in areas upstream of the geotechnically suitable areas designated for the construction of the M-X cluster group.

The computer simulation results clearly illustrate the blocking effect of the construction of the M-X cluster group (on designated geotechnically suitable areas) on the rainfall-runoff flow conditions in Dry Lake Valley (Figs. 6a and 6b).

The comparison of the results for depth of accumulated water for altered rainfall-runoff flow conditions (Fig. 6b) with the results for depth of accumulated water for natural rainfall-runoff flow conditions (Fig. 3b) clearly indicates that in certain areas which are at a higher elevations than the geotechnically suitable areas, the local accumulated water depth could increase more than 55 cm after the construction of the M-X cluster group in Dry Lake Valley at one hour after the start of the rainstorm (at the end of the rainstorm).

The comparisons of the results for effective-velocity vectors for altered rainfall-runoff flow conditions (Fig. 6a) with the results of for effective-velocity vectors for natural rainfall-runoff flow conditions (Fig. 3a) also indicates that the general drainage flow, from areas at higher elevation to the area of maximum depression which includes the playa, could change considerably after the construction of the M-X cluster group in Dry Lake Valley, at one hour after the start of the rainstorm (at the end of the rainstorm).

The effective-velocity vectors and constant-depth contours for accumulated water (subsurface and surface), representing the altered rainfall-runoff flow conditions, with the M-X cluster group on designated geotechnically suitable areas in Dry Lake Valley, at two hours after the start of the rainstorm (at one hour after the end of the mountain) are presented in Figs. 7a and 7b, respectively.

Again, at two hours after the start of the rainstorm, which corresponds to one hour after the end of the rainstorm, there is no inflow from the drainage areas along the assumed enclosure boundaries of the geohydrological region of Dry Lake Valley. Therefore, under natural drainage conditions, the depth of the accumulated water in areas at higher elevation near the mountains start decreasing rapidly (Fig. 4b). However, the computer simulation results for altered rainfall-runoff flow conditions (Fig. 7b) indicate that, in contrast to the natural rainfall-runoff flow conditions (Figs. 4a and 4b), the local depths of accumulated water at certain areas at higher elevation than designated geotechnically suitable areas do not decrease rapidly. Particularly, in areas located at the north end, in the major water course in the valley (Fig. 7b), for altered rainfall-runoff flow conditions with the M-X cluster group on designated geotechnically suitable areas, the depth of accumulated water remains approximately 70 cm higher than the depth of accumulated water that would have existed for the natural rainfall-runoff flow conditions, at two hours after the start of the rainstorm (one hour after the end of the rainstorm) in Dry Lake Valley (Figs. 4b and 7b). Hence, the rainfall-runoff flow conditions in the major water courses could be blocked considerably after the construction of an M-X cluster group on designated geotechnically suitable areas in Dry Lake Valley.

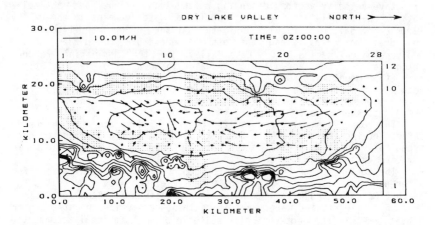

Fig. 7a. Computer simulation results for effective-velocity vectors, representing altered rainfall-runoff flow conditions in Dry Lake Valley with the M-X cluster group at 2 h after the start of the rainstorm (at 1 h after the end of the rainstorm).

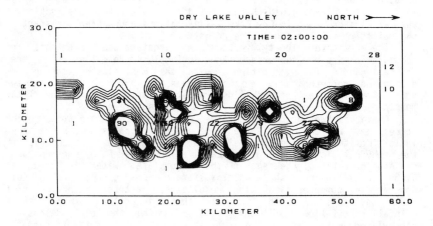

Fig. 7b. Computer simulation results for constant-depth contours for accumulated water (5 cm gradation), representing the altered rainfall-runoff flow conditions in Dry Lake Valley with the M-X cluster group at 2 h after the start of the rainstorm (at 1 h after the end of the rainstorm).

The comparison of the computer simulation results for effective-velocity vectors for altered rainfall-runoff flow conditions (Fig. 7a) with the results for effective-velocity vectors for natural rainfall-runoff flow conditions (Fig. 4a) also indicates that, at two hours after the start of the rainstorm (at one hour after the end of the rainstorm), the rainfall-runoff flow conditions in the major water courses would be generally reduced after the construction of an M-X cluster group in the designated geotechnically suitable areas in Dry Lake Valley.

Furthermore, the computer simulation results indicate that, at two hours after the start of the rainstorm (at one hour after the end of the rainstorm), the depth of accumulated water in certain areas in the vicinity of the playa would be approximately 4-8 cm lower, after the construction of the M-X cluster group (on the designated geotechnically suitable areas) (Fig. 7b) in comparison to the depth of accumulated water under natural rainfall-runoff flow conditions (Fig. 4b) in Dry Lake Valley.

The effective-velocity vectors and constant-depth contours for the accumulated water (subsurface and surface), representing the altered rainfall-runoff flow conditions in Dry Lake Valley with the M-X cluster group at four hours after the start of the rainstorm (at three hours after the end of the rainstorm) are presented in Figs. 8a and 8b, respectively.

The comparison of the computer simulation results for effective-velocity vectors for altered rainfall-runoff flow conditions (Fig. 8a) with the results for the effective-velocity vectors for natural rainfall-runoff flow conditions (Fig. 5a) indicates that, at four hours after the start of the rainstorm (at three hours after the end of the rainstorm), there would be generally reduced rainfall-runoff flow conditions in the major water course, extending from the north end toward the location of the area of maximum depression which includes the playa, after the construction of the M-X cluster group on designated geotechnically suitable areas in Dry Lake Valley.

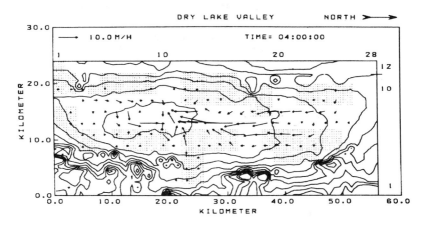

Fig. 8a. Computer simulation results for effective-velocity vectors, representing altered rainfall-runoff flow conditions in Dry Lake Valley with the M-X cluster group at 4 h after the start of the rainstorm (at 3 h after the end of the rainstorm).

The comparison of the computer simulation results for depth of accumulated water for altered rainfall-runoff flow conditions (Fig. 8b) with the results for depth of accumulated water for natural rainfall-runoff flow conditions (Fig. 5b) clearly indicates that at four hours after the start of the rainstorm (at three hours after the end of rainstorm) in certain areas, which are at higher elevations than the designated geotechnically suitable areas, the local accumulated water depth could increase more than 50 cm after the construction of the M-X cluster group in Dry Lake Valley. Particularly, in the area located at the north-east corner of the valley above the major water course, extending from the north end toward the location of the area of maximum depression which includes the playa, the construction of the M-X cluster in the designated geotechnically suitable area could result in depth of accumulated water, under altered rainfall-runoff flow conditions (Fig. 8b) to be approximately 100 cm more than the depth of accumulated water that would have existed in the same area under natural rainfall-runoff flow conditions (Fig. 5b) in Dry Lake Valley.

Furthermore, the computer simulation results for depth of accumulated water in the area of maximum depression which includes the playa indicate that, at four hours after the start of the rainstorm (at three hours after the end of the rainstorm), the depth of accumulated water under altered rainfall-runoff flow conditions, after the construction of the M-X cluster group on the geotechnically suitable areas (Fig. 8b), could be 10-40 cm lower than the depth of accumulated water that would have existed in the same area under natural rainfall-runoff flow conditions (Fig. 5b) in Dry Lake Valley.

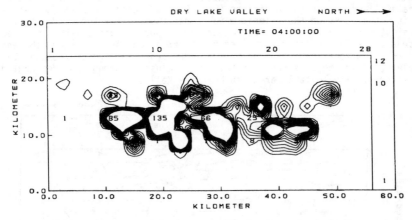

Fig. 8b. Computer simulation results for constant-depth contours for accumulated water (5 cm gradation), representing the altered rainfall-runoff flow conditions in Dry Lake Valley with the M-X cluster group at 4 h after the start of the rainstorm (at 3 h after the end of the rainstorm).

Discussions of Computer Simulation Results

Considering the natural rainfall-runoff flow conditions, the computer simulation results at four hours after the start of rainstorm (three hours after the end of the rainstorm) indicate that the major portion of the water from the precipitation on the valley surface and on the mountains is collected in three hydrological subregions which are located in the general area of maximum depression which includes the playa in Dry Lake Valley. Starting from the southern end of the valley, the three hydrological subregions have surface areas approximately equal to (1) 30 sq.km (7,400 acre), (2) 40 sq.km (9,900 acre), and (3) 30 sq.km (7,400 acre), respectively.

According to the computer simulation results for natural rainfall-runoff flow conditions (Fig. 5b), the representative depth of the accumulated water in the three hydrological subregions are approximately equal to (1) 80 cm (2.6 ft), (2) 148 cm (4.8 ft), and (3) 102 cm (3.3 ft), respectively.

Considering the altered rainfall-runoff flow conditions, after the construction of the M-X cluster group in the designated geotechnically suitable areas in Dry Lake Valley, according to the computer simulation results at four hours after the start of the rainstorm (at three hours after the end of the rainstorm), the representative depth of accumulated water in three hydrological subregions are approximately equal to (1) 85 cm (2.8 ft), (2) 135 cm (4.4 ft), and (3) 66 cm (2.2 ft).

Based on the computer simulation results for the depth of accumulated water, the differences in representative volumes of saturated ground containing water between the altered rainfall-runoff flow conditions after the construction of the M-X cluster group on designated geotechnically suitable areas (Fig. 8b) and the natural rainfall-runoff flow conditions (Fig. 5b), at four hours after the start of the rainstorm (at three hours after the end of the rainstorm), in three geohydrological subregions in Dry Lake Valley can be approximately determined as: (1) $1,500,000$ m^3 (1,200 acre-ft) increase, (2) $5,200,000$ m^3 (4,200 acre-ft) decrease, and (3) $10,800,000$ m^3 (8,700 acre-ft) decrease, respectively. Hence, the net decrease in representative volume of saturated ground containing water in three geohydrological subregions becomes approximately $14,500,000$ m^3 (12,000 acre-ft). Considering a representative value for the volumetric porosity as 0.4 (for gravel, sand, silt, clay, etc.) in three geohydrological subregions, the net decrease in the volume of water that can be collected in the general area of maximum depression which includes the playa can be approximately determined as $5,800,000$ m^3 (4,700 acre-ft).

Hence, the computer simulation results indicate that the alteration of the natural rainfall-runoff flow conditions after the construction of the M-X cluster group on designated geotechnically suitable areas could conceivable result in preventing the collection of $5,800,000$ m^3 (approximately 4,700 acre-ft) of water in the geohydrological subregions in the general area of maximum depression which includes the playa in Dry Lake Valley four hours after the start (three hours after the end) of a rainstorm, with a duration of one hour and a rainfall-rate of 2 cm/h.

The computer simulation results for altered rainfall-runoff flow conditions, during and after a hypothetical rainstorm scenario, with a rainfall duration of 1 h and rainfall rate of 2 cm/h on the entire land area of the valley and on the mountains, also indicates that geotechnically suitable areas could cause significant changes in values of local depth of accumulated water in certain areas in Dry Lake Valley. Particularly in areas at higher elevation than designated geotechnically suitable areas in major water courses, local depth of accumulated water after the construction of the M-X cluster group could be 100 cm more than the local depth of accumulated water that would have existed in the same area under natural rainfall-runoff flow conditions (Figs. 5b and 8b). The excessive accumulation of water in the major water courses due to the blockage of the rainfall-runoff flow conditions could also cause engineering design problems due to possible flooding of the roads and shelters of various proposed M-X clusters. Therefore, the results of the computer simulations indicate that careful mitigating design alternatives must be included in the construction of the proposed M-X cluster group on geotechnically suitable areas, based on careful considerations of the rainfall-runoff flow conditions in Dry Lake Valley.

43

CONCLUSIONS

A fast-transient, two-dimensional, discrete-element rainfall-runoff model and a preliminary, development-program version of its associated RAINER computer code were developed specifically for the simulation of channelized, composite subsurface-surface flow conditions in valleys with steep terrain under general rainfall conditions.

The mathematical formulation of the state-of-the-art model incorporates important unique capabilities which are not included in any other existing mathematical model or computer code that is presently being utilized for the analysis of rainfall-runoff flow conditions in flood plains.

In contrast to all existing models, the present rainfall-runoff model can realistically consider local, temporally varying flow conditions which can go through the necessary transitions between dry-ground conditions and three distinct flow regimes of rainfall-runoff which consists of (1) initial subsurface groundwater flow, (2) subsequent first-level composite flow, as the combination of subsurface groundwater flow and surface-water flow in water courses with directional channels, and (3) final second-level composite flow, as the combination of subsurface groundwater flow, surface-water flow in water courses with directional channels and surface-layer flow above the ground.

The development of the model is based on an approximate, universally applicable mathematical formulation which can consider groundwater flow and surface-water flow conditions either simultaneously or separately at different times and in different parts of a specified geohydrological region. Therefore, the model is applicable to all two-dimensional flow conditions in any geohydrological region where both groundwater and surface-water flow conditions must be considered.

The mathematical model and the preliminary development-program version of its associated RAINER computer code were applied to assessment of the potential geohydrological impact of the construction of a hypothetical (or planned) M-X cluster group on designated geotechnically suitable areas in Dry Lake Valley.

The preliminary results of the computer simulations indicated that the drainage flow conditions in the valley could be significantly altered by the construction of the M-X cluster group on designed geotechnically suitable areas due to possible blockage of the natural rainfall-runoff flow conditions in the major and/or minor water courses. The results illustrated that the blockage of the water courses could lead to excessive accumulation of water in areas at higher elevation than the designated geotechnically suitable areas.

Hence, the study concluded that, unless special mitigating design alternatives can be included in the construction, the elevated water levels under altered rainfall-runoff flow conditions could cause flooding of the roads and shelters of the M-X clusters in Dry Lake Valley.

The computer simulation results for altered rainfall-runoff flow conditions during and after a rainstorm also indicated that the construction of a hypothetical (or planned) M-X cluster group on designated geotechnically suitable areas could considerably reduce the amount of water that can be collected in the area of maximum depression which includes the playa in Dry Lake Valley.

Hence, the study also concluded that the construction of a hypothetical (or planned) M-X cluster group on designated geotechnically suitable areas in Dry Lake Valley has the potential to impact adversely the recharge characteristics of the groundwater in the aquifer systems of the geohydrological region.

ACKNOWLEDGEMENT

This study was supported by the U.S. Department of the Air Force, Ballistic Missile Office, as a part of the program for the Draft Environmental Impact Statement (DEIS) on the M-X Deployment Area Selection and Land Withdrawal/Acquisition.

REFERENCES

Bear,J. 1972. Dynamics of Fluids in Porous Media, Environmental Science Series, American Elsevier Publishing Company, Inc., New York.

Cooley, R.L. and Westphal, J.A. 1974. Application of the Theory of Ground-Water and River-Water Interchange, Winnemucca Reach of the Humboldt River, Nevada. Project Report, Desert Research Institute, Center of Water Resources Research, Reno, Nevada.

Courant, R., Friedrichs, K.O. and Lewy. H. 1938. Uber die Partiellen Differenzengleichurgen der Mathematischen Physik, Mathematische Annalen, Vol. 100, pp. 32-74.

Courant, R., Friedrichs, K.O. and Lewy, H. 1967. On the Partial Difference Equations of Mathematical Physics. IBM Journal, March 1967, pp. 215-234.

Cunningham, A.B. 1977. Modeling and Analysis of Hydraulic Interchange of surface and groundwater. Technical Report H-W 34. Water Resoures Center, Desert Research Institute, Reno, Nevada.

Eraslan, A.H. 1973. Far-Field Modeling of Thermal Discharges in Coastal and Offshore Regions. Proceedings of Conference on Environmental Impact Statements for Nuclear Power Plants. Georgia Institute of Technology. Pergamon Press.

Eraslan, A.H. 1974a. Far-Field Model for Offshore Thermal Discharges. American Nuclear Society Transactions, Vol. 18, p.55.

Eraslan, A.H. 1974b. Discrete-Element Computational Model for Predicting Transient, One-Dimensional Hydrodynamic and Water Quality Conditions in Estuaries and Coastal Rivers. 11th Annual Meeting, Society of Engineering Science, Durham, North Carolina.

Eraslan, A.H. 1975a. Transient, Two-Dimensional Discrete-Element Far-Field Model for Thermal Impact Analysis of Power Plant Discharges in Coastal and Offshore Regions. Part 1: General Description of the Mathematical Model and the Results of an Application. ORNL-4940, Oak Ridge National Laboratory, Oak Ridge, Tennessee.

Eraslan, A.H. 1975b. A transient, Two-Dimensional, Discrete-Element Model for Far-Field Analysis of Thermal Discharges in Coastal Regions. in : Schetz (Editor), Progress in Astronantics and Aeronautics, Vol. 36, AIAA publication, MIT Press, Cambridge, Massachusetts.

Eraslan, A.H. 1976. Discrete-Element Models for Thermal Impact Assessment of Large Power Plants in Estuaries and Coastal Regions. in : Transactions 12th Annual Meeting, Southeastern Seminar on Thermal Sciences, University of Virginia, Charlottseville, Virginia.

Eraslan, A.H. 1977. Applications of Unified Transport Approach Models for the Assessment of Power Plant Impact on Aquatic Environments. in: Energy and Environmental Stress in Aquatic Systems, the fifth in a series of Ecological Symposia paper No. 130, Savannah River Ecology Laboratory, Augusta, Georgia.

Eraslan, A.H. 1981a. ESTONE : A computer Code for Simulating Fast-Transient, One-Dimensional Hydrodynamic, Thermal and Salinity Conditions in Controlled Rivers and Estuaries. ORNL/NUREG-8102. Oak Ridge National Laboratory, Oak Ridge, Tennessee.

Eraslan, A.H. 1981b. A Fast-Transient, Three-Dimensional Discrete-Element Mathematical Model for Predicting Coupled Hydrodynamic, Thermal and Salinity Conditions in Lakes, Rivers, Reservoirs, Estuaries and Coastal Zones for Assessment of the Impact of Power Plant Operations. ORNL-NUREG report, Oak Ridge National Laboratory, Oak Ridge, Tennessee. (in preparation)

Eraslan, A.H. 1981c. Discrete-Element Method: An Advanced Numerical Simulation Technique for Transient Multi-Dimensional Laminar-Turbulent Free-Surface Flow Problems in Regions with Geometrically Complex Boundaries. International Conference on Numerical Methods in Laminar and Turbulent Flow, Venice, Italy, July 13-16, 1981 (to be presented).

Eraslan, A.H., Kim, K.H. and Harris, J.L. 1977. A systematic Application of Transient, Multi-Dimensional Models for Complete Analysis of Thermal Impact in Regions with Serve Reversing Flow Conditions. in: Proceedings of the First Conference on Waste Heat Management and Utilization, Miami Beach, Florida, May, 1977, p. IV-B-4.

Eraslan, A.H., Erhan, I.H. and Lin, W.L. 1980. A Supplementary Study of the Impact of Power-Plant Operation of the Hudson River with Emphesis on Bowline Point Generating Station. Appendix E, Final Environmental Statement, Bowline Point Generating Station. U.S. Army Engineer Division, New York.

Eraslan, A.H., Erhan, I.H. and Lin, W.L. 1981. A Mathematical Model for the Analysis of Potential Impact of M-X Missile System on Rainfall-Runoff Conditions in the Valleys of Great Basin Physiographic Region (report prepared for the U.S. Department of the Air Force).

Evans, M.E. and Harlow, F.H. 1967. The Particle-in Cell Method for Hydrodynamic Calculations, Los Alamos Scientific Laboratory, Rept. No. LA-2139, Los Alamos, New Mexico.

Freeze, R.A. 1972. Role of Subsurface Flow in Generating Surface Runoff 1, Base Flow Contributions to Channel Flow: Water Resources Research, Vol. 8, No. 3, pp. 609-623.

Gentry, R.A., Martin, R.E. and Daly, B.J. 1966. An Eulerian Differencing Method for Unsteady Compressible Flow Problem, Journal of Computational Physics, Vol. 1, pp. 87-118.

Goldstein, S. 1938. Modern Developments in Fluid Dynamics. Oxford Clarendon Press, London.

Harlow, F.H. and Welch, J.E. 1965. Numerical Calculation of Time-Dependent Viscous Incompressible Flow of Fluid with Free Surface. Physics of Fluids, Vol. 8, No. 12, pp. 2182-2189.

Harlow, F.H. and Welch, J.E. 1966. Numerical Study of Large Amplitude Free Surface Motions. Physics of Fluids, Vol. 9, pp. 842-851.

Holdeman, J.T. 1977 PREPR2: A Program to Aid in the Preparation of Input Data for the Farout Two Dimensional Code, ORNL/CSD/TM-19, Oak Ridge National Laboratory, Oak Ridge, Tennessee.

Irmay, S. 1938. On the theoretical derivation of Darcy and Forchheimer formulas. Trans. Amer. Geophys. Union No. 4, 39, pp. 702-707.

Leendertse, J.J., Alexander, R.C. and Siao-Kung, L. 1973. A three-Dimensional Model for Estuaries and Coastal Seas: Vol. 1, Principles of Computation. R-1417-OWRR, The Rand Corporation, Santa Monica, California.

Lin, W.L. 1981. Application of the Discrete-Element Method to Fast-Transient Three-Dimensional Incompressible Flow with Free-Surface Conditions in Regions of Arbitrary Geometry. Ph.D. Dissertation, The University of Tennessee, Knoxville, Tennessee.

Lin, W.L. and Erhan, I.H. 1978. PREPAR : Automated Discrete-Element-Generator Program for Multi-Connected Flow Regions with Geometrically Complex Shoreline Boundaries. Environmental Impact Project Report. The University of Tennessee, Knoxville, Tennessee.

Lin, W.L. and Eraslan, A.H. 1981. A Fast-Transient, Three-Dimensional Discrete-Element Hydrodynamic Model for Predicting Free-Surface Surface-Water Flow Conditions in Regions with Geometrically Complex Boundaries. ORNL-NUREG report, Oak Ridge National Laboratory, Oak Ridge, Tennessee (in preparation).

Pinder, G.F. and Sauer, S.P. 1971. Numerical Simulation of Flood Wave Modification Due to Bank Storage Effects: Water Resources Research, Vol. 7, No. 1, pp. 63-70.

Roache, P.J. 1972. Computational Fluid Dynamics. Hermosa Pulishers, Albuquerque, New Mexico.

Scheidegger, A.E. 1953. Theoretical Models of Porous Matter. Producers Monthly, Vol. 17, No. 10, pp. 17-23.

Spraggs, L.D. and Street, R.L. 1975. Three-Dimensional Simulation of Thermally-Influenced Hydrodynamic Flows TR-190, Department of Civil Engineering, Stanford University, Stanford, California.

Sundermann, J. and Holz, K.P. 1980. Mathematical Modelling of Estuarine Physics. Lecture Notes on Coastal and Estuarine Studies, Springer-Verlag, New York.

U.S. Army Corps of Engineers. 1973. HEC-1, Flood Hydrograph Package. Hydrologic Engineering Center, Davis, California.

U.S. Army Corps of Engineers. 1976. Water Surface Profiles, HEC-1, Computer Program 723-X6-L202a. Hydrologic Engineering Center, Davis, California.

U.S. Army Corps of Engineers. 1978. Flood Hydrograph Package (HEC-1) Dam Safety Version. Hydrologic Engineering Center, Davis, California.

Welch J.E., Harlow, F.H., Shannon, J.P. and Daly, B.J. 1966. The MAC Method, LA-3425, Los Alamos Scientific Laboratory, Los Alamos, New Mexico.

RUNOFF MODEL BASED ON LARRIEU'S GENERALIZED UNIT HYDROGRAPH THEORY AND TWO-PHASE INFILTRATION THEORY

H. J. Morel-Seytoux,
Professor of Civil Engineering
Colorado State University

L. A. Lindell,
District Hydrologist
Medford District Bureau of Land Management

and

F. Correia,
Graduate Research Assistant
Department of Civil Engineering
Colorado State University

ABSTRACT

The watershed simulation model, referred to as WSMOD for brevity in this paper, can be used as an event simulation model or as a sequential model, as a lumped or as a fully distributed model. The user may choose from several options for computing time to ponding, infiltration rates after ponding, discrete kernels in the routing component and determining time step to be used. The results produced by each method can be compared to discover which appears to be the most realistic for a specific watershed.

The model contains four basic components: soil moisture accounting (including evaporation), time to ponding and infiltration before ponding, infiltration after ponding and surface runoff response. The soil moisture accounting component monitors the soil moisture content in the upper and lower storage zones of the soil profile. It takes into account the cumulative infiltration during periods of rainfall, the evaporation from the soil surface when no rain is occurring, and the drainage of water from the upper zone into the lower zone storage. The time to ponding and infiltration components are subdivided into various options. One of the options is the Soil Conservation Service method which is based on the concept of the curve number. The other options break down into different combinations of ponding and post-ponding formulations. Ponding time can be calculated optionally from two formulae. Both formulae are derived from the two-phase theory of infiltration but with different simplifying assumptions. Similarly following ponding three different formulae may be used to calculate cumulative infiltration depth at various discrete times and infiltration increments for each time step. From these values an average infiltration rate over the time step is obtained and by difference with the rainfall rate the average excess rainfall rate can be determined for the time interval.

The surface runoff component uses Larrieu's routing approach to generate a runoff response for the excess rainfall calculated. Larrieu's routing method divides the watershed into subareas, called isochrones, centered about a line of constant travel time for water to reach the outlet. The computer model allows the user to further divide the watershed basin into subsections within each isochrone, thus enabling the watershed parameters to take on a spatial variability. The rainfall which occurs may be variable with respect to time and space. That is, the rainfall may vary between the isochrones and subsections within the watershed for each time step. The resulting streamflow values are obtained by means of superposition in time and space of the individual runoff responses. The unit hydrograph ordinates to be used in the runoff response may be (1) estimated from field data using any type of identification procedure such as the constrained least-squares technique, (2) obtained from a conceptual unit hydrograph model or (3) simply given by a synthetic unit hydrograph if no data is available.

The estimation of the parameters of the model is much simplified because of the physical meaning and known numerical ranges of most parameters. An example of parameter calibration is presented. Finally the importance of using a proper infiltration method in watershed modeling is emphasized through an example comparing the effect of using the infiltration approach versus the ϕ-index method for purpose of unit hydrograph identification.

1. INTRODUCTION

The role of the infiltration process in the hydrologic cycle is an important one, particularly in determining the distribution, magnitude and timing of surface runoff. Therefore the accuracy of surface runoff predictions is very dependent on reliable estimates of infiltration. An accurate estimate of the time at which ponding first occurs is essential in the prediction of surface runoff and it must be considered in watershed modeling.

The amount of infiltration also has an effect on recharge to both streams and groundwater tables. Therefore dependable infiltration equations need to be incorporated in groundwater models as well as watershed models.

Even for the strict purpose of unit hydrograph identification, the use of a proper method for computing time to ponding and infiltration rates after ponding may be crucial in getting satisfactory results. In fact, an excess rainfall hyetograph that is very different from the real one may be responsible for unrealistic and eventually jagged unit hydrographs. WSMOD encompasses various methods for calculating the infiltration of water through the soil surface. The desired infiltration method is selected by the user.

The surface routing is based on Larrieu's generalized unit hydrograph. Larrieu's routing method divides the watershed into subareas, called isochrones, centered about a line of constant travel time for water to reach the outlet. If desired the isochrones may be divided into subsections to account for the spatial variability in watershed parameters. The rainfall may be variable in time and space. That is, the rainfall may vary between various isochrones and isochrone subsections for each time step.

The soil moisture accounting component monitors the soil moisture content in the upper and lower storage zones of the soil profile. It takes into account the cumulative infiltration during periods of rainfall, the evaporation from the soil surface when no rain is occurring, and the drainage of water from the upper zone into the lower zone storage.

The characteristics of the model give it a great deal of flexibility. It may be used either as an event simulation model or as a sequential model. It may be utilized as a lumped model or as a distributed model. The time step can be of any length but larger than one minute. Different time steps may be used within a given set of data, a feature which is particularly convenient when the model is used as a sequential model and there is no rainfall for long periods.

In section 2, the model is described. An example of parameter calibration for a lumped system is presented in section 3. Finally in section 4, the effect of using a proper infiltration method on unit hydrograph identification is discussed.

2. DESCRIPTION OF WATERSHED MODEL WSMOD

2.1. General Characteristics of the Model

WSMOD may be classified as both an event simulation model and a sequential model. It also may be used as a lumped or a distributed model and variable time increments are permitted which allows implementation of small time intervals during each rainfall event and large time intervals for the periods of no precipitation between events. Input to the model consists of: (1) the hyetograph of precipitation for the rainfall event which took place over every subsection of every isochrone or, in the lumped version, over the entire watershed, (2) the daily pan evaporation data for the time period involved, measured on or near the watershed, (3) the area of the drainage basin and the percent area of every subsection, (4) the infiltration parameters for every subsection, and (5) unit hydrograph ordinates for the entire watershed or parameters allowing their internal computations. Further details regarding the input to the model and the calibration of the watershed parameters may be found in the user's manual written for this watershed model (Lindell and Morel-Seytoux, 1979).

Table 1 presents a list of lumped and distributed input variables. Those actually used in the model depend on the selected options for a particular run.

The general structure of WSMOD displayed on Figure 1 is similar to the U. S. Geological Survey one (Dawdy et al., 1972). However, WSMOD uses several optional and more sophisticated formulas to compute time to ponding and infiltration rates after ponding instead of the simple, traditional and limited, Green and Ampt equation. In this way WSMOD takes advantage of recent progress made in the field of infiltration theory (Morel-Seytoux, 1973a and later).

Essentially, WSMOD calculates the excess rainfall which occurs as a result of a rainfall event over a watershed basin. The excess rainfall is routed to the basin outlet allowing prediction of streamflows at this point. These tasks are accomplished by four basic components: soil moisture accounting and evaporation, ponding time, infiltration, and surface routing.

2.2. Soil Moisture Accounting

The soil moisture accounting component monitors the water content

51

Table 1. List of lumped and distributed input variables in the
rainfall-runoff model WSMOD.

Lumped Input Variables	Distributed Input Variables
Evaporation Constant to Adjust Pan Evaporation	Precipitation Rate
Time Step	Hydraulic Conductivity at Natural Saturation
Exponent of Normalized Water Content in the Expression of k_{rw}	Effective Capillary Drive
	Water Content at Natural Saturation
Memory Time of the Watershed (Number of Non-Zero Unit Hydrograph Ordinates)	Initial Water Content of the Upper Zone
Watershed Time Constant (To Be Used in the Larrieu's Expression for the Unit Hydrograph)	Initial Water Content of the Lower Zone
	Residual Water Content
	Rate of Replenishment of Lower Zone Storage
	Depth of Upper Zone Storage
	Depth of Lower Zone Storage
	Ultimate Viscous Correction Factor
	SCS Curve Number
	Relative Area of Every Isochrone to Total Watershed Area
	Relative Area of Every Subsection to Total Isochrone Area

in the upper and lower storage zones of the soil profile. This routine
is not called into service when the precipitation is continuous over
the entire rainfall episode. However, for intermittent rainfall events,
the soil moisture accounting component is summoned at the end of each
time interval during which there is no precipitation. At the end of
the first time step with no rainfall, immediately following a time
step during which rainfall occurred, the new soil moisture content of
the upper zone is calculated by the following equation:

$$\theta_{iu}^{\nu} = \theta_{iu}^{o} + \frac{W^o}{D_u} - \frac{q\Delta t}{D_u} - \frac{e_p \Delta t}{D_u} \qquad (1)$$

where θ_{iu}^{ν} is the new initial moisture content in the upper zone of
storage, θ_{iu}^{o} is the old initial moisture content in the upper zone of
storage used during the last period of continuous rainfall, W^o is the
total cumulative infiltration depth during the last continuous rainfall
period, D_u is the depth of the upper storage zone, q is the drainage rate of water from the upper to the lower zone, Δt is the length
of the time step, and e_p is the potential rate of evaporation. For
consecutive time steps with no rainfall, the new soil moisture content

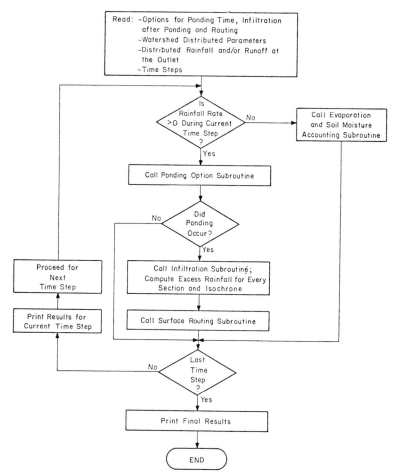

Fig. 1. Flowchart of watershed model WSMOD.

for the upper zone is obtained from Eq. (1) excluding the term for cumulative infiltration depth.

Equation (1) assumes that the cumulative infiltration becomes immediately uniformly distributed throughout the upper zone of storage. Part of this water which accumulates in the upper zone evaporates or drains to replenish the lower zone storage.

If the newly calculated value of the water content in the upper soil zone is greater than the residual water content, a new value of the effective capillary drive must be obtained since it is a function of water content. Equation (2) (Morel-Seytoux, 1978a) is used to reduce the initial value of the effective capillary drive for the following time step:

$$H_c(\theta_{iu}) = H_{cM} \left[1 - \left(\frac{\theta_{iu}^{\nu} - \theta_r}{\tilde{\theta} - \theta_r} \right)^6 \right] \tag{2}$$

where $H_c(\theta_{iu})$ is the effective capillary drive as a function of θ_{iu}, H_{cM} is the maximum effective capillary drive, θ_r is the residual water content and $\tilde{\theta}$ is the water content at natural saturation. The moisture content of the lower zone of storage in the soil profile is also calculated at the end of each time step during which there is no rainfall according to Eq. (3):

$$\theta_{i\ell}^{\nu} = \theta_{i\ell}^{o} + \frac{q\Delta t}{D_\ell} \tag{3}$$

where $\theta_{i\ell}^{\nu}$ is the new initial soil moisture content in the lower storage zone, $\theta_{i\ell}^{o}$ is the old initial moisture content in the lower storage zone, and D_ℓ is the depth of the lower zone of storage.

2.3. Calculation of Ponding Time

Three options are available for the computation of time to ponding. The simplest one is based on the Soil Conservation Service curve number method (Morel-Seytoux, 1978b; Morel-Seytoux and Verdin, 1980; Verdin and Morel-Seytoux, 1980).

The second option calculates the time of ponding by the following formula (Morel-Seytoux, 1978b):

$$t_p = t^o + \frac{1}{r}\left[\frac{(\tilde{\theta}-\theta_{iu}^{o})H_c(\theta_{iu})}{\left(\frac{r}{\tilde{K}}-1\right)} - W^o\right] \tag{4}$$

where t_p is the time of ponding, t^o is the time at the beginning of the current time step, r is the rainfall rate during the current time step, $\tilde{\theta}$ is the water content at natural saturation, θ_{iu}^{o} is the initial moisture content in the upper zone, $H_c(\theta_{iu})$ is the effective capillary drive, $\tilde{K}$ is the hydraulic conductivity at natural saturation and W^o is the cumulative infiltration since the beginning of the last continuous rainfall.

The third option uses the more complicated formula (Morel-Seytoux, 1978a), namely:

$$t_p = t^o + \frac{T^{\nu}}{r\ast}\left[\exp\left(\frac{1-\theta_0^\ast}{r\ast\beta^{\nu}-1}\right) - 1\right] \tag{5}$$

where t_p and t^o were defined previously, $r\ast = r/\tilde{K}$ is the normalized rainfall rate, T^{ν} is defined in Eq. (6), β^{ν} is the viscous correction factor for the current time step, and $\theta_0^\ast$ is the normalized water content at the soil surface at time t^o defined as $(\theta_u^o-\theta_r)/(\tilde{\theta}-\theta_r)$. The composite parameter T^{ν} is defined as:

$$T^{\nu} = \frac{(\bar{\theta}-\theta^0_{iu})H_c M}{\tilde{K}} + \frac{W^0}{\tilde{K}} \tag{6}$$

2.4. Calculation of Infiltration after Ponding

Four options are available for the computation of infiltration after ponding. The simplest option is again the use of the Soil Conservation Service Method based on the curve number (US SCS 1972; Viessman et al., 1977).

The other two methods, based on the modern two-phase theory of infiltration, use formulas derived in recent years (Morel-Seytoux, 1978b). In one option the cumulative infiltration is given by the expression:

$$W^{\nu} = W_p + S(P,\theta_{iu})\{[t^{\nu}-t_p+t_r]^{\frac{1}{2}} - t_r^{\frac{1}{2}}\} + \frac{\tilde{K}}{\beta}(t^{\nu}-t_p) \tag{7}$$

where W^{ν} is the cumulative infiltration at the end of the current time step, W_p is the cumulative infiltration at the time of ponding t_p , $S(P,\theta_{iu})$ is the *rainfall sorptivity* (Morel-Seytoux, 1978b) defined by the relation:

$$S(P,\theta_{iu}) = \sqrt{\frac{2\tilde{K}(S_f + W_p)^2}{\beta\left[S_f + W_p\left(1 - \frac{\beta_p}{\beta}\right)\right]}} \tag{8}$$

where $S_f = (\bar{\theta}-\theta_{iu})H_c(\theta_{iu})$ is the storage suction factor (Morel-Seytoux, 1978b), β is the ultimate viscous correction factor (Morel-Seytoux and Khanji, 1974), β_p is the viscous correction factor obtained at the time of ponding and t_r , a time, is given by the expression:

$$t_r = \frac{1}{2}\frac{\tilde{K}(S_f + W_p)^2}{\beta\left(r_p - \frac{\tilde{K}}{\beta}\right)^2\left[S_f + W_p\left(1 - \frac{\beta_p}{\beta}\right)\right]} \tag{9}$$

The last option is based on a relation developed from infiltration theory (Morel-Seytoux, 1973a; Morel-Seytoux and Khanji, 1974; Morel-Seytoux, 1976, 1978a,b) for time and cumulative infiltration:

$$\frac{\tilde{K}}{\beta}(t^{\nu}-t_p) = W^{\nu}-W_p-[S_f+W_p(1 - \frac{\beta_p}{\beta})]\ln\left\{\frac{S_f + W^{\nu}}{S_f + W_p}\right\} \tag{10}$$

where all symbols have been defined previously. The cumulative infiltration in Eq. (10) is not written explicitly in terms of time but vice versa. Therefore values of w^{ν} must be chosen and substituted into the equation until the correct t^{ν} is obtained.

An alternative is to expand the logarithmic term in Taylor's series, truncate after the second order term and solve the quadratic equation for $\Delta W = W^{\nu} - W^0$ in terms of $\Delta t = t^{\nu} - t^0$, with the result:

$$W^{\nu} = \left\{\frac{2\tilde{K}}{\beta}\,(S_f + W_o)\Delta t + \left[W_o - W_p\left(1 - \frac{\beta_p}{\beta}\right) - \frac{\tilde{K}\Delta t}{2\beta}\right]^2\right\}^{\frac{1}{2}} + \frac{\tilde{K}\Delta t}{2\beta} + W_p\left(1 - \frac{\beta_p}{\beta}\right) \qquad (11)$$

which is valid provided Δt is not too large. For practical purposes Eq. (11) is adequate with the small time steps used during rainfall events.

2.5. Surface Routing

The excess rainfall resulting from the infiltration component is used to generate a runoff response determined by Larrieu's generalized instantaneous unit hydrograph method. This method divides the watershed into subareas, called isochrones, centered about a line of constant travel time for water to reach the outlet. A unit impulse of excess rainfall which falls over a subarea (χ) at a given time, τ , reaches the outlet at the later time, $\tau + \chi$, and then generates a runoff response. The overall runoff response for a general rainfall over the entire watershed is obtained by superposition in time and space (Morel-Seytoux, 1981).

When it is not desired to divide the watershed into isochrones, the basin is regarded as one single isochrone with a delayed response time equal to zero. Only one value of excess rainfall is calculated for each time step and it applies to the entire watershed area. The runoff is generated by the discrete convolution relation involving the excess rainfall and the discrete kernels (unit hydrograph ordinates).

In addition to dividing the watershed into isochrones, the computer model has the capability of subdividing each isochrone into subsections. Both the rainfall intensity and the watershed parameters may vary between isochrones and subsections, while the rainfall intensity may also be variable and intermittent with respect to time.

The values of the discrete kernels may be internally computed or given by the user. In the first case any conceptual model for the unit hydrograph may be used. The expression proposed by Larrieu is currently implemented, namely:

$$\delta(n) = \exp\left[-(n-1)^2/2K^2\right] - \exp\left[-n^2/2K^2\right] \qquad (12)$$

where $\delta(n)$ is the nth unit hydrograph ordinate and K is the watershed time constant, to be supplied by the user and such that:

$$\delta(K+1) > \delta(n) \qquad n = 1,2,\ldots,K,K+2,\ldots M$$

where M is the memory time of the watershed, which is also, of necessity, the number of nonzero values of the unit hydrograph ordinates. A plot of discrete kernels (δ) versus time periods (n) for various values of the watershed time constant is presented on Figure 2.

The use of a conceptual model for the discrete kernels is not necessary. Alternately values of the discrete kernels may be read as input. These values may be obtained from a synthetic unit hydrograph or may be estimated from real data by calibration. In the examples presented in section 3, a constrained least-squares technique (Morel-Seytoux et al., 1980) was used with very good results.

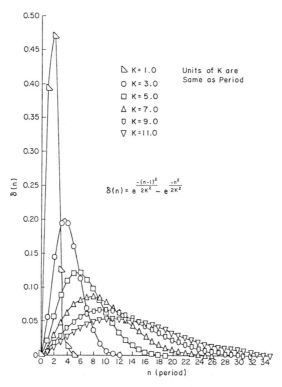

Fig. 2. Plot of the discrete kernels (δ) versus time periods (n) for various values of the watershed time constant (K) .

2.6. Calibration of Parameters

The set of infiltration parameters to be provided by the user to make runs with WSMOD consists of: H_c(cm) the effective capillary drive, $\tilde{K}$ (cm/hr) the hydraulic conductivity at natural saturation, $\tilde{\theta}$ (dimensionless) the water content at natural saturation, θ_{iu} (dimensionless) the initial water content of the upper zone, $\theta_{i\ell}$ (dimensionless) the initial water content of the lower zone, θ_r (dimensionless) the residual water content, D_u (cm) the depth of the upper zone storage, D_ℓ (cm) the depth of the lower zone storage, q (cm/hr) the rate of replenishment of lower zone storage, β (dimensionless) the ultimate viscous correction factor and CN , the curve number (only if the SCS method is used). The calibration of these parameters is not overly cumbersome because they have a very precise physical meaning, an extremely helpful fact to estimate reasonable initial values in the calibration stage. Tables 2 and 3 provide guidelines for initial estimates of the parameters.

Table 2. Range of values which may be used to estimate the watershed parameters.

Parameter	Numerical Range	Descriptive Range	Source
$\tilde{K}$	0.0 cm/hr - 5.0 cm/hr	tight clay - coarse sand	Morel-Seytoux (1978b), Chap. 3, p. 13
H_c	5 cm - 30 cm	coarse sand - tight clay	Ibid., Chap. 3, p. 13
β	1 - 1.7		Ibid., Chap. 3, p. 14
$\tilde{\theta}$	0.25 - 0.35	clay soil - sandy soil	Ibid., Chap. 3, p. 13
θ_{iu}^o	$0 - \tilde{\theta}$	completely dry - saturated soil	Ibid., Chap. 3, p. 13
θ_{il}^o	$0 - \theta_r$		
D_1	5 cm - 50 cm		Viessman et al. (1977), p. 421
D_u	$0.06\ D_1$	steep slopes, limited vegetation, low depression storage	Ibid., p. 421
	$0.08\ D_1$	moderate slopes, moderate vegetation, moderate depression storage	Ibid., p. 421
	$0.14\ D_1$	heavy vegetation or forest cover, soils subject to cracking, high depression storage, very mild slopes	Ibid., p. 421
q	0 cm/hr - $\tilde{K}$		

A trial-and-adjustment method for calibration has been utilized so far because an "automatic" calibration mode has not been fully implemented yet. A good practice is to use rainfall events that are preceded by a long, dry period. The initial moisture content can then be assumed to be at wilting point value.

Experience in dealing with the model, although not very extensive, has shown that if reasonable initial values are considered for all the parameters, then calibration on $\tilde{K}$ and/or H_c and the unit hydrograph ordinates is enough to get good results. An example is provided. It is also advisable to start calibrating only with one isochrone and one subsection and if this lumped model proves not to be good then proceed considering increasingly distributed models unless one has a very good reason for considering a distributed model from the very beginning.

3. EXAMPLE OF CALIBRATED RESULTS FOR THE SITUATION OF A SINGLE ISOCHRONE

3.1. Field Data Used in Calibration

Data obtained from the U.S.G.S. for the Red Rock Canyon watershed for two rainfall-runoff events occurring on July 2, 1972 and August 7, 1971 were used for calibration.

Table 3. Typical values of the differences $\tilde{\theta} - \theta_{fc}$ and $\theta_{wi\ell}$ (θ_{fc} is field capacity water content $\theta_{wi\ell}$ is wilting point water content) for various soils.[1]

Soil Type	$\tilde{\theta} - \theta_{wi\ell}$	$\tilde{\theta} - \theta_{fc}$
Coarse sand	0.24	0.18
Sand	0.32	0.19
Loamy fine sand	0.33	0.27
Sandy loam	0.31	0.19
Fine sandy loam	0.36	0.24
Very fine sandy loam	0.33	0.21
Loam	0.30	0.14
Clay loam	0.25	0.13
Silty clay loam	0.23	0.08
Sandy clay	0.19	0.12
Silty clay	0.21	0.09
Clay	0.18	0.07

[1] Adapted from C. B. England, "Land Capability: A Hydrologic Response Unit in Agricultural Watersheds," U. S. Department of Agriculture, ARS 41-172, September 1970. After H. N. Holtan et al., "USDAHL-74 Revised Model of Watershed Hydrology," U. S. Department of Agriculture, ARS Tech. Bulletin No. 1518, Washington, D.C., 1975.

Red Rock Canyon Basin, in Colorado, is a small watershed with 4.27 square miles. The elevation ranges from a height of 5400 feet at the stream origin to a height of 4920 feet at the basin outlet. Although the watershed has steep slopes at the stream origin, the major part of the basin consists of a gradually sloping surface which encourages a large portion of the rainfall to infiltrate into the ground rather than to immediately become surface runoff. Because the surface is protected by a vegetative cover this will result in less compaction of the soil surface and a greater permeability of the soil. Typically, the soil found in this area is a silty clay loam, calcareous and moderately alkaline. The rate of water intake ranges from moderate to moderately slow. The underlying bedrock in the basin varies between sandstone, shale and limestone. The small area of the watershed and the homogeneity of its characteristics makes particularly meaningful the consideration of only one isochrone.

The rainfall and runoff data for the two events that were considered are presented in Tables 4 and 5.

Table 4. Rainfall and runoff data for rain of July 2, 1972.

Time (hr/min)	Rainfall (cm/hr)	Runoff (m³/s)	Time (hr/min)	Rainfall (cm/hr)	Runoff (m³/s)	Time (hr/min)	Rainfall (cm/hr)	Runoff (m³/s)
18.45	7.62	0.283	19.50	0.30	14.159	20.55	0.00	2.549
18.50	14.02	0.565	19.55	0.00	11.893	21.00	0.00	2.265
18.55	14.94	1.416	20.00	0.00	10.194	21.05	0.00	1.982
19.00	6.10	3.398	20.05	0.00	8.778	21.10	0.00	1.699
19.05	3.35	5.663	20.10	0.00	7.546	21.15	0.00	1.699
19.10	2.13	8.495	20.15	0.00	6.795	21.20	0.00	1.416
19.15	1.52	11.893	20.20	0.00	5.947	21.25	0.00	1.133
19.20	0.00	15.858	20.25	0.00	5.097	21.30	0.00	0.850
19.25	0.61	20.388	20.30	0.00	4.248	21.35	0.00	0.708
19.30	0.61	26.618	20.35	0.00	3.681	21.40	0.00	0.566
19.35	0.30	33.131	20.40	0.00	3.398	21.45	0.00	0.283
19.40	0.00	24.919	20.45	0.00	3.115	21.50	0.00	0.142
19.45	0.00	17.557	20.50	0.00	2.832	21.55	0.00	0.000

Table 5. Rainfall and runoff data for rain of August 7, 1971.

Time (hr/min)	Rainfall (cm/hr)	Runoff (m³/s)	Time (hr/min)	Rainfall (cm/hr)	Runoff (m³/s)	Time (hr/min)	Rainfall (cm/hr)	Runoff (m³/s)
8.20	0.91	0.000	9.50	0.00	4.955	11.20	0.00	0.538
8.25	5.49	0.000	9.55	0.00	4.134	11.25	0.00	0.510
8.30	9.75	0.000	10.00	0.30	3.851	11.30	0.00	0.453
8.35	11.28	0.396	10.05	0.00	3.511	11.35	0.00	0.396
8.40	11.58	2.095	10.10	0.00	2.747	11.40	0.00	0.340
8.45	3.35	6.116	10.15	0.00	2.577	11.45	0.00	0.311
8.50	0.91	8.778	10.20	0.00	2.379	11.50	0.00	0.283
8.55	0.30	11.242	10.25	0.00	2.209	11.55	0.00	0.283
9.00	0.00	16.141	10.30	0.00	1.926	12.00	0.00	0.255
9.05	0.00	18.123	10.35	0.00	1.699	12.05	0.00	0.227
9.10	0.00	20.105	10.40	0.00	1.444	12.10	0.00	0.227
9.15	0.00	21.634	10.45	0.00	1.303	12.15	0.00	0.142
9.20	0.00	16.990	10.50	0.00	1.161	12.20	0.00	0.085
9.25	0.61	11.553	10.55	0.00	0.991	12.25	0.00	0.057
9.30	0.30	9.826	11.00	0.00	0.851	12.30	0.00	0.028
9.35	0.30	8.155	11.05	0.00	0.765	12.35	0.00	0.000
9.40	0.00	6.456	11.10	0.00	0.680			
9.45	0.00	5.692	11.15	0.00	0.595			

3.2. Calibration Procedure

Calibration was considered in two stages. First the infiltration parameters were calibrated with the objective of getting a match between computed and observed total discharge depth. Then the unit hydrograph ordinates were estimated from the computed excess rainfall hyetograph and the observed hydrograph. One could certainly go back and recalibrate the infiltration parameters so that an optimum prediction of ponding time or an optimum shape of the excess rainfall hyetograph were obtained. However, in the cases that were worked out, when a good prediction of total infiltration depth was obtained, the ponding time and the shape of the excess rainfall hyetograph were also predicted well. This is the result and the advantage of using a sound physically based model. In fact the parameters used in computing cumulative infiltration are mostly the same ones used in computing time to ponding. Therefore if a set of parameter values gives optimum results in one case, it is likely to give at least reasonable values in the other case.

For the rainfall-runoff event taking place on July 2, 1972 (Example 1), the initial values of the parameters were the following: $\tilde{K} = 1.38$ cm/hr , $H_c = 26.00$ cm , $\beta = 1.2$, $\tilde{\theta} = 0.26$, $\theta_{iu} = 0.00$, $\theta_{i\ell} = 0.00$, $\theta_r = 0.00$, $q = 0.01$ cm/hr , $D_u = 4.00$ cm , and $D_\ell = 22.00$ cm. These values, that were based on the characteristics and the antecedent condition of the soil, were kept constant except for $\tilde{K}$. For every value of $\tilde{K}$ an error in total infiltration depth was computed and a new value of $\tilde{K}$ was considered in order to drive the error to zero according to a linear relationship between $\tilde{K}$ and the error. This method was found to be quite efficient because for an extremely high accuracy (0.0005 cm) it converged in six steps. The successive values of $\tilde{K}$ and the error are presented in Table 6. The best value of $\tilde{K}$ was 1.94 cm/hr .

Alternative methodologies were tried keeping all parameters fixed except H_c or θ_{iu} in two separate cases. A very fast convergence to a zero error in the computed infiltration depth was also verified in these cases. For a simple event, like Example 1, it was noticed that when a good accuracy in the computation of the total infiltration depth was attained, a good or a very acceptable match was also obtained in the other variables of interest, namely time to ponding and excess rainfall time distribution. For more complicated situations it would be necessary to carry a sort of tuning with the other parameters.

For Example 2 (rain on August 7, 1971), the same procedure was used, calibrating on $\tilde{K}$ and using the same values of Example 1 for the other parameters. The best value for $\tilde{K}$ was found to be, in this case, $\tilde{K} = 1.88$ cm/hr .

When a good match in infiltration depth is obtained, Stage 1 of calibration is completed. Stage 2 is the estimation of the unit hydrograph ordinates (discrete kernels). This estimation is straightforward since a computed excess rainfall hyetograph and an observed discharge hydrograph are available. A constrained least-squares technique was used in this estimation (Morel-Seytoux et al., 1980). The memory time

Table 6. Calibration on $\tilde{K}$ (Example 1: July 2, 1972 Rainfall-Runoff Event)

$\tilde{K}$ (cm/hr)	Error (cm) (Computed Infiltration Depth Less Observed Infiltration Depth)	Algorithm Path
1.38	-0.413	increase $\tilde{K}$
1.81	-0.135	increase $\tilde{K}$
2.02	0.096	decrease $\tilde{K}$
1.93	-0.013	increase $\tilde{K}$
1.94	-0.003	increase $\tilde{K}$
1.95	0.007	decrease $\tilde{K}$
1.943	0.000	best value

of the watershed was found to be 35 in Example 1 and 47 in Example 2. Although the values of the discrete kernels are slightly different the shape of the unit hydrograph is similar. Having the discrete kernels, the computed discharge may be calculated and compared with the observed discharge. The comparison is displayed on Figure 3 for Example 1 and on Figure 4 for Example 2. If the results of Stage 2 of calibration were not satisfactory, and if it was due to the excess rainfall pattern, a return to Stage 1 would be necessary to improve the estimates of the infiltration parameters. This was not the case in both examples.

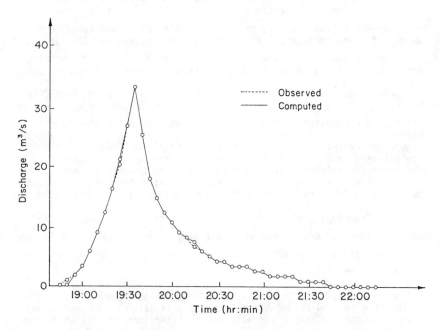

Fig. 3. Example 1 with best infiltration parameters and discrete kernels identified by a constrained least-squares technique.

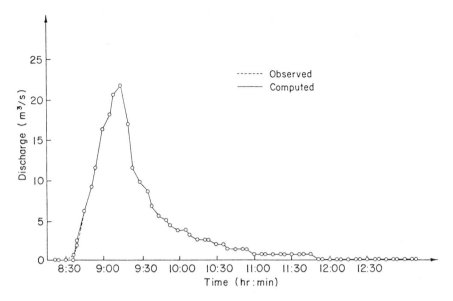

Fig. 4. Example 2 with best infiltration parameters and discrete kernels identified by a constrained least-squares technique.

3.3. Validation

As a validation procedure, "cross" computations were performed. The results show how the parameters calibrated for one case may be reliable for another situation. A complete validation procedure would

Table 7. Results when using in one example the best parameters for the other example.

		"Best" Parameters	"Cross" Parameters	Joint Estimation of Parameters
Example 1	Error in Computed Infiltration (cm)	.000	-.064	-.032
	MAD* (m^3s^{-1})	.532	11.036	3.956
	SAD (m^3s^{-1})	1.912	94.689	32.079
	SSD (m^3s^{-1})^2	.409	599.478	74.885
Example 2	Error in Computed Infiltration (cm)	.000	.074	.037
	MAD (m^3s^{-1})	.178	8.328	5.316
	SAD (m^3s^{-1})	.237	65.113	43.739
	SSD (m^3s^{-1})^2	.033	318.373	132.346

* MAD: maximum absolute deviation; SAD: sum of absolute deviations; SSD: sum of squares of deviations.

require more data and more examples to be worked out, corresponding to different antecedent conditions. Nevertheless the results obtained for Example 1 and 2 were encouraging. These results are summarized in Table 7. Some of the calculated results are presented on Figures 5 and 6. From these results it can be noticed that the infiltration results are quite reliable. In fact the differences in computed and observed discharges are mainly due to the use of a "non-optimum" unit hydrograph in the cross computations.

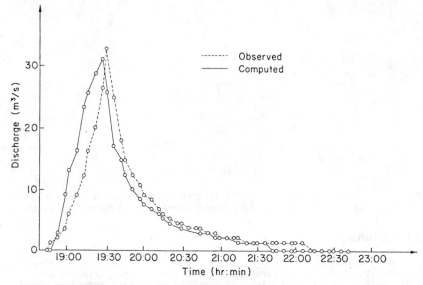

Fig. 5. Example 1 computed with best parameters of Example 2.

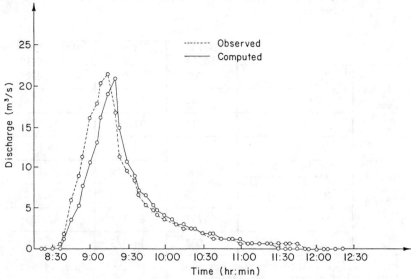

Fig. 6. Example 2 computed with best parameters of Example 1.

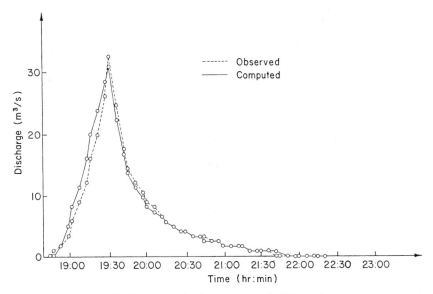

Fig. 7. Example 1 computed with average infiltration parameters and jointly identified unit hydrograph.

A joint estimation of the parameters for the two examples was also performed. Actually the average of the infiltration parameters was used and a joint estimation of the unit hydrograph was performed by the constrained least-squares technique assuming an average memory time of 41 periods. The results were obviously much better than when the "cross" parameters were used. These results are presented on Figures 7 and 8.

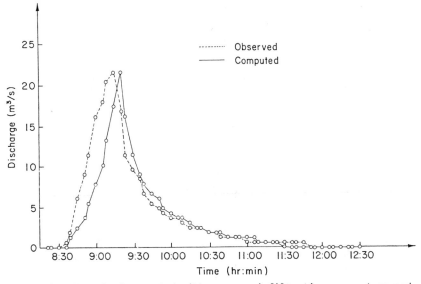

Fig. 8. Example 2 computed with average infiltration parameters and jointly identified unit hydrograph.

4. EFFECT OF PROPER INFILTRATION METHOD ON UNIT HYDROGRAPH IDENTIFICATION

The use of a proper method for computing time to ponding and infiltration rates after ponding may be crucial at securing a good estimate of the discrete kernels. An example which supports this statement was worked out using the rainfall-runoff event of July 2, 1972 (Example 1). When Eq. (5) for ponding time and Eq. (10) for infiltration after ponding were used, with the calibrated parameters presented in Section 3, the unit hydrograph estimated by constrained least-squares has the smooth realistic shape presented on Figure 9. The values of these unit hydrograph ordinates are presented in Table 8. When the ϕ-index method is used (ϕ = 10.03 cm/hr) the unit hydrograph estimated by the same method is jagged and unrealistic. These results are shown in Table 8 and on Figure 10.

Table 8. Discrete kernel estimates when using infiltration theory approach and ϕ-index method.

Time Period	Discrete Kernels Infiltration Theory	Discrete Kernels ϕ-Index Method
1	.005295	.005654
2	.012521	.004867
3	.020788	.022234
4	.031157	.018881
5	.043597	.045903
6	.058115	.040139
7	.074590	.078946
8	.097500	.068154
9	.122759	.130944
10	.091116	.107215
11	.063930	.070465
12	.051794	.055855
13	.043553	.046444
14	.037365	.039617
15	.032185	.034358
16	.028044	.029340
17	.024961	.026401
18	.021854	.023121
19	.018743	.020333
20	.015610	.016884
21	.013538	.014289
22	.012530	.012871
23	.011494	.012337
24	.010458	.010711
25	.009423	.010422
26	.008383	.008506
27	.007348	.008553
28	.006285	.006237
29	.006341	.006771
30	.005275	.006125
31	.004241	.004651
32	.003192	.004164
33	.002687	.002497
34	.002179	.003367
35	.001149	.001193
36	.000000	.001549

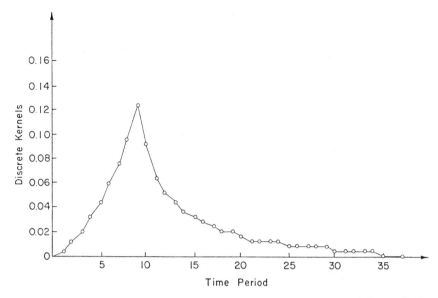

Fig. 9. Unit hydrograph ordinates when infiltration Eqs. (5) and (10) are used.

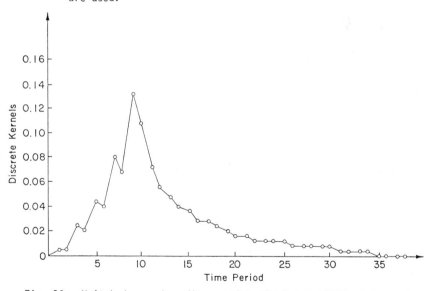

Fig. 10. Unit hydrograph ordinates when the φ-index method is used.

5. ACKNOWLEDGMENTS

The work upon which this paper is based was supported in part by funds provided by the Ministry of Agriculture and Water, Kingdom of Saudi Arabia under USDA Agreement 58-319R-8-134 and by the Office of Water Research and Technology, U.S. Department of Interior, as authorized under the Water Resources Research Act of 1964, project Colorado B-199, pursuant to grant agreement 14-34-0001-9109.

6. REFERENCES

Dawdy, D. R., R. W. Lichty and J. M. Bergmann, 1972. A Rainfall-Runoff Simulation Model for Estimation of Flood Peaks for Small Drainage Basins , U.S. Geological Survey Professional Paper, No. 506-B.

Lindell, L. A. and H. J. Morel-Seytoux, 1979. User's Manual for a FORTRAN IV Program: A Rainfall-Runoff Model for Flood Prediction, HYDROWAR Program, Colorado State University, 51 pages.

Morel-Seytoux, H. J., 1973a. Two-Phase Flows in Porous Media, Chapter in Advances in Hydroscience, V. T. Chow, Editor, Vol. 9, 1973 (August), Academic Press, pp. 119-202.

Morel-Seytoux, H. J., 1973b. Systematic Treatment of Infiltration with Applications, Environmental Resources Center, Colorado State University, Fort Collins, Colorado, Completion Report No. 50, 64 pages.

Morel-Seytoux, H. J., 1976. Derivation of Equations for Rainfall Infiltration, Journal of Hydrology, 31:203-219.

Morel-Seytoux, H. J., 1978a. Derivation of Equations for Variable Rainfall Infiltration, Water Resources Research Journal, Vol. 14, No. 4, August 1978, pp. 561-568.

Morel-Seytoux, H. J., 1978b. Abstractions, Excess Rainfall and Direct Runoff, Hydrology for Transportation Engineers, Chapter 3, T. G. Sanders, editor. Prepared by Colorado State University, Department of Civil Engineering, for the U.S. Department of Transportation, Federal Highway Administration, 734 p.

Morel-Seytoux, H. J., 1981. Engineering Hydrology: Ensemble of Lectures and Handouts, HYDROWAR Program, Colorado State University, January 1981, 250 pages.

Morel-Seytoux, H. J. and J. Khanji, 1974. Derivation of an Equation of Infiltration, Water Resources Research Journal, Vol. 10, No. 4, August 1974, pp. 795-800.

Morel-Seytoux, H. J. and J. P. Verdin, 1980. Extension of the Soil Conservation Service Rainfall-Runoff Methodology for Ungaged Watersheds, HYDROWAR Program, Federal Highway Administration Report No. FHWA-RD-80-173, CER80-81HJM-JPV14, October 1980, 75 pages.

Morel-Seytoux, H. J., J. Kimzey and F. Correia, 1980. User's Manual for UNIT: A FORTRAN IV Program to Identify Unit Hydrograph Ordinates Using a Constrained Least-Squares Approach, HYDROWAR Program, Colorado State University, Fort Collins, Colorado, CER-81HJM-JK-FC16, November 1980, 43 pages.

U. S. Soil Conservation Service, 1972. National Engineering Handbook, Section 4, Hydrology, Washington, D.C.

Verdin, J. P. and H. J. Morel-Seytoux, 1980. User's Manual for XSRAIN, a FORTRAN IV Program for Calculation of Flood Hydrographs for Ungaged Watersheds, HYDROWAR Program, CER80-81JPV-HJM17, November 1980, 176 pages.

Viessman, W., Jr., Knapp, J. W., Lewis, G. L. and T. E. Harbaugh, 1977. Introduction to Hydrology, IEP a Dun-Donnelley Publisher, New York, 704 p.

CREAMS HYDROLOGY MODEL—OPTION ONE

J. R. Williams
Hydraulic Engineer
U.S. Department of Agriculture
Grassland, Soil & Water Research Laboratory
P. O. Box 748
Temple, TX 76501, U.S.A.

A. D. Nicks
Agricultural Engineer
U.S. Department of Agriculture
Southern Great Plains Watershed Research Center
P. O. Box 400
Chickasha, OK 73018, U.S.A.

INTRODUCTION

CREAMS (Chemicals, Runoff, and Erosion from Agricultural Management Systems (Knisel, 1980) is a model for simulating agricultural contributions to water pollution. It is useful in evaluating agricultural management practices designed to prevent water pollution. The model consists of three major components--hydrology, erosion-sedimentation, and chemistry. Although hydrology is only one component of the total system, water is the principle element--it causes erosion, transports chemicals and sediment, and is an uncontrolled natural input. Two hydrology options (one uses daily rainfall, the other breakpoint rainfall) are provided for greater user convenience and general applicability. The development and testing of the CREAMS daily rainfall hydrology model (CDRHM) is reported here.

The objective in developing the CDRHM was to evaluate the effect of management decisions on runoff volume and peak runoff rate with reasonable accuracy for small ungaged watersheds throughout the U.S. To satisfy the objective the model had to be: (a) physically based and use readily available inputs (calibration is not possible on ungaged watersheds); (b) capable of computing the effects of management changes on outputs; (c) computationally efficient to allow simulation of a variety of management strategies without excessive cost; and (d) capable of simulating long periods for use in frequency analysis.

The major processes included in the model are surface runoff, percolation, and evapotranspiration. The SCS curve number technique (Soil Conservation Service, 1972) is used to predict runoff volume because: (a) it is a reliable procedure that has been used for many years in the U.S.; (b) it is computationally efficient; (c) the required inputs are generally available; and (d) it relates runoff to soil type, land use, and management practices. The use of readily available daily rainfall is a particularly important attribute of the curve number technique. For many locations, rainfall data with time increments less than one day are not available. Also, daily rainfall data manipulation

69

and runoff computation are more efficient than similar operations with shorter time increments.

Traditionally, the SCS has used an antecedent rainfall index to estimate three antecedent soil moisture conditions (I--dry, II--normal, III--wet). In reality, soil moisture varies continuously and thus curve number has many values instead of only three. Runoff prediction accuracy was increased by using a soil moisture accounting procedure (Williams and LaSeur, 1976) to estimate the curve number for each storm. Although the soil moisture accounting model is superior to the antecedent rainfall method, it does not maintain a water balance and requires calibration with measured runoff data.

Here the curve number technique was linked with evapotranspiration (ET) and percolation models to overcome these deficiences. Calibration is not necessary because CDRHM is more physically based--the soil water balance is related directly to curve number. Besides predicting daily runoff volume, an equation was also developed for predicting peak runoff rates. Tests with data from watersheds in Texas, Nebraska, Georgia, Ohio, Oklahoma, Arizona, New Mexico, West Virginia, Mississippi, Iowa, and Montana indicate the CDRHM is capable of simulating long-term runoff, ET, and percolation realistically.

MODEL DESCRIPTION

Runoff Volume

The CDRHM is based on the water balance equation

$$ST_t = ST + P - Q - ET - O \tag{1}$$

where ST is the soil water content at the beginning; ST_t is the soil water t days later; P is the amount of rainfall; Q is the amount of runoff; ET is the amount of evapotranspiration; and O is the amount of percolation below the root zone during the t day period.

Runoff is predicted for daily rainfall using the SCS equation (Soil Conservation Service, 1972)

$$Q = \frac{(P - 0.2s)^2}{P + 0.8s} \tag{2}$$

where Q is the daily runoff; P is the daily rainfall; and s is a retention parameter.

The retention parameter s is related to soil water content with the equation

$$s = s_{mx}\left(\frac{UL - ST}{UL}\right) \tag{3}$$

where ST is the plant available soil water content in the root zone, UL is the upper limit of plant available soil water storage in the root zone, and s_{mx} is the maximum value of s. The plant available water storage is computed with the equation

$$ST = (\theta - BR15) (RD) \tag{4}$$

where θ is the daily simulated volumetric fractional water content, BR15 is the volumetric fractional water content at 15 bars tension, and RD is the root zone depth. Similarly the upper limit of plant available water storage is computed with the equation

$$UL = (\phi - BR15) (RD) \tag{5}$$

where ϕ is the soil porosity in the root zone. The maximum value of s is estimated with the I moisture condition CN using the SCS equation (Soil Conservation Service, 1972)

$$s_{mx} = 254 \left(\frac{100}{CN_I} - 1 \right) \tag{6}$$

where CN_I is the I moisture condition CN. The constant value, 254, in Eq. 4 gives s_{mx} in mm. Thus, s, UL, SM, P, Q, ET, and O are expressed in mm. An estimate of the II moisture condition CN can be obtained easily for any watershed using the SCS Hydrology Handbook (Soil Conservation Service, 1972). The corresponding CN_I values are also tabulated. For computing purposes CN_I was related to CN_{II} with the polynomial

$$CN_I = -16.91 + 1.348(CN_{II}) - 0.01379(CN_{II})^2 + 0.0001177(CN_{II})^3. \tag{7}$$

If soil water is distributed uniformly in the soil profile, Eq. 3 should give a good estimate of the retention parameter and thus the runoff. However, if the soil water content is greater near the surface, Eq. 3 would tend to give low runoff predictions. Conversely, runoff would be overpredicted if the soil water content was greater in the lower root zone. To account for the soil water distribution, a weighting technique was developed. The root zone was divided into seven layers and weighting factors (decreasing with depth) were applied. The depth-weighted retention parameter is computed with the equation

$$s = s_{mx} \left(1. - \sum_{i=1}^{7} W_i \frac{ST_i}{UL_i} \right) \tag{8}$$

where W_i is the weighting factor, ST_i is the water content, and UL_i is the upper limit of water storage in storage i. The weighting factors decrease with depth according to the equation

$$W_i = 1.016 \left(\exp (-4.16 \frac{D_{i-1}}{RD}) - \exp (-4.16 \frac{D_i}{RD}) \right) \tag{9}$$

where D_i is the depth to the bottom of storage i, and RD is the root zone depth. Equation 9 assures that

$$\sum_{i=1}^{7} W_i = 1.$$

Peak Runoff Rate

Peak runoff rate is predicted with the equation

$$q_p = 160(DA)^{-0.3} (CS)^{0.159} (\frac{Q}{25.4})^{0.836(DA)^{0.0166}} (LW)^{-0.187} \qquad (10)$$

where q_p is the peak runoff rate mm/h; DA is the drainage area in ha; CS is the mainstem channel slope in m/m; Q is the daily runoff volume in mm; and LW is the length-width ratio of the watershed. Data from 304 storms that occurred on 56 watersheds located in 14 states were used to develop Eq. 10. Watershed areas ranged from 70 to 6200 ha. Since these areas are larger than what is usually considered field scale, the equation has variable exponents for DA and Q to accommodate areas down to one ha or less. These variable exponents simply prevent unreasonably high predictions for small areas (without the variable exponents, the the q_p values could exceed normal peak rainfall intensities in some areas).

Evapotranspiration

The evapotranspiration (ET) component of CDRHM is Ritchie's ET model (Ritchie, 1972). To compute potential evaporation, the model uses the equation

$$E_o = \frac{1.28 \; \delta \; H_o}{\delta + \gamma} \qquad (11)$$

where E_o is the potential evaporation; δ is the slope of the saturation vapor pressure curve at the mean air temperature; H_o is the net solar radiation; and γ is a psychrometric constant. The value of δ is computed with the equation

$$\delta = \left(\frac{5304}{T^2}\right) \exp (21.255 - 5304/T) \qquad (12)$$

where T is the daily temperature in degrees Kelvin. H_o is calculated with the equation

$$H_o = \frac{(1 - \lambda)(R)}{58.3} \qquad (13)$$

where R is the daily solar radiation in Langleys and λ is the albedo for solar radiation.

The model computes soil and plant evaporation separately. Potential soil evaporation is predicted with the equation

$$E_{so} = E_o \exp(-0.4 \text{ LAI}) \tag{14}$$

where E_{so} is the potential evaporation at the soil surface in mm and LAI is the leaf area index defined as the area of plant leaves relative to the soil surface area. Actual soil evaporation is computed in two stages. In the first stage, soil evaporation is limited only by the energy available at the surface and, thus, is equal to the potential soil evaporation. When the accumulated soil evaporation exceeds the stage one upper limit, the stage two evaporative process begins. Here the stage one upper limit is estimated with the equation

$$U = 9(\alpha - 3)^{0.42} \tag{15}$$

where U is the stage one upper limit in mm and α is a soil evaporation parameter dependent on soil water transmission characteristics (ranges from about 3.3 to 5.5 mm/d$^{1/2}$). Ritchie (1972) suggests a value of 4.5 for loamy soils, 3.5 for clays, and 3.3 for sands.

Stage two soil evaporation is predicted with the equation

$$E_s = \alpha[t^{1/2} - (t - 1)^{1/2}] \tag{16}$$

where E_s is the soil evaporation for day, t, and t is the number of days since stage two evaporation began.

Plant evaporation is computed with the equations

$$E_p = \frac{(E_o)(\text{LAI})}{3}, \qquad 0 \le \text{LAI} \le 3 \tag{17}$$

$$E_p = E_o - E_s, \qquad \text{LAI} > 3 \tag{18}$$

If soil water is limited, plant evaporation is reduced with the equation

$$E_{PL} = \frac{(E_p)(\text{ST})}{.25FC}, \qquad \text{ST} \le 0.25FC \tag{19}$$

where E_p is plant evaporation; E_{PL} is plant evaporation reduced by limited ST; and FC is the field capacity of soil in mm. Evapotranspiration, the sum of plant and soil evaporation, cannot exceed E_o. Field capacity is computed with the equation

$$FC = (\text{BR3} - \text{BR15})(\text{RD}) \tag{20}$$

where BR3 is the volumetric fractional water content at 0.3 bars tension.

Once the total ET is computed for a particular day, it must be distributed properly in the soil layers. A model for simulating root growth (Williams and Hann, 1978) is used for this purpose. Root depth is calculated with the equation

$$RD_n = 2.5 \ \frac{\sum\limits_{i=1}^{n} E_{Pi}}{\sum\limits_{i=1}^{N} E_{Pi}} \tag{21}$$

$$0 \le RD_n \le 1.0$$

where RD_n is the fraction of the root zone that contains roots on day n and N is the number of days in the growing season.

The water use rate as a function of root depth is expressed by the equation

$$v = v_0 \ \exp \ (-4.16 \ RD) \tag{22}$$

where v is the water use rate by the crop at depth, RD; and v_0 is the rate at the surface. The total water use within any depth can be computed by integrating Eq. 22 to obtain

$$ET = \frac{v_0}{4.16} \ (1 - \exp \ (-4.16 \ RD)) \tag{23}$$

The value of v_0 is determined for the root depth each day and the water use in each storage is computed with the equation

$$uw_i = \frac{v_0}{4.16} \ \left(\exp \ (-4.16 \ RD_{i-1}) - \exp \ (-4.16 \ RD_i) \right) \tag{24}$$

where uw_i is the water use in storage i (in mm) and RD_{i-1} and RD_i are the depths at the top and bottom of storage, i (recall i = 1, 7).

Percolation

The percolation component of the CDRHM uses a storage routing technique (Williams and Hann, 1976) to predict flow through the root zone. The root zone is divided into seven layers or storages for routing. The top storage is 1/36 of the root zone depth, the second storage is 5/36, and the five lower storages are 1/6 each. The top two layers are thinner than the lower layers because the topsoil has a greater effect on runoff and ET and thus needs better definition. The routing equation is

$$0 = \sigma \left(F + \frac{ST}{\Delta t} \right), \qquad \left(F + \frac{ST}{\Delta t} \right) > FC \tag{25}$$

where F is the infiltration or inflow rate; σ is the storage coefficient; and Δt is the routing interval (1 day). If inflow plus storage does not exceed field capacity, percolation cannot occur. The storage coefficient is a function of the travel time through the storage expressed by the equation

$$\sigma = \frac{2\Delta t}{2t + \Delta t} \qquad\qquad (26)$$

where t is the travel time through a storage. Travel time is estimated with the equation

$$t = \frac{ST - FC}{f_c} \qquad\qquad (27)$$

where f_c is the saturated conductivity of the soil.

MODEL TESTING AND EVALUATION

The CDRHM has been tested on basins in Texas, Ohio, Georgia, Oklahoma, Nebraska, Arizona, New Mexico, West Virginia, Mississippi, Iowa, and Montana. Results of the tests are shown in Tables 1-4. Table 1 shows that the model generally approximates long-term water yield (average annual runoff) well. Also, average annual ET and percolation predictions seem realistic. The monthly R^2 values shown in Table 1 were obtained by comparing measured and predicted monthly runoff. Table 2 contains statistics obtained by comparing measured and predicted individual runoff events. Although some of the R^2 values are lower than desirable, the standard deviations of the measured and predicted runoff are similar. This indicates that the model simulates runoff with a frequency distribution similar to that of the measured runoff, although the measured record is not duplicated precisely. There are many reasons for prediction errors. Some of the most important reasons are: (a) The curve number system's inability to consider rainfall intensity, duration, or distribution; (b) The use of average values for temperature, solar radiation, and leaf area index instead of actual values (Daily temperature and solar radiation is computed with Fourier analysis of average monthly values. Thus the same values are used for each year. Although LAI values are input each year, they are not driven by crop growth); (c) Lack of information on planting and tillage dates and incomplete soils descriptions; and (d) Rainfall and runoff data errors.

Table 3 shows a comparison of measured and predicted percolation for watershed Z at Tifton, GA. The measured values are actually subsurface flow measured at the watershed outlet. Of course, the predicted percolation is the amount of water that flows downward below the root zone. Considering these differences, the test can only indicate that the percolation model gave reasonable results.

Table 4 contains measured and predicted percolation and evapotranspiration for watershed 115 and lysimeter Y103A near Coshocton, Ohio. The measured values were obtained from the lysimeter. Both the watershed and the lysimeter grew the same crop each year. Close comparisons between measured and predicted values indicate satisfactory test results. Limited data prohibit percolation and ET model tests as extensive as those of the runoff model.

75

Watershed location	Drainage area (ha)	Length of record (yr)	Measured P (mm)	Measured Q (mm)	Average annual Predicted Q (mm)	ET (mm)	Percolation (mm)	Monthly R^2
SW-2, Riesel, TX	1.04	4	938	159.5	196.3	707	29.0	0.75
SW-12 "	1.29	9	967	229.9	163.8	778	26.7	0.72
Y-6 "	6.47	9	967	145.3	186.4	765	20.3	0.86
Y-8 "	8.29	9	967	170.7	156.0	793	17.3	0.65
21-H, Hastings, NE	1.55	13	580	86.4	93.7	487	0.8	0.41
3-H "	1.55	14	591	132.6	134.9	458	0.3	0.66
3-H "	1.55	9	596	120.6	137.4	457	0.3	0.68
PO-1, Watkinsville, GA	2.59	3	1205	221.5	210.8	839	176.5	0.46
PO-2 "	1.29	2	1124	150.9	164.1	740	236.2	0.53
104, Coshocton, OH	0.52	8	996	8.9	15.5	823	120.6	0.39
104 "	0.52	4	899	22.4	29.0	728	120.1	0.92
129 "	1.04	34	908	21.1	21.3	757	130.8	0.33
130 "	0.78	33	902	24.1	21.3	777	106.2	0.45
132 "	0.26	21	897	52.8	55.4	729	116.1	0.51
115 "	0.78	30	942	49.0	59.2	793	89.7	0.56
110 "	0.52	29	899	43.2	45.2	783	70.1	0.43
118 "	0.78	33	928	51.1	56.6	786	85.1	0.53
106 "	0.52	31	879	52.3	43.9	767	66.8	0.33
192 "	3.11	28	882	66.3	47.8	767	72.9	0.48
R-5, Chickasha, OK	9.58	8	766	44.7	49.5	694	17.8	0.73
R-7 "	7.77	8	766	151.9	134.6	614	10.4	0.86
C-4 "	12.17	9	818	87.6	70.9	745	2.5	0.59
C-5 "	5.18	9	697	51.3	48.8	643	2.3	0.35

Table 1. Runoff Model Test Results (Annual-Monthly).

Watershed location	Drainage area (ha)	Length of record (yr)	Measured P (mm)	Measured Q (mm)	Predicted Q (mm)	Predicted ET (mm)	Predicted Percolation (mm)	Monthly R^2
W-6, Cherokee, OK	0.78	19	603	84.6	90.9	509	4.3	0.45
W-7 "	0.78	19	603	91.2	91.2	509	4.3	0.53
W-13 "	0.78	7	553	42.2	53.6	497	0.5	0.59
#2, Guthrie, OK	1.29	10	711	106.2	95.0	593	29.7	0.85
W-1 "	1.04	7	696	17.0	22.6	649	39.9	0.46
W-1, Vega, TX	52.32	5	465	11.4	7.1	456	0.0	0.12
W-2	38.85	5	471	24.6	20.3	455	0.0	0.27
W-1, Spur, TX	4.66	19	510	49.0	52.1	458	0.0	0.67
W-2 "	3.88	19	510	68.1	65.0	445	0.0	0.70
W-3 "	4.66	18	511	39.4	43.7	466	0.0	0.72
63105, Lucky Hills, AZ	0.26	10	283	29.0	25.4	262	0.0	0.84
01, Ft. Stanton, NM	9.84	10	366	0.5	1.3	380	0.0	0.24
02 "	12.95	10	372	0.0	1.5	379	0.0	0.001
66001, Moorefield, WV	3.37	9	766	73.7	70.4	659	23.6	0.51
62014, Holly Spr., MS	0.52	3	1155	383.0	405.9	710	46.2	0.80
62014	0.52	3	857	240.8	245.1	615	30.2	0.65
22003, Guthrie, Ctr., IA	4.92	4	617	29.5	27.9	574	7.1	0.74
Z, Tifton, GA	0.34	6	1287	75.2	77.2	1048	182.1	0.26

Table 1. Continued.

Watershed location	Drainage area (ha)	Length of record (yr)	Average annual					Monthly R^2
			Measured		Predicted			
			P (mm)	Q (mm)	Q (mm)	ET (mm)	Percolation (mm)	
A, Sidney, MT	0.78	3	368	43.2	33.5	347	0.0	0.72
W-3, Garland, TX	4.14	8	1042	232.2	226.6	779	21.8	0.84
W-1 "	10.10	8	1073	129.8	163.1	805	88.9	0.86
W-3, Tyler, TX	3.11	9	1076	33.3	45.5	811	208.3	0.36
W-5 "	0.78	9	1056	209.0	184.1	785	89.4	0.58
W-4 "	24.09	11	1042	193.8	175.3	770	89.4	0.60

Table 1. Continued.

Watershed location	R^2	Runoff volume Standard deviation		Peak runoff rate			
				Mean		Standard deviation	
		Measured	Predicted	Measured	Predicted	Measured	Predicted
SW-2, Riesel, TX	0.85	18.80	18.80	2.19	1.77	3.17	2.43
SW-12 "	0.69	18.80	13.97	1.75	1.48	2.57	2.38
Y-6	0.90	17.27	21.34	5.43	5.45	7.60	8.92
Y-8	0.64	16.26	12.70	5.99	5.85	8.90	8.42
21-H, Hastings, NE	0.46	10.67	9.40	1.91	1.39	2.70	2.27
3-H "	0.65	11.94	10.41	3.11	1.53	4.11	2.49
3-H "	0.55	13.97	14.22	3.11	3.01	4.11	4.71
PO-1, Watkinsville, GA	0.60	15.49	12.19	4.54	3.06	6.83	5.02
PO-2	0.64	11.43	11.18	1.91	1.50	2.67	2.53
104, Coshocton, OH	0.28	2.54	2.79	0.49	0.28	0.76	0.44
104 "	0.88	10.67	9.14	0.49	0.46	0.76	1.11
129 "	0.24	6.35	4.32	0.58	0.68	1.07	1.32
130 "	0.29	6.60	4.06	0.34	0.34	0.75	0.59
132 "	0.46	8.38	6.10	0.08	0.09	0.11	0.16
115 "	0.55	8.13	7.37	0.74	0.58	1.33	1.18
110 "	0.37	8.13	5.84	0.35	0.46	0.83	0.87
118 "	0.52	7.37	5.84	0.60	0.61	1.12	1.04
106 "	0.31	5.59	4.83	0.55	0.50	1.17	0.88
192 "	0.41	9.40	6.35	1.07	1.26	2.93	2.45
R-5, Chickasha, OK	0.72	8.89	8.13	5.78	5.48	10.15	8.28
R-7 "	0.86	11.43	11.18	7.77	6.70	11.68	10.98
C-4 "	0.64	10.67	9.14	3.81	1.72	3.80	2.79
C-5	0.46	8.13	7.37	1.47	1.09	1.50	2.03

Table 2. Runoff Model Test Results (Events).

Watershed location	R^2	Runoff volume Standard deviation Measured	Runoff volume Standard deviation Predicted	Peak runoff rate Mean Measured	Peak runoff rate Mean Predicted	Peak runoff rate Standard deviation Measured	Peak runoff rate Standard deviation Predicted
W-6, Cherokee, OK	0.35	11.18	10.41	1.34	1.50	1.71	2.32
W-7 "	0.42	12.45	10.67	1.50	1.37	1.90	2.12
W-13 "	0.59	7.87	8.38	1.35	1.23	1.77	1.98
#2, Guthrie, OK	0.67	9.65	9.14	1.68	1.64	2.49	2.12
W-1 "	0.15	5.59	3.81	0.45	0.76	0.56	0.90
63105, Lucky Hills, AZ	0.64	5.84	4.32	0.29	0.13	0.47	0.20
01, Ft. Stanton, NM	0.003	0.51	3.05	1.45	1.87	1.05	2.27
02 "	0.10	0.00	3.30	1.02	1.51	0.37	2.53
66001, Moorefield, WV	0.71	17.78	13.72	0.49	0.65	0.45	1.13
62014, Holly Spr., MS	0.82	18.80	16.26	0.90	1.41	1.23	1.68
62014 "	0.62	14.48	12.70	0.90	1.02	1.23	1.41
22003, Guthrie Ctr., IA	0.44	4.06	4.32	0.64	0.97	0.28	1.30
Z, Tifton, GA	0.08	5.84	5.84	0.49	0.62	0.55	0.78
A, Sidney, MT	0.68	8.64	7.11	0.49	0.39	0.89	0.60

Table 2.   Continued.

Year	Annual percolation (mm)	
	Measured	Predicted
1970	454.7	501.4
1971	234.4	400.6
1972	297.7	308.1
1973	442.5	324.6
1974	213.6	269.2
1975	376.7	249.2
Mean	336.5	342.1
ST. DEV.	103.9	94.0

Month	Ave. monthly percolation (mm)	
	Measured	Predicted
1	54.6	67.8
2	74.9	67.8
3	38.6	40.1
4	64.5	48.0
5	19.8	26.2
6	14.5	18.0
7	10.2	3.3
8	24.9	18.3
9	18.0	6.1
10	0.0	10.7
11	0.0	6.3
12	18.3	29.5
Mean	28.2	28.4
ST. DEV.	24.6	22.9

Table 3. Percolation Model Results at Tifton, GA, Watershed Z.

Table 4. Percolation and Evapotranspiration Model Results at Coshocton, OH, Watershed 115.

| | | Annual | | | | | Ave. monthly | | | |
| | | Percolation (mm) | | ET (mm) | | | Percolation (mm) | | ET (mm) | |
Year	Crop	Measured[1]	Predicted	Measured[1]	Predicted	Month	Measured[1]	Predicted	Measured[1]	Predicted
1944	Meadow	128.5	72.4	771.4	731.5	1	28.2	35.3	19.6	19.3
1945	Corn	279.4	294.1	850.6	830.3	2	31.0	32.3	30.5	24.4
1946	Wheat	126.7	108.7	972.8	930.4	3	45.7	44.4	56.1	43.7
1947	Meadow	253.2	163.1	858.3	914.1	4	34.0	23.1	80.5	86.4
1948	Meadow	160.0	159.0	904.7	811.3	5	10.4	8.4	138.4	131.8
1949	Corn	140.7	214.9	887.7	834.9	6	6.1	7.9	137.2	126.0
1950	Wheat	327.2	264.7	926.1	954.5	7	1.0	4.1	140.2	122.9
1951	Meadow	348.0	303.8	885.2	801.4	8	0.8	0.5	117.1	108.5
1952	Meadow	232.2	258.3	887.2	787.9	9	1.0	0.5	74.4	60.7
1953	Corn	63.5	82.0	763.3	710.9	10	0.8	1.5	48.0	50.5
1954	Wheat	16.3	18.5	779.8	769.9	11	1.3	2.5	24.6	33.3
1955	Meadow	133.9	203.5	898.1	756.2	12	7.9	12.2	21.3	23.6
1956	Meadow	161.8	197.4	1005.1	1031.2	Mean	14.0	14.5	73.9	69.3
1957	Corn	181.4	245.1	852.9	790.4	ST. DEV.	16.3	15.5	48.3	43.4
1958	Wheat	94.0	33.8	940.3	993.6					
1959	Meadow	129.8	246.1	985.0	862.8					
1960	Meadow	98.0	172.7	980.7	913.4					
1961	Corn	183.6	141.2	859.5	802.1					
1962	Wheat	132.8	106.2	860.8	806.4					
Mean		167.9	173.0	888.0	843.8					
ST. DEV.		86.4	85.9	70.4	89.2					

[1] Measured percolation is from a nearby lysimeter (Y103A) with land use same was watershed 115 each year. Period of record is 1944-1962.

SUMMARY AND CONCLUSIONS

The CDRHM (CREAMS daily rainfall hydrology model) was developed as one component of a field scale model for simulating agricultural contributions to water pollution. The CREAMS model provides the user two hydrology models to accommodate a wide range in problem solving needs. Since the CDRHM uses readily available daily rainfall, it gives CREAMS greater user convenience and general applicability. The CDRHM is physically based (does not require calibration); capable of computing the effects of agricultural management on outputs; and computationally efficient (allows simulation of a variety of management strategies without excessive cost). Major processes included in CDRHM are surface runoff, percolation, and evapotranspiration.

Although the CDRHM should be useful in providing long-term runoff estimates for small watershed water resources planning, it's most important application is furnishing inputs to the other CREAMS components (erosion-sedimentation and chemistry). Runoff volume and peak runoff rate are important inputs to the erosion-sedimentation component and runoff volume, soil water content, evapotranspiration, and percolation are important to the chemistry component.

Tests with data from small watersheds in Texas, Nebraska, Georgia, Ohio, Oklahoma, Arizona, New Mexico, West Virginia, Mississippi, Iowa, and Montana indicate that CDRHM is capable of simulating long-term runoff, ET, and percolation realistically.

REFERENCES

Knisel, W.G. 1980. CREAMS, A Field Scale Model for Chemicals, Runoff, and Erosion from Agricultural Management Systems. USDA Conservation Research Report No. 26, 643pp.

Ritchie, J.T. 1972. A Model for Predicting Evaporation from a Row Crop with Incomplete Cover. Water Resources Research, Vol. 8, No. 5, pp. 1204-1213.

Soil Conservation Service, USDA. 1972. National Engineering Handbook, Hydrology, Section 4, Chapters 7-10.

Williams, J.R., and Hann, R.W. 1978. Optimal Operation of Large Agricultural Watersheds with Water Quality Constraints. Texas Water Resources Institute, TR-96, Texas A&M University, College Station, Texas. 152pp.

Williams, J.R., and LaSeur, W.V. 1976. Water Yield Model Using SCS Curve Numbers. Journal of the Hydraulics Division, ASCE, Vol. 102, No. HY9, pp. 1241-1253.

NOTATION

$BR3$ = volumetric fractional water content at 0.3 bars tension.

$BR15$ = volumetric fractional water content at 15 bars tension.

CN_I = SCS curve number for antecedent moisture condition I.

CN_{II} = SCS curve number for antecedent moisture condition II.

CS = mainstem channel slope in m/m.

D = depth to the bottom of each soil layer in mm.

DA = drainage area in ha.

E_o = potential evaporation in mm.

E_p = plant evaporation in mm.

E_{PL} = plant evaporation reduced by water stress in mm.

E_s = soil evaporation in mm.

E_{so} = potential evaporation at the soil surface in mm.

f_c = saturated conductivity of the soil in mm/h.

F = infiltration rate to top storage and inflow rate to lower storages in mm/hr.

84

FC = field capacity of the soil in mm.

H_o = net solar radiation in mm.

LAI = leaf area index (area of plant leaves relative to the soil surface area).

LW = length-width ratio of the watershed.

N = number of days in the crop's growing season.

O = amount of percolation below the root zone in mm.

P = daily rainfall amount in mm.

q_p = peak runoff rate in mm/h.

Q = daily runoff volume in mm.

R = daily solar radiation in Langleys.

RD = root zone depth in mm.

S = SCS curve number retention parameter in mm.

S_{mx} = maximum value of S (value for I antecedent moisture condition) in mm.

ST = plant available soil water content in the root zone in mm.

t = time in days.

T = daily average temperature in degrees Kelvin.

U = soil evaporation upper limit for stage one in mm.

UL = upper limit of plant available water storage in the root zone in mm.

uw = daily water use in each storage in mm.

v = water use rate for the crop at a given depth in mm/d.

v_o = water use rate for the crop at the soil surface in mm/d.

W = soil water-depth weighting factor for estimating the curve number retention parameter.

α = soil evaporation parameter.

γ = psychrometric constant.

δ = slope of the saturation vapor pressure curve at the mean air temperature.

Δ = increment of time.

θ = daily simulated volumetric fractional water content.

λ = albedo for solar radiation.

σ = storage coefficient for computing percolation.

ϕ = soil porosity in the root zone.

THEORY AND PRACTICE OF THE SSARR MODEL AS RELATED TO ANNALYZING AND FORECASTING THE RESPONSE OF HYDROLOGIC SYSTEMS

David M. Rockwood, PE
Formerly, Chief, Water Control Branch, North Pacific Division
U.S. Corps of Engineers (Retired). Presently, Consulting Engineer

ABSTRACT

The SSARR (Streamflow Synthesis and Reservoir Regulation) model has been developed progressively over the past 25 years to provide a generalized computer simulation technique for analyzing and forecasting various types of hydrologic systems. Although the model was first developed in connection with planning, designing, forecasting and managing the water resources of the Columbia River System (USA and Canada), it has subsequently been applied to many other regions for a variety of hydrologic and water management regimes in each of the major continents of the world. The model consists of various elements to simulate the processes of watershed hydrology, river system response, channel routing effects, and operational management of reservoirs or other water control facilities. The data processing system encompassed in the model provides for continuous simulation of a complex hydrologic system as a once-through process and as an efficient and easy to use processor. The model has been applied to numerous types of studies for planning and designing water control systems as well as to the operational forecasting of rivers and management of water control system on a real time basis.

The hydrologic methods utilized in the model are based on standard and generally accepted concepts of the hydrologic cycle and river analysis used in water management practice, and the formulation of the SSARR model has been described in several technical journals, manuals, and reports published previously. Accordingly, it is not the intent of this paper to repeat detailed descriptions of the algorithms and data processing techniques used in the model. It is considered important, however, to present at this time an overall synopsis of the principles of design and utilization of the model from the viewpoint of many years of experience gained in continued development and refinement. The first section of the paper deals with an overall summary of design concepts in relation to hydrologic and hydraulic theory of each of the various elements used in the model. The paper closes with a summary of examples of specific use of the model for simulating the hydrologic and/or reservoir control conditions for a complex river system or for a single river, reservoir, or tributary. The examples include those for planning, designing, or managing water control systems, together with the day-to-day forecasting of natural or regulated streamflows in a river system. This section also provides a summary of use of the SSARR model on a world wide basis.

INTRODUCTION

Symposia of the type for which we are gathered today are usually organized to present new and unpublished techniques related to scientific knowledge for a particular field and proceedures which may be used to apply that knowledge for practical application. This paper has a somewhat different point of view, it that it deals with a basic modeling proceedures which is well established. Furthermore, a significant amount of writing has been published in technical journals and manuals over the past 20 years which describe technical details related to the development of the model. Nevertheless, there is a definite need at this time not only to assess and summarize the theory in light of many years practice, but also to relate how the progressive development has been intertwined with the practical needs in applying the model to many types of hydrologic systems. Specifically, this paper summarizes the concepts used in the development and use of the SSARR (Streamflow Synthesis and Reservoir Regulation) model. Included in the SSARR model are elements related to nearly all of the topics being discussed at this seminar.

Purpose and Scope

The main purpose of this paper is to present a verbal description of the hydrologic and hydraulic principles used in model design, together with other design aspects important in utilizing the model in practical application. Reference is made to other publications which contain the detailed descriptions of the algorithms and data processing techniques used in the model. The paper also discusses the types of applications that have been made by using the model in analyzing a variety of river systems. This section includes a summary of design criteria that are related to the practical application as important requirements in addition to the hydrologic theory, in order to achieve a practical and workable system.

PRINCIPLES USED IN SSARR MODEL FORMULATION

The development of the SSARR model resulted from necessity for river system analysis in real-time, initially in connection with the operation and management of the Columbia River System by the U.S. Federal agencies involved in forecasting and management of the system. The principles used in design of the model were to provide for simulating the response of the hydrologic and water control elements, in a theoretically sound and operationally practical system. The modelling system, through evolution by practical experience in many applications, has become a highly generalized and flexible computer based proceedure. The detailed descriptions of the algorithms and data processing techniques are contained in the SSARR and COSSARR user manuals, and the reader is referred to those publications [1][2] for this information. The following section describes the principles in formulation of the watershed, river system and water control elements of river system simulation in the model.

Hydrologic Representation of the Watershed Model

General principles

The matter of hydrologic representation of runoff processes in a watershed model is highly subjective. No two hydrologists look at watershed runoff processes in exactly the same light. Nevertheless, there are some underlying principles that must be preserved in the formulation of a deterministic hydrologic watershed model. These include the logical accounting of each of the basic elements in the hydrologic cycle (rain-

fall, snowmelt, interception, soil moisture, interflow, groundwater recharge, evapotranspiration, and the various time delay processes), together with the ability to maintain continuity of each of the processes
and to represent each by objective functions, which relate them to observed hydrometeorological parameters. The differences between model
representations are in the various complexities that are incorporated
to represent a particular process.

Proper judgment must be used in determining the degree of refinement that is warranted in a model used for hydrological forecasting. In a
real-world situation, the limitations of availability of basic data --
even looking out well into the future -- are an overriding constraint
which negate the value of an overly complex representation of the hydrologic processes. Indexes of each of the processes are necessary, and
these indexes should be as simple as possible in order to provide a
workable solution for day-to-day application. Specific overriding considerations including quantifying true basin precipitation from rain-
gauge networks, the ability to handle data on real-time basis, the effect
of unforecastable meteorological elements, quantitative precipitation
forecast (QPF), etc., negate the value of overly complex hydrologic
models used for this type of work.

The balance between an empirical and completely rigorous approach
is a matter of judgment. The watershed model portion of the SSARR model
is designed to give the hydrologist as much flexibility as possible in
the way the model is applied to a particular case. The degree of the
complexity of the representation can be varied by the user to best fit
a set of circumstances. The model is designed to be as general and
flexible as possible, so that it will be a convenient tool for the hydrologist to adjust or modify characteristics in order for him to provide,
in his judgment, the best representation of the hydrologic processes.
In this sense, then, the model becomes a tremendous hydrologic bookkeep-
er and data processor for hydrologists to work with. The hydrologists
who uses it can become truly creative in developing a process which
combines both the art and the science of hydrology for his best use in a
particular case. Again, in this sense, the SSARR watershed model (which
is really only a small segment of the model as a whole) is not intended
to be a single representation of the hydrologic processes, but rather
a data processor which will provide the framework structure for a multi-
plicity of representations, depending upon the initiative of the user.

Water balance representation

The basic concept in the formulation of the SSARR watershed model
involves a closed hydrologic system in which the water accountability is
defined by inputs from rainfall and/or snowmelt, outflows from runoff as
measured at stream gaging stations, losses by the process of evapotranspiration and losses resulting from changes in soil moisture in the surface or sub-surface layers of the soil mantle. So called losses by infiltration to the ground water aquifers are accounted for by an infiltration index defining that portion of the water input which contributes
to ground water or "base flow". The water entering the deep ground water
system is not subject to losses by evapotranspiration. Continuity of the
volume of ground water is maintained throughout the simulation process,
and very significant time delays to runoff resulting from ground water
contribution are accounted for in the ground water storage and outflow
relationship used in the model. In the application of the model, it is
generally assumed that no significant ground water flow bypasses the
river flow as measured at the stream gaging station. Surface and subsurface runoff from the watershed are accounted for individually in

their contribution to the flows. Moisture inputs from rainfall and/or snowmelt are considered to be those contributing at the ground surface. Interception loss is implicitly accounted for as a portion of the evapotranspiration loss, and daily variation in the amounts of interception are accounted for in the basic soil moisture, infiltration, and surface runoff indexes used in the model. Thus, the watershed model is simplified but rigorously applied representation of the standard hydrologic cycle. The model parameters may be manipulated to represent special or unusual conditions of basin runoff or geology, as, for example, large deep ground water aquifers, unusually diverse soil mantle conditions, or varying amount of impermeable areas encountered in urban hydrology. In the SSARR model as a whole (beyond the use of just the watershed model), the effect of irrigation diversions and return flows, channel bypasses, reservoir regulation or other man-caused effects on the hydrology of river system may be incorporated in the simulation process.

Discussion of watershed model parameters

a. General   The rainfall-runoff parameters used in the SSARR watershed model are described in Section 2.01 through 2.03 the Program Desription and Users' Manual. The methods of computing and maintaining continuity of each of the indexes or parameters are fully described therein. The following paragraphs discuss the hydrologic significance of each of the indexes or parameters.

b. Definition of components   Because of the varied interpretation of the components of the water balance, each is specifically defined as used in this paper. Observed runoff is defined as the gauged volume of runoff passing a gauging station on a river or stream. Generated runoff is the observed runoff corrected for transitory storage in the soil, ground, and stream channels. Total basin precipitation is defined as the hypothetical precipitation which falls above the tree crown level. Net precipitation is the part of the total precipitation which reaches the ground. Loss is defined as that part of the net precipitation which is permanently lost to runoff by evaporation, transpiration, or soil moisture increase. Soil moisture storage differs from ground or channel storage in that water stored as soil moisture can be removed only by evaporation and transpiration, whereas water in ground or channel storage is temporarily stored in transit and will ultimately appear as runoff. Infiltration to the deep ground water aquifers is not considered to be a loss in that the water which enters these zones eventually appears as runoff. The delay to flow through ground or channel storage is accounted for by recession analysis or by the storage routing computations.

c. Net precipitation input (WP)   The average net precipitation value for a drainage basin or hydrologic unit is derived as a weighted daily or period amount from a series of individually reported or observed values. The weighting is generalized, and each station may be assigned its individual weighting value. The station weights may be determined on the basis of previously derived relationships between station and basin normal annual or normal seasonal precipitation; or the weighting values may be derived on an areal basis using the Thiessen Polygon. It is also possible to derive the individual station weights by trial and error reconstitution studies of historical streamflow data, whereby the effectiveness of each station in representing basin precipitation can be determined as an index value in relationship to other stations being used. Average basin precipitation may also be estimated indirectly from reports obtained from weather radar or from satelite observations.

In actual hydrologic simulation practice, the determination of basin precipitation is one of the least well defined of any of the processes involved. Precipitation measurements in themselves are subject to numerous sources of error, such as **gage exposure, wind, variability of** actual rainfall in time and space, the paucity of reporting networks in representing amounts of rainfall over large areas, and countless difficulties in obtaining reliable and accurate reports from relatively untrained observers. Difficulties in estimating basin rainfall are accentuated in areas where storms are primarily of a convective nature, as is usually the case in the tropics. Nevertheless, any attempt at hydrologic simulation of rainfall-runoff processes must be based on use of whatever basic rain data are available, but conceptually the computed rainfall amounts should be considered as a relatively unreliable index value of the "true" basin amount. It is important, however, to normalize insofar as possible the computed basin amounts in order to estimate values for the basin, based on long-term studies of normal annual or normal seasonal amounts.

The period distribution routine in the model provides a convenient method for computing period (usually 3 hours) rainfall amounts from daily observations. Although the model is designed to handle any designated period rainfall from observed amounts for periods ranging from one-hour to 24-hour duration, it is usually not practical to assemble a long series of rainfall data inputs for many stations, with one-hour inputs, for a long-term (many year) reconstitution study. Daily values, however, are generally available; and the distribution routine may be used to distribute rainfall from daily amounts. Diurnal rainfall patterns may be easily represented by this method with reasonable assurance that they represent realistic period values.

d. Snowmelt functions   The calculation of snowmelt by the SSARR model is a major element of the watershed modelling process. It is not within the scope of this paper to describe these functions, but reference is made to "Snow Hydrology"[3] and the SSARR user manual for back ground information and detailed descriptions of the particular methods used in applying snowmelt parameters to the computation of water inputs resulting from snowmelt. The following paragraphs outline only the broad and general aspects in the use of these functions, where snowmelt is involved in watershed runoff simulation.

Basically, there are two general options for the calculation of snowmelt, namely: (1) the temperature index method; and (2) the use of generalized equations of snowmelt as determined by the thermal budget of heat loss and gain to the snowpack. The temperature index method is usually used for daily forecasting applications, whereas the more detailed energy budget approach is more appropriate for design flood derivations. By either method, daily or period values of effective snowmelt runoff values are computed as a time series, as a function of appropriate meteorological values. The programme also provides for distributing daily amounts into period values for representing the diurnal fluctuations of snowmelt. In addition to the two methods for calculating snowmelt, two options are available to evaluate snowmelt runoff from a watershed, where the snowpack accumulates and ablates seasonally over the watershed. The first option describes the snow cover depletion by use of a function which relates the snow-covered area to the accumulation of runoff, in proportion to the total seasonal runoff volume. The second option provides for the capability to sub-divide the watershed area into "bands" of equal elevation, as determined from area-elevation relationships. Each "band" is treated separately in the determination of snow accumulation and snowmelt, as inputs to the water balance for the watershed as a whole. The North-Pacific Division Office of the Corps of Engineers has recently

completed a major revision of the watershed portion of the model, whereby the "snow band" option for computing snowmelt runoff has been reprogrammed to provide a workable and efficient system for operational use
of the model in this mode.

e. Soil Moisture Index (SMI)   The soil moisture index used in the SSARR
model represents a weighted mean value of the water stored in the soil
mantle that can be removed by plant roots through transpiration and also
by natural evaporation. It does not include the part of the soil moisture content that exists at the permanent wilting point. The computation
of the changes of soil moisture index values are based on the increases
resulting from rainfall or snowmelt, and the decreases by the evapotranspiration process. Increases in the soil moisture index values result in
a "permanent" loss to runoff in the water balance for the basin as a
whole. The upper limit of the soil moisture index is considered to be
its field capacity, which is equivalent to the capillary moisture holding
capacity, or the total amount of water which can be held under the force
of gravity under natural conditions. Thus, the soil moisture index is
a continuously varying parameter that may range from a value of zero when
the soil moisture has been reduced to the "wilting point" by the evapotranspiration process,to a maximum value represented by the field capacity of the soil for the basin as a whole.

     The soil moisture index function (SMI function) is a means for
distributing the soil moisture loss as a function of the relative wetness
of the basin. Theoretically, if the soil mantle in a drainage basin were
completely homogeneous, the soil moisture increase would occur uniformly
over the basin area, and no runoff would be generated until the soil
moisture deficit had been satisfied. When the soil mantle had then reached its field capacity, any further moisture input would be fully effective in the generation of runoff. Such a generalized, theoretical concept of runoff does not occur in natural river system, because of the
widely varying conditions of the soil types, slopes, depression storage,
and conditions of porosity in the basin as a whole. The SMI funtion is
a means to account for the diversity of soil conditions as variable per
cent of the basin area that contributes to permanent loss by soil moisture increases, as a function of the relative wetness of the basin.

f. Evaptranspiration Index (ETI)   The evapotranspiration index (ETI)
used in the SSARR model is a weighted basin mean daily value of the water
lost to the atmosphere by the evapotranspiration process. Transpiration,
soil evaporation and evaporation of free water from the plant or forest
cover are considered to act together to produce the losses by evapotranspiration. Inasmuch as the evapotranspiration loss is physically the
result of change of state of water from the liquid to vapor phase, the
process of evapotranspiration requires energy for the transformation,
and is therefore dependent upon a source of energy from the atmosphere.
This energy may be derived through solar or terrestrial radiation, or
from wind and vapor pressure gradients in the lower layers of the atmosphere, whereby there is a net exchange of energy between the plant and
soil surfaces and the atmosphere to cause the change of state of liquid
water to the vapor phase. Most commonly the source of energy is solar
radiation, but the rate of evaporation is highly dependent upon the vapor
pressure differences between the atmosphere and that of the free water
available in the soil, plant, and forest mantle.

     The rate of evapotranspiration in a particular hydrologic regime
varies widely on a daily, monthly, or seasonal basis. The computation
of daily basin-wide evapotranspiration amounts on a rigorous basis would
require observations of solar radiation, longwave radiation, atmospheric

moisture content (dew point, vapor pressure or relative humidity), and wind movement. These types of data which would represent basin mean values are normally not available. Therefore, evapotranspiration amounts are usually computed by some form of an index. One particularly useful index is the measurement obtained of daily pan evaporation, whereby the estimated basin evapotranspiration is a direct function of the daily pan evaporation amounts. Other indexes may be based on air temperature, dew point temperature, or solar radiation amounts. Evapotranspiration rates are computed as "potential" amounts, considering that adequate supply of water in the soil mantle is available to support the transpiration process. As the soil moisture decreases to the "permanent wilting capacity", there is no longer the ability to supply the water necessary for transpiration, and the actual transpiration approaches zero.

In the model, the potential evapotranspiration may be computed by either of two basic methods, namely: (i) mean daily amounts based on mean monthly values which are typical for a given hydrologic regime; or (ii) mean daily amounts based on observed values of daily pan evaporation, mean daily air temperature or dew point temperatures, or daily solar radiation amounts. The daily computed amounts are adjusted in either case by a function to account for daily or period rainfall which would reduce the potential rate of evapotranspiration.

In the application of the SSARR model, the mean daily ET amounts computed through use of mean monthly amounts are most commonly used. Although the actual amounts vary widely from day to day in a particular hydrologic regime, the variations generally tend to balance out over a period of several weeks. The normal seasonal variation is, of course, accounted for in the specific monthly mean values. If, however, it is possible to assemble more detailed daily data (such as pan evaporation, solar radiation, etc.) these values may be used to compute the daily amounts. The refinements that are achieved by so doing may have significant value under certain climatic regimes, but normally, the differences are insignificant for the relatively wet, tropical climates.

g. Baseflow Infiltration Index (BII)   Baseflow Infiltration Index used in the SSARR watershed model provides a means for computing the relative proportion of the water available in the surface layers of the soil mantle that enters the ground water aquifers as deep percolation. Under the principle of "generated runoff", as defined in subparagraph (b), above, all water which is not lost to the atmosphere by evapotranspiration, or the permanent loss by soil moisture increase, is available to runoff as a time delay function. Conceptually, the model considers the time delay to occur in three zones, namely surface, sub-surface and baseflow. The long time delay caused by base flow infiltration represents that portion of the water which is in transitory storage for several months (or possibly years under certain circumstances). To some hydrologists, this would appear to constitute a loss; but in the overall long term simulation, the continuity of water in base flow is maintained throughout the simulation process. It is, therefore, not a loss in terms of permanent losses to runoff, but it does constitute an apparent loss to immediate runoff in flood hydrology.

h. Surface - Sub-surface Flow Index (S-SS)   In the model, the surface-sub-surface(S-SS) flow separation index deals with the water excess which is generated from the residual after soil moisture and transpiration losses and base flow infiltration have been satisfied. Normally, the "direct" runoff is considered to be the result of the percolation of "free" water through the upper layers of the soil mantle (say,in the zone up to a maximum of depth of 50 cm below the ground surface) termed as sub-surface flow. When the water input rate exceeds the capacity of sub-surface zone to transmit water under gravitational force, the residual water excess amount is considered to occur directly on the ground surface or the upper few cms of the soil mantle. The surface-sub-surface flow separation is a means for defining the relative portion of the direct runoff that contributes to each portion, as a function of input rate.

The S-SS function is usually specified as a nonlinear function, whereby the lower rates of input provide water excess primarily in sub-surface zone, while high input rates are predominantly on the surface runoff. The time delay functions for routing surface and sub-surface are specified to represent the difference in storage times for each of the two zones.

i. Watershed outflow transformation by polyphase routing for each flow component input   The water excess values computed for each time period in each of the three flow components (surface, sub-surface and baseflow) must be transformed from values computed as input rates to time-distributed values of streamflow. In the SSARR model, this transformation is accomplished by polyphase routing, whereby the input rates expressed as cm per period are converted to equivalent values of steady-state outflow, expressed as cubic meters per second for the particular drainage area. The routing is performed through the use of the flow continuity equations set forth in the SSARR users manual.

The use of polyphase routing for this type of transformation has several advantages for computerized simulation of streamflow from watersheds. These includes: (i) simplicity of computation; (ii) ease of application in trial and error reconstution studies for determining basin runoff characteristics; (iii) the relatively small amount of information required to store in the computer, in order to represent the basin runoff characteristics; (iv) the completely flexible means for representing time delays to runoff either for short term flood runoff on relatively small tributaries, or for long-term base flow on ground water discharge for large river system; (v) the convenient means for preserving the continuity of flow at any specified point in time, for "stop action" or "instant replay" capabilities, particularly in its use for day-to-day streamflow forecasting; (vi) the flexibility of providing virtually any desired shape of runoff characteristics, (as, for example, a unit distribution of known characteristics), (vii) the assured preservation of continuity of flow for any computed runoff excess, (Reference is made to section 2.06 of the SSARR user manual for information in determining polyphase routing coefficients) and (viii) the ability to represent nonlinear response for runoff from a watershed.

While unit hydrographs have been developed and used for numerous hydrologic studies, the principles upon which the unit hydrograph theory was applied were based on application to flood hydrology for simulating flows for short time period (generally several days to a week or so). Polyphase routing is a much more feasible way to transform rainfall runoff excesses for hydrograph synthesis, in that it performs the same functions as a unit hydrograph but has many decided advantages in application to computerized streamflow synthesis, as listed above.

j. Watershed model processing   Each of the basic parameters or functions discussed above is processed in the model as a period by-period computation. The model may operate on any specified time period, ranging from 0.1 hour to 24 hours. Also, the length of time period may be varied for a particular simulation on the basis of specified values

94

individually for each of four portions of the total time of the simulation. This feature of the model is particularly useful in making simulations for forecasting streamflow and reservoir regulation, whereby the first several days of the forecasting period is computed on a detailed short-time period basis, followed by a long-range outlook computed with a longer time period.

The continuity of each of the rainfall runoff parameters (SMI and BII) is computed for each elemental time period, and the functions for relating the parameters to the runoff elements or flow components are used by the programme for each incremental period computation. Similarly, the flow transformation by polyphase routing is accomplished as a period-by-period process for each of the flow elements used in the model.

Optimisation of parameters

In general, the optimisation of the various parameters is normally accomplished by trial-and-error reconstitution studies of historical streamflow data using observed values of index of water input. Such studies are generally performed for several years of historical data, whereby the various parameters are tested to achieve the best fit of computed and observed streamflow. It is normally assumed that the physical factors affecting runoff are nonchanging over a period of years, so that the parameters and functions used in the model are fixed as a given set of values for the entire study period. The degree of fit between computed and historical streamflow is determined either visually by inspection of graphical plots of the data or by computation of errors of estimate by standard least-squares determinations. The principle objective is to achieve consistency over a wide range of hydrologic conditions to eliminate bias between high and low periods of streamflow and to achieve relatively uniform consistency for the years being studied.

The overall water balance for particular study areas should represent as closely as possible the known or expected values of precipitation, evapotranspiration, soil moisture and ground water condition that are characteristics for the climatological and hydrological regime of the area. The main objective of reconstitution studies is first to achieve a water balance by adjusting the following parameters: (i) precipitation weighting for estimating basin precipitation from index station values; (ii) SMI function, in terms of total soil moisture index values. and the shape of the SMI function; (iii) ETI values, based on observed or estimated amounts which properly reflect the seasonal of daily variation; and (iv) BII function, representing the portion of water input which contributes to base flow. When the overall water balance is achieved, the refinements in timing can be taken into account by adjusting the polyphase routing parameters for each components flow (surface, subsurface, and base flow), and by adjusting the S-SS flow separation function.

While the above description is only a very brief description of the methods used for adjusting and optimizing parameters, it provides the principles by which the hydrologist can develop his own expertise in streamflow simulation by use of the model. The efficient use of the model and concerted efforts in optimizing parameters can best be achieved by actual use in a creative sense by the hydrologist in fitting data for a particular hydrologic regime.

Automatic optimization of the parameters has been achieved by Dr. Bolyvong Tanovan in connection with studies made for the Mekong River in Indochina. An auxilary programme, designated MEKSSA, was written which provided the means for objectively determining the best fit of the

SSARR parameters through a method of convergence. This type of analysis provides a means for optimizing the functions through objective "best fit" determinations. In principle, it is an improvement on the trial-and-error approach recommended herein, but there are practical consider-ations that must be considered before making a judgment on the use of this method. In particular, the hydrologist should, through use of the trial-and-error approach, become thoroughly familiar with the basin hydrologic characteristics and recognize any deficiency in the water accounting that may result from unusual conditions reflected by the model parameters. Secondly, it often develops that inconsistencies in basic data are not readily apparent unless the results of the studies are inspected indi-vidually through subjective analysis. The so-called automatic adjust-ment procedures would not provide this aspect of knowledge of model per-formance and data deficiencies.

Hydraulic Representation of Unsteady Flows in River Channels

There are many methods commonly used for simulating the response of river systems subjected to unsteady inflows from natural or man-caused fluctuations. Streamflow routing procedures range from simple, empirical methods for translating and computing the attenuation of the unsteady flow fluctuations, to highly complex and completely rigorous computerized solutions of the unsteady-state flow equations. Each has its use for particular types of applications, depending on the type of river system, the general ranges of flow variations normally experienced, the effects of variable backwater conditions caused by tides or reservoir operation, the overall accuracy of the computed fluctuations in relation to the needs for a particular application, the time and effort that can be expended in the solution for timely use, and the availability of basic data required for application.

The streamflow routing functions contained in the SSARR model provide a generalized system for solving the unsteady flow conditions in river channels where streamflow and channel storage effects are related either at one point or at a series of points along a river system. In principle, the method involves a direct solution of storage-flow rela-tionship involved in maintaining continuity of streamflow and storage in each element of the river, through use of a procedure which solves the relationships in finite elements of time and river reach. This involves a completely general and flexible method for solving the flow routing equations which can be applied in many ways depending upon the type of basic data available, and the conditions of the river system with respect to backwater effects from variable stage discharge effects, such as tidal fluctuations or reservoir fluctuations. The model provides a two-dimensional analysis of unsteady flows in a river system, in the sense that the changes in flow are determined only in y and t coordinates: that is, streamflows and water levels may be computed for each element of distance along the axis of the river, and in each element of time used in the solution of the routing equations. There is no intent to solve the varying conditions of velocity in the cross-section of the river. In the model, the conditions of flow at a point are a function of hydraulic parameters which relate the total flow at a point along the river channel to either the water surface elevation at that point, or as a function of the gradient of the water surface as measured at upstream or downstream points. Inasmuch as the changes in momentum which take place in the unsteady flow of water in river channels are usually negligible, there is no attempt to account for the continuity of changing momentum in the SSARR model channel routing procedure.

One of the basic concepts in the formulation of the SSARR model is

96

to utilize in so far as possible "predigested" data in expressing the
storage and flow relationship in the analysis of water systems. Re-
cognition of this principle is very important in understanding the reasons underlying the mathematical flow-discharge relationships used in
the model. Countless thousands of repetitive solutions of the flow-
routing are required in solving the unsteady state flow equations required for a complete analysis of a hydraulic system. It is senseless
to solve the full equations commonly used to express the flow in channels,
(for example, by use of Manning's equation) for the iterations that are
required in unsteady flow computations, when these flow and storage relationships may be "predigested" to provide simplified and easy to use
routing coefficients for a particular case. Accordingly, it is much
more feasible to develope an efficient and easy to use flow routing
procedure which utilizes simplified expressions relating streamflow,
water surface elevation, and channel storage at each point in the river
channel. The concept of the use of a constant or variable time of
storage (T_s) does explicitly provide for such a solution. This is one
of the basic elements in the formulation of the routing method used in
the SSARR model. In the case where backwater effects must be considered
in the flow routing equations, the model utilizes generalized 3-variable
functions which relate flow between water surface elevations successivelly at
downstream points. Again, the concept "predigested" data becomes important, in that these functions can be easily defined from observed data,
or from backwater computations based on steady-state flows.

In actual practice, the necessary functional relationships, or
flow routing coefficients for the algorithms used in the model can be
derived, by a number of different procedures, depending upon the availability of basic data, and other considerations as to type of application and the necessity for refinement in the procedure as a whole. The
simplest technique is to develop channel routing coefficient utilizing
observed streamflow data at various points in the river system and general knowledge of the channel characteristics. From this knowledge,
repetitive trial and error solutions of historical streamflow conditions,
utilizing the model in connection with deriving local inflows as a time
series, provides a technique for refining the routing coefficients and
determining the time of storage variation which best expresses the
"time of travel" in streamflow for a particular reach. Another method
for developing routing coefficients is to analyse the streamflow-
storage-elevation relationships for points where discharge measurements
are made. If the cross-section data for such points are representative
of the channel reach as a whole, considering both normal and overbank
flow conditions, these provide a discreet evaluation of T_s per km of
channel length, which may be used in solving the incremental storage
functions for that reach. A third method, which is used for conditions
where backwater significantly affects the flow relationships at a point,
the effect of variable channel slope on discharge at a point may be developed from simultaneously observed water-surface and streamflow data
for the range of conditions that are normally experienced.

The fourth and most rigorous method for developing the routing
characteristics for the model is that of computing steady-state water
surface profiles for a range of flows and elevations at downstream control points which are to be expected. From these steady-state flow
evaluations, data may be enterred into the model which represents the
3-way variability of streamflow to elevation for each successive point
in the river system. This relationship, when combined with the variable
increment per unit of reach, provides a completely flexible method for
routing streamflows where backwater is a significant factor.

97

From the above discussion, it can be understood that the simulation of river system response as   computed by the SSARR model is not a single method for routing streamflows in a river channels, but rather offers a variety of methods for application for solving the continuity equations required for flow routing. In all cases, the continuity of streamflow is completely preserved in each routing reach. The importance of the principles involved in the development of the procedure lies in the flexibility in application and efficiency of operation with respect to computer utilization. Accordingly, it may be applied easily to many types of applications for hydrologic studies required in project planning, designing and operation, or for streamflow forecasting and real-time management of water control systems.

Reservoir Regulation Functions

The SSARR model is designed to include the effects of reservoirs or other water control elements within the streamflow simulation processing. Reservoirs may be described for any location in a river system, whereby inflows are defined from single or multiple tributaries, derived either from watershed simulation for river basins upstream, or from specified flows as a time series, or combination of the two.

Outflows are determined on the basis of specified operating conditions. The processing of the hydraulic conditions at reservoirs is performed sequentially with all other elements in the river basin simulation, in order to provide a once-through process for the system as a whole, including all natural and man-caused effects.

Specification of reservoir characteristics

Reservoirs are specified in a manner to provide the model with the hydraulic characteristics of the combined water-passage facilities which affects the reservoir levels, outflows, and tailwater conditions. The basic elements include the normal operating (maximum and minimum) reservoir levels, and tables which relate reservoir storage and reservoir outflows as a function of water surface elevation in the reservoir. The elevation-outflow relationships are normally specified for the combined flow of regulating outlets, spillways, hydraulic turbines or other water passage facilities. These values are usually specified as their combined flow, thereby representing the maximum flow that can be passed within the normal operating constraints of the project. This requirement is particularly important in evaluating the limiting outflows that can be passed during the use of surcharge storage or as a limiting constraint which the model checks against when specifying reservoir regulation for a particular operating function. In addition to the hydraulic characteristics or the projects, the specifications include information for the logical processing the reservoir in relation to the other elements in the river system, and other data processing information.

Specification of reservoir regulation

The regulation of hydraulic conditions at reservoir may be simulated by various modes, including: (1) "free-flow", (2) outflows specified as a time series; (3) reservoir elevations specified as a time series; (4) change in storage, expressed as a flow rate, and specified as a time series; (5) change in storage elevation per day, specified as a time series, and (6) change in storage volume per day, specified as a time series. Within a particular simulation, any one or all of the above types of specifications may be interspersed within the period of simulation. Also, point values for a particular specification may be designated at irregular intervals, and the programme automatically determines intermediate values, based on lineal interpolation between the designated values. This is important in reducing the volume of input data required for a particular simulation, and for ease in data preparation. If the specification of reservoir operation results in exceeding the upper limits of the normal operating pool level or the discharge capacity for the particular reservoir level, the model will automatically process outflows as "free-flow" routing until such time as the operating specifications result in operation below those limits. Also, in connection with other computer programmes developed in the North Pacific Division of the Corps of Engineers, the reservoirs may be regulated to simulate the hydraulic conditions required for at-site or system hydro-power requirements, or for flood control requirements at downstream control points in a river system.

River Model Options for Maintaining Storage and River Continuity

Additional basic requirements for a river system simulation model are to provide results of the simulation at intermediate points within the system, and maintain complete continuity of all water gain or loss within the system, whether it be large or relatively small. There are several options in the SSARR model which can be utilized for these purposes, and also for simplifying the total processing requirements for a system simulation. These include: (1) A completely general and flexible method for processing and summing simulated flows, in downstream order, for providing discharges and/or water levels at any desired point in the system; (2) An "Adjacent Basin" capability, for calculating the intermediate or "local" runoff from minor drainage areas contributing to the system, as a direct or functional relationship of flows as synthesized for an adjacent major drainage; (3) A means for conviently deriving "local" inflows for ungaged areas, by direct water continuity computations between upstream and downstream points; (4) A means for computing overbank flows which may be lost to the system; and (5) A means for accounting for irrigation diversions and return flows.

Computer Utilization

The successful application of a general system model of this type is perhaps as much dependent upon the data processing elements as the mathematical formulation. Usually a large mass of data must be gathered, processed, and checked for adequacy. Burdensome data preparation may overwhelm the individual user. Numerous trial studies must usually be performed to assure the adequacy of basin runoff characteristics, etc. Convenience in operating the system from previously stored data, by use of on-line computer terminals is of upmost importance in large-scale studies. It is also important to be able to analyze results in a convenient manner, whereby graphical displays may be made immediately available on interactive output terminals. Response time from time of submission to completion of results, available to the user is of particular significance to river forecasting and real-time management of water control systems.

In summary, it is essential that the data-processing system be simple to use from the user point of view, for the methods of preparation of input data, on-line terminal operation, and outputs provided in a timely and usable form.

The design of the SSARR model has progressively been developed with these requirements in mind. The computer based data handling

systems are complex from a computer programming point of view, in order
to achieve the overall requirement for generality and flexibility, in
application and operation. This complexity is invisible to the user,
however, and he simply utilizes the power of the system in a prescribed
manner which may be easily learned. In this sense, the model may be
thought of as a major data processor which may be operated upon to achieve
the goals of hydrologic system analysis. The SSARR model requires the use
of a moderate to large scale computer system. There is, however, scaled
down version of the model termed COSSARR, which can be processed on a
small scale computer. This form of the model performs essentially the
same hydrologic and hydraulic analytical functions as the large-scale
SSARR model, but is much more limited in data processing capability with
respect to size of system that may be processed, internal data processing
operation, time of processing, and interactive control.

UTILIZATION OF THE SSARR MODEL

Historical Background

As indicated in the introduction, the SSARR model was initially
developed beginning in 1956 in order to meet the needs of U.S. Corps of
Engineers in analyzing and forecasting the hydrologic and reservoir
systems in the North Pacific Division area of the Corps of Engineers which
camprises the Columbia River System in the U.S. and Canada, Coastal rivers
in Western Oregon and Washington, and the State of Alaska. As the model
was being further developed and refined in the 1960's, other U.S. Federal
Agencies participated in its development and utilization. The Portland
River Forecast Center of the U.S. National Weather Service was and is now
highly involved in its application to river forecasting in the Pacific
Northwest, and performed many studies which advanced the model capabilities in the forecasting mode. A formal agreement between the Washington
offices of the two agencies (Corps of Engineers and National Weather
Service) was consumated in 1963 whereby the two agencies agreed to co-
operate by use of the software and hardware computer systems available
in the Corps of Engineers, together with the river forecasting personnel
and facilities available in the Portland River Forecast Center. This
agreement has continued through the years, and was augmented by agreement
with the Bonnville Power Administration, U.S. Department of Energy, to
participate in the continued development and application of the hydrologic procedures as described herein. This summary points the important
consideration that the model is not a static, but a dynamic system that
is continually being upgraded through experience gained in its utilization in many types of studies.

Familiarity with the model has come about primarily from its use
in day-to-day management of water control systems and river forecasting.
The use of the model as a tool in the derivation of design floods, reservoir operating plans, and basic hydrologic studies in connection with
project planning is not as well known. These applications include major
water resource studies, such as the development of the Probable Maximum
and Standard Project Floods for the Columbia River System, and for many
individual project studies in the Columbia Basin, for both Federal or
Non Federal Projects. Similar project studies have been completed for
several major projects in Alaska by the Alaska District of the Corps
of Engineers.

Because of the nature of the model as a generalized hydrologic
processor, and the fact that the utilization of computerized technique
in hydrologic analysis and river forecasting received considerable attention starting in the mid-1960's, a large interest developed at about

that time in application of the model to areas outside of the North Pacific Division area of the Corps of Engineers. This came about by technical exchanges both in the United States and abroad. The major areas of interest inside the United States included several universties, Federal and State agencies, and private engineering companies. Several provincial and federal agencies in Canada pursued the use of the model in streamflow forecasting and hydrologic analysis. Significant advances in the use of the model in Canada were made in connection with analysis of rivers in Alberta, Ontario, and New Brunswick. Outside of the North American Continent, the model has been supplied to many countries and organizations, through programmes of the U.S. Agency for International Development, and the United Nations and upon individual requests. Most notable examples are: (1) The use of the model in analyzing and forecasting the Mekong River in Indochina; (2) general use of the model in analyzing rivers in Indonesia; (3) the use in analyzing and forecasting rivers in Brazil; (4) the application to the Morava River in Yugoslavia; and (5) the application of the model to 7 river systems that were analyzed on a worldwide basis through the UN/World Meteorological Organization Programme for testing conceptual hydrologic models used in river forecasting. It is interesting to note the principles contained in the model have been applied by independent investigators in several countries for developing similar hydrologic processors.

Summary of Utilization

From the above description, it is seen that the model has been transferred to many locations throughout the world. There is no way to evaluate the over-all effectiveness in its utilization outside of the area from which it was developed. In many cases, the model was supplied with little or no training, but in other cases, the technological transfer was incorporated in a training programme when there was considered "feed back" from the experience gained in application to other river systems. In several cases, the model users have creatively extended its use into areas that were not envisioned in its conception. It is far beyond the scope of this paper to summarize these experiences. However, a summary of Technical Reports [4] in the form of abstracts was published by the North Pacific Office of the Corps of Engineers in 1976, which summarizes 44 technical papers which at that time were known to have been prepared on subjects related to the development and application of the SSARR model. Obviously, this is not a complete summary of all applications, but only those for which technical documentation was available at that time. Finally, it is noted that, as of 1976, the model had been supplied to 32 agencies or organizations in the United States, and 33 organizations outside of the United States.

Design "Feed Back" from Application

The question now is, what have we learned from the 25 years of experience in the development and application of this model. The validity of design principles has been confirmed by numerous types of applications. But the principles of design were by no means fully developed at the initiation of the work. On the contrary, many of them came about from the experience gained in application, as a continuity process of "feed back" in the several major phases of development. Annex I lists the types of application, together with the principles of design used in the continued development of the SSARR model. This annex summarizes what we now consider to be the important design principles incorporated in the present formulation of the model. From our experience, we feel that these elements are all important characteristics that are necessary for the successful operation of hydrologic system processor.

A Look into the Future

The use of this system has provided experience far beyond what was dreamed of when the programme was first being developed back in the 1950's. At that time, computerized continuous simulation for hydrologic analysis was an exciting frontier, just as the case of the development of commercial aviation when it was first actively pursued in the early 1930's. It is unlikely that anyone at that time (speaking of aviation) could foresee the modern air transport system that we have today. But each step in its development came along with greater advances. So it is with the subject I am dealing with today. Somewhat of a plateau has now been reached in this development, from a hydrologic point of view, but further refinements will be in order when more adequate hydrologic data justifies them. In the meantime, further use will be made of advanced large scale computer systems as well as minicomputers, to provide more efficient and timely simulation of hydrologic systems through use of the SSARR model. The really productive advances will come from those hydrologists who look at this model as a tool which they may use as a processor to be applied creatively to achieve their goals. It will certainly not come from those who look at it as a "cook-book" and problem solver, without using the creativity of the human mind and common sense in its application. The person who uses the model should accept full responsibility for its application. The credit or blame for the results depend as much on the skill and intelligent application by the user as the formulation of the model itself. Finally, let us be humble in our approach, recognizing the deficiencies of any system to fully represent all details of analysis of the hydrologic processes. But let us procede with caution in incorporating additional refinements without considering the basic limitations of quality of data, data processing requirements, and practical considerations of day-to-day application.

REFERENCES

Technical Reports

1) U.S. Army Engineer Division, North Pacific, Portland, Oregon, "Program Description and Users Manual for SSARR Model," September 1972 (Revised June 1975)

2) U.S. Army Engineer Division, North Pacific, Portland, Oregon, "User Manual for COSSARR Model," January 1972

3) U.S. Army Engineer Division, North Pacific, Portland, Oregon, "Snow Hydrology," 30 June 1956

4) U.S. Army Engineer Division, North Pacific, Portland, Oregon, "SSARR Model, Summaries of Technical Reports," January 1976

5) Ecole Polytechnique Federale de Lausaune, Department
de Genie Civil, Thesis No. 224 (1975) "Contribution
a L'Etude de L'Ecoulement Pluvial dans de Bassin du
Mekong," par Bolyvong Tanovan

ANNEX I

GENERAL SUMMARY OF APPLICATIONS
AND PRINCIPLES OF DESIGN USED IN THE CONTINUED
DEVELOPMENT OF THE SSARR MODEL

1. Application of SSARR model to hydrologic and reservoir system simulations

1.1. Generalized hydrologic simulation model and system processor, for
use in many types of applications for analyzing river and reservoir
systems.

1.2 Types of applications to which SSARR model has been applied

    (a) Large scale systems

        (i) Complex hydrologic systems ($200,000 - 1,000,000$ km^2)

        (ii) Complex reservoir systems

        (iii) Combination of systems involving many tributaries and
projects

    (b) Small scale systems

        (i) Small experimental watershed areas (as small as $5 - 10$ km^2)

        (ii) Single river or tributary streams ($500 - 10,000$ km^2)

        (iii) Single project studies

    (c) Long period studies

        (i) $20 - 30$ years of synthetic record

(ii) Extending streamflow records

(iii) Performing basic long term hydrologic studies for runoff characteristics of rivers

(d) Short term studies

(i) Single flood occurance, 1 - 5 days

(ii) Forecast 1 - 20 days

(iii) Design flood studies for many purposes

(e) Applied to planning or designing studies related to water control projects

(i) Detailed studies of synthetic floods, reservoir operating plans, or hydrologic studies for determining water supply and utilization.

(ii) Application to many types and sizes of river systems, for a variety of hydrologic regimes (including rainfall and/or snowmelt runoff), on a world-wide basis.

(f) Applied to forecasting and water management

(i) Time of completion of the essence

(ii) Objective forecasting procedure for river forecasting and management of water control systems for large rivers, single tributaries, or combination.

(iii) Application to real-time forecasting and water management for a variety of systems on a world-wide basis.

(iv) Application to derivation of natural flows on a "real time" basis, for assessing effects of reservoir regulation on a current basis.

2. Principles of design used in the continued development of the SSARR Model

2.1 Hydrologic methods as a balance between highly theoretical techniques and practical needs for use in applied hydrology, considering limitations of availability of basic data, requirements for normal hydrologic investigations, and needs for streamflow forecasting.

2.2 Model is completely generalized, whereby a particular condition of hydrology, river system, reservoir system, or combination thereof is defined through use of characteristics, which, when properly calibrated, become fixed data which define the particular river or reservoir system.

2.3 When fixed data are established, the model operates from time-variable data, as time series including hydrometeorological variables, such as rainfall, evapotranspiration, streamflow, snowmelt functions, reservoir operations, limiting tidal fluctuations, or in any combination thereof, as necessary.

2.4 Comprehensive model for simulating all river sub-basins, channels, and reservoir operation as a single once-through operation and processor for a given river system.

2.5 Ease in application; also ease in training in its use for application by others.

2.6 Efficiency in operation, both in preparation of input data and in processing algorithms through computer, particularly for forecasting and water management.

2.7 Developing basin characteristics (rainfall - runoff, etc) by easily performed trial and error reconstitution studies (developed by using model) and ease of adjusting characteristics in the trial and error process.

2.8 Flexible method for performing various water related elements that affect streamflow in the simulation process

    (a) Irrigation depletions and return flow

    (b) Over bank flow conditions

    (c) Bi-furcation of channels

    (d) "Adjacent basin" capability, to relate minor drainage runoff to a major drainage, on basis of statistically derived correlation, for convenience in forecasting.

    (e) Water requirements for hydro-power, irrigation, navigation, flood control, water quality, etc.

2.9 Providing means of regulating reservoirs by flexible methods to represent all requirements for simulating water use of reservoir projects.

2.10 Use of watershed polyphase storage routing technique employing time of storage concept (T_s) for transforming water inputs to runoff components, as improved method for maintaining continuity of streamflow, particularly to provide for "stop action," "instant replay'" and "back-up period" adjustments for streamflow forecasting, and ease in application generally.

2.11 Generalized and easily applied channel routing procedure, which includes evaluation of variable backwater conditions in rivers or channels.

2.12 Continued development of the model from its use in may applications whereby the needs for refinement become apparent, in various types of applications for planning, designing, and forecasting water resource systems.

2.13 Standard formats for various types of data, for ease in systematizing all data inputs (punch cards or computer terminals).

2.14 Outputs in a form that are easily interpreted, in the form of listings of numerical values, or as plotted values in a time sequence.

2.15 For either line printer or cathode ray display terminals, provide the capability for interactive response between the computer and the user, through use of a computer terminal, for instant response, and thereby minimize "turn-around time."

2.16 Use of table-functions, as well as mathematical functions, for ease in defining various characteristics used in the simulation processing.

2.17 Generalized time routine, which can be applied flexibly for determining incremental times used in the simulation processing.

2.18 Application to medium or large scale computers (SSARR) or small scale computers (COSSARR).

A SYSTEMS APPROACH TO REAL TIME RUNOFF ANALYSIS WITH A DETERMINISTIC RAINFALL-RUNOFF MODEL

Robert J. C. Burnash
Hydrologist in Charge
National Weather Service
California-Nevada River Forecast Center
Room 1641, 1416-9th Street
Sacramento, California 95814

R. Larry Ferral
Deputy Hydrologist in Charge
National Weather Service
California-Nevada River Forecast Center
Room 1641, 1416-9th Street
Sacramento, California 95814

ABSTRACT

The responsibility for providing the United States with warnings of river conditions was assigned to the National Weather Service in 1890. This requirement has led to the development of a systems approach to hydrologic data collection, runoff computation, and forecast production which is being developed and applied in the service area of the California-Nevada River Forecast Center. This system is focused on producing information on the future distribution of water in time and space which effects the safety, welfare and economic well being of the nation and its inhabitants.

A principal element of any hydrologic warning system is an effective rainfall-runoff model. The authors have constructed such a model, which is now being generally applied by the National Weather Service. It has been our goal to construct a physically realistic and understandable model which, although necessarily parametric, would describe the runoff generation process in a manner which was consistent with the physics of plant, atmosphere, and soil water interaction. As a basic part of this effort, it was considered necessary to model the percolation process in a manner which retained the desirable infiltration characteristics described by Horton and others and to recognize and improve upon the limitations which such systems possessed.

The formulation of such a system did not, however, solve the rainfall-runoff analysis problem. As the runoff equations were developed to a high level of effectiveness, they led to an ever increasing appreciation of the deficiencies in available rainfall data. The magnitude of the rainfall data problem, coupled with the authors' responsibilities for real-time river forecasting, led to an additional effort to improve the effectiveness, timeliness and stability of precipitation measurement in order to provide better data input to the runoff analysis process.

These efforts have recently led to the completion of highly cost-effective computerized techniques for real-time data collection and analysis. The basic measuring device utilized for such data collection is a self-reporting gage which sends each increment of data via an event activated radio system.

Data storage, analysis and the forecasting of hydrologic responses may be accomplished instantaneously with such a system. More commonly, however, a hydrometeorologist maintaining close surveillance by interacting with a powerful minicomputer, can incorporate site-specific weather forecasts into a local hydrologic analysis. This makes it possible for the first time to provide site-specific, quantitative and reliable rainfall-runoff analysis, i.e.: flood forecasts suitable for effective warning, evacuation or flood fighting in even the smallest basins.

Thus, a systems approach to real-time runoff modeling includes not only an effective deterministic rainfall runoff technique which allows rational analysis of the physical processes: it must also include effective automated, dimensionally stable inputs, reliable communications, and refined short-term hydrometeorologic forecasts. The development of such an approach provides important feedback mechanisms which provide conjunctive improvements in both the real-time meteorological and hydrological analyses.

INTRODUCTION

In 1890 the United States Congress established a new agency, now known as the National Weather Service, to provide public weather and hydrologic warnings. Recognizing the significant interdependence of hydrologic and meteorologic warnings, the National Weather Service has attempted to meet its assigned responsibility through the design, testing and application of equipment and procedures which contributed to the conjunctive success of this dual mission.

At the present time, thirteen National Weather Service River Forecast Centers carry the primary responsibility for those hydrologic warnings which can be quantitatively produced. For those locations where discrete warning techniques have not yet been developed, general warnings are provided by the nation's many Weather Service offices. Such warnings, based upon an interpretation of hydrologic and meteorologic data, lack the specificity which can be obtained from discrete analysis. They do, however, provide important information to locations where more discrete technology is inappropriate or is not yet available.

The process of producing discrete hydrologic warnings affecting the safety, welfare and economic well being of the nation and its inhabitants is now being advanced through the application of a highly cost effective technology. Although this automated technology has been applied only over a limited portion of the United States, the potential benefits which can be gained from it are enormous. The analysis of flood threats and the evaluation of potential river flows and surface water availability all stem from the same data collection and analysis techniques. As a consequence, similar systems are being planned for many areas in the United States. The completion of these systems is expected to contribute well over a billion dollars a year in economic benefits to the nation through reduced flood damages and increased effectiveness of water management decisions.

The systems which hold such substantial promise have already produced benefits well in excess of development and operational costs in the

limited areas where they have been applied. These benefits will become
available to a growing portion of the United States as the resources become available to allow implementation of a systems approach. At the
present time, areas of the western United States are enjoying the first
fruits of these system concepts (Bartfeld and Taylor 1980).

RAINFALL-RUNOFF MODELING

A key element in the technology of automated hydrologic warnings has
been the development and application of an effective rainfall runoff
methodology. After extensive testing nationally and internationally,
(World Meteorological Organization 1975) the National Weather Service
selected the Sacramento Rainfall Runoff Model (Burnash, Ferral and
McGuire 1973) for this purpose. The Sacramento model is a deterministic
generalized hydrologic model which is based upon a parametric concepu-
alization of percolation, soil moisture storage, drainage and evapotranspiration characteristics. Each variable required in the model is intended
to represent a discrete and recognizable characteristic required for
effective real time hydrologic analysis.

The definition of model parameters is achieved by establishing a
soil moisture computation which allows the determination of basin streamflow from basin precipitation. Effective moisture storage capacities in
the soil profile are estimated not by sampling of the soil profile, but
by inference from the rainfall and discharge records. The five basic
soil moisture components of the model are upper zone and lower zone
tension water storages, which are filled preferentially by infiltrated
water, and three free water storages. Upper zone free water storage
supplies water for percolation to lower zones and for interflow. The two
lower zone free water storages fill simultaneously from percolated water
and drain independently at different rates, giving a variable groundwater
recession. These storages are diagrammed in Fig. 1.

Rainfall occurring over the basin is considered as falling on two
basic areas, 1) a permeable portion of the soil mantle, and 2) a portion
of the soil mantle covered by streams, lake surfaces, marshes, or other
impervious material directly linked to the streamflow network. The first
area produces runoff when rainfall rates exceed percolation rates, while
the second area produces direct runoff. In the permeable portion of the
basin, the model visualizes an initial soil moisture storage identified
as Upper Zone Tension which must be totally filled before moisture becomes available for other purposes. This represents that volume of
precipitation which would be required under dry conditions to meet all
interception requirements and to provide sufficient moisture to the upper
soil mantle so that percolation to deeper zones and sometimes horizontal
drainage could begin. When the Upper Zone Tension volume has been satisfied, excess moisture above the Upper Zone Tension water capacity is
temporarily accumulated in Upper Zone Free Water. Upper Zone Free Water
is that volume of moisture in the upper level soil from which lateral
drainage, appearing as streamflow, is observable. This form of lateral
drainage is identified as interflow. Upper Zone Free Water not only has
the horizontal potential to generate interflow, but more significantly has
a vertical potential. Interflow is produced whenever the Upper Zone Free
Water has a residual moisture content after meeting the requirements for
vertical drainage. The residual moisture content produces interflow by a
simple storage depletion term UZK such that:

FLOIN = UZK * UZFWC.

where:

109

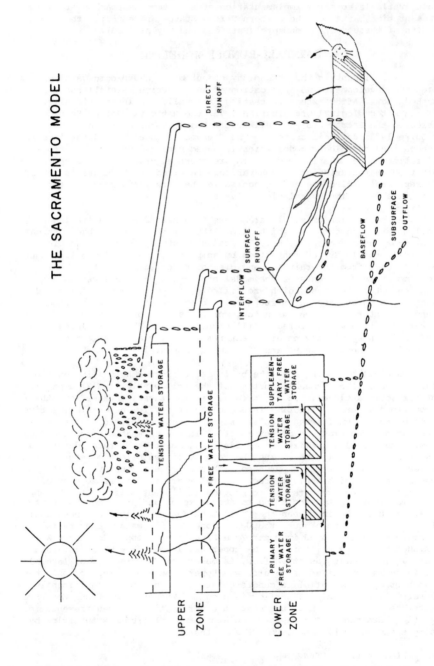

Figure 1. Sacramento Model. A conceptual real-time deterministic
hydrologic model.

FLOIN is the generated interflow,
UZK is the lateral drainage coefficient of UZFWC,

and

UZFWC is the free water content of the upper zone after
percolation to the lower zone has occurred.

The rate of vertical drainage is controlled by the contents of the Upper
Zone Free Water and the deficiency of lower zone moisture volumes. The pre-
ferred path for moisture in Upper Zone Free Water is considered to be vertical. Only when the rate of infiltrated precipitation exceeds the rate at
which vertical motion can take place from the upper zone, does horizontal
flow in the form of interflow occur. If the precipitation rate exceeds the
percolation rate and the maximum interflow drainage capacity, then surface
runoff occurs. Under this system, surface runoff is a highly rate-dependent
volume with the rate of runoff being determined by the rate of precipitation
application and the degree of dryness of the different zones. Surface
runoff is consequently a residual volume which occurs whenever the upper
zone storages are met at a rate which exceeds the ability of the upper zone
free water to convey water both horizontally as interflow and vertically
through percolation.

The percolation mechanics which represent the transfer of moisture from
the upper soil to the lower soils have been designed to correspond with observed characteristics of the motion of moisture through the soil mantle.
The mechanics are intended to provide a parametric parallel of the formation
and transmission characteristics of the wetting front in the soil mantle.
Water in excess of Upper Zone Tension requirements can percolate to a deeper
portion of the soil mantle through transfer by Upper Zone Free Water. The
mechanics of transfer from upper zone to lower zone volumes is based upon
the computation of a lower zone percolation demand. When the lower zone is
totally saturated, the percolation into the lower zone must be limited to
that water which is draining out of the lower zone. This limiting drainage
rate, representing saturated conditions, is computed as the sum of the products
of each lower zone free water storage capacity and its drainage rate. This
limiting rate of drainage is defined as PBASE, such that:

$$PBASE = (LZFSM*LZSK) + (LZFPM*LZPK)$$

where:

LZFSM = the maximum capacity of the quicker draining
         lower zone free water storage,
LZSK = the drainage rate of LZFSM,
LZFPM = the maximum capacity of the slower draining
         free water storage,

and

LZPK = the drainage rate of LZFPM.

The saturated drainage rate PBASE, which represents the minimum potential for percolation, is used to define a drainage rate which represents
the maximum potential for percolation. This maximum rate, which corresponds
to the driest conditions in the lower zone, is larger than PBASE by a quantity which may be defined as Z*PBASE. Thus, the percolation demand under
all circumstances can be estimated by evaluating the change from PBASE*(1+Z),
the driest percolation condition, to PBASE, the saturated percolation condition. An exponential relationship defined by the exponent REXP, provides

111

a curvilinear percolation relationship capable of reproducing widely varying percolation characteristics reported in the literature.

The percolation demand, PD, varies from PBASE to (1+Z)*PBASE by the relationship:

$$PD = PBASE\left(1+Z\left(\frac{\Sigma\ (lower\ zone\ capacities\ less\ contents)}{\Sigma\ (lower\ zone\ capacities)}\right)^{REXP}\right)$$

The lower zone capacities consist of:

LZTWM - the lower zone tension water capacity,

as well as LZFSM and LZFPM described earlier.

The computation of a percolation demand is not, however, adequate to define actual percolation, for actual percolation is influenced by the degree of wetness in the upper soil profile such that:

$$Percolation = PD * \frac{UZFWC}{UZFWM}$$

provided, however, that actual percolation cannot exceed the volume available in UZFWC.

The sums of lower zone capacities and contents include both tension water and free water. Thus percolation is defined by an interrelationship between soil drainage characteristics and soil moisture conditions. The volume which is percolated to the lower zone is divided among three significant soil moisture storages. The first of these, lower zone tension, represents that volume of moisture in the lower zone soil which will be claimed by dry soil particles when moisture from a wetting front reaches that depth. The tension water capacities defined in this model are capacities for change. In the lower level this is the difference between that water held against gravity after wetting and that remaining after plant roots have extracted all that they are capable of withdrawing. In the upper zone some additional water is lost and the tension water capacity is enlarged by direct evaporation from the soil. Tension water deficiencies would absorb all percolated water until these deficiencies are satisfied. However, variations in soil conditions and rainfall amounts over a drainage basin cause variations from this condition. The effect of these variations is approximated by diverting a fraction of the percolated water into lower zone free water storages before tension water deficiencies are fully satisfied. The free water storages in the lower zone represent those storages which generate horizontal flow generally observable as increases in base flow at the gaging point. As the tension water storage is completely filled, all percolation is diverted to free water storages. At all times the distribution of percolated water between free water storages is a function of their relative ratios of contents to capacity.

If the natural boundary conditions of the basin should allow all applied moisture to leave the basin either at the gaging point or through evapotranspiration, these soil moisture divisions would be adequate. However, subsurface drainage bypasses the gaging site in many basins. In order to approximate this effect within a particular basin, it is assumed that those soils in areas draining in a direction or to a depth away from the stream channel have the same basic drainage characteristics as those soils which drain to the stream channel. Thus the volume of such subsurface flows can be expressed as a fraction of the volumes integrated from the surface outflow hydrographs. This volume exists within the basin in addition to the volumes which will be observed through the surface outflow hydrograph.

Streamflow is thus the result of processing precipitation through an algorithm representing the uppermost soil mantle and lower soils. This algorithm produces runoff in five basic forms: 1) direct runoff from permanent and temporary impervious areas, 2) surface runoff due to precipitation occurring at a rate faster than percolation and interflow can take place when both upper zone storages are full, 3) interflow resulting from the lateral drainage of a temporary free water storage, 4) supplementary base flow, and 5) primary base flow. Runoff forms one and two have similar drainage characteristics while the drainage of each of the remaining components corresponds to observed streamflow features with uniquely different characteristics.

It should be noted that the proportion of impervious runoff, i.e., direct runoff, does not remain a constant with this model. It has been observed in many basins that upon filling the tension water storages, an increasing area assumes impervious characteristics. This, the additional impervious area, provides a useful representation of the filling of small reservoirs, marshes, and temporary seepage outflow areas which achieve impervious characteristics as the soil mantle becomes wetter.

An examination of Fig. 1 indicates that water percolating from the upper zone free water to the lower zone may go totally to tension water or some fraction of the percolated water may be made available to the primary and supplementary storages. At any time that the lower zone tension storage becomes filled, continued percolation is divided between the two lower zone free water storages. At all times water made available to primary and supplementary storages is distributed between them in response to their relative deficiencies.

Evaporation from water and phreatophyte surfaces is computed at the potential rate. Over other portions of the soil mantle, evapotranspiration is treated as the only process which depletes tension storage. As the soil mantle dries from evapotranspiration, moisture is withdrawn from the upper zone at the potential rate multiplied by the proportional loading of the upper zone tension water storage. In the lower zone evapotranspiration takes place at a rate determined by the unmet potential evapotranspiration times the ratio of the lower zone tension water content to total tension water capacity. If evapotranspiration should occur at such a rate that the ratio of contents to capacity for available free water exceeds the ratio of contents to capacity of tension water then water is transferred from free water to tension water and the relative loadings balanced in order to maintain a moisture profile that is logically consistent. Depending upon basin conditions, some fraction of the lower zone free water is considered to be below the root zone and therefore unavailable for such transfers. Various algorithms have been utilized to compute evapotranspiration demand. Hounam (1971) has documented many of the procedures intended for such purposes and indicated many of the problems associated with them. The authors are presently utilizing either daily mean values of evapotranspiration varying with day of the year and defined by model optimization techniques or redimensioned computations of daily evaporation based upon the work of Kohler, Nordenson and Fox (1955). If computed values are used, they are adjusted by a coefficient that varies with day of the year.

Although the system mechanics of the generalized hydrologic model are simplified approximations of natural processes, the total effect is consistent with observations of the soil moisture profile made by experimental studies such as those by Green et al (1970) and Hanks, Klute, and Brestler (1960).

The application of these mechanics has been made on areas ranging in size from less than ten square kilometers to over one hundred thousand square kilometers. As with any analysis system, the effectiveness of the solution depends upon the ability to obtain effective inputs and the homogeneity of the area and the data set.

Tests of model sensitivity have demonstrated that the single most important determinant in defining system performance is the precipitation dimension, while the second most important determinant is the evapotranspiration dimension. A proportional change in these basic inputs generates a shift in model performance which is approximately an order of magnitude greater than a similar proportional change in any model parameter.

Versions of the model have been prepared which operate with inputs as fine as twelve minutes or as coarse as twenty-four hours. In order to maintain reasonable sensitivity of the rate functions in the model, the precipitation is analyzed in increments of 5 millimeters, utilizing a temporal distribution function whenever the input volume exceeds the 5 millimeter criteria.

Although this system is intended for flood forecasting, the water balance characteristics necessary for effective determination of rainfall-runoff conditions, emphasize the input deficiencies of precipitation measurement during major storms when winds are likely to have their greatest impact on rain gage catch. This input problem is frequently most dramatic at the same time as discharge values are most difficult to measure. As a consequence, the authors believe that the correlation coefficient of monthly flow (produced by summing all simulated and observed flows during the month) is generally a better indicator of system performance than any term which is more sensitive to short-term input difficulties.

Monthly correlation coefficients representing model performance for various analyses across the United States demonstrate a high level of performance in widely diverse geographic and climatologic regimes.

Basin	Size KM^2	Number of Months Analyzed	Correlation Coefficient of Computed to Observed
South Yamhill, Oregon	1300	120	.996
Santa Ynez, California	1031	108	.986
Merrimac, Missouri	2023	96	.985
Bird Creek, Oklahoma	2344	84	.996
French Broad, North Carolina	176	60	.984

Much of the model's performance is due to the flexibility with which the model represents a variety of drainage and infiltration characteristices.

INFILTRATION COMPARISONS

Horton (1939) suggested an infiltration equation which has become quite famous as a technique for computing infiltration. The equation:

$$f = f_c + (f_o - f_c) e^{-kt}$$

is actually a simple decay curve where

   f = infiltration rate at time t

f_c = a minimum infiltration rate

f_o = the infiltration rate at t = $\emptyset$

e = the Napierian base

k = a decay constant

t = time

A plot of the log of the derived infiltration versus time is a straight line with a negative slope. Computation of continuous infiltration capacity requires that rainfall be in excess of infiltration throughout the time period. As written, the equation is not applicable for intermittent rainfall, with alternate wetting and drying periods, or for rainfall that does not continuously exceed f. Infiltration capacity almost always exceeds rainfall rates early in a storm, resulting in actual infiltration rates equal to rainfall rates, as described and explained by Mein and Larson (1973). Holtan (1961) and Bauer (1974) proposed modifications and elaborations of Horton's equation for application to intermittent rainfall. They approached this by making infiltration a function of moisture in the soil, not time as such.

The Sacramento Model, though not designed as an infiltration model, provides an indication of the vertical distribution of water in the soils. The modeled components of runoff, and by implication net infiltration as the difference between rainfall and runoff, are computed as functions of moisture contained in the soil. Though designed to be applied for real time hydrologic analysis of runoff conditions with intermittent and variable natural rainfall, system mechanics can be applied to continuing rainfall in excess of infiltration to provide an interesting comparison with Horton's equation. Some similarities and differences can be seen:

1) Observed infiltration curves show very high rates after a short time period, with initial rates at time zero undefined. Examples are seen in Rubin (1966), Mein and Larson (1973), Linsley, Kohler and Paulhus(1975), and Green et al (1970). The Sacramento model similar to observed conditions provides for very rapid initial infiltration. This takes place as void spaces in the upper soil levels are being filled. Infiltration then drops to a much lower rate which diminishes slowly with time, as rainfall in excess of infiltration capacity continues. When rainfall stops, infiltration capacity increases very quickly as moisture in upper soil levels drains away, then more slowly as lower soils drain and moisture is removed by evapotranspiration. These features are all compatible with natural conditions and are necessarily included in the model in order to provide an effective real time rainfall runoff transfer function for intermittent as well as continuous rain conditions.

2) Observed infiltration curves in Rubin (1966) p746, Green et al (1970) p869, Mein and Larson (1973) p390, and Linsley et al (1975) p263, when replotted in a log f vs t form, are concave upward. A typical plot is shown in Figure 2a. After the rapid initial infiltration rate, the Sacramento Model's subsequent infiltration, with some parameter combinations, can give an infiltration-time curve very similar to Horton's, i.e.: linear on a log f vs t plot. More commonly, the Sacramento Model gives a log f versus t curve that is concave upward, equivalent to a Horton's k that diminishes with time. A typical plot is shown in Figure 2b. Other parameter combinations, much less common, can give a concave log f vs t curve, equivalent to a Horton's k that increases with time.

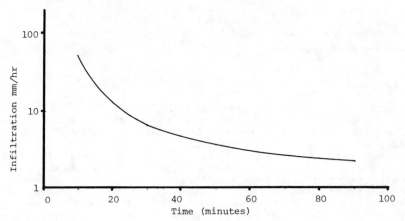

Figure 2a. A typical observed infiltration curve plotted as log f vs time. (Replotted from Linsley, Kohler and Paulhus, Houston black loam.)

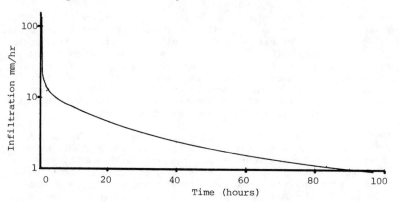

Figure 2b. An infiltration curve generated by the Sacramento Model, plotted as log f vs time. (Parameters fitted to Arroyo Seco, California. Excess water applied after wetting and prolonged drying.)

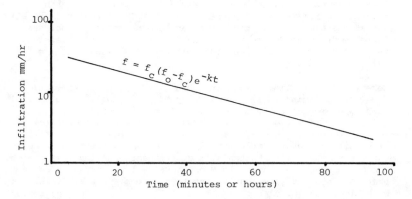

Figure 2c. Horton's equation plotted as log f vs. time.

Thus it appears that the Sacramento Model has substantially greater flexibility in modeling the diverse observed infiltration curves than is possible with the classical Horton equation. In addition, it provides a reasonable estimate of initial wetting conditions which are not adequately described by Horton's equation.

DATA COLLECTION

This modeling process, in order to be effectively applied for purposes of hydrologic warnings, must necessarily be supported by an effective real time data collection system. Manual observation techniques rarely provide sufficiently timely or dimensionally stable data for adequate effectiveness in hydrologic warnings. As a consequence, the application of the Sacramento Model with data systems, which in the past were considered acceptable, has led to numerous problems. The model sensitivity, a requirement for more effective analysis, exposed the inadequacy of many existing data base systems.

Very few manual observation techniques can maintain a measurement accuracy of five percent over a prolonged period of time. The growth of vegetation, the construction of buildings, slight changes in equipment or exposure all contribute to a lack of consistency.

If such inputs are used in a sensitive model, the impacts upon runoff projections can be quite large. As an example of the sensitivity problem, Figure 3 illustrates the effect of the change in forecast runoff which is produced by a 5% change in the precipitation input. This problem has led to the development and application of fully automated data collection systems which were more appropriate to the needs of an operational warning system.

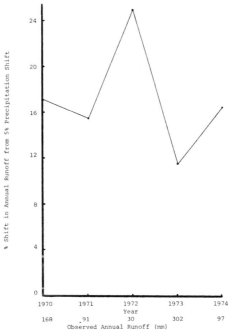

Figure 3. Computed change in runoff produced by a five percent change in precipitation input. Sespe Creek, Ventura County, California (655 sq. km.)

That system of automated data which has demonstrated the most effective benefit cost ratio is based upon installing totally self-contained sensors at those locations where data is required. Such sensors contain their own power supply and communication equipment. Whenever there is a change in the sensor value, the units transmit a self-initiated radio message. Under conditions when the sensors are not producing changes, reports are sent at periodic intervals to verify system operation. To date this technology has been applied to precipitation, snowpack, temperature, and water levels. A remarkable auxiliary benefit of automated data collection was a reduction in the true cost of collecting data (Burnash & Bartfeld 1980).

Of primary interest to most hydrologists are the precipitation gages. Three types of precipitation gages are utilized. They are, 1) a simple rain gage for relatively snow free areas, 2) a modified design for areas where snow occurs but where the winter precipitation does not exceed one hundred centimeters of water content, and 3) the deep snow gage which may be used in areas where the snow depth can reach as much as seven meters. See Figure 4. All sensors send brief radio signals, less than one-quarter second in duration, which place a very small load on the power supply. A single four kilogram battery has adequate reserves to power a precipitation gage for over a year in the wettest areas of the world. Such gages have been installed utilizing basically line-of-sight radio transmission paths. Data is acquired by a minicomputer which monitors a radio receiver. Where direct radio transmission to a base station is not feasible, radio relays have been utilized. At some locations data is received and interpreted by local microcomputers. These microcomputers meet local requirements for data and site-specific hydrologic warnings. The microcomputers are programmed to allow telephone polling by the RFC computer. An example of various communications paths is contained in Figure 5.

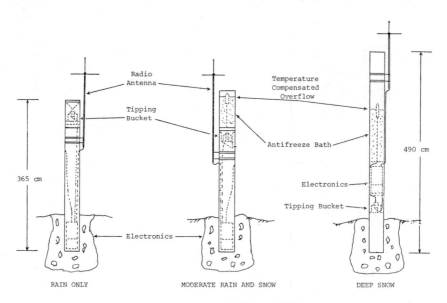

Figure 4. Self-contained event reporting precipitation gages designed for various site conditions.

Inasmuch as the RFC computer can be set to interrogate local data collection minicomputers on the basis of satellite imagery or other meteorological data, the telephone polling can be kept at a cost effective level. The frequency of interrogation is dependent upon the significance of the event. At the RFC, the data is analyzed by a powerful minicomputer which has in storage the hydrologic characteristics of all areas for which forecast service is provided. The RFC computer evaluates the precipitation input for the area and through the Sacramento model produces forecasts of runoff and streamflow. Upon completion of the discharge analysis, the computer evaluates significant stage conditions associated with the discharge forecast and prepares an English language statement of river conditions. Based upon the significance of the analysis, the system then determines the appropriate routing of the warning message and the information is routed to the appropriate local office.

At the present time, this developing technology is being applied to a limited but growing number of locations. Although the entire process of data collection, model application, forecast generation and distribution are not yet instantaneous, they can all be completed within minutes. This new dimension in timeliness is not restricted to an arbitrary data collection time, for data is constantly available and is always current. Until very recently, the lack of timely data limited site-specific flood warnings to large, relatively slow-rising rivers. This limitation has been eliminated by the continuous collection of real-time data, the use of this data to improve short-term quantitative rainfall forecasts, and the automation of hydrologic data analysis and forecasting. It is now possible to provide effective flood warning systems where some of the greatest flood hazards exist - along small, fast-rising rivers, creeks and arroyos.

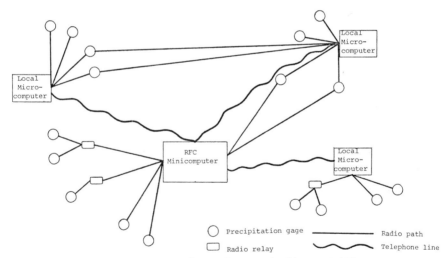

Figure 5. Typical data collection and communications system.

CONCLUSIONS

The combination of the rainfall-runoff model with the other technologies we have discussed has resulted in an information and processing system that not only solves old problems in a more efficient manner, it adds a new dimension to the supporting capability of the meteorologist. The data collected by these systems describes storm movements and intensity changes in a manner which allows a substantial improvement in the ability to evaluate precipitation which is likely to occur during the next few hours. The feedback of such determinations into the hydrologic analysis allows real-time warnings to be generated for areas where the time from the slackening of heavy rain to crest conditions is in the scale of minutes.

Thus real-time automated systems based upon effective data collection, continuous meteorologic and hydrologic analysis and automated forecast generation and distribution provide the potential for a remarkable improvement in flood warning programs.

REFERENCES

Bartfeld, I. and Taylor, D.B. 1980. A Case Study of a Real-Time Flood Warning System, Sespe Creek, Ventura County, California. Symposium on Storms, Floods and Debris Flows in Southern California and Arizona in 1978 and 1980. Environmental Quality Laboratory, California Institute of Technology.

Bauer, S.W. 1974. A Modified Horton Equation for Infiltration During Intermittent Rainfall. Hydrologic Sciences Bulletin, Vol. 19, pp. 219-225.

Burnash, R.J.C. and Bartfeld, I. 1980. A Systems Approach to the Automation of Quantitative Flash Flood Warnings. Second Conference on Flash Floods, Preprints, pp. 216-226. American Meteorological Society.

Burnash, R.J.C., Ferral, R.L. and McGuire, R.A. 1973. A Generalized Streamflow Simulation System. Joint Publication, National Weather Service and California Department of Water Resources, Sacramento, California, 204 pages.

Green, D.W., Dabri, H., Weinaug, C.F. and Prill, R. 1970. Numerical Modeling of Unsaturated Groundwater Flow and Comparison to a Field Experiment. Water Resources Research, Vol. 6, pp. 862-874, 1970.

Hanks, R.J., Klute, A. and Brestler, E. 1969. A Numerical Method for Estimating Infiltration, Redistribution, Drainage and Evapotranspiration of Water from Soil. Water Resources Research, Vol. 5, pp. 1064-1069.

Holtan, H.N. 1961. A Concept for Infiltration Estimates in Watershed Engineering. Agr. Res. Service, U.S.D.A., ARS 41-51.

Horton, R.E. 1939, Analysis of Runoff Plot Experiments with Varying Infiltration Capacity. Transactions American Geophysical Union, Vol. 20, pp. 693-711.

Hounam, C.E. 1971. Problems of Evaporation Assessment in the Water Balance. WMO/IHD Report No. 13, WMO No. 285.

Linsley, R.K., Kohler, M.A. and Paulhus, J.L.H. 1975. Hydrology for Engineers, 2d. ed, McGraw-Hill Book Co., New York, p. 263.

Mein, R.G. and Larson, C.L. 1973. Modeling Infiltration During a Steady Rain. Water Resources Research, Vol. 9, pp. 384-394.

Rubin, J. 1966. Rainfall Uptake. Water Resources Research, Vol. 2, pp. 739-749.

World Meteorological Organization. 1975. Intercomparison of Conceptual Models Used in Operational Hydrological Forecasting. Operational Hydrology Report No. 7, WMO No. 429.

THE NEW HEC-1 FLOOD HYDROGRAPH PACKAGE

Arlen D. Feldman, Chief Research Branch
Paul B. Ely, Hydraulic Engineer
David M. Goldman, Hydraulic Engineer

Hydrologic Engineering Center
U. S. Army Corps of Engineers

ABSTRACT

HEC-1 is a mathematical watershed model containing several methods with which to simulate surface runoff and river/reservoir flow in river basins. The hydrologic model together with flood damage computations (also included in the model) provide a basis for evaluation of flood control projects. HEC-1 was developed by the Hydrologic Engineering Center (HEC), U.S. Army Corps of Engineers, in the late 1960's; a new version of the model, with greatly expanded capabilities, was released in 1980 and is described in this paper. The capabilities of the new HEC-1 Flood Hydrograph Package include: simulation of rainfall and/or snowmelt runoff from subbasins and flow through a stream network, simulation of flows in urban areas, hydrologic calculations for dam safety and dam failure studies, and economic calculations for planning flood control systems.

HEC-1 simulates a stream network using four components: 1) runoff from a subbasin, 2) hydrograph routing, 3) combining of hydrographs, and 4) flow diversion. Most complex, branching stream networks can be simulated with the model. The various options for watershed runoff calculation are described including: precipitation, interception/ infiltration, precipitation excess-to-runoff transformation, river routing, and flow through reservoirs. Diversions and multistage pumping plants capabilities are also described. Flow in urban areas can be simulated using kinematic wave routing of rainfall excess along a path which includes overland flow elements, collector channels, and a main channel to a subbasin outlet. A special routing routine is described for simulating flow through a dam and spillway, over the top of dam, or through a dam breach. This can be used in conjunction with other stream network modeling capabilities to determine potential hazards from dam overtopping or failure. This capability has been frequently used in the U.S. National Non-Federal Dam Safety Inspection Program.

In addition to its hydrologic capabilities, HEC-1's application to economic evaluation of flood hazards and flood control systems is presented. Expected annual flood damage is computed using the watershed model results together with flood-frequency and flood-damage data. Flood damage may be calculated for any locations in the river basin and for existing and alternative flood control projects. When damage estimates are combined with cost data for the projects and a systematic search procedure, the model can provide an estimate of the optimal size of the flood control projects based on maximum net benefits. This enables a planner to select the most desirable flood control scenario.

INTRODUCTION

History of HEC-1

The HEC-1, Flood Hydrograph package, computer program was originally developed in 1967 by Leo R. Beard and other members of the Hydrologic Engineering Center staff to simulate flood hydrology in complex
river basins. The first package version represented a combination of
several smaller programs which had previously been operated independently to simulate various aspects of the rainfall/snowmelt process.
In 1973, the program underwent a major revision. The computational
methods used by the program remained basically unchanged; however, the
input and output formats were almost completely restructured. These
changes were made in order to simplify input requirements and to make
the program output more meaningful and readable.

The present program (HEC, 1981a) again represents a major revision
of the 1973 version of the program. The program input and output
formats have been completely revised and the computational capabilities
of the dam-break (HEC-1DB), project optimization (HEC-1GS) and kinematic wave (HEC-1KW) programs have been combined in the one program.
The new program gives the powerful analysis features available in all
the previous programs, together with some additional capabilities, in a
single easy-to-use package.

Purpose of HEC-1

The HEC-1 model is designed to simulate the surface runoff
response of a river basin to precipitation by representing the basin
with interconnected hydrologic and hydraulic components. It is primarily applicable to flood simulation. English or metric units may be
used. Each component models an aspect of the precipitation-runoff
process within a portion of the basin, commonly referred to as a subbasin. A component may represent a surface runoff entity, a stream
channel, or a reservoir. Representation of a component requires a set
of parameters which specify the particular characteristics of the
component and mathematical relations which describe the physical processes. The result of the modeling process is the computation of
streamflow hydrographs at desired locations in the river basin.

The flood hydrograph information provided by HEC-1 has been extensively used in flood plain information studies and flood control project evaluations. The interconnection of HEC-1's hydrologic outputs
with water surface profile and reservoir operation models and flood
damage analyses was described by Feldman (1981). The other water
resources system simulation models of the Hydrologic Engineering Center
are also described in that publication.

COMPONENTS OF THE MODEL

The stream network model is the basic foundation capability of the
HEC-1 program. All other program computation options build on this
option's capability to calculate flood hydrographs at desired locations
in a river basin. This section discusses: the conceptual aspects of
using the HEC-1 program to formulate a stream network model from basic
river basin data; model formulation as a step-by-step process; and the
functions of each component in representing individual characteristics
of the river basin.

Stream Network Model Development

A river basin is subdivided into an interconnected system of stream network components using topographic maps and other geographic information. A basin schematic diagram (e.g., figure 1) of these components is developed by the following steps:

(1) The study area watershed boundary is delineated first. In a natural or open area this can be done from a topographic map. However, supplementary information, such as municipal drainage maps, may be necessary to obtain an accurate depiction of an urban basin's extent.

(2) Segmentation of the basin into a number of subbasins determines the number and types of stream network components to be used in the model. Two factors impact on the basin segmentation: the study purpose and the hydrometeorological variability throughout the basin. First, the study purpose defines the areas of interest in the basin, and hence, the points where subbasin boundaries should occur. Second, the variability of the hydrometeorological processes and basin characteristics impact greatly on the number and location of subbasins. Each subbasin is intended to represent an area of the watershed which, on the average, has the same hydraulic/hydrologic properties. Further, the assumption of uniform precipitation and infiltration over a subbasin becomes less accurate as the subbasin becomes larger. Consequently, if the subbasins are chosen appropriately, the average parameters used in the components will more accurately model the subbasins. The number of subbasins used also has a direct effect on the cost of the model. Consequently, it pays to be as economical as possible with the number of subbasins.

(3) Each subbasin is to be represented by a combination of model components. Subbasin runoff, river routing, reservoir and diversion and pump components are available to the user.

(4) The subbasins and their components are linked together to represent the connectivity of the river basin. HEC-1 has available a number of methods for combining or linking together outflow from different components. This step finalizes the basin schematic.

Land Surface Runoff Component

The subbasin land surface runoff component, such as subbasins 10, 20, 30, etc. in figure 1, is used to represent the movement of water over the land surface and in stream channels. Inputs to this component can be a precipitation hyetograph and a soil water infiltration rate function. Note that the rainfall and infiltration are assumed to be uniform over the subbasin. The infiltration losses are subtracted from the rainfall and the resulting rainfall excesses are then routed by the unit hydrograph or kinematic wave techniques to the outlet of the subbasin producing a runoff hydrograph. The unit hydrograph technique produces a runoff hydrograph at a discrete point, usually the most downstream point in the subbasin. If this location for the runoff computation is not appropriate, it may be necessary to further subdivide the subbasin or use the kinematic wave method to distribute the local inflow. The kinematic wave rainfall excess-to-runoff transformation allows for the uniform distribution of the land surface runoff along the length of the main channel. This uniform distribution of local inflow (subbasin runoff) is particularly important in areas where many lateral channels contribute flow along the length of the main channel.

River Routing Component

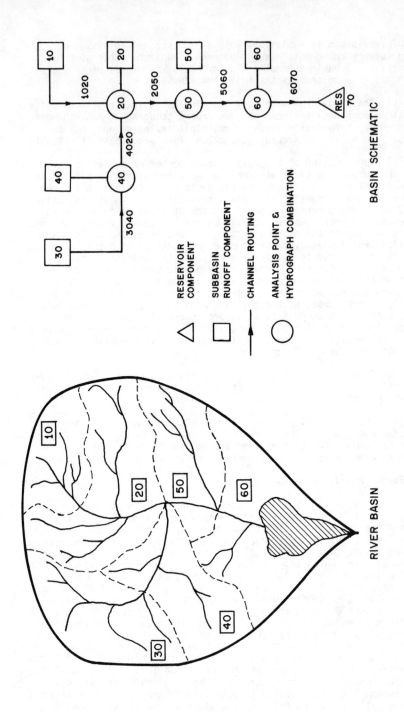

Figure 1 HEC-1 DEPICTION OF A RIVER BASIN

124

A river routing component, element 1020, figure 1, is used to simulate the flow of water in a river channel. The input to the component is an upstream hydrograph resulting from subbasin runoff, river routings or combinations of both. If the kinematic wave method is used, the local subbasin distributed runoff is also input to the main channel and combined with the upstream hydrograph as it is routed to the end of the reach. The hydrograph is routed to a downstream point based on the characteristics of the channel.

Combined Use of River Routing and Subbasin Runoff Components

Consider the use of subbasin runoff components 10 and 20 and river routing reach 1020 in figure 1. The runoff from component 10 is calculated and routed to control point 20 via routing reach 1020. Runoff from subbasin 20 is then calculated and combined with the outflow hydrograph from reach 1020 at control point 20. Note that this method of adding flows approximates the addition of lateral inflow to reach 1020. The runoff from subbasin 20 could be calculated directly at control point 20 in a unit hydrograph subbasin runoff calculation, or it could have been uniformly distributed along reach 1020 in a kinematic wave subbasin runoff calculation. A suitable combination of the subbasin runoff component and river routing components can be used to represent the intricacies of any rainfall-runoff and stream routing problem. The connectivity of the stream network components is implied by the order in which the data components are arranged. Simulation must always begin at the uppermost subbasin in a branch of the stream network. The simulation (succeeding data components) proceeds downstream until a confluence is reached. Before simulating below the confluence, all flows above that confluence must be computed and routed to that confluence. The flows are combined at the confluence and the combined flows are routed downstream. In figure 1, all flows tributary to control point 20 must be combined before routing through reach 2050.

Reservoir Component

The reservoir component can be used to represent the storage-outflow characteristics of a reservoir, lake, detention pond, highway culvert, etc. The reservoir component application is similar to that of the river routing component. Upstream inflows are routed through a reservoir based on the specified storage outflow characteristics as is the case in some river routing options. Consequently, the same flood routing methods can be applied for either component.

Diversion Component

The diversion component is used to represent channel diversions, stream bifurcations, or any transfer of flow from one point of a river basin to another point in or out of the basin. The diversion component receives an upstream inflow and divides the flow according to a user-prescribed rating curve.

PRECIPITATION-RUNOFF SIMULATION

The HEC-1 model components are used to simulate the precipitation-runoff process as it occurs in an actual river basin. The model components function based on simple mathematical relationships which are intended to represent individual meteorologic, hydrologic and hydraulic processes which comprise the precipitation-runoff process. These processes are separated into precipitation, interception/infiltration, transformation of precipitation excess to subbasin outflow, addition of baseflow and flood hydrograph routing.

Precipitation

A precipitation hyetograph is used as the input to all runoff calculations. The specified precipitation is assumed to be a subbasin average (i.e., uniformly distributed over the subbasin). Any of the model options used to specify precipitation will eventually result in a hyetograph. The hyetograph represents subbasin average precipitation depths over a computation interval. Precipitation data for an observed storm event can be supplied to the program by either of two methods: subbasin-average, or gages and weightings.

There are three methods for generating synthetic storm distributions: standard project, probable maximum, and specific frequency storms. The Standard Project Storm (Corps of Engineers, 1952) has a duration of 96 hours. The percentages of the index precipitation falling during each 24-hour period of the storm are automatically calculated by HEC-1 according to the Corps criteria. Probable Maximum Precipitation (National Weather Service, 1956) may be simulated for a minimum of 24 hours and up to 96 hours. The day with the largest amount of precipitation is preceded by the second largest and followed by the third largest. The fourth largest precipitation day precedes the second largest. The distribution of 6-hour precipitation during each day is according to standard criteria of the Weather Service or the Corps.

A synthetic storm of any duration from 5 minutes to 10 days can be generated based on given depth-duration data (National Weather Service, 1961). Depth for 10-minute and 30-minute durations are interpolated from 5-, 15- and 60-minute depths using equations from HYDRO-35 (National Weather Service, 1977). Cumulative precipitation for each time interval is computed by log-log interpolation of depths from the depth-duration data. Incremental precipitation is then computed and rearranged so the second largest value precedes the largest value, the third largest value follows the largest value, the fourth largest precedes the second largest, etc.

Snowfall and Snowmelt

Where snowfall and snowmelt are considered, there is provision for separate computation in up to ten elevation zones within a subbasin. These zones may be of any equal increments of elevation with a corresponding air temperature lapse rate per zone. The input temperature data are those corresponding to the bottom of the lowest elevation zone. Temperatures are reduced by the lapse rate in degrees per increment of elevation zone. The base temperature at which melt will occur, must be specified because variations from $0^\circ C$ ($32^\circ F$) might be warranted considering both spatial and temporal fluctuations of temperature within the zone. Precipitation is assumed to fall as snow if the zone temperature is less than the base temperature plus 2 degrees. Melt occurs when the zone temperature is equal to or greater than the base temperature. Snowmelt is subtracted from and snowfall is added to the snowpack in each zone. Snowmelt may be computed by the degree-day or energy-budget methods. The basic equations for snowmelt computations are from EM 1110-1-1406 (Corps, 1960b). The energy-budget equations have been simplified for use in this program.

Interception/Infiltration

126

Land surface interception, depression storage and infiltration are referred to in the HEC-1 model as precipitation loss rate computations. Interception and depression storage are intended to represent the surface storage of water by trees or grass, local depressions in the ground surface, in cracks and crevices in parking lots or roofs, or in a surface area where water is not free to move as overland flow. Infiltration represents the movement of water to areas beneath the land surface.

Two important factors should be noted about the precipitation loss computation in the model. First, precipitation which does not contribute to the runoff process is considered to be lost from the system. Second, the equations used to compute the losses do not provide for soil moisture or surface storage recovery (the Holtan loss rate option is an exception in that soil moisture recovery occurs by percolation out of the soil moisture storage). This fact dictates that the HEC-1 program is a single-event-oriented model.

The precipitation loss is considered to be a subbasin average (uniformly distributed over an entire subbasin). For the kinematic wave runoff transformation separate precipitation losses can be specified for two types of overland flow planes. The losses are assumed to be uniformly distributed over each overland flow plane. In some instances, there are negligible precipitation losses occurring for a portion of a subbasin. This would be true for an area containing a lake, reservoir or impervious area. In this case, precipitation losses will not be computed for a specified percentage of the area labeled as impervious.

There are four methods that can be used to calculate the precipitation loss. Using any one of the methods, an average precipitation loss is determined for a computation interval and subtracted from the rainfall/snowmelt hyetograph. The resulting precipitation excess is used to compute an outflow hydrograph for a subbasin.

An initial loss (units of depth) and a constant loss **rate** (depth/hour) is the first option. All rainfall is lost until the volume of initial loss is satisfied. After the initial loss is satisfied, rainfall is lost at the constant rate. The second method is the HEC Exponential Loss Rate Method. This is an empirical method which relates loss rate to rainfall intensity and accumulated losses. Accumulated losses are representative of the soil moisture storage. Estimates of the parameters of the exponential loss function can be obtained by employing the HEC-1 parameter optimization option described in a later section. A similar loss rate function is used for snowmelt.

The Soil Conservation Service (SCS), U.S. Department of Agriculture, has instituted a loss rate technique which relates the drainage characteristics of soil groups to a curve number, CN (SCS, 1965 and 1975). The SCS provides information on relating soil group type to the curve number as a function of soil cover, land use type and antecedent moisture conditions. Precipitation loss is calculated based on supplied values of CN and an initial surface moisture storage capacity in units of depth. Since the SCS method gives total excess for a storm, the incremental excess (the difference between rainfall and loss) for a time period is computed as the difference between the accumulated excess at the end of the current period and the accumulated excess at the end of the previous period.

The fourth loss rate option is a method developed by Holtan et al.

(1975). It computes loss rate based on the infiltration capacity given by the formula:

$$f = f_c + G*a*s^b \quad . . . . \quad (1)$$

where f is the infiltration capacity in inches per hour, G is a growth index representing the relative maturity of the ground cover, a is the infiltration capacity in inches per hour per (inch of available storage)^b, s is the equivalent depth in inches of pore space in the surface layer of the soil which is available for storage of infiltrated water, f_c is the constant rate of percolation of water through the soil profile below the surface layer, and b is an empirical exponent, typically taken equal to 1.4.

Precipitation Excess-to-Runoff Transformation

HEC-1 provides two methods for transforming rainfall/snowmelt excesses into runoff: unit hydrograph and kinematic wave. The unit hydrograph technique has been discussed extensively in the literature (Linsley et al., 1975, and Viessman et al., 1977). This technique is used in the subbasin runoff component to transform rainfall/snowmelt excess to subbasin outflow. A unit hydrograph can be directly input to the program or a synthetic unit hydrograph can be computed from user supplied parameters. The parameters for the synthetic unit hydrograph can be determined from gage data by employing the parameter optimization option described in a later section. Otherwise, these parameters can be determined from regional studies or from guidelines given in references for each synthetic technique. There are three synthetic unit hydrograph methods available in the model. The synthetic techniques compute the unit graph for whatever computational time interval is being used in the simulation.

The Clark method (1945) requires three parameters to calculate a unit hydrograph: the time of concentration for the basin, a storage coefficient, and a time-area curve. The time-area curve defines the cumulative area of the watershed contributing runoff to the subbasin outlet as a proportion of the time of concentration. In the case that at time-area curve is not supplied, the program utilizes a synthetic elliptical time-area curve. The Snyder method (1938) determines the unit graph peak discharge, time to peak, and widths of the unit graph at 50 and 75% of the peak discharge. The method does not produce the complete unit graph required by HEC-1. Thus, HEC-1 uses the Clark method to produce a Snyder unit graph. The Soil Conservation Service, SCS, dimensionless unit hydrograph method (1965) uses a single parameter, which is equal to the lag (hours) between the center of mass of rainfall excess and the peak of the unit hydrograph. Peak flow is computed using subbasin area and time to peak. The unit hydrograph ordinates are computed from a dimensionless graph using the peak flow and time to peak.

The kinematic wave subbasin runoff method in HEC-1 (HEC, 1979b) is composed of three elements: overland flow planes, collector channels, and a main channel, figure 2. Through these elements, the kinematic wave technique transforms rainfall excess into subbasin outflow. This simulation may be done on a detailed street-by-street basis in an urban area or set up to simulate representative drainage systems within a

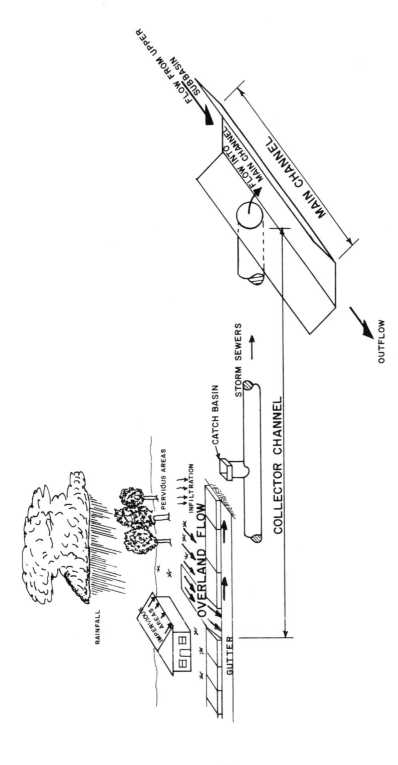

Figure 2   DEPICTION OF KINEMATIC WAVE RUNOFF
(figure 1.5: HEC, 1979b)

129

subbasin. If a representative system is used, the program automatically computes the total subbasin runoff as a function of the area of the representative system and the total area of the subbasin.

In the kinematic wave interpretation of the equations of fluid motion the momentum equation is reduced to a stage-discharge relation. The wave characteristics of a flood are then described solely by the continuity equation. HEC-1 solves the kinematic wave equations using a finite difference algorithm based on the same method developed for the MITCAT simulation model (Harley, 1975). Detailed development of the specific finite difference equations, the coding procedures and boundary requirements can be found in the following references: Harley, 1975; and Hydrologic Engineering Center, 1979b.

Base Flow

Two distinguishable contributions to a flood hydrograph are direct runoff (described earlier) and base flow which results from releases of water from surface and subsurface storage. The HEC-1 model provides means to include the effects of base flow on the streamflow hydrograph as a function of three input parameters: starting flow, a recession threshold, and a recession rate as shown in figure 3. Both the initial

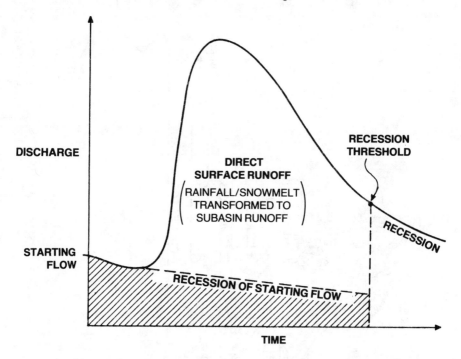

Figure 3 HEC-1 SUBBASIN RUNOFF SIMULATION

and base flow recession occur at an exponential decay rate, which is assumed to be a characteristic of the basin. The rising limb of the streamflow hydrograph is adjusted for base flow by adding the recessed starting flow and computed direct runoff flows. The falling limb is determined in the same manner until the computed flow is determined to be less than the threshold. From this time on, the streamflow hydrograph is computed using the recession equation unless the computed flow rises above the base flow recession. This is the case of a double-peaked streamflow hydrograph where the rising limb of the second hydrograph is computed as before, using the recessed starting flow and the computed direct runoff.

Flood Routing

Flood routing is used to simulate the outflows from river reaches and reservoirs. Most of the flood routing methods available in HEC-1 are based on the continuity equation and some relationship between flow and storage or stage. These methods are Muskingum, kinematic wave, modified Puls, working R and D, and level-pool reservoir routing. In all of these methods, routing proceeds on an independent reach basis from upstream to downstream; backwater effects are not considered. These methods cannot simulate discontinuities in the water surface such as jumps or bores. These methods should, however, give good results for routing floods through channels on moderate to steep slopes and through reservoirs. There are also two routing methods in HEC-1 (Tatum and Straddle-Stagger) which are based on lagging averaged hydrograph ordinates. These methods are not physically based, but have been used on several rivers with good results. Channel infiltration losses may be simulted. Hydrographs are adjusted for losses after routing for all methods except modified Puls; for modified Puls, losses are computed before routing.

The Muskingum method (Chow, 1964) computes outflow from a reach as a function of the current period inflows and the previous period inflow and outflow. The routing procedure may be repeated for several sub-reaches. The total travel time through the reach and the size of the wedge storage coefficient are checked by the program for physical and computational constraints.

Storage routing methods in HEC-1 are those methods which require data about the storage characteristics of a routing reach or reservoir. These methods are: modified Puls, working R and D, and level-pool reservoir routing. These methods also require outflow data which is related to storage. There are three methods for determining routing reach storage in HEC-1: (1) direct input, (2) surface area and elevation for reservoirs (conic method), and (3) channel cross-section and reach length (normal-depth channel flow). Outflow characteristics can be computed from: direct input, normal-depth channel flow, weir equation (spillway), critical depth (trapezoidal spillway), or ogee spillway data. Whenever storage and outflow data are computed from methods other than direct input, elevation (stage) data must be supplied so the relation between storage and outflow can be determined. If the storage routing procedure used is the modified Puls (given storage versus outflow or computed by normal-depth channel flow), the working R&D, or the trapezoidal spillway critical depth method, a storage versus outflow relationship is first computed from the input data and then used in all time-interval computations. If the level-pool reservoir routing with low-level orifice and weir spillway outflows is used, storage and outflow are computed from the current reservoir water surface elevation in each time interval.

Storage and outflow data for use in storage routing may be computed from channel characteristics. The program uses an 8-point cross section which is representative of the routing reach. Outflows are computed for normal depth using Manning's equation. Storage is cross-sectional area times reach length. Storage and outflow values are computed for 20 evenly-spaced stages beginning at the lowest point on the cross section to a specified maximum stage. The cross section is extended vertically at each end to the maximum stage.

The modified Puls routing method (Chow, 1964) is a variation of the storage routing method described by Henderson (1966). A storage indication function is computed from given storage and outflow data. The outflow at the end of the time interval is interpolated from a table of storage indication versus outflow. Storage is then computed from a continuity relationship. When stage data are given, stages are interpolated for computed storages. Initial conditions can be specified in terms of storage, outflow, or stage. The corresponding value of storage or outflow are computed from the given initial value. The working R and D method (Corps of Engineers, 1960a) is a variation of modified Puls method which accounts for wedge storage as in the Muskingum method.

Level-pool reservoir routing assumes a level water surface in a reservoir. It is used in conjunction with the pump option described subsequently and with the dam-break calculation described in a later section. Using the principle of conservation of mass, the change in reservoir storage for a given time period is equal to average inflow minus average outflow. An iterative procedure is used to determine end-of-period storage and outflow. Pumps may be included as a part of level-pool reservoir routing. The program checks the reservoir stage at the beginning of each time period. If the stage exceeds the "pump-on" elevation the pump is turned on and the pump output is included as an additional outflow term in the routing equation. When the reservoir stage drops below a "pump-off" elevation, the pump is turned off. Several pumps with different on and off elevations may be used. Each pump discharges at a constant rate. Pumped flow is lost from the system and is not available for any further calculations.

Reservoir outflow for storage routing may be computed from a description of the outlet works (low-level outlet and spillway). There are two subroutines in HEC-1 which compute outflow rating curves. The first uses simple orifice and weir-flow equations while the second computes outflow from specific energy or design graphs and corrects for tailwater submergence. An outflow rating curve is computed for 20 elevations which span the range of elevations given for storage data. Storages are computed for these outflows and this storage versus outflow relation is used for modified Puls or working R and D routing. For level-pool reservoir routing outflows are computed for the orifice and weir equations for each routing interval. Trapezoidal and ogee spillways (Corps of Engineers, 1963) may also be simulated using appropriate pier and abutment losses.

Kinematic wave channel routing can be utilized independently of the other elements of the subbasin runoff. In this case, upstream inflow is routed through a reach (independent of lateral inflows) using the previously described kinematic wave methods.

PARAMETER OPTIMIZATION

Calibration and verification are essential parts of the modeling process. Rough estimates for the parameters in the HEC-1 model can be obtained from the literature, however, the model should be calibrated to observed flood data whenever possible. HEC-1 provides a powerful optimization technique for the estimation of some of the parameters when gaged precipitation and runoff data are available. By using this

technique and regionalizing the results, rainfall-runoff parameters for
ungaged areas can also be estimated (HEC, 1981b). A summary of the
HEC's experience with automatic calibration of rainfall-runoff models
is given by Ford et al. (1980).

The parameter optimization option has the capability to automatically determine a set of unit hydrograph and loss rate parameters that
"best" reconstitute an observed runoff hydrograph for a subbasin. The
data which must be provided to the model are: basin average precipitation; basin area; starting flow and base flow parameters; and the
outflow hydrograph. Unit hydrograph and loss rate parameters can be
determined individually or in combination. Parameters that are not to
be determined from the optimization process must be estimated and
provided to the model. Initial estimates of the parameters to be
determined can be input by the user or chosen by the program's optimization procedure.

The runoff parameters that can be determined in the optimization
are the unit hydrograph parameters of the Snyder, Clark and SCS methods
and loss rate parameters of the exponential, Holtan, SCS, and initial/
constant methods. The melt rate and threshold melt temperature can
also be optimized for snow hydrology studies.

The "best" reconstitution is considered to be that which minimizes
the weighted squared difference between the observed hydrograph and the
computed hydrograph. Presumably, this difference will be a minimum for
the optimal parameter estimates. The sum of the weighted squared
differences STDER objective function is defined as follows:

$$\text{STDER} = \sum_{i=1}^{n} (QOBS_i - QCOMP_i)^2 * WT_i/n \quad \ldots \ldots \ldots \ldots (2)$$

where $QCOMP_i$ is the runoff hydrograph ordinate for time period i
computed by HEC-1, $QOBS_i$ is the observed runoff hydrograph ordinate
i, n is the total number of hydrograph ordinates, and WT_i is the
weight for the hydrograph ordinate i computed from the following equation:

$$WT_i = (QOBS_i + QAVE) / (2*QAVE) \quad \ldots \ldots \ldots \ldots \ldots \ldots (3)$$

where QAVE is the average computed discharge. This weighting function
emphasizes accurate reproduction of peak flows rather than low flows by
biasing the objective functions. Any errors for computed discharges
that exceed the average discharge will be weighted more heavily, and
hence the optimization scheme should focus on reduction of these errors.

The minimum of the objective function is found by employing the
univariate search technique (Ford et al., 1980). The univariate search
method computes values of the objective function for various values of
the optimization parameters. The values of the parameters are systematically altered until STDER is minimized. The range of feasible
values of the parameters is bounded because of physical limitations on
the values that the various unit hydrograph, loss rate, and snowmelt
parameters may have, and also because of numerical limitations imposed
by the mathematical functions. The optimization procedure does not
guarantee that a "global" optimum (or a global minimum of the objective
function) will be found for the runoff parameters; a local minimum of

133

the objective function might be found by the procedure. To help assess the results of the optimization, HEC-1 provides graphical and statistical comparisons of the observed and computed hydrographs. From this, the user can then judge the accuracy of the optimization results.

HEC-1 may also be used to automatically derive routing criteria for certain hydrologic routing techniques. Criteria can be derived for the Tatum, straddle-stagger and Muskingum routing methods. Observed hydrographs are reconstituted to minimize the squared sum of the deviations between the observed hydrograph and the reconstituted hydrograph. The procedure used is essentially the same as for the unit hydrograph and loss rate parameter optimization.

MULTIPLAN-MULTIFLOOD ANALYSIS

The multiplan-multiflood simulation option allows a user to investigate a series of floods for a number of different characterizations (plans) of the watershed in a single computer run. The advantage of this option is that multiple storms and flood control projects simulations can be performed in a single computer run and the results compared with a minimum of effort by the user.

The multiflood simulation allows the user to analyze several different floods in the same computer run. The floods are specified as fractions of a base event (e.g., .5, 1.0, 1.5, etc.) which may be of either precipitation or runoff. In the case of rainfall, each ordinate of the input base-event hyetograph would be multiplied by a ratio and a stream network rainfall-runoff simulation carried out for each ratio. This is done for every ratio of the base event. In the case of runoff ratios, the ratios are applied to the computed or direct-input hydrograph and no rainfall-runoff calculations are made for individual ratios.

The multiplan option allows a user to conveniently modify a basin model to reflect desired flood control projects and changes in the basins's runoff response characteristics. This is useful when, for example, a comparison of flood control options or the effects of urbanization are being analyzed. The user designates PLAN 1 as the existing river basin model, and then modifies the existing plan data to reflect basin changes (such as reservoirs, channel improvements, or changes in land use) in PLANS 2, 3, etc. If the basin's rainfall-runoff response characteristics are modified in one of the plans, precipitation ratios and not runoff ratios must be used. Otherwise, ratios of hydrographs should be used. The program performs a stream network analysis, or multiflood analysis, for each plan. The results of the analysis provide flood hydrograph data for each plan and each ratio of the base event. The summary of the results at the end of the program output provides the user with a convenient method for comparing the differences between plans (alternative flood control systems).

DAM SAFETY/FAILURE ANALYSIS

The dam failure analysis capability was added to the HEC-1 model to assist in studies required for the United States National Non-Federal Dam Safety Program. This option uses simplified hydraulic techniques to estimate the potential for and consequences of dam overtopping or structural failures on downstream areas.

A dam failure analysis utilizes the network modelling techniques with some added capabilities for reservoir routing. These additional reservoir routing capabilities calculate flow through low-level outlets, spillway, over the top of the dam, and through a breach. The dam failure simulation differs from the previously described reservoir routing in that the stage-outflow relation is computed by determining the flow over top of the dam (dam overtopping) and/or through the dam breach (dam break) as well as through other reservoir outlet works, figure 4. The stage-outflow characteristics are then combined with the level-pool storage routing to simulate a dam failure.

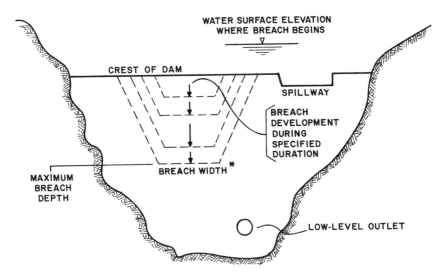

* BREACH SHAPE MAY BE TRAPEZOIDAL,
  RECTANGULAR, OR TRIANGULAR.

Figure 4   COMPONENTS OF NORMAL AND BREACH FLOW
THROUGH A DAM

The discharge over the top of the dam is computed by a weir flow equation. Spillway discharges continue to be computed by the spillway equation even as the water surface elevation exceeds the top of the dam elevation. The weir flow for dam overtopping is added to the spillway and low-level outlet discharges. Critical flow over a non-level dam crest is computed from crest length and elevation data. A dam crest is divided into rectangular and trapezoidal sections and the flow is computed through each section. When a dam is breached the width of the breach is subracted from the crest length beginning at the lowest portion of the dam.

Dam breaks are simulated using the methodology proposed by D. L. Fread (1979). Structural failures are modeled by assuming certain geometrical shapes for the dam breach. The variables used in the analysis, as well as the dam breach shapes available in the program, were shown in figure 4.

Flow through a dam breach is computed as weir flow using progressively larger weirs as the breach develops. The breach is initiated when the water surface in the reservoir reaches a specified elevation. The breach begins at the top of the dam and expands linear-

135

ly to the bottom of the breach and to its full width in a specified
time. The failure duration is divided into 50 computation intervals.
These short intervals are used to minimize routing errors during the
period of rapidly changing flows when the breach is forming. Downstream routing methods in HEC-1 use a time interval which is usually
greater than the time interval used during breach development. The
program output shows the short-interval failure hydrograph and the
location of the regular HEC-1 time intervals. It is important to be
sure that the breach hydrograph is adequately described by the HEC-1
end-of-period intervals or else the downstream routings will be
erroneous.

The dam-break simulation assumes that the dam-break hydrograph
will not be affected by tailwater constraints and that the reservoir
pool remains level. Also, HEC-1 hydrologic routing methods are assumed
appropriate for the dynamic flood wave. Under the appropriate conditions, these assumptions will be approximately true. However, care
should be taken in interpreting the results of the dam-break analysis.
If a more accurate analysis is needed, then an unsteady flow model,
such as the National Weather Service's DAMBRK (Fread, 1979), should be
used.

PRECIPITATION DEPTH-AREA RELATIONSHIP SIMULATION

One of the most difficult problems of hydrologic evaluation is
that of determining the effect that a project on a remote tributary has
on floods at a downstream location. A similar problem is that of
deriving flood hydographs, such as the standard project floods or
100-year exceedence interval floods, at a series of locations throughout a complex river basin. Both problems could require the successive
evaluation of many storm centerings upstream of each location of
interest.

Since the average depth of precipitation over a tributary area for
a storm generally decreases with the size of contributing area, it
would ordinarily be necessary to recompute a decreasing consistent
flood quantity contributed by each subarea to successive downstream
points. In order to avoid the proliferation of hydrographs that would
ensue, the depth area calculation of HEC-1 makes use of a number of
hydrographs (termed "index hydrograph") computed from a range of
precipitation depths throughout the river basin. The index hydrographs
are computed from a set of precipitation depth-drainage area (index
area) values, a time distribution of rainfall, and appropriate loss
rate and unit hydrograph parameters. A consistent hydrograph is that
which corresponds to the appropriate precipitation depth for the
sub-basin's drainage area. The consistent hydrographs are determined
by interpolating between the two index hydrographs bracketing the
subareas drainage area. The stream system procedure of generating
index hydrographs, interpolating, routing and interpolating, is
repeated throughout a river basin for as many locations as are desired
as described in the HEC-1 Users Manual (HEC, 1981a).

FLOOD CONTROL BENEFIT ANALYSIS

Flood control planning requires the ability to rationally assess
the economic consequences of flood damage. The HEC-1 benefit analysis
option provides the capability to assess flood damage and explore the
economic benefits provided by alternative flood control measures. The
benefit due to the implementation of a flood control plan is determined

by computing the difference between damage occurring in a river basin with the flood control plan and without the plan. River basin damage is determined by summing the damage computed for particular areas or reaches of the basin.

Expected annual damages (EAD) are computed as the sum of the damages weighted by a frequency of occurrence. This sum can be thought of as the average yearly damage that can be expected to occur in the reach over an extended period of time. The basic assumption of the EAD analysis is that the damage frequency curve can be obtained by combining damage versus flow (stage) and flow (stage) versus frequency relations which are characteristic of the area that the damage reach represents. The damage versus flow (stage) relation ascribes a dollar damage that occurs in an area to a level of flood flow. The flow (stage) versus exceedence frequency relation ascribes an exceedence frequency to the magnitude of flood flow. By combining this information, the damage versus frequency curve and, hence, the EAD for a reach can be determined. By comparing river basin EAD with and without flood control projects, benefits are computed as the reduction in damages.

There are two basic computations in a benefit calculation: exceedence frequency curve modification and EAD calculation. Structural flood control measures (e.g., reservoirs and channel improvements) affect the flow-frequency relationship. Nonstructural measures (e.g., flood proofing and warning) do not usually have much impact on the flood-frequency relationship but do modify the stage-damage relationship.

Frequency Curve Modification

The flow-exceedence frequency data provided for damage reaches refer to PLAN 1 or the base plan of the multiplan-multiflood model. Implementation of structural flood control measures will change this exceedence frequency relation. HEC-1 computes modified frequency relationships using the following methodology. A multiflood analysis is performed for PLAN 1 to establish the frequency of the peak discharge of each ratio of the design event. The peak-flow frequency for each ratio of the design event is interpolated from the input flow-frequency data tables for a damage reach. A stage-frequency curve is established in essentially the same manner as for flows when stage-frequency data are specified for a damage reach.

A multiflood simulation is now performed for the flood control plans. The peak discharges (stages) are computed at each damage reach for each ratio of the design event. HEC-1 assumes that the frequency of each ratio remains the same as computed for the base case above; and only the peak flows associated with each ratio change for different plans. In this manner, the modified flow-frequency curve is computed for all ratios as shown in figure 5. The assumption inherent in this procedure is that the event ratio-frequency relation is not affected by basin configuration.

Expected Annual Damage (EAD) Calculation

EAD is calculated by combining the flow or stage-frequency curve and the flow- or stage-damage for each PLAN and damage reach (HEC, 1979a). The flow-frequency curve is used in conjunction with the flow-damage data to produce a damage-frequency curve as shown in figure 6. The area under the damage-frequency curve is the EAD for the reach. This area is computed using a three point Gaussian Quadrature

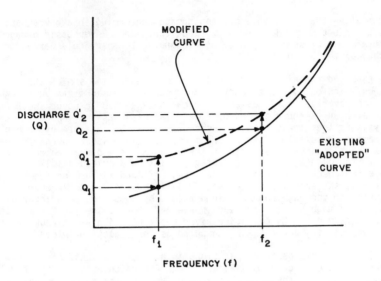

Q$_i$ = PEAK DISCHARGE FOR DESIGN STORM ℓ

f$_i$ = FREQUENCY OF Q$_i$ FROM GIVEN CURVE

Q'$_i$ PEAK DISCHARGE FROM DESIGN STORM ℓ UNDER MODIFIED
WATERSHED CONDITION

Figure 5   MODIFYING FREQUENCY CURVES

formula. If more than one damage category is specified for a reach,
the above steps are repeated for each land use. The EAD is summed for
all the land uses to produce the EAD for the reach. The benefit
accrued due to the employment of a flood control plan is equal to the
difference between the PLAN 1 EAD and the flood control plan EAD. The
model performs this computation for all plans in the
multiplan-multiflood analysis.

FLOOD CONTROL SYSTEM OPTIMIZATION

The flood control system optimization option is used to determine
optimal sizes for the flood control components in a river basin flood
control plan (Davis, 1974). The optimization model is an extension of
the flood damage model previously described. The optimization model
utilizes a two-plan damage analysis: PLAN 1 is the base condition of
the existing river basin and PLAN 2 is the flood control plan being
optimized. Data on the costs of various sizes of flood control
projects are required, otherwise the formulation of the optimization
model is essentially the same as in the flood damage model case. The
flood control components that can be optimized as part of the flood
control system are as follows: reservoirs, diversions, pumping plants,
and local protection projects (levees, etc.).

The storage of a reservoir may be optimized by determining the
elevation of the reservoir spillway, thus defining the point where
reservoir outflows are uncontrolled. The low-level outlet characteristics of the reservoir are fixed by input. Flow diversions, such as
described for the stream network simulation, may have their channel
capacity optimized. The diverted flow may be returned to another

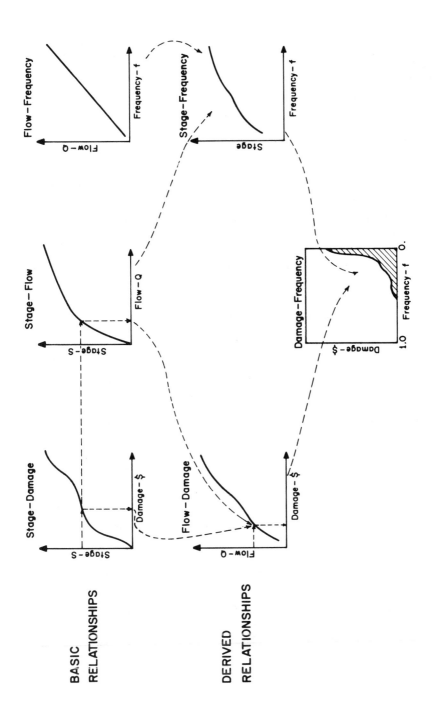

Figure 6   EXPECTED ANNUAL DAMAGE CALCULATION
(figure 1 : HEC, 1979a)

139

branch of the stream network or simply lost from the system. Pumping plants may be located virtually anywhere in a stream network and their capacity may be optimized. The pumped water is considered lost from the system and cannot be returned to another branch of the stream network. A local protection project can be used to model a channel improvement or a levee. This component can only be used in conjunction with the damage analysis of a reach because it only modifies the damage function. The local protection project analysis requires capacity and cost data together with pattern damage tables for maximum and minimum sizes of the project. Damage functions are interpolated for project sizes between these maximum and minimum design values.

The flood control component optimization model requires data as described for the flood damage model plus information about the capital and operating costs of the projects and about the objective function for the flood control scheme. The data for the various types of flood control components are essentially the same and may be separated into cost data, capacity constraints, and optimization criteria. Minimum and maximum capacity must be specified for each flood control component. An initial estimate of the size of the flood control component is also required to give the optimization procedure a starting point.

Two types of data are supplied to the program which are used to calculate the total annual cost of a flood control component. First, capacity versus capital cost tables are required to determine the capital cost for any capacity of the flood control component. A capital recovery factor is also required so that equivalent annual costs for the capital investments can be computed. Second, operation and maintenance costs are computed as a proportion of the capital cost. For pumping plants, average annual power costs for various pump capacities are required. Pump operation costs are computed in proportion to the volume pumped. Capital and operating costs for non-optimized components of the system may also be considered.

The optimization methodology can operate on maximum net benefits and/or flow targets criteria. Maximum net benefits are computed using the cost and flood damage data previously described. Desired streamflows may also be specified at any point downstream of a flood control project. These streamflow limitations, referred to as "flow targets" are specified as the flow (stage) which is desired to occur at a given frequency. For example, it may be desired to have the 5% flood at a particular location be 1,000 m^3/s. The input data for flow targets are the discharge or stage and the frequency.

The model determines an optimal flood control system by minimizing a system objective function. The system objective function is the sum of flood control system total annual cost and the expected annual damage occurring in the basin. If flow targets are specified, then the previous sum is multiplied by a penalty factor which increases the objective function proportionately to deviations from the target. Note that the minimization of the objective function leads to the maximization of the net benefits accrued due to the employment of the flood control system. Net benefits are equal to the difference between the EAD occurring in PLAN 1 and the sum of the system costs and EAD occurring in PLAN 2.

An initial system configuration is analyzed by the program based on capacities specified by the user. The model performs a stream network simulation and expected annual damage calculation for the base condition, PLAN 1, without the proposed flood control measures. The stream network and expected annual damage calculations for the initial sizes of the proposed flood control system are then calculated and the initial value of the objective function is determined. The model then uses the univariate search procedure to estimate a minimum value for the objective function. The search proceeds by using the stream network and EAD calculation to generate points on the system objective function for various flood control system capacities. These capacities are systematically altered by the procedure until an optimum is reached. As in the river basin parameter optimization, a global optimum can not be guaranteed (in fact there maybe many alternative optimal solutions). However, by inspecting the resulting net benefits provided by the system, the desirability of the optimal system can be assessed.

PROGRAM USAGE

This section describes the general organization of the input data, program output, example problems, and computer requirements.

Input Data

There are two general types of data cards for HEC-1: input control and river basin simulation data. The input control cards tell the program the format of the river basin data as well as controlling certain diagnostic output. The river basin simulation data are all identified by a unique two-character alphabetic code in card columns one and two. These codes serve two functions: they identify the data to be read from the card; and they activate various simulation options. The first character of the code identifies the general data category and the second character identifies a specific type of data within a category. The data may be input in a free or fixed format. The stream network structure can be protrayed diagrammatically. This option causes the program to search the input data deck and determine the job step computations. A flow chart of the stream network simulation is printed.

The user may enter time series data, either hyetographs or hydrographs, at time steps other than the computation interval of the simulation. This option is convenient when entering data generated by another program or in a separate HEC-1 simuldtion. In many instances, certain physical characteristics are the same for a number of subbasins in the stream network model (for instance, infiltration characteristics). Further, in a multiplan analysis, much of the PLAN 1 subbasin data remains unchanged in subsequent plans. The HEC-1 program input conventions make it unnecessary to repeat much of this information in the data deck.

Program Output

A large variety and degree of detail in the printer output are available from HEC-1. The output may be categorized in terms of input data feedback, intermediate simulation results, summary results, and error messages. The degree of detail of virtually all of the program output can be controlled by the user. The input data file for each job is read and converted from free format to fixed format and a sequence number is assigned to each line. The reformatted data can be printed so the user can see the data which are going into the main part of the program.

The data used in each hydrograph computation can be printed as well as the computed hydrograph, rainfall, storage, etc. as applicable. The sources of these data are indicated by the card identification code and input line number printed on the left side of the page. Hydrographs may be printed in tabular form and/or graphed (printer plot) with the date, time, and sequence number for each ordinate. For runoff calculations, rainfall, losses, and excesses are included in the table and plot. For snowmelt calculations, separate values of loss and excess are printed for rainfall and snowmelt. For storage routings, storage and stage (if stage data are given) are printed/plotted along with discharge.

The program produces hydrologic and economic summaries of the computations throughout the river basin. The standard program hydrologic summary shows the peak flow (stage) and accumulated drainage area for every hydrograph computation in the simulation. Economic summary data show the flood damages and benefits (also costs for project optimization) for each damage reach and for the river basin. The river basin damage/benefit results may also be summarized by two locational descriptors (e.g., river name and county name) if desired. The user can also choose time series data at selected stations to be displayed in tables at the end of the job. Hyetographs, losses, excesses, stages, storages, and hydrographs can be printed in these tables in any desired order as specified by input control.

Example Problems

The HEC-1 Users Manual (HEC, 1981a) contains several test problems which serve both as illustrative examples of various capabilities of HEC-1 and as benchmark tests to verify that the program is working correctly. The first three example problems illustrate the most basic river basin modeling capabilities. Following these, specialized capabilities of HEC-1 are added to the basic model. The last four examples are a sequence of steps necessary to perform multiflood, multiplan, flood damage, and flood control project optimization analyses.

Computer Requirements and Support

HEC-1 requires a FORTRAN IV compiler and up to 16 input/output scratch (tape, disk, etc.) files. The computer memory required on the CDC 7600 is 115,000 words. It requires approximately 7 seconds to compile on that machine. The program has been tested on several major computers and the machine dependent code removed whenever possible. The users manual and programmers supplement describe detailed program characteristics and modifications necessary to run the program on different computer systems and to reduce memory requirements. The HEC provides user support for HEC-1 and other programs (Eichert, 1978). The program and documentation may be obtained from the HEC for the cost of reproduction and handling.

REFERENCES

Chow, V. T. 1964. Handbook of Applied Hydrology. McGraw-Hill, New York.
Clark, C. O. 1945. Storage and the unit hydrograph. Transactions of the American Society of Civil Engineers 110, pp. 1419-1446.
Corps of Engineers 1952. Standard Project Flood Determinations. Engineering Manual 1110-2-1411, U.S. Army, Washington, D.C.
Corps of Engineers 1960a. Routing of Floods through River Channels. Engineering Manual 1110-2-1408, U.S. Army, Washington, D.C.

Corps of Engineers 1960b. Runoff from Snowmelt. Engineering Manual 1110-2-1406, U.S. Army, Washington, D.C.

Corps of Engineers 1963. Hydraulic Design of Reservoir Outlet Structures. Engineering Manual 1110-2-1602, U.S. Army, Washington, D.C.

Davis, D. W. 1974. Optimal sizing of urban flood control systems. Journal of the Hydraulics Division 101, pp. 1077-1092, American Society of Civil Engineers.

Eichert, B. S. 1978. Experiences of the Hydrologic Engineering Center in Maintaining Widely-Used Hydrologic and Water Resources Models. Technical Paper No. 56. Hydrologic Engineering Center, U.S. Army Corps of Engineers, California.

Feldman, A. D. 1981. Water Resources System Simulation. in: V. T. Chow (Editor), Advances in Hydroscience, Vol. 12, pp. 297-423, Academic Press, New York.

Ford, D. T., Morris, E. C. and Feldman, A. D. 1980. Corps of Engineers' experience with automatic calibration of a precipitation-runoff model. In "Water and Related Land Resource Systems" (Y. Haimes and J. Kindler, eds.). Pergamon Press, New York.

Fread, D. L. 1979. DAMBRK: The NWS Dam-Break Flood Forecasting Model. Technical Paper, Office of Hydrolgy, National Weather Service, Silver Spring, Maryland.

Harley, B. M. 1975. MITCAT Catchment Simulation Model, Description and Users Manual, Version 6, Resource Analysis Corporation, Massachussetts.

Henderson, F. M. 1966. Open Channel Flow, Macmillan Co., New York, pp. 356-362.

Holtan, H. N., Stitner, G. J., Henson, W. H. and Lopez, N. C. 1975. USDAHL-74 Revised Model of Watershed Hydrology. Technical Bulletin No. 1518, Agricultural Research Service, U.S. Department of Agriculture, Washington, D.C.

Hydrologic Engineering Center 1979a. Expected Annual Flood Damage Computation. Program Users Manual, U.S. Army Corps of Engineers, California

Hydrologic Engineering Center 1979b. Introduction and Application of Kinematic Wave Routing Techniques Using HEC-1. Training Document No. 10, U.S. Army Corps of Engineers, California.

Hydrologic Engineering Center 1981a. HEC-1 Flood Hydrograph Package (preliminary). Program Users Manual, U.S. Army Corps of Engineers, California.

Hydrologic Engineering Center 1981b. Hydrologic Analysis of Ungaged Watersheds with HEC-1 (preliminary), U.S. Army Corps of Engineers, California.

Linsley, R. K., Kohler, M. A. and Paulhus, J. L. 1975. Hydrology for Engineers, 2nd edition, McGraw-Hill Co., New York.

National Weather Service 1956. Seasonal Variation of Probable Maximum Precipitation East of the 105th Meridian for Areas from 10 to 1,000 Square Miles and Durations of 6, 12, 24 and 48 Hours. Hydrometeorological Report No. 33, U.S. Department of Commerce, Washington, D.C.

National Weather Service 1961. Rainfall Frequency Atlas of the United States. Technical Paper No. 40, U.S. Department of Commerce, Washington, D.C.

National Weather Service 1977. Five to 60-Minutes Precipitation Frequency for the Eastern and Central United States. Technical Memo NWS HYDRO-35, National Oceanographic and Atmospheric Atmospheric Administration, U.S. Department of Commerce, Maryland.

Snyder, F. F. 1938. Synthetic unit hydrographs. Transactions of the American Geophysical Union, Vol. 19, Part 1, pp. 447-454.

Soil Conservation Service 1965. Computer Program for Project
    Formulation Hydrology. Technical Release No. 20, U.S. Department of
    Agriculture, Washington, D.C.
Soil Conservation Service 1975. Urban Hydrology for Small Watersheds.
    Technical Release No. 55, U.S. Department of Agriculture,
    Washington, D.C.
Viessman, W. Jr., Knapp, J. W., Lewis, G. L. and Harbaugh, T. E. 1977.
    Introduction to Hydrology, Dun-Donnelley Co., New York.

RAINFALL/RUNOFF MODEL USING
BASIN WATER HOLDING CAPACITY

Ming C. Shiao
Tennessee Valley Authority
Water Systems Development Branch
Norris, Tennessee 37828

Walter O. Wunderlich
Tennessee Valley Authority
Water Systems Development Branch
Norris, Tennessee 37828

ABSTRACT

Experiences are described with a simple conceptual rainfall/ runoff model for streamflow forecasting in gaged headwater catchments. The model uses basin water holding capacity for soil moisture accounting. The basin water holding capacity (WHC) is determined as a function of baseflow. Infiltration is determined as a function of basin water holding capacity and rainfall intensity. For the time distribution of surface runoff, a set of unit responses is used which is derived from the area-distance distribution of the drainage basin and water travel time. The model has very moderate computer requirements and low computational cost. It was tested on two watersheds with drainage areas of 137 and 764 square miles. In both cases, the results reproduced satisfactorily the observed data.

INTRODUCTION

This paper describes experiences with a relatively simple conceptual model for streamflow forecasting in gaged headwater catchments. The model is suitable for streamflow computation for a specific historical storm event as well as for operational day-to-day streamflow forecasting. For applications to specific events, updating of the model to current basin conditions is possible without extensive "warm-up" preparations. For day-to-day operational applications, the model, due to its simplicity, is computationally efficient. The model simulates a simplified hydrologic process which consists of only the principal hydrologic components: surface runoff, infiltration, base flow, and soil moisture storage. A particular feature of the model is the use of observed streamflow to estimate the basin water holding capacity which in turn is used to determine infiltration and base flow.

In the following, the formulation of the model is described. The model is applied and tested on two tributary watersheds in the Tennessee River basin. The results are discussed and recommendations for further developments are given.

MODEL FORMULATION

The WHC Model

The hydrologic process in a catchment which converts rainfall into runoff consists of many components, such as areal precipitation, interception, evaporation, transpiration, infiltration, depression storage, soil moisture storage, base flow (interflow + groundwater discharge) and surface runoff. A schematic diagram relating these components to each other is shown in Figure 1.

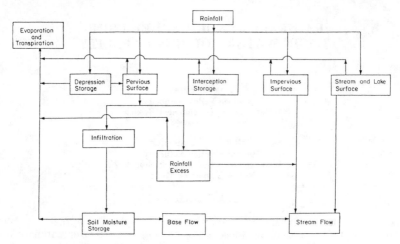

FIGURE 1. SCHEMATIZED CONCEPTUAL RAINFALL-RUNOFF PROCESS

The described model simulates only those components of the rainfall-runoff process which are considered most important in a head-water basin (source area). The hydrologic process in the catchment is assumed to have one input (precipitation), and two outputs (streamflow and evapotranspiration). Flow contributed by snow is not included. The process is characterized by the basin water holding capacity (WHC) which is defined as the additional amount of water that can be absorbed by the soil before it reaches saturation (i.e., the soil moisture deficit). The model is hereafter called the WHC model. The basin WHC is increased by evaporation from soil and transpiration from plants (collectively referred to as evapotranspiration) and by discharge of base flow. It is reduced by precipitation. The basic concept of the model is expressed mathematically by the mass conservation equation of the hydrologic system:

$$P_t = F_t + R_t \tag{1}$$

$$(S_t - S'_t)/\Delta t = (W'_t - W_t)/\Delta t = F_t - G_t - E_t \tag{2}$$

with $W_t = SM - S_t$

where P_t, F_t, R_t, G_t, E_t are precipitation, infiltration, rainfall excess (direct runoff), base flow, and evapotranspiration, respectively, during the t-th time interval, Δt, SM is the saturation soil moisture storage, S'_t and S_t are the soil moisture storages at the beginning and at the end of the t-th time interval, respectively, and W'_t and W_t are the basin water holding capacities at the beginning and at the end of the t-th time interval, respectively.

A schematic diagram of the WHC model is shown in Figure 2. Rainfall is divided into infiltration and rainfall excess. Infiltration replenishes soil moisture storage which produces base flow and allows for evapotranspiration. Base flow and rainfall excess are combined to produce streamflow.

Not explicitly included in the model are interception, depression storage, and precipitation falling directly on streams, lakes and impervious surfaces. Both interception and depression storage are components of a transient nature. They eventually either add to soil moisture storage through infiltration or return to the atmosphere as evaporation. Thus, interception and depression storage are included in the model as part of the infiltration and evapotranspiration components. For small storm events, they can be a significant percentage of precipitation. However, their accurate estimation is difficult. In practical applications, a deduction from the precipitation to account for these components can be made. The deduction is site dependent and can be

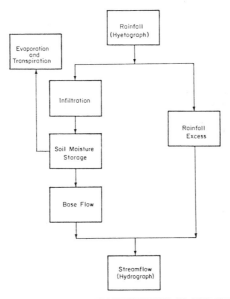

FIGURE 2.   SCHEMATIC REPRESENTATION OF THE WHC MODEL

either a constant or a function of time of the year. The deduction can best be determined through application of the model to the specific basin using events occurring in different seasons.

Mean Areal Precipitation and Rainfall Hyetograph

The accuracy that is justifiable for a rainfall-runoff model is predicated on the accuracy with which the areal precipitation and its temporal distribution (i.e., rainfall hyetograph) can be determined. The described model uses mean areal precipitation over the entire watershed. Techniques which have been used to compute the mean areal precipitation are the representative point method, arithmetic mean method, polygon method, isohyetal method, etc. The Thiessen polygon method is used in this study.

The temporal areal rainfall distribution is determined by distributing the daily precipitation at each station in the basin over

time in proportion to the rate of rainfall recorded at the nearest recording station. A rainfall hyetograph is thus constructed at each station
in the basin. For each time period, the mean areal rainfall is obtained
by summing over all stations the products of the rate of rainfall and
their corresponding Thiessen weights (representative of the fraction of
the drainage area attributed to each station). This procedure requires
that, for each storm event, at least one recording station is available in
or near the basin which provides the necessary information on temporal
rainfall distribution. In general, the reliability of the computed areal
rainfall and the time-distributed areal rainfall hyetograph improves with
the number of rain gages (especially recording gages) available in the
basin.

Basin Water Holding Capacity

The soil moisture storage at the beginning of a storm event
plays an important role in determining rainfall excess, base flow and
their time distributions. Suppose there were two hydrographs resulting
from the same storm, one with high initial soil moisture (wet soil) and
the other with low initial soil moisture (dry soil). The former would
produce a higher peak flow and greater runoff volume than the latter.
Also, the basin lag defined as the time from the centroid of the rainfall
volume to the peak of the hydrograph, would be shorter for the high
soil moisture case than for the low soil moisture case.

Direct measurement of soil moisture storage in a basin at the
beginning of the storm is practically impossible. It has been suggested
that base flow at the beginning of the storm can be used as a good
index of initial soil moisture conditions if supplemented by information
on preceding rainfall (Linsley, et al., 1958). It is assumed that the
initial basin water holding capacity can be used as a complement of the
initial soil moisture storage. In the following, an approach is described
which determines the initial basin WHC using the antecedent base flow.
As a first approximation, the influence of preceding rainfall on soil
moisture storage is neglected.

Substituting equation (1) into (2) yields

$$\Delta S_t = S_t - S_t' = (P_t - R_t - G_t - E_t) \, \Delta t \tag{3}$$

Integrating equation (3) over the duration of precipitation, T_d, results
in

$$\sum_{t=1}^{N} \Delta S_t = \left(\sum_{t=1}^{N} P_t - \sum_{t=1}^{N} R_t - \sum_{t=1}^{N} G_t - \sum_{t=1}^{N} E_t \right) \Delta t \tag{4}$$

where $N = T_d / \Delta t$.

To obtain an estimate of basin infiltration, it is assumed that, for an
isolated short duration storm event, base flow and evapotranspiration
during the rainfall period (G and E in Equation 4) are relatively small
and can be neglected. For a storm event whose rainfall is large enough
to satisfy the soil moisture capacity, equation (4) can be written as

$$\sum_{t=1}^{N} \Delta S_t \approx \left(\sum_{t=1}^{N} P_t - \sum_{t=1}^{N} R_t \right) \Delta t = \left(\sum_{t=1}^{N} F_t \right) \Delta t$$

$$= S_N - S_o = SM - S_o = W_o \tag{5}$$

with indexes o and N referring to the beginning and the end of the rain period. The initial basin WHC, W_o, is therefore equaled to the total storm infiltration.

To develop the relationship between the initial basin WHC and the antecedent base flow, a number of storm events which are isolated and of short duration are selected. Each storm has a relatively low initial soil moisture storage; in other words, the influence of preceding rainfalls is minimal. The total infiltration or initial water holding capacity, W_o, is obtained as the difference between the total precipitation and the volume of direct runoff. The latter is obtained by subtracting base flow from the observed hydrograph. For base flow separation, a composite recession curve is used for the falling limb and a third order polynomial is used for the rising limb of the base flow. The total infiltration is then plotted versus the antecedent base flow. Such a plot is shown in Figure 3. Not every storm event selected has

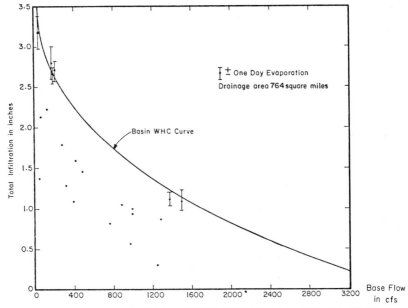

FIGURE 3. TOTAL INFILTRATION VS ANTECEDENT BASE FLOW FOR THE EMORY RIVER BASIN

enough rainfall to satisfy the soil moisture capacity. Of all events with the same antecedent base flow, only the one with the highest total infiltration satisfies or comes close to satisfying equation (5). The envelope curve which connects those points with highest total infiltrations is defined as the basin WHC curve. For a given antecedent base flow, the ordinate of a point on the basin WHC curve represents the amount of infiltration required to fill the soil moisture capacity. This is equivalent to the basin water holding capacity for the given antecedent base flow. A one-day evaporation estimate is used in Figure 3 to show the effect of evapotranspiration on the computed total infiltration.

The basin WHC curve can be expressed as:

$$W_t = SM - \alpha \, q_t^\beta \tag{6}$$

where SM is the soil moisture capacity in inches and q_t is the baseflow yield of the drainage basin in cfs per square mile. SM can be estimated from available soil moisture characteristics (Carlson, 1959; Longwell, et al, 1963), or by curve fitting as the vertical intercept of the basin WHC curve; α and β are coefficients which can be determined by a nonlinear least-square method.

The basin WHC curve can be used to determine the basin WHC at the beginning of a storm. The basin WHC thus derived tends to overestimate the amount of water that can be absorbed by the soil. For an event strongly affected by a previous storm, the initial soil moisture storage can be high, and an adjustment (reduction) in the initial basin WHC may be necessary. This adjustment varies as a function of groundwater table and preceding rainfall. As a first approximation no adjustment is included here. The rainfall events selected for the derivation of the basin WHC curve were isolated single peak events of relatively short rainfall durations and large intensities. However, once the basin WHC curve is established, it does not limit the application of the model to such events.

Equation (6) can be rearranged into a flow vs. storage equation for base flow computation using W'_t, the water holding capacity at the beginning of the t-th time step:

$$G_t = \alpha' \ A \ (SM - W'_t)^{\beta'} \tag{7}$$

where G_t is the base flow during the t-th time interval in cfs, A is the drainage basin area in square miles, W'_t is the basin water holding capacity at the beginning of the t-th time interval in inches, and α' and β' are empirical coefficients resulting from rearranging Equation (6).

Infiltration

As a next step, an infiltration model is developed which computes the rate of infiltration during the rainfall-runoff process. In general, the infiltration rate is dependent on physical soil properties, vegetative cover, antecedent soil moisture conditions, rainfall intensity and the slope of the infiltrating surface. A detailed discussion of these factors can be found in Linsley et al. (1958). In the past, extensive efforts have been made to develop the mathematical theory of the infiltration process and the subsequent movement of the infiltrated water within the soil (Green and Ampt, 1911; Hillel, 1971). However, due to the spatial non-uniformity of soil and the difficulty of measuring the required physical parameters, the application of theoretically based equations is very limited.

Various empirical formulations were proposed for infiltration computation. Horton (1940) treats the infiltration rate as a function of time but provides no provision for a recovery of the infiltration capacity during periods of low or no rainfall. Holtan (1961) proposed an empirical equation which relates the infiltration rate to the available soil water storage capacity. Holtan's model is able to describe the recovery of infiltration capacity during periods of low or no rain. Aron, et al. (1977) developed a model which relates infiltration to available soil water storage capacity and cumulative rainfall.

In this study, since measurements of infiltration are not available, the (apparent) infiltration for each selected storm is first estimated (heuristically) so that the observed surface runoff is best reproduced. These (apparent) infiltrations and their corresponding basin WHC and rainfall increment are then fitted for all selected storms with various mathematical functions using a nonlinear least-square method. The best fit of the data was achieved with the infiltration model:

$$F_t = (a \ P_t + b) \ W'^{(c \ P_t + d)}_t \tag{8}$$

where a, b, c and d are empirical coefficients determined by a non-linear least-square method. Normally, heavier rainfall saturates the surface soil layer faster than lighter rainfall. The saturated layer has a low infiltration capacity which results in runoff rates being near the rainfall rates (Hann and Barfield, 1978). Equation (8), however, states that the infiltration rate increases with the rainfall rate. The (apparent) infiltration used in the model is actually a combination of infiltration directly from rainfall and from the depression storage. The contribution from depression storage increases as the rate of rainfall increases until depression storage reaches its capacity. The applicability of equation (8) is limited to storm events with rainfall rates within the range and for the time step (two hours) used in deriving the model. Adjustments of the infiltration model may be necessary as more storm events are being tested.

Evapotranspiration

There are two ways of evaluating evapotranspiration for use in a rainfall-runoff model. One is calculating evapotranspiration with a mathematical model such as the Penman-Monteith model (Monteith, 1965) or the thermal budget method (TVA, 1978). This approach requires meteorological inputs such as air temperature, dew point temperature, wind speed, cloud cover, etc., as well as information on surface vegetation. The second approach approximates the actual evapotranspiration with historical pan evaporation data available at land-pan stations in or near the basin. As a first approximation (and also to reduce data requirements in real time operation), the second approach is used here.

Surface Runoff Routing

In a headwater basin, runoff translation is accomplished by converting the basin rainfall excess hyetograph into a runoff hydrograph at the basin outlet. The runoff hydrograph is the time distribution of the flow rate that results from a particular rainfall excess hyetograph. It is mainly a function of the basin surface drainage characteristics. By definition, the volume of the rainfall excess is equal to the total surface runoff volume.

Numerous methods for surface runoff routing have been proposed. The most commonly used methods are the unit hydrograph method (or convolution method) and the kinematic flow approximation method. The selection of a routing method depends on many factors, such as available computer resources, quantity and quality of input data, required modeling accuracy, etc. River channel geometries are not available in most headwater basins. In addition, most of them have only one flow gage at the basin outlet. The method of the unit hydrograph is therefore used in this study. The method is based on several assumptions. They include uniform (or a specified non-uniform) spatial distribution of rainfall excess over the watershed and uniform rate of rainfall excess during the computation time step. However, the most important assumption is that of linearity, meaning that the time distribution of the runoff flow rate is proportional to the rainfall excess volume of the computation time step.

Rainfall excess is translated to the basin outlet using a set of unit responses derived from the basin area-distance distribution. The area-distance distribution is a plot of accumulated drainage area versus channel distance from the mouth. The distribution can be constructed using an area topomap. Figure 4a shows such an area-distance distribution for a hypothetical drainage basin.

For a given water travel speed, v, the area-distance distribution can be converted to an area-time distribution by dividing the distance by v (see Figure 4b). The area-response of an area-time distribution is obtained by plotting the gradients of the area-time distribution versus time. The area-response is normalized so that it has an area of one. This produces an instantaneous unit response (IUR) with ordinates in units of time^{-1} shown by the dashed line in Figure 4c. The IUR can be used for runoff translation if all the water

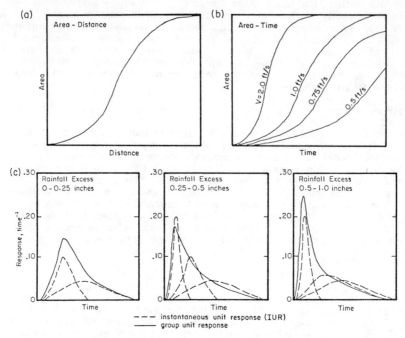

FIGURE 4. (a) AREA - DISTANCE DISTRIBUTION
(b) AREA - TIME DISTRIBUTION
(c) GROUP UNIT RESPONSES

travels with the same speed. Actually, basin runoff consisting of flows varying from laminar overland flow to turbulent channel flow does not travel with the same speed. So, instead of one single IUR, the sum of several IUR's each with different speed and weighted by the fraction of the total flow volume attributed to that speed is used. The resulting group unit response is shown by the solid line in Figure 4c. The share of flow volume attributed to each speed is determined by trial-and-error until the group unit response best reproduces the (calibration) storm hydrographs.

One well-recognized problem with using unit response (or unit hydrograph) for runoff translation is its inability to account for wave celerity changes with flow, especially when a wide range of flow is involved in the application. To circumvent this problem, a multiple linear method similar to that developed by Keefer and McQuivey (1974) is used which divides rainfall excesses by size into several groups. Within the group, linearity is assumed and a group unit response is identified. In Figure 4c, three groups each consisting of a different combination of IUR's are defined. For the group with large rainfall excess, a large portion of the flow will travel with high speed while, for the group with small rainfall excess, most of the flow will travel with low speed. The runoff hydrograph at the basin outlet is computed by convoluting rainfall excesses and group unit responses.

USE OF THE WHC MODEL

General

The first step is the computation of the areal rainfall hyetograph for the basin by the Thiessen Polygon Method. Then, the antecedent base flow is estimated from the observed streamflow. In case of an extended recession, the antecedent baseflow is approximately equal to the observed streamflow. In cases where flow is influenced by a prior storm, the antecedent baseflow is estimated from the observed streamflow using a baseflow separation technique. The initial basin WHC is then computed from the antecedent base flow using equation (6). Successive base flows and infiltrations are determined by equations (7) and (8), respectively. As a first approximation, the rate of evapotranspiration is assumed to be equal to the monthly mean of historical pan evaporation. Changes in the basin WHC and rainfall excess are computed for successive time increments by means of equations (1) and (2). The two-hourly rainfall excess hyetograph is then transformed into a surface runoff hydrograph at the basin outlet using a family of unit response functions described previously. Due to lack of knowledge about the time distribution of base flow, the computed base flow for each time period is used without adjustment. The flow hydrograph is obtained as the sum of surface runoff and base flow.

Operational Use

For operational flow forecasting, the basin WHC is adjusted at the beginning of each forecast using the most recently observed (or estimated) base flow by equation (6). This procedure enables the operator to tune the model to the current soil moisture conditions. A flowchart which illustrates the operation of the model is shown in Figure 5.

Application to the Emory River Basin

The WHC model was tested for predicting surface runoff and streamflow hydrographs for two river basins in the Tennessee Valley region. The first is the Emory River basin with a catchment area of 764 sq miles (1979 km^2). As shown in Figure 6, the available rain gage network consists of five daily reporting (nonrecording) stations and one radio gage (reporting every two hours). Streamflow is monitored at Oakdale, the basin outlet.

The basin WHC curve, Figure 3, and the basin (apparent) infiltration model (see Table 2) were derived from 21 storm events (see Table 1) ranging from 1.10 inches (27.9 mm) to 4.58 inches (116.3 mm). The (apparent) infiltration rate is bounded by the rate of precipitation. The WHC model was tested on a separate set of data containing nine storms with rainfalls from 1.48 inches (37.6 mm) to 3.09 inches (78.5 mm). Storms with rainfall outside this range remain to be tested. The initial soil moisture storage was determined using the observed antecedent base flow. The computed and observed flow volumes for all test events are given in Table 2. Except for storm 5, the WHC model performed satisfactorily. The average error (excluding storm 5) in terms of total flow volume is about ±6 percent. The model tends to underestimate the total flow volume and peak flow for storms which generate high flow rates. This is probably due to the time step (two hours) used for the input rainfall which is not detailed enough to

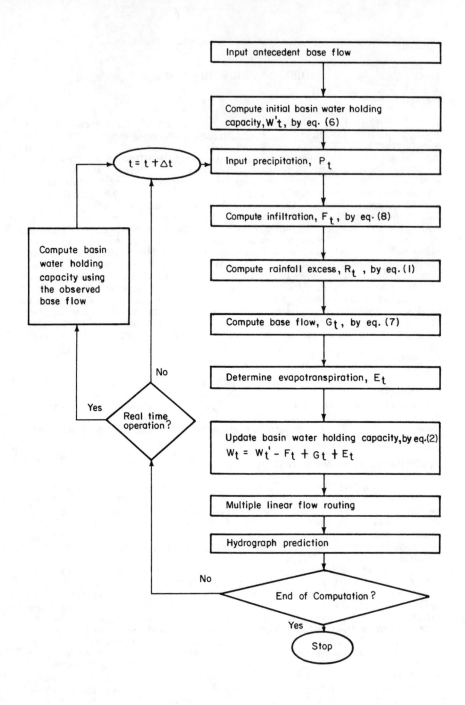

FIGURE 5. FLOWCHART OF THE WHC MODEL

154

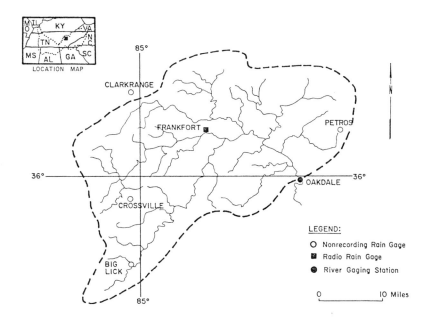

FIGURE 6.   THE EMORY RIVER CATCHMENT

Table 1.   Calibration Storm Events For The Emory River Basin

Storm No.	Date	Rainfall Duration (hours)	Calculated Basin Rainfall Depth (in)	Observed Peak Flow (cfs)
1	May 18, 1961	7	1.24	5850
2	Jan. 5-6, 1962	19	1.25	9700
3	Apr. 28-29, 1963	47	2.54	7200
4	Aug. 16, 1964	9	2.53	4600
5	Apr. 12-13, 1966	19	2.19	22040
6	Dec. 8-9, 1966	26	2.41	23500
7	Jan. 26-27, 1967	9	1.36	9850
8	June 3-4, 1967	19	1.37	3900
9	Mar. 11-12, 1968	19	2.50	35000
10	Mar. 21-22, 1968	23	1.77	13700
11	June 23, 1969	8	4.58	32500
12	Apr. 1-2, 1970	19	2.32	24000
13	Oct. 14, 1970	7	1.50	3600
14	Jan. 4, 1971	15	1.39	13820
15	Feb. 4-5, 1971	13	1.73	27100
16	Oct. 18-19, 1972	23	2.60	15600
17	Sept. 22-23, 1975	35	3.60	25040
18	Apr. 22-23, 1977	24	2.35	14700
19	Dec. 24, 1977	1	1.32	9200
20	Jan. 25, 1978	7	1.10	32600
21	June 8, 1978	13	2.18	15150

Table 2. Summary of Model Testing for the Emory River Basin

Storm No.	Storm Date	Calculated Basin Rainfall Depth (in)	Total Basin Flow Obs. (in)	Total Basin Flow Comp. (in)	Peak Flow Obs. (cfs)	Peak Flow Comp. (cfs)	Error Total Flow (%)	Error Peak Flow (%)	Hydrograph Comparison VRE[1]	Hydrograph Comparison REM[2]	Hydrograph Comparison ABSEM[3]
1	May 15-25, 1975	3.09	1.51	1.67	10670	14721	10.6	38.0	0.54	0.11	0.30
2	Oct. 16-25, 1975	2.81	1.94	1.85	34600	27900	-4.6	-19.4	0.27	-0.05	0.16
3	Nov. 12-19, 1975	2.58	2.23	2.15	43000	30572	-3.6	-28.9	0.39	-0.04	0.18
4	Mar. 12-19, 1977	2.00	1.89	1.65	28000	18628	-12.7	-33.5	0.53	-0.13	0.31
5	Sept. 26-Oct. 2, 1977	2.61	0.68	0.98	5200	10768	44.1	107.1	1.18	0.44	0.72
6	Nov. 20-27, 1977	2.81	2.46	2.45	22600	20367	-0.4	-9.9	0.22	0	0.17
7	May 12-22, 1978	1.84	1.54	1.46	9400	11817	-5.2	25.7	0.52	-0.05	0.36
8	Mar. 3-11, 1979	1.48	1.66	1.64	18500	18495	-1.2	0.0	0.34	-0.01	0.25
9	Apr. 12-19, 1979	1.61	1.63	1.67	10600	14705	2.5	38.7	0.48	0.03	0.36

[1]Coefficient of variation of residual of errors
[2]Ratio of relative error to mean
[3]Ratio of absolute error to mean

Water holding capacity vs. base flow relationship:

$$W_t = 4.0 - 2.2\, q_t^{0.37}$$

$$G_t = 0.12\, A\, (4.0 - W_t')^{2.7}$$

with W_t in inches; q_t in cfs/sq miles; G_t in cfs; A in sq miles.

Basin (apparent) infiltration model:

$$F_t = (0.34\, P_t + 0.02)\, W_t\, (0.14\, P_t + 1.14)$$

with F_t and P_t in inches/hour.

156

simulate the short duration downpour that may have caused the peak flow.

The 44 percent error in flow volume of storm 5 was further examined. The rainfall data revealed that storm 5 was a convective-type storm with heavy rainfall on one part of the basin and light or no rainfall on the rest. Since the Thiessen Polygon Method makes the assumption of uniform rainfall distribution in each polygon according to its representative station, the entire area of each polygon is assumed to receive precipitation and to contribute runoff. In reality, only a portion of the basin may have actually contributed. Thus, areal rainfall was probably overestimated for storm 5. Consequently, the model overestimated the flow volume.

Observed and predicted flow hydrographs of three selected events are shown in Figures 7-9. The figures permit the model performance to be judged visually. In addition, three verification criteria for evaluating the model's accuracy with respect to the observed data were computed:

Coefficient of variation of residual errors

$$VRE = [\Sigma(y_c-y_o)^2/n]^{1/2}/\bar{y}_o \qquad (10)$$

Ratio of relative error to mean

$$REM = \Sigma(y_c-y_o)/(n\bar{y}_o) \qquad (11)$$

Ratio of absolute error to mean

$$ABSEM = \Sigma|y_c-y_o|/(n\bar{y}_o) \qquad (12)$$

where y_c is the computed flow, y_o is the observed flow, $\bar{y}_o$ is the mean of the observed flows ($=\Sigma y_o/n$), and n is the total number of observa-

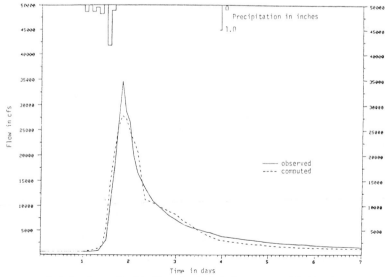

FIGURE 7. HYDROGRAPH FOR THE EMORY RIVER AT OAKDALE, OCTOBER, 16-25, 1975

157

tions.   These criteria are listed in Table 2.   The numbers are similar for storms 1 through 9 with the exception of storm 5.   Storm 5 stands out as the worst fit.   Figures 7 and 8 show a high flow and a medium flow storm (storms 2 and 6), respectively, with two distinct rainfall patterns.   Storm 5 is shown in Figure 9.   As discussed before, both the areal rainfall and its temporal distribution were probably miscalculated for this storm.

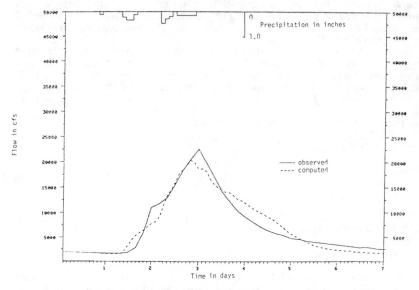

FIGURE 8,   HYDROGRAPH FOR THE EMORY RIVER AT OAKDALE, NOVEMBER 20-27, 1977

Application to the Doe River Basin

The second catchment tested was the Doe River basin.   It has a catchment area of 137 square miles (355 km^2).   The raingage network consists of three daily reporting stations and one radio gage, Figure 10.

The basin WHC curve, Figure 11, and the basin (apparent) infiltration model (see Table 4) were developed based on 26 storm events (see Table 3) ranging from 0.8 inches (20.3 mm) to 4.1 inches (104.1 mm).   A separate set of 10 storm events was used for model testing.   They range from 0.75 inches (19.1 mm) to 3.65 inches (92.7 mm).   The observed and computed flow volumes and the goodness-of-fit criteria for hydrograph comparison are presented in Table 4.   The average total flow volume error (excluding storms 4 and 8) is approximately ±10 percent.   The large volume errors for storms 4 and 8 (40 percent and 43 percent, respectively) are again attributed to the non-uniform areal rainfall distribution of convective-type storms.   The size of the convective storm error is of the same magnitude as that of the Emory River basin.   This is somewhat unexpected since the polygon size of each individual rain gage is much smaller in the Doe River basin (average approx. 35 sq miles [91 km^2]) than in the Emory River basin (average approx. 130 sq miles [337 km^2]).   A comparison of the goodness-of-fit criteria computed for both basins indicates that they are of the same magnitude.   The results obtained so far indicate that the

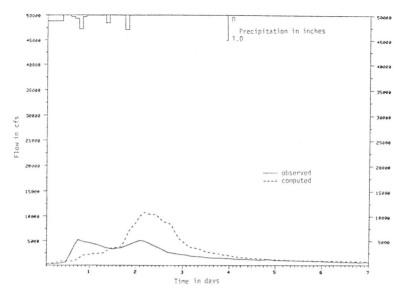

FIGURE 9.   HYDROGRAPH FOR THE EMORY RIVER AT OAKDALE,
SEPTEMBER 26-OCTOBER 2, 1977

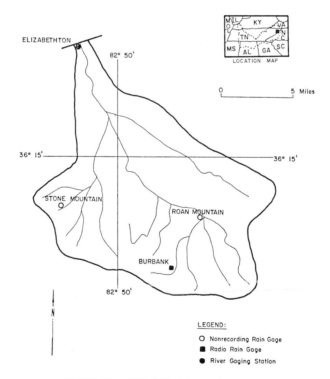

FIGURE 10.   THE DOE RIVER CATCHMENT

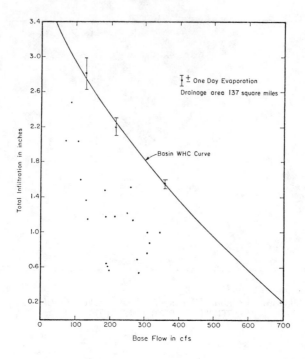

FIGURE 11.   TOTAL INFILTRATION VS ANTECEDENT BASE FLOW
FOR THE DOE RIVER BASIN

size of the catchment has no discernible effect on the performance of
the model.
      For visual comparison, four predicted and observed hydro-
graphs are shown in Figures 12-15. The agreement between the observation and prediction in Figures 12 and 14 is satisfactory. Again, the
effect of the convective storm on the computation of areal rainfall and
its temporal distribution can be seen in Figures 13 (storm 4) and 15
(storm 8). The model results are based on the temporal rainfall distribution recorded at the radio gage at Burbank. Given its eccentric
location with respect to the basin (see Figure 10) and the possibility of
local effects (mountains), it is quite possible that sometimes rainfall
recorded at the radio gage does not adequately represent the temporal
rainfall distribution for the basin.
      In both Emory and Doe River applications, the computed flow
in the recession portion of the hydrograph is frequently lower than the
observed one. The discrepancies in the observed and computed
recession hydrographs are small and are not considered critical for the
use of the model.

Table 3. Calibration Storm Events For The Doe River Basin

Storm No.	Date	Rainfall Duration (hours)	Calculated Basin Rainfall Depth (in)	Observed Peak Flow (cfs)
1	July 31-Aug. 1 1961	9	1.74	5600
2	Nov. 9-10, 1962	24	2.29	1375
3	Feb. 2, 1963	11	0.80	1095
4	Mar. 5-6, 1963	21	2.27	5390
5	Jan. 24-25, 1964	16	0.91	993
6	Apr. 7, 1964	11	1.38	2010
7	Oct. 16, 1964	21	3.77	5460
8	Mar. 25-26, 1965	26	2.71	7479
9	Feb. 12-13, 1966	21	2.48	4510
10	Mar. 3-4, 1966	6	0.92	1260
11	July 30, 1966	5	1.40	1700
12	Oct. 18-19, 1966	15	1.81	1341
13	Feb. 17-18, 1967	25	1.70	1233
14	Mar. 6-7, 1967	7	1.08	1760
15	Dec. 22, 1967	9	0.81	1186
16	Mar. 31-Apr. 1, 1968	9	1.72	2372
17	Oct. 19, 1968	7	1.72	1106
18	Apr. 1-2, 1970	22	1.40	2660
19	Apr. 27-28, 1970	24	2.08	1800
20	Jan. 22-23, 1971	16	1.66	2069
21	Jun. 20-21, 1972	27	3.46	3695
22	Mar. 15-17, 1973	34	4.10	5545
23	Dec. 27-28, 1974	15	1.31	1981
24	Mar. 12-14, 1975	38	2.90	4757
25	Mar. 12, 1977	1	1.38	2624
26	Sept. 7-8, 1977	10	2.48	1910

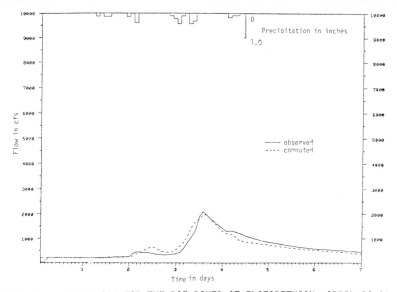

FIGURE 12. HYDROGRAPH FOR THE DOE RIVER AT ELIZABETHION, APRIL 24-30, 1973

161

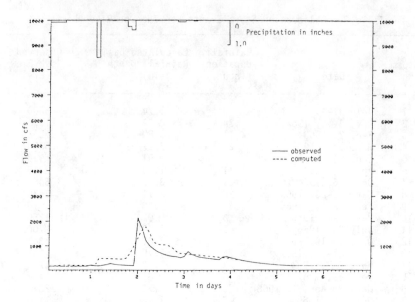

FIGURE 13.   HYDROGRAPH FOR THE DOE RIVER AT ELIZABETHION,
JULY 9-12, 1971

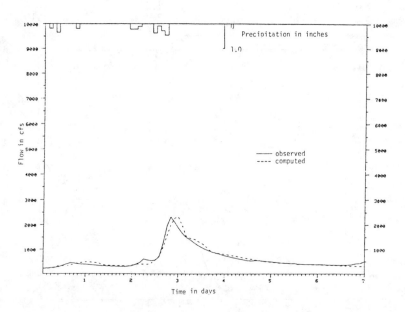

FIGURE 14.   HYDROGRAPH FOR THE DOE RIVER AT ELIZABETHION,
MARCH 10-16, 1968

Table 4. Summary of Model Testing for the Doe River Basin

Storm no.	Storm Date	Calculated Basin Rainfall Depth (in)	Total Basin Flow Obs. (in)	Total Basin Flow Comp. (in)	Peak Flow Obs. (cfs)	Peak Flow Comp. (cfs)	Error Total Flow (%)	Error Peak Flow (%)	Hydrograph Comparison VRE[1]	Hydrograph Comparison REM[2]	Hydrograph Comparison AMSEM[3]
1	Feb. 2-8, 1975	2.28	1.24	1.34	1122	1540	8.1	37.3	0.49	0.08	0.32
2	May 29-June 3, 1974	1.91	1.01	1.09	1398	1881	7.9	34.5	0.36	0.08	0.29
3	Apr. 24-30, 1973	2.86	1.19	1.23	2102	2262	3.4	7.6	0.21	0.03	0.14
4	July 9-12, 1971	2.71	0.50	0.70	2135	1882	40.0	-11.9	0.65	0.38	0.46
5	Mar. 10-16, 1968	2.39	1.14	1.20	2324	2408	5.3	3.6	0.20	0.05	0.13
6	Aug. 22-26, 1966	0.75	0.58	0.51	944	746	-12.1	-21.0	0.37	-0.12	0.24
7	Apr. 6-14, 1965	2.63	2.13	1.84	2005	2670	-13.6	33.2	0.59	-0.14	0.37
8	May 15-24, 1963	3.65	1.32	1.89	1821	2871	43.2	57.7	0.77	0.43	0.44
9	Jan. 22-Feb. 5, 1962	3.38	2.94	2.47	1580	1782	-16.0	12.9	0.28	-0.16	0.18
10	Feb. 20-Mar. 1, 1961	2.96	1.92	2.10	1710	2400	9.4	40.4	0.37	0.09	0.24

[1] Coefficient of variation of residual of errors

[2] Ratio of relative error to mean

[3] Ratio of absolute error to mean

Water holding capacity vs. baseflow relationships:

$$W_t = 4.0 - 1.3 \, q_t^{0.66}$$

$$G_t = 0.67 \, A \, (4.0 - W_t')^{1.51}$$

with W_t in inches; q_t in cfs/sq miles; G_t in cfs and A in sq miles.

Basin (apparent) infiltration model:

$$F_t = (0.34 \, P_t + 0.04) \, W_t^{(1.04 \, P_t + 0.31)} \quad \text{with } F_t \text{ and } P_t \text{ in inches/hour.}$$

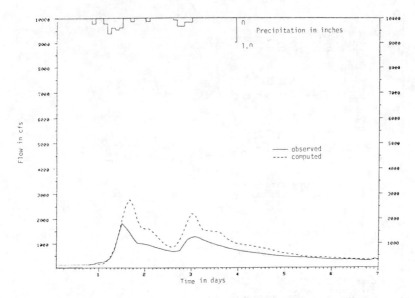

FIGURE 15.   HYDROGRAPH FOR THE DOE RIVER AT ELIZABETHION,
MAY 15-24. 1963

CONCLUSIONS AND RECOMMENDATIONS

A simple conceptual model was developed for operational streamflow prediction. The model uses the basin water holding capacity (WHC) for soil moisture accounting. The basin soil moisture storage at the beginning of a storm, represented by the initial basin WHC, is determined by the antecedent base flow estimated from the observed streamflow. Both infiltration and baseflow are computed as functions of the basin WHC. The runoff hydrograph is obtained by using a of set group unit responses—one for each range of rainfall excesses. The model can be used for streamflow computation for a specific historical storm event as well as for operational day-to-day streamflow forecasting.

The model inputs are two-hourly precipitation and mean monthly pan evaporation. The model is tailored to the basin for which streamflow forecasting is required. This requires developing the basin WHC curve, the basin infiltration equation, and a set of unit response functions for the basin. However, once they were developed, they should remain usable unless there is a major physical change within the basin, such as a large scale land cover or land use change.

The model was applied to two watersheds in the Tennessee River basin: the Emory River basin with an area of 764 sq miles (1979 km^2), and the Doe River basin with an area of 137 sq miles (355 km^2). The test data contained 9 rainfall events for the Emory River basin and 11 rainfall events for the Doe River basin. Despite the simplifying assumptions, the model performed satisfactorily in both basins except for three events dominated by local storms. The average error in terms of total flow volume is about ±6 percent in the Emory River basin and about ±10 percent in the Doe River basin. In the case of local storm dominated events, the errors amounts to about 40 percent.

Some recommendations for further development of the model are as follows:

1. A procedure to include preceding rainfall in addition to antecedent base flow in soil moisture storage estimation should be developed.

2. A procedure of assessing the seasonal variation of interception and depression storage should be developed.

3. For real time operational streamflow forecasting, a method which provides a reliable estimate of baseflow from the rising hydrograph should be developed.

4. For ungaged watersheds, e.g., direct reservoir drainage areas where the basin runoff drains into a reservoir, the relationship between the basin WHC and base flow cannot be established directly. A method which allows the transfer of basin WHC using results from nearby gaged basins should be developed.

5. The (apparent) infiltration model should be expanded to storms with rainfalls higher than those of calibration storms.

6. To improve the performance of the model for convective storms, subbasin rainfall/runoff computation should be investigated.

7. The sensitivity of the hydrograph response for similar storm events but various antecedent base flows should be tested.

8. The use of a different time step for excess rainfall calculation and basin WHC updating should be investigated.

REFERENCES

Aron, G., A. C. Miller, Jr., and D. F. Lakatos, 1977. Infiltration Formula Based on SCS Curve Number, ASCE Journal of Irrigation and Drainage, No. IR4, pp. 419-427.

Carlson, C. A., 1959. Approximation of the Field Maximum Soil Moisture Content, Soil Science Society of America Proceedings, Vol. 23, No. 6, pp. 403-405.

Green, W. H., and G. A. Ampt, 1911. Studies of Soil Physics I - The Flow of Air and Water Through Soils, Journal of Agricultural Science, Vol. 4, pp. 1-24.

Hann, C. T., and B. J. Barfield, 1978. Hydrology and Sedimentology of Surface Mined Lands, Office of Continuing Education and Extension, College of Engineering, University of Kentucky, Lexington, Kentucky.

Hillel, D., 1971. Soil and Water: Physical Principles and Processes, Academic Press Book Co., New York.

Holtan, N. H., 1961. A Concept for Infiltration Estimates in Watershed Engineering, Agricultural Research Service, ARS 41-51, 25 pp., U.S. Department of Agriculture.

Horton, R. E., 1940. Approach Toward a Physical Interpretation of Infiltration Capacity, Soil Science Society of America Proceedings, Vol. 5, pp. 339-417.

Keefer, T. N., and R. S. McQuivey, 1974. Multiple Linearization Flow Routing Model, ASCE Journal of Hydraulics, Vol. HY7, pp. 1031-1046.

Linsley, R. K., Jr., M. A. Kohler, and J.L.H. Paulhus, 1958. Hydrology for Engineers, McGraw-Hill Book Co., New York, pp. 170-172.

Longwell, T. J., W. L. Parks, and M. E. Springer, 1963. Moisture Characteristics of Tennessee Soils, Agricultural Experiment Station, Bulletin 367, 46 pp., University of Tennessee, Knoxville, Tennessee.

Monteith, J. L., 1965. Evaporation and Environment, in G. E. Fogg (Editor), The State and Movement of Water in Living Organisms, Symposia for the Society for Experimental Biology, Vol. 19, pp. 205-234.

Tennessee Valley Authority, WRMM Staff, 1978. Computation of Evapotranspiration, Water Systems Development Branch, Report No. WM28-2-500-112, 12 pp., Knoxville, Tennessee.

SMAP - A SIMPLIFIED HYDROLOGIC MODEL

João Eduardo G. Lopes
Engineer
Dept. de Águas e Energia Elétrica
R. Riachuelo, 115 - 30 and.
01007 - São Paulo - SP - Brazil

Benedito P. F. Braga Jr.
Assistant Professor
Escola Politécnica - USP
05508 - São Paulo - SP - Brazil

João Gilberto L. Conejo
Chief Engineer
Dept. de Águas e Energia Eletrica
R. Riachuelo, 115 - 30 and.
01007 - Sao Paulo - SP - Brazil

ABSTRACT

In developing nations the scarcity of hydrologic information poses a difficult problem to the hydrologist who is interested in the generation of streamflow series through the use of a sophisticated rainfall-runoff model. The difficulty is mainly related to the time interval required for the calculations in detailed conceptual models. In general the time interval used in these models is the hour or even a fraction of the hour. The availability of information in such detail is extremely rare and most of the time daily streamflow records are the finest time interval operationally available to the hydrologist.

This paper presents a simplified rainfall-runoff model whose parameters are related to the physical characteristics of the watershed. The model works on a daily basis which is consistent with the availability of data in less developed countries. The number of parameters that depends on the calibration procedure is kept to a minimum to avoid problems associated with the short length of the streamflow records that are generally available.

The model uses the concept of curve numbers from the Soil Conservation Service procedure to define the soil retention capacity of the unsaturated zone. Moisture accounting is continously performed from initial conditions given to the model. At each day moisture updating of the unsaturated zone is done by computing infiltration through the SCS runoff equation. Recharge to the saturated zone is done using the concept of field capacity. The advantage of the model is that the parameters involved in the updating procedure are easily defined in terms of the soil and vegetation cover characteristics alone.

An application is described for three rural watersheds of different characteristics located in southern Brazil. The results have indicated that the methodology is adequate. The error in terms of annual and monthly volumes for the calibration period of 4 years was in the range of 20 percent. The model was applied to watersheds with drainage areas smaller than 3,500 square kilometers.

INTRODUCTION

In recent years the hydrologic community has been flooded by a countless number of hydrologic continous simulation models. In general these models have two basic disadvantagens when applied to less developed countries. The first is their complexity which requires a large number of parameters to be estimated from a small record of observed flows and the second is the time step (hour or fraction of the hour) used in the computations which is not easily found in operational form. Some sophisticated models poses both problems (Crawford and Linsley, 1966; U.S. Army, 1972, EPA, 1971), others less sophisticated in terms of time steps still have many parameters (Mero, 1969). Kraeger (1971) proposed a method to deal with the problem of small time steps through disaggregation of daily precipitation into hourly amounts. This statistical procedure gives good results when precipitation is relatively homogeneous as in the case of temperate climates. In the tropics, however, the predominance of convective rainfall poses some difficulties in the application of the approach.

In this paper is presented an alternative hydrologic model which has a relatively simple structure and operates with daily rainfall and mean monthly potential evapotranspiration inputs. The model was developed in such a way to minimize the subjectivity in the calibration process. Actually there is just one parameter that does not have a direct physical estimator. This parameter is associated with the groundwater recharge and some indications for a first guess are shown in the case study presented at the end of the paper.

The model utilizes a soil moisture accounting procedure that is based on two linear reservoirs representing the unsaturated and saturated zones. Infiltration is taken into account by the Soil Conservation Service (SCS) runoff equation. In the following sections a brief review of the SCS method is presented together with a description of the accounting procedure. Finally, three applications of the proposed model to rural watersheds in the State of São Paulo, Brazil are discussed.

MODEL DEVELOPMENT

The Soil Conservation Service Technique

The Soil Conservation Service (SCS) runoff curve number technique allows the obtaining of runoff volumes from the equation below (Soil Conservation Service, 1975):

$$Q = (P - IA)^2/(P - IA + S) \qquad (1)$$

where Q is the direct runoff, P is precipitation, IA is the initial abstraction and S is the potential abstraction, all in milimeters.

Potential abstraction (in mm) is calculated by

$$S = 25.4 \left[(1,000/CN) - 10\right] \qquad (2)$$

with CN defined as the curve number, related to the soil, vegetation, cover and antecedent moisture conditions of the watershed. Estimates of CN can be obtained for different antecedent soil moisture conditions (dry, normal and wet), through the use of tables (Soil Conservation Service, 1975). Setzer and Porto (1979) in a recent study for the State of São Paulo, Brazil has adapted those tables for the semi-tropical conditions of watersheds of this part of the world. Initial abstraction depends on the land cover conditions and estimates for different situations are easily found elsewhere (e.g. Haan, 1976).

For a given precipitation over the watershed the use of the SCS runoff curve number technique produces direct runoff volumes. This method is event based, restricted to three basic soil moisture conditions and to obtain the event hydrograph one has to add the base flow component. Despite these limitations, it has been used worldwide with acceptable results. Its simplicity and suitability for application to practical problems suggests the use of the SCS concepts to generate direct runoff in hydrologic continuous simulation models. To calculate base flow and update soil moisture conditions the soil and groundwater components have to be simulated.

Soil Moisture Accounting Procedure

Continuous hydrologic simulation models are usually a combination of transfer functions and reservoirs that simulates the natural phenomena. They differ in the transfer functions and (or) the number or type (linear, non linear) of reservoirs. In this paper, the choice was to use three linear reservoirs representing the surface, soil and aquifer retention characteristics (Figure 1). Transfer functions are treated preferentially as linear functions to save computer time whenever the assumption of linearity is acceptable.

A fraction of precipitation (P) is conveyed as direct runoff (DR) by the SCS runoff curve numbers technique (equation 1), to a linear reservoir that routes this water to the basih outlet (RDR). The remaining water depth (P - DR) is depleted at the potential evapotranspiration rate (PE). Excess water (P - DR - PE) enters a linear reservoir representing the upper soil horizon (unsaturated zone). Moisture is lost from this zone at an evapotranspiration rate (AE) proportional to the rate of filling (RF) of the reservoir (actual level divided by the maximum level) times the potential evapotranspiration (PE).

The output from this reservoir is the recharge to the groundwater reservoir. The concept of field capacity (FC) is used in this transfer function. If the actual level (RSOIL) of the soil reservoir is higher than field capacity, recharge (REC) occurs and is equal to:

$$REC = (RSOIL - FC) \times RF \times CREC \qquad (3)$$

where RF = RSOIL/SAT is the rate of filling of the soil reservoir, SAT is the maximum soil water content (saturation capacity) and CREC is a recharge coefficient, an input parameter representing the rate of depletion of the soil reservoir. Water transferred to the aquifer below is regulated by this parameter.

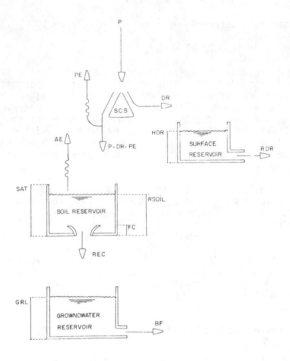

Figure 1. The Soil Moisture Accounting Model.

The groundwater reservoir is another linear reservoir whose depletion characteristic is that of the observed base flow hydrograph. Discharge is obtained as the addition of the routed direct runoff and the base flow component (BF) (output of the groundwater reservoir).

Following the above procedure the levels in each reservoir are continuously updated for each time step.

This updating enables the use of a variable potential abstraction (S) in Equation 1. Accordingly, S is computed by substracting from the saturation capacity the actual soil reservoir level at the time step being considered (Eq. 4). It should be noticed that RSOIL is the actual reservoir level that is used to calculate the potential abstraction (S) for the next time interval.

MODEL PARAMETERS AND OPERATION

As daily data is the most frequent available hydrological information, specially in less developed countries, the model was structured in such a way that the computations are performed on a daily basis. Basic input data to the model are daily precipitation, mean monthly evapotranspiration by month and daily streamflow for the calibration period.

The following ten parameters used in the model are either obtainable directly from data or based on watershed characteristics described below.

AREA (SqKm) - drainage area of the watershed

PCOF - rainfall weighting coefficient relating point rainfall to watershed rainfall, obtainable from Thiessen or isoietal maps.

CN - Soil Conservation Service (SCS) runoff curve number estimated from watershed soil and land cover characteristics and soil moisture condition at the starting time of simulation (obtainable from tables: SCS, 1975: Setzer and Porto, 1979)

IA (mm) - initial abstraction, as defined in Equation 1, is a function of watershed land cover type, including interception and surface retention losses.

FC (percentage) - field capacity of the upper soil horizon,estimated for the whole basin. Can be determined experimentaly from soil characteristics.

CREC - recharge coefficient. A parameter related to the movement of water in the unsaturated soil zone, and therefore a function of soil type, determined during calibration through successive approximations.

K1 (1/day) - base flow recession constant determined from hydrograph analysis (Linsley et al., 1975, pp. 225 - 230).

K2 (1/day) - direct runoff recession constant obtainable from hydrograph separation (Linsley et al., 1975, pp. 225 - 230).

SOLIN (percentage) - initial soil moisture content.

BASIN (CMS) - base flow at starting time of simulation.

As mentioned before the model is composed of three linear reservoirs with separation of direct runoff and infiltration performed by the SCS runoff curve number technique (Equation 1), with S being updated at each time interval by:

$$S = SAT - RSOIL \qquad (4)$$

SAT is the soil saturation capacity. This constant is calculated once for all at the beginning of simulation. At this time RSOIL is exactly equal to SOLIN (the initial soil moisture in percentage) times SAT and S is directly obtained from CN (input parameter) by using Equation 2. From Equation 4 above the saturation capacity (SAT) is obtained by dividing S by (1 - SOLIN). The value of SAT thus obtained is used throughout the simulation as a constant.

Depletion of direct runoff reservoir (RDR) will occur as

$$RDR = (1 - K2) \times HDR \qquad (5)$$

where K2 is the direct runoff recession constant and HDR is the actual level of the surface reservoir. The same procedure is used to deplete the groundwater reservoir, substituting K2 by K1 and HDR by GRL (see Figure 1). The initial groundwater reservoir level (GRL) is obtained through the conversion of the initial base flow input (BASIN) to depth in milimeters as

$$GRL = 86.4 \times BASIN/ \left[A (1 - K1) \right] \qquad (6)$$

171

where BASIN is the given initial base flow in CMS, A is the drainage
area in Square meters and K1 is the base flow recession constant.

The model was designed to continuously simulate the daily mean
flow but can also be used as an event based model. It was developed to
run on small to medium size computers and the present FORTRAN IV version
of the model requires 32 K bytes of core storage.

MODEL APPLICATION

The model was applied to three watersheds of the State of São Paulo,
Brazil: the Pinheirinho river basin with 113 km^2, the Camanducaia river
basin with 928 km^2 and the Jaguari river basin with 3,399 km^2. Figure 2
shows schematically the shape and location of the hydrologic gage
stations. Available data are daily precipitation, mean daily flows and
class A pan evaporation records. Input data for the calibration period
is daily rainfall, average daily streamflow and mean monthly
evapotranspiration by month. Table 1 shows the details of the hydrologic
gage stations used in the simulation.

Vegetation cover of the watersheds is mainly grassland and the soil
type is classified as groups A and B according to the Soil Conservation
Service tables given by Setzer (1979). Hydrometeorological factors show
marked seasonality, with dry winter (June, July) and rainy summer
(December, January, February). Mean annual potential evapotranspiration
varies from 1250 mm (Camanducaia and Jaguari river basin) to 1550 mm
(Pinheirinho river basin). The minimum mean monthly evapotranspiration
is around 2 mm by day (Camanducaia and Jaguari) in September and the

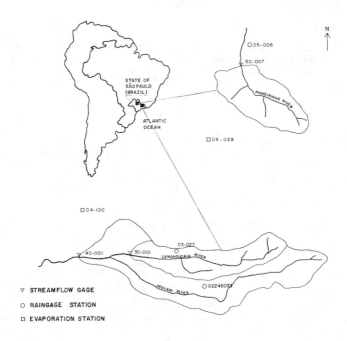

Figure 2. Location of the Watersheds Used in Simulation Runs.

TABLE 1 - Gage Stations Used in the Simulation Runs

Streamflow	Precipitation	Evaporation (Class A)	Drainage Area (km^2)	Calibration period
3D-001 (Camanducaia at Fazenda Barra)	D3-027 (Monte Alegre do Sul)	D4-100 (Campininha)	928	Oct, 1963 to Sept, 1968
4D-001 (Jaguari at Usina Ester)	02246033 (Bragança Pau lista)	D4-100 (Campininha)	3,400	Oct, 1958 to Sept, 1963
5D-007 (Pinheirinho at Usina Três Saltos)	D5-006 (Torrinha at Usina Três Saltos)	D5-028 (Barra Boni-ta)	113	Oct, 1940 to Sept, 1947

maximum mean is about 6,0 mm by day (Pinheirinho) in February. Average annual rainfall is about 1400 mm for the Camanducaia River Basin and for the Jaguari River basin and 1300 mm for the Pinheirinho River basin. In a rainy month the total rainfall can amount to more than 450 mm and in a dry month it could be zero.

Calibration of model parameters is done according to the description given in the previous section. The set of parameters that gave the best results for the calibration period are given in Table 2.

TABLE 2 - Final Model Parameters Used in Simulation

River Basin Station Code	3D-001	4D-001	5D-007
Rain Gauge Station Code	D3-027	02246033	D5-006
Pan Evaporation Station Code	D4-100	D4-100	D5-028
Rain Coefficient (PCOF)	1.0	1.0	1.0
Drainage Area (Area-sgkm)	928.	3399.	113.
SCS Curve Number (CN)	42.	45.	45.
Initial Abstraction (IA-mm)	5.	5.	5.
Field Capacity (FC-%)	0.35	0.40	0.35
Recharge Coefficient (CREC)	0.012	0.009	0.013
Base flow recession (K1-1/day)	0.992	0.993	0.993
Direct runoff recession (K2-1/day)	0.800	0.800	0.700
Initial soil moisture (SOLIN-%)	0.37	0.42	0.38
Initial base flow (BASIN-cumecs)	3.70	33.0	1.00

Results are presented in Figures 3, 4 and 5 in the form of mean monthly flows for the calibration period. Most of the differences between monthly observed and simulated flows do not depart more than 20 percent, which is almost the usual level of dispersion found in basic hydrological data in developing countries. Daily flows have greater discrepancies despite the good agreement in the recession part of the hydrograph between observed and simulated flows, and some peak flows hydrographs that are well close to the observed. The most important factor that explains these discrepancies is the non representativeness of the spatial rainfall over the watershed for many high flows events. Spatial rainfall variability is significant in this area of the world, due to the characteristics of the weather, with cold fronts and convective effects. Since the model uses just one precipitation station the best results were those from the Pinheirinho River Basin which is the smallest one (113 km^2).

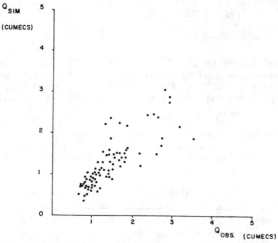

Figure 3. Observed (Q_{OBS}) and Simulated (Q_{SIM}) Mean Monthly Flows for the Pinheirinho River (A = 113 km^2).

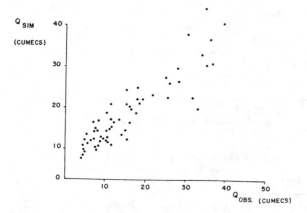

Figure 4. Observed (Q_{OBS}) and Simulated (Q_{SIM}) Mean Monthly Flows for the Camanducaia River (A = 928 km^2).

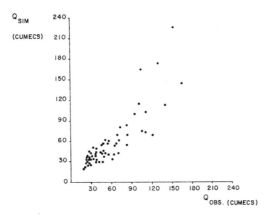

Figure 5. Observed (Q_{OBS}) and Simulated (Q_{SIM}) Mean Monthly Flows for
the Jaguari River (A = 3,400 km^2).

Figure 6 shows the results for the Pinheirinho River Basin in terms
of mean daily flows for a typical year of the simulation period. It can
be observed that the model is adequate and even the peak flows are well
represented.

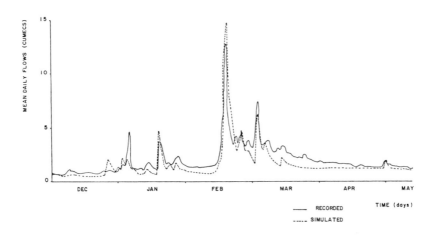

Figure 6. Observed and Simulated Mean Daily Flows for the Pinheirinho
River from Dec, 1946 to May, 1947.

CONCLUSIONS

A model has been developed that is suitable for application when
hydrologic data is scarce. The scarcity of data seriously limits the
application of complex rainfall - runoff models to watersheds in less

175

developed countries. The model presented here can successfully simulate daily flows from daily precipitation and monthly potential evapotranspiration. The hydrologic data used in the applications come from regular stations of the hydrologic network of the State of São Paulo, Brazil. This was done to avoid using experimental basins that are well measured and in general are not representative of the natural environment found in practical engineering applications.

Although no extensive sensitivity analysis has been developed, from the case study presented, it can be seen that the spatial variability of rainfall will limit direct applications of the model to small watersheds (up to 900 km^2). This limitation can be circunvemted, however, by subdividing the watershed into small segments and then apply a convenient routing procedure to route the flows from one segment to the other.

REFERENCES

Crawford, N.H. and Linsley, R.K. 1966. Digital Simulation in Hydrology. Technical Report no. 39, Stanford University, Stanford, California.

Environmental Protection Agency. 1971. Storm Water Management Model. U.S. Government Printing Office, Washington, D.C.

Hann, C.T. 1976. Mini-Course 3 - Urban Runoff Hydrographs. National Symposium on Urban Hydrology, Hydraulics and Sediment Control. Kentucky.

Kraeger, B.A. 1971. Stochastic Monthly Streamflow by Multistation Daily Rainfall Generation. Technical Report no. 152. Stanford University, Stanford, California.

Linsley, R.K.; Kohler, M.A. and Paulhus, J.L. 1975. Hydrology for Engineers. McGraw Hill. New York.

Mero, F. 1969. Application of the Groundwater Depletion Curves in Analysing and Forecasting Spring Discharges Influenced by Well Fields. I.A.H.S. publication no. 63. Sumposium on Surface Water, pp. 107-117.

Setzer, J. and Porto, R.L.L. 1979. Tentativa de Avaliação de Escoamento Superficial de Acordo com o Solo e seu Recobrimento Vegetal nas Condições do Estado de São Paulo. Boletim Técnico no. 2, vol. 2, Departamento de Águas e Energia Elétrica, São Paulo.

Soil Conservation Service. 1975. Urban Hydrology for Small Watersheds. Technical Release no. 55. U.S. Department of Agriculture.

U.S. Army Engineer Division. 1972. Program Description and user Manual for SSAR - Streamflow Synthesis and Reservoir Regulation, U.S. Army, Portland, Oregon.

DAILY SNOWMELT RUNOFF DURING PREMONSOON MONTHS IN BEAS BASIN USING LIMITED DATA

S.M. Seth
Scientist E
National Institute of Hydrology
Roorkee, India

ABSTRACT

The demand for water coupled with construction of reservoirs in Himalayan catchments has increased the requirements of accurate stream-flow forecasts. Usually limited data of daily precipitation and temperature are available at base stations situated at lower elevations. In these catchments the snow accumulation season is from beginning of November to March and the snowmelt season for melting of temporary snowcover is from mid March to May, which is also the premonsoon season. The present study deals with development of a snowmelt runoff model using information regarding the areal extent of permanent and temproary snow covers by comparison of satellite imageries; and observed data of daily precipitation (rain and snow) and daily temperature for premonsoon season. Three years (1977, 1978, 1979) of data from the Beas river catchment upto Manali gauge site (1829 m above m.s.l.) has been utilized in the study for verification of the model. The catchment is divided into four elevation zones at 610 m intervals for temporary snow covered area of 287.68 km^2 up to 4269 m elevation. The remaining area of 81.73 km^2 lying above 4269m constitutes a permanent snow covered area. The altitudinal effect on temperature has been considered by lapse rate. The orographic effect on precipitation has been considered by adding an incremental value of 5 percent for each 305 m rise in elevation, to recorded precipitation at Manali station. The daily snowfall data for November to mid March provides depth of snow cover data and the melt is computed by degree day method. The model considers melt due to rain, losses from meltwater, effect of rain falling on non-snow covered area, and uses simple routing relationship for obtaining daily streamflow at the catchment outlet. The eight parameters representing degree day factors for two parts of the season, losses from snowmelt, rain on snowcovered area, rain on non-snow covered area, lapse rate, melt due to rain, and recession factor are estimated by a pattern search optimization technique using least squares criteria. The model gives encouraging results, as indicated by comparison of observed and computed flows as well as sensitivity analysis. However, due to limitation of data the model could not be verified using a split sample. Use of cross correlation analysis is suggested for examining reasons for good or bad reproduction of observed flows.

INTRODUCTION

Snow is an important part of the hydrological cycle. In many areas it is the dominant source of streamflow. The deposition of snow during the winter months of November to mid-March and the melting of snow during subsequent months resulting in streamflow, forms one of the most important phases of the hydrologic cycle in northern Himalayan region of India. The snowmelt runoff estimation is vitally important in forecasting seasonal water yields regulating the storage reservoirs, estimating design floods, etc. Despite its importance, snow hydrology has not developed to a level where daily streamflow forecasts could be made for snowmelt runoff. However, with the increased demand for water and considerable construction activity of reservoirs in Himalayan catchments, the requirement of accurate forecasts of streamflows has considerably increased, particularly for premonsoon months.

Since the data availability is very limited, using existing level of technology in this area, suitable methodology has to be developed for estimation of daily snowmelt runoff using limited data. Generally data of daily precipitation and temperature are available at base stations situated at lower elevations, where the streamflow is also gauged. In such Himalayan catchments the snow accumulation season is during winter months of November to mid-March. However, the records of snowfall at lower elevations could only provide an index for total snow accumulation, because of orographic effect. Moreover in these catchments some portion at higher altitudes always remains under permanent snowcover, whereas most of area at lower elevations comes under snowcover temporarily during winter months. Due to temperature rise in premonsoon months as well as some rains, the temporary snowcover starts melting gradually from lower elevations. Due to great variation in elevation, it is thus, necessary to consider lapse rate of temperature. Besides these factors, the melt due to rain falling on snow as well as losses from snowmelt and rain on snowcovered areas have to be considered. When temporary snow cover melts, loss from rain on non-snow covered area becomes important.

The degree day method is widely used for snowmelt computations because of its simplicity. When air temperature measurements are the only relevant data available, it is the only method for computation of snowmelt runoff and is appropriate for forested areas. The studies of U.S. Army Corps of Engineers (1956) have shown that basinwide snowmelt increases as the melt season progresses due to heat storage in snowpack and other effects. Recently, with the availability of satellite imageries, the areal extent of permanent snow cover and temporary snow cover can be ascertained by comparison of imageries for beginning of November, mid-March and end of May. Viessman et al (1977) have discussed in detail point and areal snow characteristics, Martinec (1976) has stressed the need for considering differences in temperature and in the snow cover for basins with a great elevation range. He suggests use of a constant recession coefficient for routing the snowmelt runoff. Rango and Martinec (1979) report the application of a snowmelt runoff model using satellite data. They consider lapse rate, runoff from precipitation, simple recession coefficient, division into elevation zones and degree day approach. However, the effect of melt from rain and rain falling on non-snow covered area are not covered.

For the Himalayan catchments a model was developed by Thapa (1980) for his Master's disseration at University of Roorkee. It was tested with data of Beas basin and by trial and error calculations, values of parameters were evaluated. The present study is essentially an extension of this work using general computer program with optimization technique. The main features of the study are discussed in the following paragraphs.

THE MODEL

As discussed in detail by Thapa (1980) the model is based on the following important assumptions (i) division of catchment into altitudinal zones to take into account the orographic effect on precipitation and the lapse rate of temperature, (ii) contribution of snowmelt from both the temporary and permanent snow covers (iii) accumulation season from November to mid-March (or so) and the premonsoon snowmelt season from mid-March (or so) to May, (iv) negligible snowmelt during accumulation season and depletion of snow in different altitudinal zones only during melt season, (v) losses occur from snowmelt as well as rain falling on snow covered area by evaporation and infiltration, (vi) the predominant factors affecting snowmelt rate are the air temperature and rainfall, (vii) there is increase of degree day factor in later part of melt period in premonsoon months, (viii) the melt water can be routed by simple recession relation, and (ix) data available only at base station.

Divisions of Catchment Area

The catchment area is divided into the permanent snow covered zone, temporary snow covered zone and the non-snow covered zone using information from satellite imageries or aerial photographs. Further subdivisions are made for temporary snow covered area and non-snow covered area depending upon elevation differences.

Temperature Change with Elevation

Since the model assumes availability of data only at base station, the temperature at mean heights of each elevation zone is computed by using lapse rate relationship.

$$T_i = T_b - (X2) \ (H_i - H_b) \tag{1}$$

where T_i is the daily mean temperature in $^{\circ}C$ at mean height of ith zone and T_b is that for base station, H_i and H_b are elevations in m and (X2) is lapse rate ($^{\circ}C$ per 305 m).

Orographic Effect on Precipitation

The increase in precipitation with elevation due to orographic effect is considered by adding an incremental value of five percent for each 305 m rise in elevation. Thus knowing the precipitation depth for base station, the same could be computed for mean height of every elevation zone. The value of five percent has been assumed only for Beas basin on the basis of information available.

Degree Day Method

The daily snowmelt depth M_i in metre for elevation zone i is given by

$$M_i = (X3) \ (T_i) \tag{2}$$

179

for first part of premonsoon melt season, and

$$M_i = (X5) \ (T_i) \tag{3}$$

for second part of premonsoon melt season. Where X3 and X5 are parameters representing degree day factors, and T_i is mean temperature in $^\circ$C for ith zone.

Snowmelt due to Rain

The depth of water from snow D_i (m) melted by rain is given by

$$D_i = (X4) \ (R_i) \ (T_i) \tag{4}$$

where R_i is rainfall in the zone in m, and X4 is the parameter.

Daily Snowmelt Discharge for a Zone

Considering the loss from melt water by a runoff coefficient X6, the daily snowmelt discharge Q_i (m^3/sec) for ith zone of area A_i (m^2) will be given as

$$Q_i = (X6) \ (T_i A_i / \ 86400) \ (X3 + X4 \ (R_i)) \tag{5}$$

for first part of premonsoon melt season, and

$$Q_i = (X6) \ (T_i A_i / \ 86400) \ (X5 + X4 \ (R_i)) \tag{6}$$

for second part of premonsoon melt season

Daily Snowmelt Discharge for N Zones

Using recession constant method of routing, the daily snow melt discharge from N zones fully covered by snow is given by

$$(Q_s)_j = \sum_{i=1}^{N} Q_i \ (1-X1) + X1 (Q_s)_{j-1} \tag{7}$$

Where $(Q_s)_j$ and $(Q_s)_{j-1}$ are daily snowmelt discharges in m^3/sec at day j and j-1 respectively and X1 is recession constant.

Contribution to Discharge from Rainfall

The rainfall on temporary snow covered area as well as on non-snow covered area will contribute to runoff at outlet and the parameters X7 and X8 are runoff coefficients respectively for these two cases. If rainfall depth is R_i (m) for zone i, having area A_i (m^2) under temporary snow cover, then discharge in (m^3/sec) is given by

$$(Qts)_i = (X7) \ (R_i) \ (A_i / \ 86400) \tag{8}$$

and if area A_i is non-snow covered, then

$$(Qns)_i = (X8) \ (R_i) \ (A_i / \ 86400) \tag{9}$$

the discharge $(Qts)_i$ for temporary snowcover will be added with Q_i in equation (7). For small catchments the storage effect is negligible for the discharge $(Qns)_i$ for non-snow covered area and it could be added to final $(Qs)_j$ to give total direct runoff at outlet.

THE DATA

Three years (1977, 1978, 1979) date of Beas river catchment up to Manali gauge site, having an elevation of 1829m above mean sea level has been utilised in the study. The satellite imageries indicated an area of 81.73 km^2 lying above 4269 m elevation to be under permanent snow cover. The remaining area of 287.68 km^2 has been divided into four elevation zones at 610 m interval. The sketch of catchment area and area distribution in different elevation zones are given in Fig. 1. The snow accumulation data regarding daily mean temperature TC, rainfall and snowfall was available for

    (i)   3rd December 1976 to 27th March 1977
    (ii)   3rd November 1977 to 12th March 1978
    (iii) 25th November 1978 to 17th March 1979.

The premonsoon snowmelt period of these three years has been taken as follows after considering the distribution of daily mean temperature.

    (i)   28th March 1977 to 30th April 1977 (Ist part)
           Ist May 1977 to 31st May 1977 (2nd part)
    (ii)   13th March 1978 to 30th April 1978 (Ist part)
           Ist May 1978 to 31st 1978 (2nd part)
    (iii) 18th March 1979 to 20th April 1979 (Ist part)
           21st April 1979 to 31st May 1979 (2nd part)

For premonsoon snowmelt season divided into two parts for each of three years the baseflow has been separated from observed runoff and only direct runoff in m³/sec is used for optimization of eight parameters of the model. (Thapa, 1980). The water equivalent of snow has been assumed as 0.10 through the study.

OPTIMIZATION

For optimization of eight parameters X1, X2, X3, X4, X5, X6, X7, and X8 in equations (1) to (9), Pattern Search optimization technique described by Monro (1971) has been used. The programme was written in Fortran IV and run of DEC 2050 computer of Roorkee University Regional Computer Centre. The upper and lower limits for parameter values and initial starting values were adopted on the basis of initial results obtained by Thapa (1980). These are given in Table 1. The objective criteria used for optimization is sum of squares of differences between observed and computed direct runoff.

The objective function FCS is given by

$$FCS = \sum_{I=1}^{M} \; Q\,O(I) - QC(I) \quad ^2 \qquad (10)$$

where QO and QC are observed and computed direct runoffs in m³/sec, I is an index and M is total number of data values. The function FCS1 and FCS2 refer to Ist part of melt season and total melt season respectively,

TABLE 1

STARTING PARAMETER VALUES AND LIMITS

Parameter	Lower Limit	Upper Limit	Starting Value
X1	0.7000	1.0000	0.90000
X2	1.4000	1.6500	1.53000
X3	0.0015	0.0035	0.00180
X4	0.0110	0.0140	0.01250
X5	0.0015	0.0035	0.00315
X6	0.0000	1.0000	0.90000
X7	0.0000	1.0000	0.59500
X8	0.0000	1.0000	0.27800

LEGEND
RIVER
CONTOURS
GAUGING SITE

A. CATCHMENT AREA

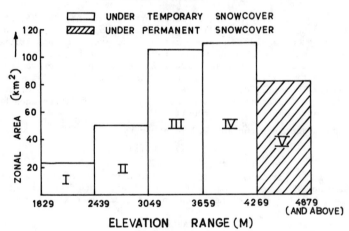

B. AREAS OF ELEVATION ZONES

FIG. 1.   BEAS RIVER BASIN UPTO MANALI GAUGE SITE

SUMQO1, SUMQC1, and SUMQO2, SUMQC2 are corresponding values for sums of observed and computed direct runoffs in m^3/sec.

The optimised parameter values and functions for three data sets are given in Tables 2 and 3.   Their comparison indicates stability except for parameters X6, X7, X8 which represent runoff factors for snowmelt, rain on snow covered area and rain on non-snow covered area respectively.   The observed and computed hydrographs of direct runoff are shown in Figs. 2 and 3.   For 1978 and 1979 the hydrographs compare well.   The mean values of observed and computed flows also agree.   Only for year 1977 the simulation is not that good.   Some of the possible reasons are the effects of baseflow separation, part coverage of catchment by snow at

TABLE 2

OPTIMISED PARAMETER VALUES

Parameter		Optimised values	
	1977 data	1978 data	1979 data
X1	0.8930	0.94500	0.94500
X2	1.4500	1.60650	1.45350
X3	0.0020	0.00153	0.00171
X4	0.0110	0.01128	0.01128
X5	0.0030	0.00347	0.00246
X6	0.3630	0.99225	0.99225
X7	0.3990	0.58235	0.66655
X8	0.1350	0.37279	0.15909

TABLE 3

OPTIMISED FUNCTION VALUES

Function		Values of function	
	1977 data	1978 data	1979 data
FCS1	1299.38	1548.84	913.35
SUMQO1	543.05	718.32	896.07
SUMQC1	492.31	916.57	895.95
FCS2	4237.99	8476.38	2847.75
SUMQO2	1312.93	2818.62	2921.93
SUMQC2	981.27	2583.14	2781.86

the beginning of melt season and relatively more amounts of rainfall during melt season. Further study is needed for going into these aspects.

However, due to limitation of data the model could not be verified using a split sample. It will be necessary to test the performance of the model by verification with independent data from same watershed or a similar watershed. Further data is being collected for this purpose.

The results of sensitivity study changing one parameter value at a time as given in Table 4, shows high sensitivity of results to change in X1 which is recession constant. Further studies are planned with different sets of starting parameter values and some modifications in procedure for accounting of losses from rainfall as well routing of runoff from rainfall on non-snow covered areas.

The results of cross correlation analysis for 1978 data presented in Table 5, indicate possible lines of approach. It is clearly seen that for this data there is strong correlation between observed daily temperature $TC(^{o}C)$ and observed direct runoff $QO(m^3/sec)$. The correlation between TC and computed runoff QC is slightly higher. It could be possible to include correlation between some data values in optimization criteria. Further work is proposed in this direction.

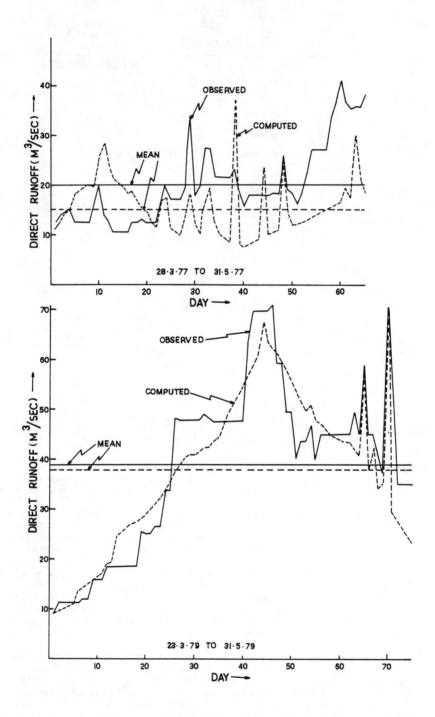

FIG. 2.   COMPARISON OF OBSERVED AND COMPUTED FLOWS FOR 1977 AND 1979

184

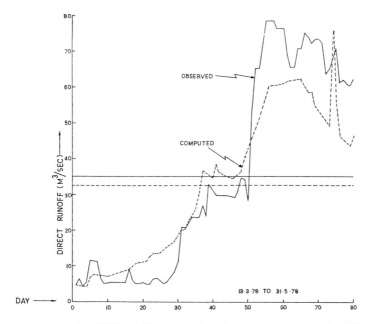

Fig. 3. COMPARISON OF OBSERVED AND COMPUTED FLOWS FOR 1978.

TABLE 4

SENSITIVITY OF OPTIMISED PARAMETER VALUES

Parameter	Values of function FCS2		
	1977 data	1978 data	1979 data
For (+) Five Percent Change			
X1	6312.0	75526.5	40958.7
X2	6658.3	8744.8	2907.8
X3	6678.6	9328.9	2907.9
X4	6480.2	8464.2	2849.8
X5	6136.2	7878.4	2951.6
X6	6529.4	7445.2	2794.2
X7	6462.0	8445.5	2857.9
X8	6489.1	8448.8	2848.4
For (-) Five Percent Change			
X1	6654.1	11816.7	8293.4
X2	6057.3	8761.7	3193.0
X3	6341.0	7736.9	2905.9
X4	6481.9	8421.9	2845.7
X5	6844.5	9063.9	2807.5
X6	6442.8	9743.2	3260.3
X7	6504.2	8447.0	2840.5
X8	6480.6	8441.5	2856.4

185

TABLE 5

RESULTS OF CROSS CORRELATION ANALYSIS FOR 1978 DATA

Correlation Coefficient	Log	Between TC & QO	Between TC & QC	Between QC & QO
(0)	0	0.885	0.904	0.960
(1)	1	0.884	0.903	0.953
(2)	2	0.883	0.902	0.942
(3)	3	0.878	0.908	0.935
(4)	4	0.864	0.905	0.923
(5)	5	0.864	0.901	0.904

CONCLUSIONS

By considering division of catchment area into permanent snow cover and temporary snow cover, effect of rainfall on snow covered and non-snow covered areas, orographic effect on precipitation and division of melt season in premonsoon in two parts, quite encouraging results have been obtained. For all three years, there is good reproduction of observed flows inspite of limitations of data. However, further verification tests with independent data will be needed for establishing suitability of the model for general use. This model provides a suitable methodology for prediction of daily snowmelt during premonsoon season, in small Himalayan catchments using limited precipitation and temperature data and appropriate information from satellite imageries.

ACKNOWLEDGEMENTS

The author gratefully acknowledges useful discussions with Mr. K. P. Sharma, Reader in Civil Engineering, University of Roorkee and Mr. K. B. Thapa, Trainee Officer, School of Hydrology, University of Roorkee. He is also highly grateful to Dr. S. Ramaseshan, Director, National Institute of Hydrology, for constant encouragement.

REFERENCES

Martinec, J. 1976. Snow and Ice. in; J. C. Rodda (Editor), Facets of Hydrology, pp. 86-115, John Wiley, London

Monro, J. C. 1971. Direct Search Optimization in Mathematical Modelling and a watershed model application. NOAA Technical Memorendum NWS HYDRO-12, Silver Springs, Maryland.

North Pacific Division, Corps of Engineers, U.S. Army, Portland, Oregon. 1956 Summary Report of the Snow Investigations Snow Hydrology

Rango, A. and Martinec, J. 1979, Application of a Snowmelt runoff model using lands at data, Nordic Hydrology, Vol. 10, pp. 225-238.

Thapa, K.B. 1980 Analysis for Snowmelt runoff during premonsoon months in Beas basin using satellite imageries. Unpublished M.E. Dissertation, School of Hydrology, University of Roorkee, Roorkee.

Viessman, W. (Jr.), Knapp, J.W., Lewis, G.L. and Harbaugh, T.E. 1977, Introduction to Hydrology, IEP a Dun Donnelley, Publisher, New York.

Section 2
EVALUATION OF CATCHMENT MODELS

A TAXONOMY FOR EVALUATING SURFACE WATER QUANTITY MODEL RELIABILITY

L. Douglas James, Professor
Utah Water Research Laboratory
and Department of Civil and Environmental Engineering

David S. Bowles, Research Associate Professor
Utah Water Research Laboratory
and Department of Civil and Environmental Engineering

and

Richard H. Hawkins, Professor
Department of Forestry and Outdoor Recreation
(Watershed Science Unit)
and Department of Civil and Environmental Engineering

Utah State University
Logan, Utah 84322

ABSTRACT

Rainfall-runoff and other surface water quantity models provide information needed to solve a wide variety of water resources problems including flood forecasting and control, drought and low flow river forecasting, streamflow regulation encompassing reservoir operation and other efforts to satisfy such instream flow needs as fish and wildlife, recreation, navigation, and hydroelectricity. Models range from very simple formulas used with a hand calculator for a quick approximation to large and complex digital computer programs.

Some available models cover many possible applications (with varying degrees of reliability) while others concentrate on one or two. Some make approximate estimates from limited information, and others require a great deal of descriptive data and use a large amount of computer time in detailed computations. To the hydrologist or water resources planner, it is often unclear what type or complexity of model will satisfactorily supply the information he needs. The purpose of this paper is to 1) develop and present a system of model classification based on user needs and modeling approaches, 2) use a matrix built from this two-way classification for evaluating the reliability of the current state of the modeling art for meeting specific user needs, and 3) also use the matrix to formulate a taxonomy for general model assessment.

User needs encompass nine basic types of information which can be supplied by models. The models use one or more of ten structural components to replicate physical processes (evapotranspiration, soil moisture storage, infiltration, snow accumulation and melt, subsurface flow,

flow through streams, and flow through lakes) or to preserve relationships in observed data (of input to output, among locations over a
region, or over time at a given location.

It was found that the matrix of nine information types versus
ten model components could be used to evaluate the reliability of the
current state of the modeling art represented by each element. Reliability is determined on the bases of 1) agreement with known scientific
principles, 2) ability to match measured occurrences in controlled
situations, and 3) reasonableness in terms of the expectations of experienced people. Evaluation of the current state of the modeling art
also needs to consider 1) accuracy achievable in various applications,
2) restrictions imposed by deficiencies in scientific understanding and
data availability, and 3) priorities for supplying new information
needs.

The model taxonomy used for general assessment identified 24
model species which fall into 8 generic types. With respect to these
8 genera, continuous simulation physical watershed process models were
found to be a powerful tool for replicating measured flows and also for
modeling watersheds with sparse or nonexistent streamflow records.
Provided accurate information on channel and flood plain geometry is
available, channel process models give excellent flow rates and water
surface profiles for reservoir operation and flood plain management.
Flood statistical relationship models were widely used in pre-computer
days and are being phased out as computerized simulation and information
storage systems become widely available. Although they are not yet
widely used, urban runoff models are very valuable tools for urban storm
water control system design and operation and urban flood plain management. Flood and drought frequency analysis provide reasonable to good
results for most applications although these results are best when the
choice of probability distribution is not restricted. Results from
alluvial process (erosion, transport, and sedimentation) models are very
approximate as a great deal more work is needed to make them reliable.
Ice prediction models are still at the developmental stage. Soil-water
interaction and snow-pack models represent efforts to provide a scientific basis for water conservation through better management of agricultural, forest, and range lands and offer major opportunities to do for
the better.

The evaluation and taxonomy are useful for selecting a model to
provide hydrologic information needed for dealing with a particular water
management problem. The evaluation is useful for assessing the reliability of results that can be obtained from a particular type of model.
In addition, poor evaluations suggest research is needed to improve model
reliability.

INTRODUCTION

Sound water programs are formed from reliable technical information.
For example, for the rainfall-runoff and other surface water quantity
models encompassed within the scope of this paper, the formulation of
plans maximizing net benefits from building new or operating old facilities to harness water resources, requires reliable prediction of the
consequences of each alternative. Similarly, land use planning programs
need information on water requirements and on how land use affects
flood problems and water supply downstream. In the private sector,
irrigators seek technical advise on how to ration water wisely.

Any calculation scheme to provide needed technical information can be called a model. Models range from simple formulas applied with a hand calculator to complex digital computer programs requiring man years of effort in assembling the data and years of training for effective use.

The ultimate surface water model would be comprehensive in its coverage of the physical factors influencing surface water movement and consequently reliable in its quantitative estimation of stream flow and in its prediction of how flows would be affected by proposed designs or operating procedures. We do not now have such a comprehensive model because we lack the required understanding of the interacting physical relationships and the descriptive data to verify relationships thought to be true.

Available models are neither completely comprehensive nor fully reliable. Consequently they can be classified according to the components they cover and rated as to the reliability of their coverage. Model reliability is demonstrated by 1) ability to match measured occurrences, 2) agreement of the relationships used with known scientific principles (particularly important if the model is used to estimate quantities outside the measured range), and 3) reasonableness of estimated quantities in terms of the expectations of experienced people.

In one sense, reliability is with respect to the desired application, whether for water policy formulation, project design, or system operation. In evaluating models to meet a given need (flood peak estimation for example), one finds that a given model may be well suited for one application (selecting a flow for culvert design) but totally inappropriate for another (real time control of an urban storm water system). This evaluation, however, is not designed to evaluate given models (the Stanford Watershed Model, the log-Pearson Type III distribution for flood frequency estimation, the Muskingum approach to channel routing, etc.) but rather the present state of the art in modeling to meet given needs.

This limitation to the scope of this evaluation reduces the task to one of manageable size. We are not evaluating all models applicable to meeting a given need (flood peak estimation in our example), but rather making a judgment (based on information on the models referenced in Appendix I) as to the most appropriate available model type for major user needs and evaluating that modeling approach as the most reliable the state of the art has yet produced.

Also, the viewpoint is taken that all users need reliability but some require much more detail than others. In the above example, one needs complete flood hydrograph detail at specific sites for operating an urban storm water system, only the flood peaks at those sites for culvert sizing, and only a summary statistic on the aggregate flood problem over the area for setting policy on budgeting a storm water program. While the mass of desired flood data is greatly reduced as one goes from system operation applications to the need for a single measure of overall problem magnitude, all applications need reliability in the estimates they obtain; and the reliability of what can be provided depends in all cases on what the state of the art of hydrologic modeling can currently produce. In summary, limiting this evaluation to the best available models of a given type and viewing reliability in absolute rather than relative terms have been used to keep the scope of this paper within reasonable range.

This perspective should not be taken as an attempt to discourage the development of new models. One often finds new surface water models being developed for new applications, a situation favored because of the communications and training barriers to gaining familiarity with previous work by others. From one perspective, new models mean duplication; but in reasonable amounts, multiple modeling efforts capture independent thinking, enhance interaction among experts, produce theoretically sounder models as differences are reconciled, and broaden the experience base required to make models more comprehensive and reliable.

CLASSIFICATION FOR RELIABILITY EVALUATION

Two perspectives should be considered in evaluating model reliability. The water resources planner perceives reliability in terms of how well a model estimates flood peaks, low flows, annual runoff volumes or other physical quantities required for system design or operation. The scientific hydrologist perceives reliability more in terms of how well a model replicates measured processes such as channel routing or infiltration. These two dimensions of reliability are of course interrelated. Reliablity for planning estimation is achieved through reliability in hydrologic process replication. For estimating a given physical quantity, however, some physical processes are much more important than others.

These two contrasting perspectives can be considered as two dimensions of reliability and represented in a two-dimensional matrix. Each element in such a matrix focuses on a reliability evaluation of how satisfactorily the current state of the art in replicating of a given hydrologic process serves planning and operation needs for estimating a given physical quantity.

THE USER DIMENSION

Users want surface water models to organize information for the design of cost effective water resources development, the effective operation of systems already in place, or guidance in resolving various policy issues. Applications address the four hydrologic issues shown in Table 1. Each application requires several types of specific information. For example, flood forecasting and control requires estimates of the magnitude of flood peaks for channel or bridge design as shown by Item 1a (Table 1), flood hydrographs for reservoir design and operation as shown by Item 1b, etc. The user's information need is determined by his desired application. Accordingly, each information requirement item on Table 1 is listed on at least two lines. The first line names the information. The second line states the application for the information. In all, Table 1 lists 26 information requirements for the four issue areas.

Examination of these 26 information requirements reveals considerable commonality. For example, a model that does a good job of estimating low river flows for fish and wildlife management (4aa) will also do a good job for recreation (4ba) or navigation (4ca). Separate evaluations for the three would be repetitious. Accordingly, the 26 information requirements on Table 1 are collapsed to the 9 information types numbered in the right-hand column. These 9 are defined in Table 2 and provide the user dimension for surface water quantity model reliability evaluation.

THE SCIENTIFIC DIMENSION

Surface water flow models are built from components which can be subdivided into two broad groups. Physical process components

Hydrologic Issue	Information Required	Applications	Information Need (Table 2)
Flood forecasting and control	Flood peaks	Channel and bridge design	1
	Flood hydrographs	Reservoir design and operation	2, 6
	Simultaneous flood hydrographs	Storm water disposal system design and operation	2J, 6J*
	Flood depth mapping	Floodplain land use planning	9F
	Effects of land use on down-stream flows	Upstream land use planning	1, 2
	Flood peaks after dam failures	Emergency preparedness planning	3
	Soil moisture conditions	Land drainage design	8
Drought and low flow river forecasting	Low river flows	Offstream uses	4
	Timing of drought sequences	Cumulative economic impact	4, 6
	Soil moisture conditions	Precipitation-supplied uses	8
Streamflow regulation (including reservoirs)	Runoff volume	Maximum obtainable yield	5
	Runoff time patterns (within and among years)	Reservoir sizing	6
	Simultaneous runoff volumes in regional streams	Regional water supply planning	6J
Instream flow needs a. Fish and wildlife	Low river flows	Fish support potential estimating	4
	Within-year timing of low flows	Fish life cycle matching	6
	Timing of drought sequences	Reservoir or lake low level estimating	6
	Flow velocities within streams	Estimating effects on fish species	2, 9
b. Recreation	Low river flows	Sustaining recreation capacity and aesthetic appeal	4

Table 1. Issue relationships used in model evaluation.

Hydrologic Issue	Information Required	Application	Information Need (Table 2)
	Timing of flow sequences	Matching with recreation periods	6
	Runoff time patterns (within and among years)	Estimating lake level fluctuation impact	6
c. Navigation	Low river flows	Determining waterway capacity	4
	High river flows	Determining navigation interference	2
	Formation of surface ice	Determining navigation interference	7
d. Hydroelectricity	Timing of flow sequences	Run-of-the-river generating capacity estimation	6
	Runoff time patterns (within and among years)	Streamflow regulation design	6
	Simultaneous runoff volumes in regional streams	Regional generating system planning	6J

*2J Information need 2 (Table 2 with respect to model component J (Table 3)

Table 1. (Continued)

Table 2. List of surface water flow information needs.

  I. High flows

      1. Peaks
      2. Hydrographs and longer high flow sequences (durations shorter than a month)
      3. Dam breaks

 II. Low flows

      4. Low flow sequences (durations shorter than a month)

III. Flow volumes

      5. Amounts (over time periods of a month or longer)
      6. Volume sequences (from one time period to the next)

 IV. Special hydrologic information

      7. Ice formation and breakup
      8. Source area moisture and runoff conditions
      9. Channel and inundated area geometry

estimate flow or use rates from formulas developed by scientific methods to represent water movement and storage over, into, or through the soil or in natural or manmade channels. Statistical (empirical) relationship components estimate flow or use rates from associations in historical flow or use data. These two groups can be subdivided according to the physical process involved or the type of association sought in the historical data (Table 3). The ten resulting components provide the

Table 3. List of components used in surface water model building.

  I. Physical processes

      A. Evaporation from water/surfaces and transpiration from plants
      B. Soil moisture changes (wetting or drying)
      C. Infiltration (water losses between precipitation and runoff)
      D. Snowpack accumulation and melt
      E. Subsurface flow into stream (baseflow)
      F. Flow through stream channels
      G. Flow through lakes and reservoirs

 II. Statistical Relationships

      H. Associations from site data (input-output or black box type models)
      I. Associations from regional data
      J. Associations from time sequences (data generation, operational hydrology, frequency analysis)

needed scientific dimension for model reliability evaluation. A given model can be classified according to the components it uses or assigned a complexity index according to the number of components included and user needs satisfied. Many sophisticated computer models incorporate all seven physical processes. Most incorporate several.

AN ASIDE ON MODEL STRUCTURE

Models also vary in their structural characteristics as noted in Table 4. Specifically, models vary according to the sorts of data used,

Table 4. Structural characteristics of surface water models.

A. Data considered
   a. Weather data (precipitation, evaporation, temperature)
   b. Watershed surface descriptors
   c. Historical streamflows and storage levels
   d. Flow channel and storage area characteristics
   e. Subsurface soil and geological data

B. Spatial resolution
   a. Entire area considered as a single unit or subdivided into smaller units
   b. Size of subareas (square feet to many square miles)
   c. Number of locations where flows are provided

C. Temporal resolution
   a. Continuous representation on a short time scale such as hourly
   b. Long time scale totals (or highs or lows) such as monthly or annually
   c. Amounts defined by frequency, such as 100-year floods

D. Regional specificity
   a. Applicable nationally or internationally
   b. Applicable to specific regions such as Colorado River Basin
   c. Applicable to local areas such as the Logan City water distribution system

E. Sectoral specificity
   a. By land use (forest, range, or cultivated farms)
   b. Rural vs. urban

F. Calibration method (selection of values for model parameters to meet needs of specific applications)
   a. By judgment of user
   b. Trial and error by user to match measured data
   c. Self calibration by programmed statistical analysis

the detail with which variations in flow or use rates from location to location and time to time are replicated, the geographical area and land or water use conditions to which they are applicable, and the method used to select numerical values for the coefficients in the model formulas that best match local conditions. These structural characteristics will not be formally used in the evaluation to follow but do provide background on factors which should be kept in mind.

A MATRIX OF EVALUATIONS

The matrix used to evaluate model component types (Table 3) in meeting types of information needs (Table 2) is shown in Figure 1. Three dimensions were used in making the ratings within each element of the matrix:

   1. The importance of the physical model component in determin-

ing the information (Table 5) or the usefulness of the sort of statistical relationship (Table 6). The importance rating is based on the
sensitivity of the information need to the physical process. For
example, the N for element Al means that physical process A (evaporation
and transpiration as defined on Table 3) was judged as not applicable to
estimating information need 1 (high flow peaks as defined on Table 2).
The rating of "not applicable" does not mean that absolutely no physical
relationship exists between evaporation and flood peaks; it is a judgment
that the magnitude of the relationship is too small to be of consequence
to flood peak estimation for the design or operation of water resource
systems. The usefulness ratings for statistical relationships (Table 6)
are based on judgments on whether statistical analyses of the sort
indicated is now (U) or could potentially if developed (W) contribute
meaningfully to meeting the information need.

Table 5. Ratings of importance of physical process model components.

V   Information vital to informed planning, operation, and policy
     decisions

VL   Information important but generally not vital to informed plan-
     ing, operation, and policy decisions

L   Information applicable in a limited number of situations but
     generally not important

LN   Information applicable only in unusual special circumstances

N   Information not applicable (except indirectly or with respect to
     technicalities not within the scope of this evaluation)

Table 6. Ratings of usefulness of statistical relationships.

U   A useful and applied relationship

W   A potentially useful relationship which has not yet been
     sufficiently developed to become widely applicable

N   Relationship not known to be useful

        2. The model order (Table 7) judged as most appropriate for
meeting the need. As defined on Table 7, models vary from the very
complex digital computer programs used in continuous digital simulation of the entire runoff phase of the hydrologic cycle (water movement from the time it falls as precipitation to when it flows out of
the basin in a river or underground, hour by hour, continuously over
periods of a year or longer) to simple estimating equations (for example,
to select a peak flood flow for culvert design). In between these two
extremes are "event" models that estimate flows continuously but over
some shorter period such as a few days for a winter flood peak or a month
or two of low flow in the late summer. The models can be structured to
replicate flows at a defined point on a stream (or a set of defined
points simultaneously), or they can be structured to estimate soil
moisture and runoff conditions on a defined small plot of ground.
The judgment on the most appropriate model order was biased to select the
highest order applicable model. If all orders are applicable, for
example, the rating is based on the R model. Lower order models (for
example, Q models) are evaluated only after a judgment that the higher

Table 7. Model orders.

R   Continuous hydrologic simulation models (reproducing the entire runoff phase of the hydrologic cycle from precipitation to stream-flow (including canals, reservoirs, interbasin transfers, etc.) out of the basin over periods measured in years).

E   Event simulation models (reproducing runoff over short periods such as the duration of a storm hydrograph)

Q   Equations (estimating equations usually giving a single flow property such as the flood peak)

S   Statistical models (based on relationships found among historical flows over time and at different locations in a region)

R*   Continuous or event simulations respectively that are localized
E*   to a specific area rather than simultaneously addressing points throughout an entire watershed.

order models (R and E in the sample case) are not sufficiently cost effective to be useful in meeting the indicated information need. However, the authors recognize that models of orders other than those evaluated in a particular case are in comon use.

    3. The reliability grade assigned the model as defined on Table 8. The grades, following standard academic symbols (A through F), were assigned to the physical process components and statistical relationships on the basis of a judgment on model reliability for a) planning (selection and design of best alternative), b) operation (including resource management and regulation), and c) policy (program administration and budgeting) applications. The primary considerations were a) agreement with known scientific principles, b) ability to match measured occurrences in controlled situations, and c) reasonableness in terms of the expectations of experienced people.

Table 8. Model component grading scheme.

A   Modeling of the physical process at the current state of the art does an excellent job in supplying the needed information

B   Information between adequate and excellent

C   Modeling does an adequate job for most purposes

D   Information between unsatisfactory and adequate

F   The supplied information is generally unsatisfactory

    Each rating shown on Figure 1 includes all three dimensions along a diagonal with importance in the upper left corner of the element, the model order evaluated in the center, and the reliability grade in the lower right. For five elements, two quite different sorts of models were identified for evaluation and both are rated. Many cases were found where the physical or statistical component was not judged applicable to meeting the information need. In these cases, the single rating of N was given.

SCIENTIFIC DIMENSION
Model Components (Table 3)

		Physical							Statistical			Overall Academic Grade
		A ET	B Soil	C Infl.	D Snow	E Bsfl.	F Ch. R.	G Rs. R	H Site	I Reg.	J Time	
High Flows — D. B.	1 Pk.	N	VL R C	V R C	VL R D	LN R A	VL R B	VL R A	U Q D	USC URB	U S B	B
	2 Seq.	LN R D	V R C	V R C	VL R F	L R B	V R B	VL R Q	U E D	UED URC	W S C	C
	3	N	N	N	N	N	V E C	L E C	N	N	N	C
Lows	4 Seq.	VL R C	L R A	LN R B	L R D	V R C	L R A	VL ·R A	N	U S C	U S B	B
Volumes	5 Amnt.	V R C	VL R B	VL R B	VL R B	VL R B	N	LN R A	U Q C	U S D	U S B	A
	6 Seq.	V R C	V R C	V R C	VL R C	VL R C	N	VRB VLED	N	W S D	U S C	B
Special	7 Ice	L E F	N	N	N	N	VL E D	L E D	U Q C	N	N	D
	8 S. Area	V R D	V R C	V R D	VL R B	N	N	N	U E C	W R D	W E D	C
	9 Geom.	N	L E C	N	N	N	VEC LRF	VLRC LEC	N	U S D	W S D	C

USER DIMENSION
Information Needs (Table 2)

Note: The evaluation notation combines 1) an importance (Table 5) or usefulness (Table 6) rating in the upper left, 2) a judgment selecting the most appropriate model order (Table 7) for evaluation in the center, and 3) an academic grade (Table 8) in the lower right. Evaluations for elements containing two model types are denoted in the same order on a single line. The difference between the two models in this case is explained in the text on the evaluation of that element.

Figure 1. Matrix of evaluation: How well the components used to classify models by generic type (Table 2) do in supplying surface water flow and use information needs (Table 3).

ELEMENT EVALUATIONS

The element evaluations were based on the models reported in the literature and on the descriptions of their performance cited in Appendix 1. The evaluations are judgments originally based on the experience of the authors, modified in part in response to comments received from reviewers of an earlier version of this paper solicited by the Office of Technology Assessment, and subject to revision as additional information is found and as the topics are discussed within the profession. Additional commentary on specific evaluations follow.

Information Need 1: Peak Flows

A. Evapotranspiration

Compared to rates of flood peak flow, the rates of transpiration from watershed vegetation and stream channel evaporation are inconsequentially small. It is true that evapotranspiration dries the soil and thereby reduces soil moisture and increases infiltration rates, but this evaluation is concerned with direct effects.

B. Soil Moisture Storage

The capacity of the watershed soil to store water during a storm can greatly reduce flood peaks. This capacity, and hence the effect, is relatively less for more urbanized areas or rarer (larger) floods. The modeling of how storage capacity varies between storms and accepts water during a storm is not to the point of producing precise estimates, a problem of most concern for intermediate sized watersheds. Small watersheds produce small flood peaks. Large watersheds usually have long records of gaged flows, and a frequency analysis of the record can meet the need.

C. Infiltration

During storm events, water is retained within the basin and thus does not immediately appear as flood runoff. Estimation of these "loss" rates during flood producing precipitation is key to flood peak replication and notoriously difficult to model on a basinwide basis. A higher grade (B) can be given to modeling urban watersheds where losses are less and more predictable.

D. Snow Accumulation and Melt

A lesser importance rating was given to all snow components because they are not needed in many climates. Flood peaks can result from heavy rains on frozen ground or be coupled with rapid spring thaws. The modeling of neither process is reliable on a basinwide basis.

E. Subsurface Flow

Subsurface flow contributes a significant portion of the flood peak total only in relatively rare situations where spring or seep discharge is large in comparison with direct runoff. Continuous simulation models generally estimate base flow contributions to flood peaks quite adequately. Event or equation models, where base flow amounts are often arbitrarily chosen, would have to be graded much lower (C).

F. Channel Routing

This routing is considered to include both overland routing of flow to the nearest defined channel and movement of the water

down the channel. Overland flow routing is the much less reliable (in part because of greater data problems) of the two, but
is significant in determining flood peaks only in very small
watersheds. Routing in well defined channels is a fairly precise
art (provided the hydraulic loss coefficients can be calibrated
with real data), but the reliability is much less for cases with
significant overbank flow, a likely situation during flood peaks.
The greatest problem in channel routing is the costliness of the
surveying required to collect the channel geometry required for
precise routing. Model users need to balance cost with the
benefits of applying the information.

G. Storage Routing

Storage routing is vital to flood peak modeling wherever flows
pass through lakes or reservoirs. Continuous simulation provides
a basis for determining initial storage conditions. Given the
initial conditions, available storage routing methods are reliable and economical.

H. Site Associations

Many simple equations have been presented for estimating flood
peaks from watershed characteristics. These are useful, especially at ungaged locations, in providing rough estimates but
cannot be relied upon for design of costly structures.

I. Regional Associations

Regional flood control agencies have used statistical analyses of associations between flood peaks from gaged records
throughout their jurisdiction with watershed characteristics to
develop estimating equations for ungaged watersheds. These
generally give satisfactory results but must be watched when
applied to watersheds different than any included in the original
data base.

    An alternative type of regional flood peak model simulates
flood peaks at many points simultaneously. Such models are more
expensive to develop but are rated more highly because of their
greater power to estimate the hydrologic effects of changing land
surface and channel conditions.

J. Time Associations

Statistical methods are widely used to estimate the flood peak
for a given return period or frequency from a recorded series of
historical events. Given a reliable record, the method can
perform reasonably well and is graded B if the user is free to
select a distribution of reasonable fit. If the user is constrained to the customary Log Pearson Type III distribution fit by
the method of moments, the grade would be C on a nationwide basis.

Overall Evaluation

Given adequate data for calibration, a continuous simulation
model can generally do a rather good job of estimating flood
peaks. The tendencies in model development and calibration to
emphasize matching peak flows biases model reliability toward
peaks. For certain applications, regional relationships and time

analyses (also graded B) are superior.

Information Need 2:   High Flow Sequences

A.   Evapotranspiration

Evapotranspiration losses may in rare cases cause significant
reductions in flood flows. Streams with large lakes or desert
streams lined with phreatophytes are possible cases. Modeling
how evaporation rates vary during storm (high flow) periods is
not well developed with many models simply using average values.

B.   Soil Moisture Storage

Basin soil moisture storage capacity modeling is even more vital
to replicating high flow sequences than it is for the peaks
alone. Soil moisture storage capacity is likely to be filled
during a prolonged period of high flows whereas the peak instantaneous flow may be more influenced by peak loss rates.

C.   Infiltration

Available models have never been able to accurately portray
basinwide infiltration and associated streamflow hydrographs
throughout long wet periods. Spatial variation in both precipitation and soil conditions are key problems, and the time variation in infiltration rates required to model flow sequences makes
the problem more difficult than it is for flood peaks alone. The
situation is somewhat better defined for urban watersheds, where
the grade can be raised to a B.

D.   Snow Accumulation and Melt

Hydrographs resulting from rain or snow and combining with snowmelt or rain on frozen ground are notoriously difficult to model
because they require a methodology and data for performing energy
balance computations on the snow surface and within the snow pack.
Also when the temperature is near freezing, it is very difficult
to know whether the precipitation will be rain or snow.

E.   Subsurface Flow

The subsurface contribution is relatively larger for an entire
high flow sequence than just for the peak, making accurate base
flow modeling more vital; the modeling, however, is more difficult and less reliable because of difficulty in portraying base
flow rise during extended storm periods.

F.   Channel Routing

The discussion of Section 1F still pertains with the added
complexity that routing is always essential to developing flow
sequence hydrographs whereas nonrouting methods sometimes work
for estimating peaks alone. River reaches with large amounts of
overbank flow are the greatest problem.

G.   Storage Routing

Same discussion as for Section 1G.

H. Site Associations

This event model group generally uses empirical methods for estimating losses and distributes the flow based on hydrographs representing average flow patterns during previous storms (unit hydrographs). The methodology is rated less reliable than continuous simulation.

I. Regional Associations

Regional methods for unit hydrograph derivation are used and rated as shown and as discussed in Sections 1I and 2H.. Regional simulation models, as also described in Section 1I, are rated on their ability to provide simultaneous hydrographs at a number of locations over a region.

J. Time Associations

High flow sequences could potentially be generated by methods used in operational hydrology for annual volumes, but the method has not been tested. The WSC rating is for frequency analysis of flood volume data (for example, peak day or week). Frequency analysis was judged less satisfactory for flood volumes on the basis that appropriate distributions have been less thoroughly explored.

Overall Evaluation

A calibrated model has greater difficulty in replicating the entire high flow sequence than in matching the peak alone if for no other reason than that there are so many more points to model. The lower grades on some model components also suggest a greater problem.

Information Need 3: Dam Breaks

A. Evapotranspiration

Evapotranspiration losses are not significant in reducing downstream flows below dam breaks.

B. Soil Moisture Storage

Losses to soil moisture storage are larger than those to evapotranspiration but are still not large enough to be significant, except perhaps in estimating rates of flow dissipation.

C. Infiltration

Same discussion as for Section 3B.

D. Snow Accumulation and Melt

Accumulated snow on the ground could potentially affect the flood wave created by a dam break, but this problem is not known to have been covered by modeling.

E. Subsurface Flow

Base flow is too small to be a significant concern in routing dam break floods.

F. Channel Routing

Reliable channel routing procedures are key to modeling dam break floods. Modeling is done on a dam break event basis because there is no advantage to continuous simulation modeling. Validity checking is complicated by the lack of measured data on checking actual events, and results are much better for confined and reasonably regular channels than for unconfined flood plains where the flow spreads out in two dimensions. Hazard mapping is most difficult in large, flat urban areas.

G. Storage Routing

Storage routing of dam break floods is needed for cases where a second dam is located near enough downstream to be threatened. The dynamic routing of a dam-break flood wave passing over a lake is much less reliabe than is normal flood storage routing.

H. Site Associations

No real need exists for statistical modeling of dam break floods as the relevant hydraulic principles are well enough understood to be reliable in direct applications and data on historical events are too sparse for statistical analysis.

I. Regional Associations

Same discussion as for Section 3H.

J. Time Associations

Same discussion as for Section 3H.

Overall Evaluation

Because of the poor quality of the data available for reliability checking, it is difficult to assess exactly how well calibrated dam break models perform. Nevertheless, the hydraulic principles used are well enough established to provide reasonable confidence in the results.

Information Need 4: Low Flows

A. Evapotranspiration

Stream and phreatophyte evapotranspiration can significantly reduce low flows, particularly in arid climates. In humid climates, soil drying caused by transpiration from hillside vegetation is also sometimes significant. Losses from water sources can be modeled more reliably than are losses from vegetation suffering a water shortage.

B. Soil Moisture Storage

If watershed soils hold the water they absorb during wet periods for long durations, soil drainage may be a significant source of streamflow during subsequent dry periods. In some humid climates, low flow periods may be short enough for water stored

above the water table in the soil to drain into streams during
low flow periods. In arid climates or humid climates with
seasonal rainfall, the drainage would normally be from below the
water table; and soil moisture storage would thus be a less
important factor.

C. Infiltration

Infiltration rates are minimal during low flow periods. Prior
infiltration amounts can be important in determining how far
flows drop during short periods between storm hydrographs.

D. Snow Accumulation and Melt

Snowmelt is not a factor during the summer low flows that are
critical in most climates. In cold climates, winter cold spells
prevent snowmelt and may even freeze the stream itself with
the result that flows become extremely low and may cease al-
together. The modeling of these winter low flows has not been
refined.

E. Subsurface Flow

Subsurface sources generally supply the entire low flow.
Reliable modeling of subsurface water movement is thus vital for
estimating these flows. Most models are based on recession curve
slopes fit to historical data, and the reliability of these fits
is limited by the tendency of recession rates to become flatter
as flows decrease. Subsurface flows enter a stream from different aquifers with different transmissibilities, and this sort of
variability is seldom explicitly modeled. Three other problems
are that 1) continuous models are often designed and calibrated
to match high more than low flows and thus perform better in
their area of greater emphasis, and 2) low flows are greatly
affected by human activity (water main leaks, irrigation drainage, etc.) which is not seldom modeled explicitly, and 3) the
lack of precision in low flow measurements.

F. Channel Routing

Channel routing of low flows is sometimes required for estimating
rates of water transport below storage releases for downstream
use. This can be done on either an event or a continuous simulation basis, and available models are generally quite reliable
given adequate data on channel losses and cross sections.

G. Storage Routing

Low flows below reservoirs are largely governed by reservoir
operating policy. Storage routing through the reservoir (on a
coarse time grid) is necessary to determine what low flows can be
maintained with various operating policies without the reservoir
going dry. See also Section 6G.

H. Site Associations

Simple equations (other than those regionally derived as in
Section 4I) for estimating low flows from watershed characteristics are rarely used.

I. Regional Associations

   Regional equations for predicting low flows, analagous to those
   discussed for high flows in 1I, have been developed with reasonably satisfactory results. Regional urban simulation models are
   not known to have addressed low flows.

J. Time Associations

   Frequency analyses are made of low flows with generally satisfactory results. Zero flows and flows greatly modified by human
   activity can cause special problems.

Overall Evaluation

   A properly calibrated simulation model can generally do well at
   estimating low flows, particularly extreme lows. Just as occurs
   with flood peaks, it is more difficult to match all the intermediate lows during the year.

Information Need 5: Amount of Flow

A. Evapotranspiration

   Since all precipitation ultimately becomes either runoff or
   evapotranspiration losses from vegetation or water surfaces,
   evapotranspiration amounts must be estimated and subtracted from
   precipitation to estimate streamflow. The principal modeling
   problems are in estimating 1) the difference between potential
   (as determined by solar radiation) and actual (as limited by
   available soil moisture) losses, and 2) how this difference
   increases as soil conditions dry and vary with vegetative
   and other conditions over the drainage area. 'Model replication
   of these differences is limited by a lack of data on areal
   variation in precipitation and evaporation and by a tendency for
   modeling errors to be summed in the evapotranspiration category.

B. Soil Moisture Storage

   Water stored in the soil is delayed in movement and partially
   protected against evaporation, but its presence encourages
   greater vegetative growth and hence transpiration losses.
   Continuous simulation models generally replicate the process
   reasonably well for runoff volume estimation.

C. Infiltration

   Infiltration creates soil moisture storage and reduces direct
   runoff. Its influence in modeling volumes is between that for
   flood (Section 2C) and low (Section 4C) flows.

D. Snow Accumulation and Melt

   Snow modeling is vital in climates where snowmelt is a significant source of runoff. Available models do a reasonable job in
   estimating total volumes.

E. Subsurface Flow

Subsurface flows generally contribute a major portion of the
total runoff volume with the percentage determined by subsurface
conditions. While most models can be calibrated to match the
total runoff volume reasonably well, data and definitional
problems make it very difficult to know whether one has achieved
the proper division between surface and base flow.

F. Channel Routing

Channel routing is only relevant to estimating runoff volumes if
major storms occur so near the end of a volume period that one
needs the routing to divide the runoff between periods. Routing
for the division, when necessary, is easily performed.

G. Storage Routing

Evaporation losses from lake surfaces can significantly reduce
runoff volumes in watersheds with large storage areas. The
process is readily modeled. See also Section 4G.

H. Site Associations

Many simple equations estimate runoff volumes from watershed
characteristics as was discussed for flood peaks under 1H.
Volume equations generally perform a little better because total
runoff is more predictable than the peak.

I. Regional Associations

The most sophisticated form of regional flood volume modeling is
an operational system of total annual flow generation that
preserves the relationship statistics among two or more gaged
records and serial relationships at each one. These models most
frequently generate simultaneous streamflow sequences, but
sometimes they have been used to generate simultaneous sequences
of climatological and streamflow data such as for modeling water
level fluctuations. Satisfactory methods for calibrating these
models have only been developed for a few special cases.

J. Time Associations

The development of models to generate sequences of annual flows
for a single river location and then use of those flows for
reservoir sizing to supply water during a design drought of
specified frequency has been one of the major advances in hydrologic modeling in the last 20 years. The techniques are now
advanced to the point of giving results roughly equally reliable
to those from flood frequency analysis.

Overall Evaluation

A properly calibrated continuous simulation model should do an
excellent job of estimating annual runoff volumes. The quality
of estimation for a given year depends largely on how well the
flow conditions during that year were represented in the flows
recorded during the calibration period.

Information Need 6: Volume Sequences (the distribution of total annual
runoff among the 12 months)

A. Evapotranspiration

   Volume sequences are sensitive to evapotranspiration modeling
   and have the difficulties described in Section 5A. The difficulties are not significantly greater for matching volume sequences
   than for matching volume totals.

B. Soil Moisture Storage

   Soil moisture storage is relatively more important in distributing runoff over the year than it is in determining the annual
   total. Continuous simulation models exhibit greater difficulty
   in matching recorded distributions than they do recorded totals.

C. Infiltration

   Much the same as Section 6B.

D. Snow Accumulation and Melt

   Snowmelt models are less successful at distributing the runoff
   among months than they are at matching the total.

E. Subsurface Flow

   Subsurface flows supply most of the runoff between storm periods,
   and many simulation models exhibit difficulty in matching lows
   between peaks. These lows, however, are often not very important
   to correctly estimating monthly flow totals. See also discussion
   for Section 4E.

F. Channel Routing

   Channel routing would significantly affect volume sequences
   expressed in time intervals of a month or longer only in very
   large basins or for storm periods occurring on time unit
   boundaries.

G. Storage Routing

   The water balance models used to simulate reservoir operation are
   a special sort of storage routing basic to estimating reservoir
   yield, lake levels, etc. Specific problems occur in matching
   evaporation, precipitation on the water surface, and bank storage; and these problems are largely associated with a paucity of
   data of these sorts.

      Storage sediment routing is used in determining how rapidly
   a reservoir accumulates sediment and thereby loses useful
   storage capacity. These estimates are generally made by an
   analysis based on annual runoff totals rather than continuous

   simulation. Sediment movement is very difficult to model
   accurately.

H. Site Associations

   Historical data may be collected to determine typical monthly
   flow distributions, but the process is not really modeling.

I. Regional Associations

The data generation models described in Section 5I can be expanded to disaggregate the simulated flows into volume sequences
without much loss in reliability. Many more parameter values
must, however, be estimated. The reliability of the estimates
is reduced by short periods of record and by a lack of generally
applicable estimating procedures.

J. Time Associations

See discussion for Section 5J. Some loss in accuracy is associated with estimating volume sequences from volume totals.

Overall Evaluation

Modeling dividing the total runoff volume into sequences of
periods is inherently more difficult than just matching the
recorded annual total and becomes more unreliable as divisions
by shorter and shorter periods are sought.

Information Need 7: Ice Conditions

A. Evapotranspiration

Snow evaporation may be significant, and evaporative cooling is a
factor in waterway freezing. Models of the process are not
known.

B. Soil Moisture Storage

No significant physical relationship involved.

C. Infiltration

No significant physical relationship involved.

D. Snow Accumulation and Melt

Frozen ground reduces snowmelt from the bottom of the snowpack
as discussed in Section 4D.

E. Subsurface Flow

Large subsurface flow sources can increase winter surface water
temperatures and hence inhibit ice formation, but models of the
process were not discovered.

F. Channel Routing

Surface freezing increases channel hydraulic resistance, and
some ice particles (frazil ice) within the flow cause problems
at hydroelectric power plants. As the ice grows thicker, it
creates added pressure on emerged structures. Spring breakup
carries ice within the flow and can lead to ice jams at channel
constrictions. Both ice flows and ice jams increase flood
stages. No continuous simulation models are known that replicate
freezing and breakup processes and the associated changes in
channel hydraulics, and available event models are not reliable.

G. Storage Routing

Storage routing can be important to predicting ice formation and breakups on lakes and reservoirs in cold climates. Lake surface temperature is determined by temperature gradients within the lake and evaporation rates from the lake surface, but reliable event models of the freezing process are not known.

H. Site Associations

Site relationships are used for predicting dates for winter freezing and spring ice breakup and are useful in waterway operation. This sort of information is consistent enough to provide reasonable predictions. Site associations for predicting flood stage increases during spring ice breakup are much less reliable.

I. Regional Associations

Statistical studies for estimating freezing and breakup dates and maximum winter ice thickness on regional streams are plausible but not known as being used.

J. Time Associations

No relationships of this sort are known, but frequency analysis of the historical times (Section 7H) of ice formation or breakup would be useful.

Overall Evaluation

No reliable physically-based models of ice formation and breakup are known.

Information Need 8: Source Area Conditions (done on small areas and validated by careful instrumentation at the locations)

A. Evapotranspiration

Model reliability in replicating the drying of soil above the water table is not great because of reasons presented in Section 5A.

B. Soil Moisture Storage

Soil moisture accounting is a basic purpose of source area models and the one for which the best results are achieved. Mathematical models giving reasonably reliable predictions from detailed soil profile information have not been reduced to give reliable results over larger areas or from the soils data resolution normally available with field conditions.

C. Infiltration

Source area infiltration modeling is somewhat more difficult than soil moisture storage replication because of its greater sensitivity to the spatial and temporal variability. Soil conditions vary greatly over short distances, and infiltration rates can decrease drastically during a storm as wetting fronts contact more impermeable layers deeper within the soil.

D. Snow Accumulation and Melt

Small area snow models are generally more reliable than soil
moisture models because of the greater uniformity in snow's
physical properties. Some problem exists in replicating the
effects of drifting and of differences in exposure to radiation
in attempts to expand the modeling to watershed-size areas.

E. Subsurface Flow

Source area moisture and runoff conditions are defined here as
relating to surface and soil conditions above the water table and
not including subsurface flows that percolate down to the water
table nor overland flow runoff.

F. Channel Routing

See Section 8E.

G. Storage Routing

See Section 8E.

H. Site Associations

Models are available for estimating runoff volumes and peaks from
defined source areas for defined storm events. These are widely
used and provide reasonable results.

I. Regional Associations

Infiltration rate patterns can be defined over a watershed from a
spatially finely tuned continuous simulation model or from a
statistically derived relationship between infiltration and soil
characteristics. Such models do not have many direct user
applications but are useful in developing larger scale models for
design, operation, and policy purposes.

J. Time Associations

Many models use constant runoff coefficients and in effect assume
all precipitation to be divided between runoff and losses in the
same proportion. This assumption leads to using precipitation
frequency to estimate flood peak frequency when in fact time
varying antecedent moisture conditions cause considerable variation in the division. Resolution of the problem by frequency
analysis requires analysis of joint probabilities for storm and
soil moisture conditions. This approach is not well developed,
and the problem is better addressed by continuous simulation
modeling.

Overall Evaluation

A source area model can provide reasonable replication of moisture conditions when calibrated against measured data; the
difficulty is that the data are so site specific that extrapola-
tions to other areas are not very reliable.

Information Need 9: Channel and Inundated Area Geometry

A. Evapotranspiration

   No physical relationship exists.

B. Soil Moisture Storage

   Event models use the principles of soil mechanics to predict bank
   sloughing of farm drains, natural channels through wet soils, and
   reservoir banks during periods of rapid drawdown. The last
   application is the most important for project planning and
   operation.

C. Infiltration

   No physical relationship exists.

D. Snow Accumulation and Melt

   No physical relationship exists.

E. Subsurface Flow

   Subsurface flow into a channel or lake can contribute to bank
   sloughing, particularly when water levels fall below the elevation at which seepage is occurring. See Section 9B.

F. Channel Routing

   In stable channels, the routing provides the flood flows that can
   be translated by hydraulic analysis into areas inundated by depth
   and flow velocities for flood hazard mapping. The estimates are
   normally made for selected flood events rather than on a continuous simulation basis because that is all the detail that is
   generally useful. Model reliability is greatest (B) in humid
   climates where the flow is confined between established banks,
   less (C) where the flow spreads out over areas with manmade
   development that distorts natural patterns, and least (D) on
   alluvial flood plains where the channels are overtopped at spots
   that vary from storm to storm.

      In alluvial channels, flows change the channel geometry
   through erosion and deposition. Gulleys may erode where no
   previous channel existed. The process is continuous but more
   intense during high flow periods. The art of erosion and
   sediment transport modeling is not adequate for reliable continuous simulation.

G. Storage Routing

   Storage routing is improved by integration with modeling of the
   process of sedimentation within the lake or reservoir (see
   Section 6G) and of bank sloughing (see Section 9B). These
   processes are not modeled to great precision but adequately
   enough for most design and operation purposes.

H. Site Associations

   No models are known for estimating channel geometry from site
   characteristics. In practice one would survey the channel cross
   section.

I. Regional Associations

    Regional studies have been completed to define channel area,
    depth, and flow velocity from tributary watershed characteristics
    and are useful for estimating these parameters where the cost of
    field surveys and flow hydraulics analysis is not justified. The
    results are reasonable for predicting average channel character-
    istics over long channel reaches but are not recommended for
    application to a specific site.

J. Time Associations

    Channels are enlarged by the extra runoff that occurs with
    urbanization and filled by watershed changes that increase
    upstream sediment production. Channel erosion problems also
    occur downstream from reservoirs which collect the natural
    sediment load and cause more erosive clear water to flow through
    the reach. These processes could be simulated (Section 9F) or
    evaluated statistically. Statistical estimating procedures have
    not been sufficiently validated to generate much faith in their
    reliability.

Overall Evaluation

    Calibrated models are generally reliable except for dealing with
    the specialized problems associated with erosion-sedimentation
    processes and flooding patterns on alluvial fans. An F grade is
    assigned to these cases.

MATRIX APPLICATIONS

    One can use Figure 1 in the evaluation of a given model proposed for
a specific application. For example, one might be considering a physical
process model incorporating state of the art process representations for
soil moisture (Column B), infiltration (Column C), and channel routing
(Column F) for use to estimate peak flows for sizing a bridge opening.
One could inspect the evaluations in these three columns in the top row
of Figure 1 and would have more confidence in the model if the ratings
were high. In this case, respective ratings of C, C, and B suggest that
the model will do an adequate to excellent job (Table 8). The other
physical process columns (A, D, E, and G) rate the importance of the
omitted considerations. In this example, the failure to include evapotranspiration is indicated to be no problem (Column A). Baseflow is
indicated as a consideration to check but not a likely problem. The
importance ratings in Columns D and G provide reminders that snowmelt
sources or the effects of an upstream reservoir should be covered by the
model if the bridge is located where they are relevant.

    One finds an overall rating of B for physical process models
for flood peak estimation for bridge design in Figure 1. Columns I
and J show that a regional flood frequency study or a flood frequency
analysis, respectively, are also rated B and thus in fact are preferable
considering that less work is required in their application. The physical process model rating assumes that the available model uses state of
the art sophistication in its components, that sufficient input data
of needed quality are available and that the modeler is proficient in
calibrating and using the model. If any of these conditions are not
met, the rating would be less than B; and the advantage of the statistical alternatives would be greater. Many other possible applications are
suggested by this example.

MODEL TAXONOMY

Figure 1 also provides a basis for a useful model taxonomy for assessing the overall usability of the state of the surface water quantity modeling art and prioritizing research needs. In other words, it can be directly used for the evaluation of modeling for a specific application as outlined in the above bridge example and indirectly used for more generalized model assessment and research and development prioritization purposes. The taxonomy follows.

Figure 1 has 90 elements from 10 model components multiplied by 9 information needs. Of these 90, 29 were judged as situations where modeling would not significantly contribute to supplying information needs for water resources management. Five elements were judged as containing two diverse model types. The resulting 66 (90 - 29 + 5) cases were reduced (based primarily on similarity in the scientific principles used in the modeling) to the 24 model species indicated in the matrix of Figure 2 (specie numbers are assigned from left to right, row by row; consequently, genera groupings may not contain specie in numerical order) and the species were grouped into the 8 genera shown in Figure 3. Genera and species assessments follow; but these generalities are averages in that the complexities noted in the original evaluations are not carried forward.

I. Continuous Simulation Physical Watershed Process Models

Models of watershed 1) surface soil-water interactions, 2) snow accumulation and melt, 3) baseflow, and 10) evapotranspiration estimate flood peaks, low flows, and total flow volumes and are most useful in situations where gaged streamflow records are sparse or nonexistant. These models provide a powerful tool that come far closer to replicating measured flows than was possible before their advent.

1. Watershed surface, soil-water interaction models

Replication of how precipitation divides between runoff (both during the storm and after movement through the soil or underground aquifers) and "losses" or water that eventually evaporates is key to the reliability of continuous hydrologic models. The modeling needs to simulate water movement into and through the soil and how water entry slows as the soil becomes wetter. Water movement needs to be simulated during storm events for flood peak estimation, as annual totals for estimating water supply amounts, and during dry periods to estimate low flows. The modeling art does best at estimating low flows and total annual runoff volumes and does worse (but still acceptably well) for short term flow sequences and flood peaks.

2. Snow accumulation and melt models

For climates where snow remains on the ground for more than about a month, the processes of snow accumulation on the ground and later melt need to be simulated. A VL importance rating (Figure 1) was used because the modeling is only important in some climates. Total runoff volumes from snowmelt can be replicated quite well for projecting water availability during the following summer, but the models cannot be counted on to replicate snowmelt flood peaks (magnitudes are bad and timing is worse), because of the uncertainty of spring weather conditions.

214

3. Baseflow models

During low flow periods, except in very humid climates, water
only enters streams by seepage from aquifers or swamps into channels
intersecting the water table. Available models do not estimate flow
rates from measured aquifer characteristics but rather use observed
flow patterns during long dry periods. They do quite adequately for
estimating baseflow during floods and reasonably well for estimating
low flows.

10. Evapotranspiration models

Evaporation from water surfaces and transpiration from plants
are modeled to simulate the soil drying process. The drier the soil,
the more precipitation will infiltrate during the next storm. The
results are more reliable for basinwide average soil moisture than at
specific locations but are difficult to check because of the scarcity
of field data.

II. Channel Process Models

Models for routing 4) flows down channels, 5) flows through reservoirs, 12) flood waves below dam breaks, and 22) water spreading over a
flood plain give good to excellent results for reservoir operation and
flood plain management when supplied accurate information on channel and
flood plain geometry. Since this information is costly to collect, the
user must decide whether the additional accuracy purchased by collecting
more data is worthwhile.

4. Channel routing models

Channel shapes and flow conditions (hydraulic losses) can be
measured or estimated well enough to provide reliable estimates of
changes in flow depth and velocity as water moves downstream in a
channel so that this information can be made available for operation
of flood control facilities or warning flood plain occupants during
flood emergencies. The estimation is poorer for larger floods where
streams leave the banks and is particularly unreliable where the flow
spreads over large flat areas.

5. Lake and reservoir routing models

When flows enter a lake or reservoir they add to the water depth
and increase outflow through the spillway or other outlet control.
This process is accurately modeled for such applications as sizing a
spillway big enough to ensure dam safety or operation of the spillway
gates to minimize downstream flood damage.

12. Dam break models

Dynamic channel routing models are used to determine the areas
that would be inundated downstream from a dam break. Their reliability is difficult to check (few historical records are available) and
decreases as the flood wave decreases in height and spreads out over
a wider area.

22. Flood inundation models

The mapping of flood hazard areas provides the basic data for
flood plain management. The best results are achieved where the

flow is confined between stable channel banks, and the least reliability is achieved on broad flat floodplains where flows are deflected by small obstructions and spread in drastically different patterns from one event to the next.

III. Flood Statistical Relationship Models

Methods employing 6) flood formulas, 7) regional flood equations, 11) unit hydrographs, and 24) channel geometry equations provide approximations for estimating flood flows and areas of inundation for structural design or reservoir operation in situations where the extra effort for continuous simulation is not justified. Use of these relationship models can be expected to decrease as computerized simulation and information storage systems are advanced.

6. Flood formulas

Simple equations for estimating flood peaks from watershed characteristics have been used for years to estimate design sizes for small structures but can only be recommended for situations where construction expenditures are relatively small.

7. Regional flood formulas

Equations derived statistically for estimating flood peaks on streams throughout a hydrologically homogeneous region are much more reliable than flood formulas for flood structure planning.

11. Unit hydrograph models

An assumption that a given amount of runoff from a given watershed will always occur in a given time pattern has long been used to establish flow patterns for flood control reservoir design. The results are reasonable for estimating flood peak by frequency but are poor for matching flow patterns such as is required for reservoir operation.

24. Channel geometry equations

Regional statistics on channel size and shape and associated watershed characteristics are used to calibrate statistical models for estimating channel conditions for flood routing or flood surveying cost at locations where precision is not necessary.

IV. Urban Runoff Models

Continuous simulation models that generate simultaneous flows from many small urban watersheds and aggregate them into flood flows at downstream points are a sophisticated tool for urban storm water control system design and operation and urban flood plain management.

8. Regional flood simulation models

Continuous simulation models have been developed for some urban areas to generate simultaneous flood hydrographs for many small watersheds. This is a very valuable tool for urban drainage design, flood plain regulation, and tributary area land use controls.

V. Flood and Drought Frequency Analysis

Time series models of 9) flow frequency, 13) reservoir water
releases, 14) reservoir inflows, and 15) simultaneous inflows to several
reservoirs estimate flood and drought magnitudes by frequency of occurrence for sizing flood control or water supply projects. The concept of
selecting the frequency which gives an economically optimal size provides
the best possible project for the money. These models provide reasonable
to good results for most applications.

9. Flow frequency models

Historical sequences of flood peaks, flood volumes, or low
flows provide time series used to estimate maximum or minimum flow
magnitudes to be expected, on the average, no more than once every
10, 100, or some other period of years. Reliable results require
selection of a model (probability distribution) that matches
the historical data, and reliability improves with longer record
length.

13. Reservoir water accounting models

An accounting of inflow (stream and precipitation) and outflow
(evaporation, uncontrolled releases, and project water delivered)
over long periods is used to size reservoirs and establish operating policy. Excellent results are obtained if reliable data are
available to describe inflow and storage volume geometry.

14. Annual data generation models

Models are used to generate annual runoff sequences that match
the magnitude, probability distribution, and other patterns of
historical flows to provide inflows for reservoir sizing design.
The results are generally good for monthly, seasonal, or annual
time periods.

15. Regional data generation models

The models of Species 14 can be expanded to generate simultaneous flow sequences for many points so that the design and operation of systems of reservoirs can be coordinated. The mathematics
for properly calibrating these models is not yet satisfactory.

VI. Alluvial Process Models

Models of 16) reservoir sedimentation, 21) reservoir and channel
bank sloughing, and 23) channel erosion and deposition provide information for designing to handle these problems at minimal cost. A great
deal more work is needed to make these models reliable.

16. Reservoir sedimentation models

Processes determining how much of the sediment washed into a
reservoir settles to the bottom and diminishes its storage capacity
are modeled but not with the validity required for reliable
estimation of useful reservoir life.

21. Bank sloughing models

Stream banks slough to fill drainage ditches, and reservoir

217

banks are prone to slide during rapid water level drawdowns. Soil analyses to predict these conditions are reasonably reliable for the design of bank drains to relieve water pressures, but drainage ditch stability is more difficult to assess.

23. Channel erosion and deposition models

Flows in movable channels erode and deposit their sediment loads downstream. Problem locations can be identified, but amounts of erosion and deposition are not reliably predictable for channel design or maintenance management.

VII. Ice Prediction Models

The processes of river or lake freezing or ice breakup can be replicated from either 17) physical, or 18) statistical bases for navigation, recreation, and spring flood control applications. The modeling is still developmental.

17. Ice formation and breakup models

The process of freezing, surface ice growing thicker (stronger but also exerting greater pressure on emerged structures), spring breakup, and higher flow stages in ice laden water could beneficially be modeled for navigable waterways and certain structural designs. The models are more general and reliable than the formulas but cannot be counted as highly accurate.

18. Freezing and breakup formulas

Records of historical ice conditions have been used to estimate probable freezing and spring breakup dates on various rivers and lakes. They reasonably replicate average seasons but should not be counted on to predict extraordinary situations.

VIII. Soil-Water Interaction Models

Models following the principles used by Genera 1 can be applied to small plots to replicate 19) soil-water interactions, and 20) snow accumulation and melt. The efforts to provide a scientific basis for these decisions are making a substantial contribution to water conservation through better management and offer major opportunities to do far better.

19. Plot surface, soil water interaction models

Models of Species 1 are sometimes specialized to replicate moisture conditions on a small plot, rather than as an average over an entire watershed, for agricultural water management and land use decision making. The results are a great improvement over decision making without this information but cannot really be considered reliable.

20. Plot snow accumulation and melt models

Models of Species 2 specialized for a small plot give good results in predicting spring moisture and runoff conditions.

SUMMARY

In Table 1, 26 requirements for quantitative water availability information for water resources planning, water system operation, or water policy formulation are outlined. These are reduced to 9 user information needs in Table 2. In Table 3, the model components available for making these estimates are classified into into 10 groups according to the physical processes simulated and statistical relationships incorporated. These two classifications define 90 elements in the assessment matrix of Figure 1. Each element is assessed in terms of importance, the most relevant modeling approach, and reliability. The reliability grades assume a user obtains the relevant data and applies the model as it was intended. Severe data limitations or selection of an inappropriate model greatly reduce reliability.

In order to facilitate assessment of the state of the modeling art, 24 model species are defined in Figure 2, grouped into 8 generic types in Figure 3, and discussed in the accompanying text. Additional information on model capabilities and the successes and failures of those who have tried to use them are invited and would no doubt improve the results.

SCIENTIFIC DIMENSION
Model Components (Table 3)

		A ET	B Soil	C Infl.	D Snow	E Bsfl.	F Ch.R.	G Rs.R.	H Site	I Reg.	J Time
Peaks	1 Pk	N	1	1	2	3	4	5	6	7/8	9
	2 Seq.	10	1	1	2	3	4	5	11	11/8	9
	3 D.B.	N	N	N	N	N	12	12	N	N	N
Lows	4 Seq.	10	1	1	2	3	4	13	N	7	9
Volumes	5 Amnt	10	1	1	2	3	N	LN	6	15	14
	6 Seq.	10	1	1	2	3	N	13/16	N	15	14
Special	7 Ice	17	N	N	N	N	17	17	18	N	N
	8 Area	10	19	19	20	N	N	N	6	19	LN
	9 Geom.	N	21	N	N	N	22/23	16/21	N	24	23

USER DIMENSION
Information Needs (Table 2)

Note: If two specie numbers are shown here but only one rating appears on Figure 1, both models are rated the same.

Figure 2. Model species identification matrix.

SCIENTIFIC DIMENSION
Model Components (Table 3)

USER DIMENSION Information Needs (Table 2)			A ET	B Soil	C Infl.	D Snow	E Bsfl.	F Ch. R.	G Rs. R.	H Site	I Reg.	J Time
Peaks	D. B. Seq. Pk	1		I	I	I	I	II	II	III	III IV	V
		2	I	I	I	I	I	II	II	III	III IV	V
		3						II	II			
Volumes	Lows Seq.	4	I	I	I	I	I	II	V		III	V
	Amnt	5	I	I	I	I	I			III	V	V
	Seq.	6	I	I	I	I	I		V VI		V	V
Special	Ice	7	VII					VII	VII	VII		
	Area	8	I	VIII	VIII	VIII				III	VIII	
	Geom	9		VI				II VI	VI VI		III	VI

Figure 3. Model genera identification matrix.

Overall, surface water quantity models are providing sounder technical information as they are improved and save considerable expenditure through more cost effective water project design, more efficient water system operation, and better conceived water policy.

ACKNOWLEDGEMENT

This evaluation was performed for the Office of Technology Assessment to contribute to their overall assessment of the state of the water resources modeling art for decision making.

ILLUSTRATIVE REFERENCES

1. Amein, M. and C. S. Fang. 1970. Implicit flood routing in natural channels, Journal of the Hydraulics Division, American Society of Civil Engineers, Volume 96, HY12.

2. Anderson, E. A. 1964. The synthesis of continuous snowmelt runoff hydrographs on a digital computer. Tech. Rep. 36, Dept. of Civil Engineering, Stanford Univ., Stanford CA.

3. Anderson, E. A. 1973. National Weather Service river forecast system: Snow accumulation and ablation model. NOAA Tech. Memo. NWS HYDRO-17. US Dept. of Commerce, Silver Spring MD. 217 p.

4. Anderson, E. A. 1976. A joint energy and mass balance model of a snow cover. NOAA Tech. Rep. NWS 19. US Dept. of Commerce, Silver Spring MD. 150 p.

5. Baltzer, R. A. and C. Lai. 1968. Computer simulation of unsteady flow in waterways. Journal of the Hydraulics Division, ASCE, Volume 94, Number HY4, Proceeding Paper 6048.

6. Barnes, B. S. 1952. Unitgraph procedures. Bureau of Reclamation, U. S. Department of the Interior, Denver, Colorado.

7. Beer, C. E. and H. P. Johnson. 1965. Factors related to Gully growth in the deep loess area of western Iowa. US Department of Agriculture, Misc. Pub. 970, 37-43.

8. Betson, R. P. 1972. A continuous daily streamflow model. TVA Res. Pap. No. 8, Knoxville TN.

9. Bowles, D. S. and J. P. Riley. 1976. Low flow modeling in small steep watersheds. Journal of the Hydraulics Division, ASCE, 102 (HY9):1225-1239.

10. Brune, G. M. 1953. Trap efficiency of reservoirs. Trans. Amer. Geophys. Union. 34(3):407-418.

11. Burnash, R. J. C., R. L. Ferral, and R. A. McQuire. 1973. A generalized streamflow simulation system: Conceptual modeling for digital computers. Unnumbered Pub., California Dept. of Water Resources. 204 p.

12. Chen, C-L and J. T. Armbruster. 1980. Dam-break wave model: Formulation and verification. Proc. ASCE. 106(HY5):747-768.

13. Chow, V. T. 1964. Handbook of applied hydrology. New York: McGraw-Hill.

14. Churchill, M. A. 1948. Discussion of "Analysis and use of reservoir sedimentation data," by L. C. Gottschalk. Proc. Federal Interagency Sedimentation Conference, Denver CO. pp. 139-140. (Published by US Dept. of Interior, Bureau of Reclamation, Denver).

15. Crawford, N. H. and R. K. Linsley. 1966. Digital simulation in hydrology: Stanford watershed model IV. Tech. Rep. 39, Dept. of Civil Engineering, Stanfrod Univ., Stanford CA.

16. Dawdy, D. R., R. W. Lichty, and J. M. Bergman. 1972. A rainfall-runoff simulation model for estimation of flood peaks for small drainage basins. US Geological Survey Prof. Pap. 506-6. USGPO, Washington. 28 pp.

17. Dawdy, D. R., J. C. Schaake, Jr., and W. M. Alley. 1978. Distributed routing rainfall-runoff model. US Geological Survey Water-Resources Investigations 78-90. 146 pp.

18. DeWalle, D. R. and A. Rango. 1972. Water Resources Applications of stream channel characteristics on small forested basins. Water Resources Bulletin 8(4):697-703.

19. Dilley, J. F. 1976. Lake Ontario ice modeling. IFYGL Phase 3. General Electric Co., Philadelphia PA. NTIS PB-264 998.

20. Dooge, J. C. I. 1973. Linear theory of hydrologic systems. US Dept. of Agriculture Tech. Bulletin 1468. USGPO, Washington.

21. Eggleston, K. O., E. K. Israelsen, and J. P. Riley. 1971. Hybrid computer simulation of the accumulation and melt processes in a snowpack. Utah Water Research Laboratory Pub. PRWG65-1. 77 p.

22. Eychaner, J. H. 1976. Estimating runoff volumes and flood hydrographs in the Colorado River Basin, southern Utah. US Geological Survey Water-Resources Investigations 76-102. 18 pp.

23. Ficke, E. R. and J. F. Ficke. 1977. Ice on rivers and lakes: A bibliographic essay. US Geological Survey Water Resources Investigations Reston VA. (NTIS PB-279 528).

24. Fields, F. K. 1975. Estimating streamflow characteristics for streams in Utah using selected channel-geometry parameters. US Geological Survey Water-Resources Investigations 34-71. (NTIS PB-241 541). 19 pp.

25. Fiering, M. B. and B. B. Jackson. 1971. Synthetic streamflows. Water Resources Monograph 1, Amer. Geophys. Union, Washington, D. C. 98 p.

26. Fread, D. L. 1978. Dam-break model. Hydrologic Research Laboratory. Office of Hydrology, National Weather Service, NOAA, Silver Spring, Maryland.

27. Fread, D. L. 1978. National Weather Service operational dynamics wave model. Rockville, Maryland.

28. Fread, D. L. 1973. Technique for implicit dynamic routing in rivers and tributaries. Water Resources Research 9(4):918-926.

29. Garrison, J. M., J. P. Granju, and J. T. Price. 1969. Unsteady flow simulation in rivers and reservoirs. Journal of the Hydraulics Division, ASCE, Volume 95, No. HY5, pp. 1559-1976.

30. Gifford, G. F. and R. H. Hawkins. 1979. Deterministic hydrologic modeling of grazing system impacts on infiltration rates. Water Resources Bulletin 15(4):924-934.

31. Goldstein, R. A., J. B. Mankin, and R. J. Luxmore. 1974. Documentation of PROSPER: A model of atmosphere-soil-plant water flow. EDFB-IBP-73-9, Oak Ridge National Laboratory, Oak Ridge TN.

32. Grunsky, C. E. 1908. Rain and runoff near San Francisco, California. Trans. ASCE 1090.

33. H. G. Acres, Ltd. 1971. Review of current ice technology and evaluation of research priorities. Report to Inland Waters Branch, 105. Environment Canada. 299 pp.

34. Hailey, B. M., F. E. Perkins, and P. S. Eagleson. 1970. A modular distributed model of catchment dynamics. Ralph M. Parsons Laboratory for Water Resources and Hydrodynamics, Report 133, M.I.T., Department of Civil Engineering, Cambridge, Massachusetts.

35. Hall, F. R. 1968. Base-flow recessions: A review. Water Resources Research. 4(5):973-983.

36. Hedman, E. R. 1970. Mean annual runoff as related to channel geometry of selected streams in California. US Geological Survey Water Supply Rep. 1999-E. USGPO, Washington. 17 pp.

37. Helvey, J. D. 1967. Interception by eastern white pine. Water Resources Research. 3(3):723-729.

38. Hewlett, J. D., G. B. Cunningham, and C. A. Troendle. 1977. Predicting stormflow and peakflow from small basins in humid areas by the R-index method. Water Resources Bulletin. 13(2):231-253.

39. Hipel, K. W. and A. I. McLeod. 1978. Preservation of the rescaled adjusted range, part two, simulation studies using Box-Jenkins models. Water Resources Research. Vol. 14, No. 3, pp. 509-516.

40. Holtan, H. N. and N. C. Lopez. 1971. USDAHL-70 model of watershed hydrology. US Dept. of Agriculture Tech. Bulletin 1435. 84 p.

41. Holtan, H. N. and N. C. Lopez. 1973. USDAHL-73 revised model of watershed hydrology. US Dept. of Agriculture Plant Physiol. Inst. Rept. 1. 67 pp.

42. Hydrocomp. 1969. Hydrocomp simulation programming operations manual. Hydrocomp, Inc., Palo Alto CA.

43. Hydrocomp, Inc. 1976. FULEQ, Full equations hydraulics dam failure. Palo Alto CA.

44. Hydrocomp, Inc. 1978. Planning and modeling in urban water management. Palo Alto CA.

45. Hydrologic Engineering Centre. 1972. HEC-2, Water surface profiles: User's manual. US Army Corps of Engineers, Davis CA.

46. Hydrologic Engineering Centre. 975. HEC-5c, Reservoir system operation for flood control and conservation. US Army Corps of Engineers, Davis CA.

47. Hydrology Committee, Water Resources Council. 1976. Guidelines for determining flood flow frequency. Bulletin 17.

48. James, D. E. (Ed.) 1979. U. S. National report to the IUGG. Amer. Geophys. Union. Washington.

49. James, L. D., D. S. Bowles, W. R. James, and R. V. Canfield. 1979. Estimation of water surface elevation probabilities and associated damages for the Great Salt Lake. Utah Water Research Laboratory Pub. UWRL/P-79 03. 182 p.

50. James, L. D. and W. O. Thompson. 1970. Least squares estimation of constants in a linear recession model. Water Resources Research. 6(4):1062-1069.

51. Jaynes, R. A. 1978. A hydrologic model of aspen-conifer succession in the western United States. US Dept. of Agriculture Forest Service Res. Pap. INT213. 17 p.

52. Jennings, M. E. and N. Yotsukura. 1979. Status of surface-water modeling in the U. S. Geological Survey. US Geological Survey

Circular 809. 17 pp.

53. Larson, N. M. and M. Reeves. Undated. Analytic analysis of soil-moisture and trace contaminant transport. Oak Ridge National Laboratory Rep. ORNL/NSF/EATC-12.

54. Leaf, C. F. and R. R. Alexander. 1975. Simulating timber yields and hydrologic impacts resulting from timber harvest on subalpine watersheds. 20 p.

55. Leaf, C. F. and G. E. Brink. 1973. Computer simulation of snowmelt within a Colorado subalpine watershed. US Dept. of Agriculture Forest Service Res. Pap. RM-99. 22 p.

56. Leaf, C. F. and G. E. Brink. 1973. Hydrologic simulation model of Colorado subalpine forest. US Dept. of Agriculture Forest Service Res. Pap. RM-107. 23 p.

57. Lee, B. H., B. M. Reich, T. M. Rachford, and G. Aron. 1974. Flood hydrograph synthesis for rural Pennsylvania watersheds. Dept. of Civil Engineering, Penn. State Univ. 94 pp.

58. Leopold, L. B. and T. G. Maddock. 1953. The hydraulic geometry of stream channels and some physiographic implications. US Geological Survey Prof. Pap. 252.

59. Lowham, H. W. 1976. Techniques for estimating flow characteristics of Wyoming streams. US Geological Survey Water-Resources Investigations 76-112. 83 pp.

60. Lumb, A. M. and L. D. James. 1976. Runoff files for flood hydrograph simulation. Proc. ASCE. 102(HY10):1515-1531.

61. McCuen, R. H., W. J Rawls, G. T. Fisher, and R. L. Powell. 1977. Flood flow frequency for ungaged watersheds: A literature evaluation. US Dept. of Agriculture, Agriculture Research Service ARS-NE-86. 136 pp.

62. McLeod, A. I. and K. W. Hipel. 1978. Preservation of the rescaled adjusted range, part one: A reassessment of the Hurst phenomenon. Water Resources Research, Vol. 14, No. 3, pp. 491-508.

63. Mandelbrat, B. B. 1971. A fast fractional Gaussian noise generator. Water Resources Research. 7(3):543-553.

64. Mandelbrat, B. B. 1972. Broken line process defined as an approximation to fractional noise. Water Resources Research. 8(5):1354-1356.

65. Mandelbrat, B. B. and J. R. Wallis. 1969. Computer experiments with fractional Gaussian noises, parts 1 to 3. Water Resources Research, Vol. 5, Number 1, pp. 228-267.

66. Mejia, J. M., I. Rodriguez-Itunbe, and D. R. Dawdy. 1972. Stream-flow simulation 2: The broken line process as a potential model for hydrologic simulation. Water Resources Research. 8(4):931-941.

67. Mejia, J. M. and J. Rousselle. 1976. Disaggregation models in hydrology revisited. Water Resources Research. 12(2):185-186.

68. Muller, A. 1978. Frazil ice formation in turbulent flow. Rep. No. 214. Iowa Institute of Hydraulic Research, Iowa City IA.

69. Muller, A. and D. J. Calkins. 1978. Frazil ice formation in

turbulent flow. Proc. IAHR Symposium on Ice Problems, Lulea, Sweden. Volume II: 219-234.

70. National Environmental Research Council. 1975. Flood studies report, Volume I: Hydrological Studies. National Environmental Research Council, London, England. 550 pp.

71. National Environmental Research Council. 1975. Flood Studies report, Volume II: Flood routing studies. London, England.

72. Newton, D. W. and J. W. Vinyard. 1967. Computer determined unit hydrograph from floods. Journal of the Hydraulics Division, ASCE HY5. 75.

73. Nieber, J. L., M. F. Walter, and R. D. Black. 1975. A review and analysis of selected hydrologic modeling concepts. OWRT Completion Rep., Project No. A-061-NY. 66 pp.

74. North, M. 1980. Time dependent stochastic model of floods. Proc. ASCE. 106(HY5):649-665.

75. O'Connell, P. E. 1974. Stochastic modeling of long-term persistence in streamflow sequences. Internal Report, Hydrology Section, Civil Engineering, Department, Imperial College, London. 284 pp.

76. Ol' deKop, E. M. 1911. On evaporation from the surface of river basins. Tr. IUr'ev Obs. Leningrad. (Cited in Sellars, 1965).

77. Osterkamp, T. E. 1978. Frazil ice formation: A review. Proc. ASCE. 104(HY9):1239-1256.

78. Parzen, E., and M. Pagano. 1979. An approach to modeling seasonally stationary time series. Journal of Econometrics, Vol. 9, pp. 137-153.

79. Peck, E. L., N. A. Bochin, A. Kouzel, D. H. Lennox, and B. Sundblad. (Eds.) 1979. State of the art in transposable watershed models. Proc. Third Northern Research Basin Symposium Workshop (International Hydrological Program), Quebec City.

80. Pfankuch, D. J. 1975. Stream reach inventory and channel stability evaluation. US Dept. of Agriculture Forest Service, Region 1, Missoula MT. 26 pp.

81. Rao, D. V. 1980. Log pearson type 3 distribution: A generalized evaluation. Proc. ASCE. 106(HY5):853-872.

82. Renard, K. G., W. J. Rawls, and M. M. Fogel. 1980. Currently available models. Chapter 13 in "Hydrologic Modeling of Small Watersheds," ASAE Monograph. St. Joseph MI.

83. Riggs, H. C. 1978. Streamflow characteristics from channel size. Proc. ASCE. 104(HY1):87-96.

84. Riley, J. P. and R. H. Hawkins. 1975. Hydrologic modeling of rangeland watersheds. In Watershed Management on Range and Forest Lands, Proc. Fifth Workshop of the United States/Australia Rangelands Panel. Boise ID. Published by Utah Water Research Laboratory.

85. Rockwood, D. M. 1964. Streamflow synthesis and reservoir regulation. Engineering Status Project 171, Technical Bulletin No. 22, U. S. Army Engineer Division, North Pacific Division, Portland, Oregon.

86. Rogers, J. C. 1976. Evaluation of techniques for long range forecasting of air temperature and ice formation. NOAA Tech. Memo. ERL GLERL-8.

87. Rosgen, D. F. Mass wasting. Chapter III in WRENS: US Forest
Service Report to EPA, Agreement EPA-IAG-D6-0060 (Review Draft).

88. Ross, B. B., V. O. Shanholtz, D. N. Contractor, and J. C. Carr.
1975. A model for evaluating the effect of land uses on flood
flows. Bulletin 85, Virginia Water Resources Research Center. 137 pp.

89. Saah, A. D. et al. 1976. Development of regional regression equations
for solution of certain hydrologic problems in and adjacent to Santa
Clara County. Santa Clara Valley Water District, San Jose, CA.

90. Sellars, W. D. 1965. Physical climatology. U of Chicago Press. 272 p.

91. Sherman, L. K. 1932. Stream-flow from rainfall by the unit-graph
method. Engineering News-Record. 108:501-505.

92. Simons, D. B., R. M. Li, and M. A. Stevens. 1975. Development of a
model for predicting water and sediment yield from storms on small
watersheds. US Dept. of Agriculture Forest Service, Rocky Mountain
Forest and Range Experiment Station, Flagstaff AZ.

93. Snyder, F. F. 1938. Synthetic Unit-graphs. Transactions, American
Geophysical Union.

94. Soil Conservation Service. Computer program for project formulation-
hydrology. SCS-TR-20, Washington, D. C.

95. Solomon, R. M., P. F. Ffolliott, M. B. Baker, Jr., and J. R. Thompson.
1976. Computer simulation of snowmelt. US Dept. of Agriculture
Forest Service Res. Pap. RM-174. 8 pp.

96. Spencer, D. W. and T. W. Alexander. 1978. Technique for estimating
the magnitude and frequency of floods in St. Louis County, Missouri.
US Geological Survey Water Resources Investigations 78-139. 23 pp.

97. Staff, Hydrologic Research Laboratory. 1972. National weather
service river forecast system, forecast procedures. NOAA Tech. Memo.
NWS HYDRO-14 and HYDRO-17. US Dept. of Commerce, Silver Spring MD.

98. Stall, J. B. and Y. S. Fok. 1968. Hydraulic geometry of Illinois
streams. Univ. Illinois Water Resources Center Res. Rep. 15.

99. Swanson, J. F. and D. N. Swanston. 1973. A conceptual model of solid
mass movement surface soil erosion and stream channel erosion processes. Confierous Forest Biome Ersion Modeling Group, Biome Report #72.

100. Tao, P. C. and J. W. Delleur. 1976. Seasonal and nonseasonal
ARMA models. Journal of the Hydraulics Division, American Society
of Civil Engineers, Vol. 102, No. HY10, pp. 1541-1559.

101. Tatinclaux, J. C. 1977. Equilibrium thickness of ice jams. Proc.
ASCE. 103(HY9):959-974.

102. Tatinclaux, J. C. 1978. River ice jam models. Proc. IAHR Symposium on ice Problems. Luela, Sweden. Volume II:449-460.

103. Terstriep, M. L. and J. B. Stall. 1974. The Illinois drainage area

simulator, ILLUDAS. Illinois State Water Survey Bulletin 58. 90pp.

104. Thompson, J. R. 1964. Quantitative effects of watershed variances on the rate of Gully head advancement. Trans. ASAE. 7(1):54-55.

105. Thomsen, A. G. and W. D. Striffler. 1979. Hydrograph simulation with fall meeting, San Francisco CA. 23 pp.

106. US Army Corps of Engineers. 1966. Basin rainfall and snowmelt computation. Compute Program No. 723-G-1-L2260. The Hydrologic Engineering Center, Davis, California.

107. US Army Corps of Engineers. 1971. Monthly streamflow simulation. Computer Program 723-X6-L2340. The Hydrologic Engineering Center, Davis, California.

108. US Army Corps of Engineers. 1976. Gradually varied unsteady flow profiles. Computer Program 723-G2-L7450. The Hydrologic Engineering Center, Davis, California.

109. US Army Corps of Engineers. 1977. Scour and deposition in rivers and reservoirs. Computer Program 723-G2-L2470, HEC-6. The Hydrologic Engineering Center, Davis, California.

110. US Army Corps of Engineers. No date. Deposit of suspended sediment in reservoirs. Computer Program 23-J2-L264. The Hydrologic Engineering Center, Davis, California.

111. US Army Corps of Engineers. No date. Flood hydrograph package. Computer Program 723-X6-L2010, HEC-1. The Hydrologic Engineering Center, Davis, California.

112. USDA, Soil Conservation Service. 1972. National engineering handbook, Section 4, Hydrology. USGPO, Washington.

113. USDI, Bureau of Reclamation. 1977. Design of small dams. A Water Resources Technical Publication, Sections 135 to 139. USGPO, Washington.

114. US Water Resources Council. 1977. Guideline for determining flood flow frequency. Bulletin No. 17A of the Hydrology Committee, Washington, D. C.

115. Valencia, D. and J. C. Schaake, Jr. 1973. Disaggregation processes in stochastic hydrology. Water Resources Research. 9(3):580-585.

116. Van Haveren, B. P. and G. H. Leaveslen. 1979. Hydrologic modeling of coal lands. US Dept. of Interior, Bureau of Land Management, Denver. 13 pp.

117. Watt, W. E. 1971. A relation between peak discharge and maximum 24-hour flow for rainfall floods. Journ. Hydrology. 14:285-292.

118. Williams, J. R. and W. V. LaSeur. 1976. Water yield model using SCS curve numbers. Proc. ASCE. 102(HY9):1241-1254.

119. Wilson, T. V. and J. T. Ligon. 1979. Prediction of baseflow for piedmont watersheds. Tech. Rep. 80. Clemson Univ. Water Resources Research Institute.

120. Woodward, D. E. 1973. Hydrologic and watershed modeling for watershed planning. Transactions of the ASAE. 99(3):582-584.

121. World Meteorological Organization. 1975. Intercomparison of
      conceptual models used in operational hydrological forecasting.
      WMO No. 429, Geneva.

122. Yakowitz, S. 1979. A nonparametric Markov model for daily river
      flow. Water Resources Research, Vol. 15, No. 5, pp. 1035-1043.

123. Yen, B. C. and N. Pansic. 1980. Surcharge of sewer systems.
      Illinois Water Resources Center Res. Rep. 149.

APPENDIX I
REFERENCES ILLUSTRATING MODELS CONSIDERED IN SPECIES EVALUATIONS

TABLE OF MODEL SPECIES

General (48, 52, 61, 73, 79, 82, 116)

1. Watershed surface, soil-water interaction models (8, 9, 11, 15, 16, 17, 21, 34, 40, 41, 42, 51, 54, 56, 84, 85, 88, 92, 94, 96, 97, 103, 111, 112, 118, 119, 120, 121)

2. Snow accumulation and melt models (2, 3, 4, 11, 15, 21, 40, 41, 42, 51, 54, 56, 85, 97, 105, 109, 120, 121)

3. Baseflow models (8, 9, 11, 15, 17, 21, 35, 40, 41, 42, 50, 51, 54, 56, 84, 97, 119, 120, 121)

4. Channel routing models (1, 5, 8, 11, 15, 16, 17, 27, 28, 29, 40, 41, 42, 46, 71, 85, 88, 92, 96, 97, 103, 107, 111, 112, 121)

5. Lake and reservoir routing models (17, 42, 46, 96, 97)

6. Flood formulas (22, 32, 37, 38, 57, 70, 76, 90, 112)

7. Regional flood equations (22, 24, 47, 57, 59, 70, 83, 89, 117)

8. Regional flood simulation models (22, 44, 57, 60, 117)

9. Flow frequency models (70, 74, 81, 114)

10. Evapotranspiration models (8, 11, 15, 16, 17, 21, 31, 40, 41, 42, 51, 53, 54, 55, 56, 88, 95, 96, 97, 118, 119, 120, 121)

11. Unit hydrograph models (6, 20, 22, 38, 57, 72, 91, 93, 111, 112)

12. Dam break models (8, 12, 26, 43, 107)

13. Reservoir water accounting models (11, 97)

14. Annual data generation models (25, 39, 62, 63, 64, 65, 66, 75)

15. Regional data generation models (18, 24, 36, 49, 59, 67, 75, 78, 83, 100, 108, 115, 122)

16. Reservoir sedimentation models (10, 14, 97)

17. Ice formation and breakup models (19, 23, 33, 68, 69, 77, 86, 101, 102, 110, 123)

18. Freezing and breakup formulas (13, 19, 23, 33, 69, 77, 86, 101, 102, 123)

19. Plot surface, soil water interaction models (30, 31, 51, 53, 54, 55, 56, 84, 92, 95, 112)

20. Plot snow accumulation and melt models (31, 51, 54, 55, 56, 95)

21. Bank sloughing models (113)

22. Flood inundation models (45)

23. Channel erosion and deposition models (7, 80, 87, 99, 104)

24. Channel geometry equations (58, 98, 106)

WATERSHED EVALUATION AND RESEARCH SYSTEM APPLICATION

Thomas E. Croley, II, Ph.D., Hydrologist, U.S.
Department of Commerce, National Oceanic and Atmospheric
Administration, 2300 Washtenaw, Ann Arbor, Michigan

David G. Carvey, Ph.D., Agricultural Economist, U.S.
Department of Agriculture, Economic Research Service,
Natural Resource Economics Division, 1405 S. Harrison Rd.,
East Lansing, Michigan

James G. Robb, M.S., Research Assistant, Michigan
State University, Agriculture Hall, Michigan State
University, East Lansing, Michigan.

ABSTRACT

The integration of rainfall-runoff and economic models provides a basis for studying alternatives for adjusting and constraining resource use in accordance with social needs and the capacity of natural resources to support such activities. USDA's Economic Research Service (ERS) and the University of Iowa's Institute for Hydraulic Research (IIHR) have cooperated to develop and apply an integrated Watershed Evaluation and Research System (WATERS). WATERS incorporates rainfall-runoff modeling based on a distributed-parameter watershed model. WATERS analyses resource uses including agricultural alternatives with some soil conservation and water quality management practices. Rainfall events and resource uses are reflected through infiltration, overland runoff, sediment production, and overland transport models. Channel flow and transport are included. Economic effects are described through an agricultural budget generator. A multiple objective programming model uses hydrologic and economic information for simultaneous optimizations of generally non-commensurable objectives and related tradeoff analyses.

WATERS is applied to a small Iowa watershed to study hydrologic and economic relationships for several cropping and conservation management alternatives and a selected storm of record. Multiple optimizations were run for four objectives, of which two profit objectives are commensurable: maximize crop profits, maximize rotation profits, minimize soil loss, and minimize runoff. While maximums (minimums) generally aren't attained, analytical results suggest the following: if decision makers desire high crop profit, it will be accompanied by a low rotation profit, moderate runoff, and fairly high soil loss. Attaining a high rotation profit results in a moderate crop profit, high runoff, and high soil loss. If runoff reduction is emphasized, moderate crop and rotation profits result with fairly low soil loss. Emphasizing soil loss reduction, produces low crop and rotation profits with moderate runoff. This type of information can assist decision makers in forming value judgments concerning the relative level of importance of each objective in question thus facilitating the tradeoffs that are necessary to achieve a desired mix of objectives.

INTRODUCTION

Human activities often cause serious natural resource problems for which solutions traditionally involve the institution of controls on activities. Stronger demands for indicators of performance of natural resource programs are being made. Accountability is clearly the theme of natural resource and environmental policy in the foreseeable future. This is evidenced by legislation pertaining to the Resource Conservation Act and included Soil and Water Conservation Program, the Clean Water Act with its Rural Clean Water Program, and the National Forest Management Act. In the case of point source water pollution, evaluation of these controls is often related to monitoring results of controls. In this case, it is fairly easy to identify and analyze problems and evaluate solutions. Nonpoint source (NPS) pollution involving runoff and soil erosion from diffuse sources presents more difficulties in identifying and analyzing problems and evaluating alternative solutions. The legislation of controls such as soil loss and other pollutant limits, without an adequate system for analysis and evaluation, creates a situation of uncertainty on the part of land users and policy makers. Resource use decisions which consider productivity and related gains together with resource base conservation depend on an improved understanding of NPS problems, causes, and solutions.

The objective of this paper is to present an integrated interdisciplinary approach to the study of complex problems confronting public and private decision makers who guide watershed resource use and management. This is done by 1) describing a hydrologic-economic systems approach to watershed research and 2) analyzing hydrologic and economic implications of several agricultural land use and management alternatives.

WATERSHED EVALUATION AND RESEARCH SYSTEM OVERVIEW

The Watershed Evaluation and Research System (WATERS) was cooperatively developed by the University of Iowa's Institute of Hydraulic Research (IIHR) and the U.S. Department of Agriculture's Economic Research Service, Natural Resource Economics Division. Partial funding was provided by the Soil Conservation Service, USDA. WATERS represents the integration of a watershed simulation model (Croley et al, 1979) and an optimizing model incorporating hydrologic and economic data (Carvey and Robb, 1980). WATERS encompasses a watershed resource data base, a distributed-parameter hydrologic model, and economic and multiple objective tradeoff models. WATERS components and linkages are shown in Figure 1.

The resource data base includes many types of topographic, geometric, and land use data. Although this information allows the definition of a wide range of land uses and options, the present focus is on agricultural activities and their hydrologic and economic consequences. Cropping activities, tillages, and conservation management practices are reflected in watershed model parameters and related hydrologic effects. WATERS can accommodate competing non-agricultural activities through deletions to available cropland and through impacts on hydrologic parameters.

The watershed model includes storm event, infiltration, overland runoff and sediment production and transport, channel flow and sediment transport, and flood components. For a given storm event, the model simulates attendant runoff, sediment production and transport, and flood information.

Economic and analytical components include an agricultural budget generator which describes onsite economic consequences (costs and returns) of land use and management alternatives. A multiple objective model integrates economic and hydrologic information in an analytical framework allowing descrip-

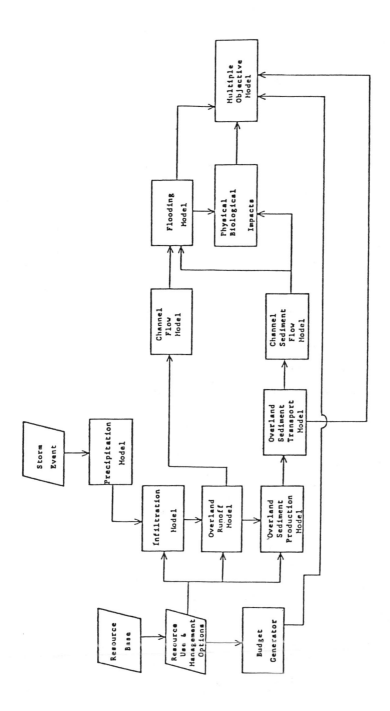

Figure 1. The Watershed Evaluation and Research System (WATERS).

231

tion of inherent physical and economic relationships and analysis of multiple, competing objectives.

WATERSHED MODEL

Description

Watershed runoff and sedimentation is represented, on a storm event basis, by a distributed parameter hydrologic model featuring detailed geometric description of watershed subareas. Outflows are sensitive to modifications in any portions of the watershed. Evaporation and transpiration are neglected while infiltration and interception are considered with the Soil Conservation Service (SCS) method (USDA Hydrology Guide). Kinematic wave models are used to determine distributions and routing of runoff and sediment produced by a known rainfall excess for each subarea. Finite difference schemes similar to the Lax-Wendroff Scheme are used in the numerical solution of the kinematic models (Croley et al, 1979). Flows at all points in the watershed, in subareas and channels, are determined at each point in time from the areal and temporal distribution of rainfall excess.

A physical model based on the bed shear stress of flow is applied to each subarea along with a modified form of the Universal Soil Loss Equation (USLE) to estimate the relative amount of sediment production. The capacity to transport sediment by overland flow is determined from flow depths and velocities available via the kinematic flow models. Sediment produced in excess of transport capacity is deposited wherever the imbalance occurs. Sediment transport at levels below transport capacity is described by using finite difference numerical solutions of sediment transport continuity equations. Sediment transport within channels presumes complete mixing where sediment enters channels. Transport in channels is described by finite difference solutions of the continuity equation in concert with the hydraulic flow solutions.

Arbitrarily fine definition can be given to any watershed where the surface is complex. This allows close approximation to spatially variable slope, roughness, infiltration, soil types, land uses, and rainfall. The model uses parameters observed from the watershed and does not rely on gage data although initial testing was performed on a gaged watershed. Most data are obtained from topographic, land use, and soils maps. Such data are used to determine nearly all geometric descriptions needed for modeling. In addition, they are useful in calculating infiltration, runoff, sediment production and transport, and hydraulic parameters.

Spatial variations in watershed parameters are incorporated into the model in a discrete, distributed-parameter approach. The hydrology of the watershed is represented by dividing the watershed into subcatchments about all tributaries of concern. Each subcatchment is further divided into sections. The boundaries of each section consist of a watershed divide as an upper bound, a drainageway as a lower bound, and the lines of steepest slope as the side boundaries. Each section is treated as a "streamtube" such that neither runoff nor sediment is considered across its side boundaries. Each streamtube is further divided into segments of sufficiently uniform characteristics, e.g., slope, land use, and cover. The result is that each segment is described by a lumped parameter model different from those of adjacent segments. Segments of each streamtube comprise a series (cascade) of elements in which each element discharges into the next. Streamtubes empty into channel segments which are also treated as a cascade.

The division of the watershed must also coincide with anticipated parameter changes for planned sub-divisions of the watershed if alternate

232

management strategies are to be analyzed. For urban and suburban settings, the watershed division may be easily accomplished since the boundaries of segments often correspond to physical entities, e.g.: paved highways, lawns, rooftops, and development plots. In rural settings, the division is not as easily made but is possible by inspecting soils, infiltration and groundwater parameters, land uses, cover, slopes, roughnesses, etc.

A network of nodes (points) is defined for ordering computations automatically and efficiently: streamtube midpoints on the channel, upstream ends of all channels, tributary junctions, points where channel characteristics change, and the watershed outlet. Each node is identified as to its type: streamtube inflow only, streamtube and channel inflows, two channel inflows, and channel inflow only. Other types of nodes with more than two inflows are represented by defining two or more nodes at the same point with zero channel length between them. Nodes are numbered in the desired computational order so that required computations are ready as needed (Croley, 1980).

Data Requirements

Geometric parameter values for every streamtube and channel segment are determined from topographic maps and by limited field survey as necessary. Manning's roughness coefficients were assigned to every streamtube segment from tables relating Manning's n to land surface types. Topographic, soils, and land use maps are used to identify surface type for all streamtube segments. A Manning's n value for each channel segment is determined from U.S. Geological Survey channel cross sections. Roughness values are computed where none are available by using onsite study of channel segments and relating values to channel depth, vegetation, obstructions, etc., with standard relations (Chow, 1964 and Barnes, 1967). An alternative is to use values from known points and interpolate with respect to distance and depth.

Thiessen weights are used to assign hyteographs to every streatube segment. Likewise, the SCS and USLE model parameters are compiled from soil and land use information for every streamtube segment. This work used 5-foot contour maps for a 3 square mile area divided into 70 subareas. Field data are required for culverts, bridge openings, and other features not present on maps. All data are assembled for streamtubes and channel segments according to nodes in the order required by the node computation algorithm. This is done by using the algorithm to output the required order of the data.

Model Calibration

Calibration begins by using total infiltration amounts for each of several test storms to compute the equivalent constant infiltration rate for the entire watershed. This is subtracted from the storm hyetograph to determine the rainfall excess intensities as a first approximation. Channel roughness values are adjusted from their observed values to calibrate timing and magnitude of peak flow rate. Adjustment is limited to varying all values proportionately for channel segments. Observed overland roughness values are not changed.

After flow parameters are determined, the sediment model is calibrated. In this model, coefficients for transport and detachment capacities and the exponent of shear stress are the only parameters not determined from work or observations in the watershed. Since the modeled watershed is steep and has fine sediment sizes, detachment is considered to control sediment movement. Therefore the transport capacity coefficient and the shear stress exponent are taken as consistent with findings of other investigators. The detachment capacity coefficient is adjusted for agreement between model estimates and observed sediment volumes.

After the first model calibration using constant infiltration rates, a modified SCS infiltration model is introduced. The SCS curve number (runoff coefficient), minimum infiltration rate, and antecedent moisture condition are determined for storms of different seasons in three steps. First, watershed runoff and rainfall volume coefficients are determined from SCS guidelines. Second, a least-squares regression analysis is used on data from several storms to determine the least-sum-of-squares error values of the curve number and minimum infiltration rate. These values are then used in a trial and error determination of the antecedent moisture condition prior to each storm for subsequent use. Third, the SCS procedure is then applied to each subarea in the distributed model. Minor adjustments of the coefficients are made for some subareas to achieve a reasonable match with selected storm hydrographs. After SCS model parameters are established, the detachment capacity coefficient is again adjusted to restore agreement with recorded sediment losses while using the better infiltration model. Ten storms were used in this determination with hourly data selectively taken from historical records to cover a range of conditions.

The hydrologic model is easily transferable, especially in the humid midwest. Storm data is used to transfer the model if infiltration and interception are considered with the SCS method. In the case of an ungaged watershed, this method is not used. New parameters for the hydrologic model are taken from SCS published guidelines.

ECONOMIC COMPONENTS

Agricultural Budget Generator

Onsite economic returns to farming due to land use and management decisions are estimated by the Multiple Objective Resource Evaluation (MORE) System Budget Generator (Economics and Statistics Service, USDA, 1980). This model combines preplant, planting, harvesting, management, and land costs with appropriate yields and prices to calculate net returns per hectare and per unit of production. This may be done for any combinations of 99 physiographic areas and 9 treatment (management) needs in any one budget generator run. Treatments may be changed for any or all areas for the purpose of analyzing a wider range of alternatives. All such changes are reflected in the hydrologic description of affected watershed subareas.

More specifically, the generator calculates onsite agricultural costs and returns for each cropping alternative, tillage, and conservation management practice selected for analysis. Preplant or operating capital costs include seed and chemical applications. Planting and harvesting costs include those for all machinery, labor, storage, and drying. A variety of methods are available for calculating management and land costs. Fuel costs may also be calculated separately. Efficiency adjustments may be made to account for impacts of conservation management treatments on time and cost requirements for production activities.

Multiple Objective Model

A traditional approach to the analysis of public and private sector resource allocation problems has been the use of a single objective model with appropriate resource constraints and activity demands. However, the solution of resource allocation problems is often subject to more than one type of consideration (or objective). Considerations such as economic efficiency, land use contributions to flooding, and soil loss may be extremely important to those with a common interest in resolving a problem. Single objective analysis requires the collapse of such noncommensurable objectives (without common units of measurement) into a single overriding objective. If economic efficiency is chosen, all solution effects of a given project must be translated into economic units of measurement. In this sense, the single objective approach requires knowledge of relationships between objectives and the definition of an economic proxy for utility prior to optimization. The difficulty with this is that often the analyst has little or no information on which to base the determination of monetary equivalents for noneconomic effects (Cohon, 1978). The result of such an approach is a single optimal solution which hides realistic relationships between economic and other related aspects of the original problem.

The approach used here encompasses noncommensurable multiple objectives in a framework of optimal targets for each objective in accordance with resource constraints. The multiple objective model does not require a pre-optimization utility approximation. It can systematically analyze a range of alternatives available to decision makers, providing them with information concerning possible tradeoffs in relevant units of measurement. This flow of information facilitates the formation of decision makers' value judgments about the relative importance of their objectives. This approach discards the concept of one optimal solution for a set of efficient solutions. This set of solutions outlines differences between objectives and shows how alternatives contribute to meeting objectives, thus providing a basis for finalizing value judgments and decisions.

The tradeoff function in Figure 2 shows the competitive range between two common agricultural objectives: minimizing soil loss and maximizing crop profits. At point A soil loss is minimized. At point B short run profits are maximized. The curve can be thought of as minimizing the disutility between resource allocations A and B for the infinite number of combinations of crop profits and soil loss. However, this illustration is not satisfactory because the two objective functions are not completely independent. Soil loss does not occur independently of income producing activities. To more clearly see the implications of tradeoffs, one must 'look behind the curve' to the activities in solution for a given point on the tradeoff function.

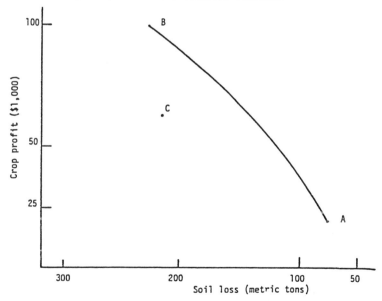

Figure 2. Estimated crop profit and soil loss tradeoff function (extreme values computed for North Branch Ralston Creek Watershed).

Multiple objective analysis is considered to be a partial analysis, meaning that the 'best planning solution may lie in the inferior region' (Brill, 1979). That is, a two objective analysis may not provide the 'best' solution if related problems exist but are not included in the analysis. For example, if a flooding problem is hydrologically related to levels of crop profit and soil loss, analytical objectives should include profit, soil loss, and flooding. Consideration of only profit and soil loss objectives will not provide information as to their most appropriate levels for obtaining some desired level of relief from flooding. In terms of Figure 2, the 'best' solution may occur at some inferior point C which reflects those levels of crop profit and soil loss which accommodate a desired solution to the flooding problem. The multiple objective approach facilitates sensitivity analyses of alternatives representing various combinations of planning objectives.

The linear multiple objective model used in WATERS has an overall objective function specifying that the model simultaneously optimize the attainment of all individual objectives. This is done in accordance with resource constraints and targets established by individual optimizations on each objective considered. Generally, maximums and minimums are not attained due to the influence of all objectives. The model uses reverse slacks and weights on objectives (Cohon). An objective's weight represents the relative level of importance placed on the achievement of that objective. A computational form of the multiple objective programming model may be expressed as:

$$\text{Minimize } z = (\lambda_1 I_1 x_{ijkl}) + (\lambda_2 I_2 x_{ijkl}) + (\lambda_3 S x_{ijkl}) + (\lambda_4 R x_{ijkl}) \qquad (1)$$

$$\text{Subject to: } Ax_{ijkl} \leq B$$

$$x_{ijkl} \geq 0$$

Where: λ_i = relative weights (objective coefficient marginals) for four objectives;

I_1, I_2 = vectors of crop and rotation profits;

x_{ijkl} = vector of crop production activities with i crops (4), j rotations (4), k tillages (2), and l residue levels (2);

S = a vector of soil loss coefficients;

R = a vector of runoff coefficients;

A = a matrix of interaction coefficients among crop production activities and available resources; and

B = a vector of available resources.

WATERS APPLICATION

Background

WATERS is applied to a 769 hectare watershed on the outskirts of Iowa City, Iowa. About 283 hectares are in corn, oats, soybeans, and alfalfa production in various rotations. Woodland covers nearly 182 hectares while permanent meadow grows on 279 hectares. Part of this meadow is contained in a nature park in the lower reaches and a golf course in the upper watershed. About 27 hectares are currently in residential development. Watershed topography is strongly rolling to rough and soils are predominantly silt loams.

For this application of multiple planning objectives in small watershed management, analyses focus on agricultural land use and conservation management activities on the 283 hectares of cropland. Land use on the remaining 486 hectares is held constant in all analyses. The watershed is divided into 70 streamtubes for modeling purposes. Cropland is found in 32 of these, primarily in the upper reaches (Map 1). The storm event selected for this application has an expected recurrence level of about 3.5 years, based on analysis of annual maximum storm volumes. Research funds limited this test application to a single rainfall event at approximately 65 percent crop cover. Plans included analysis of an expected pattern of events from preplant through harvest. However, the point of this application is not lost. Given a range of land use and conservation management alternatives, a rainfall event (or series of events) will produce related hydrologic effects. These can be jointly studied with effects of economic activities in such a way as to provide more complete information concerning the relationships between several noncommensurable planning objectives.

Based on a farm operator survey, resource uses analyzed include 18 crop activities in four rotations, two tillages, and two residue levels. Two fallow conditions representing periods of no crop growth are also studied at two residue levels. Fallow conditions provide baseline information under circumstances most suitable for runoff and sediment production and transport. These 20 resource uses are studied in terms of their hydrologic and economic impacts. They represent 20 nonstructural alternatives for controlling agricultural runoff and soil loss--agricultural contributions to flooding and nonpoint source pollution. Relationships between the economic returns to various resource uses and associated runoff and soil loss are analyzed. A dry reservoir, designed to temporarily retard storm flow, is included in the hydrologic analysis. However, a comparative economic analysis of the structure is beyond the scope of this application. The structure is designed to protect development outside the watershed and involves no opportunity costs or benefits to land use activities within the watershed.

Hydrologic Impacts

Hydrologic impacts of alternative crop activities, tillages, and residue levels for the watershed and for those streamtubes containing cropland are presented in Table 1. The least runoff was achieved with no till planting of soybeans under high residue conditions following corn (run 15). Corn crops with reduced tillage and low residue levels produced the highest runoff (runs 2,5,8, and 11). Both fallow conditions exceeded all cropping alternatives in runoff production (runs 19 and 20).

Least soil loss is achieved with hay (run 18). No till corn following hay (run 3) and no till corn before soybeans (run 6) under high residue levels were close to hay in keeping soil loss low. Greatest soil loss for cropping activities is found with reduced tillage soybeans following corn leaving low residue (run 14). Both fallow conditions far outstrip cropping activities in soil losses (runs 19 and 20).

All cropland in various corn activities, produces about 60 percent of all watershed runoff and from 50 to 85 percent of its soil loss. No till activities contribute the least runoff and soil loss. Planting all cropland in no till soybeans after corn with high residue produces 55 percent of all runoff and 58 percent of total soil loss. Reduced till soybeans after corn with low residue produces 85 percent of all soil loss. For high residue conditions,

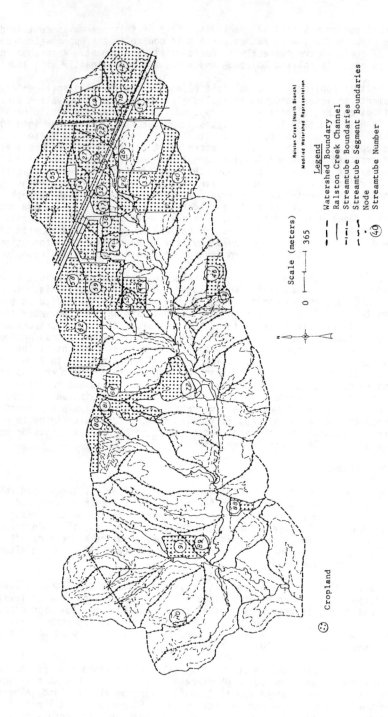

Map 1. North Branch Ralston Creek Watershed cropland by streamtube.

Legend

Watershed Boundary
Ralston Creek Channel
Streamtube Boundaries
Streamtube Segment Boundaries
Node
40 Streamtube Number

Scale (meters)

0   365

Ralston Creek (North Branch)
Modified Watershed Representation

Cropland

Table 1. North Branch Ralston Creek Watershed hydrologic data: runoff and soil loss by crop and management activity.[1]

Land Use[2]	Management[3] Tillage[3]	Residue[4]	Runoff Cropland (mil. cubic meters)	Runoff Per Hectare (cubic meters)	Cropland Runoff as a Percent of Watershed Runoff (percent)	Soil Loss Cropland (metric tons)	Soil Loss Per Hectare (metric tons)	Cropland Soil Loss as a Percent of Watershed Soil Loss (percent)
1. Corn after hay[5]	reduced	high	1.443	5,101.3	60	149.8	.53	57
2. "		low	1.547	5,466.8	60	325.2	1.15	74
3. "	no till	high	1.395	4,931.8	59	117.6	.42	51
4. Corn before soybeans[5]	reduced	high	1.443	5,101.3	60	149.8	.53	57
5. "		low	1.547	5,466.8	59	325.2	1.15	74
6. "	no till	high	1.395	4,931.8	60	117.6	.42	51
7. Corn continuous	reduced	high	1.443	5,101.3	60	223.6	.79	67
8. "		low	1.547	5,466.8	60	574.9	1.82	82
9. "	no till	high	1.395	4,931.8	59	163.5	.58	59
10. Corn after soybeans	reduced	high	1.443	5,101.3	60	514.9	1.82	82
11. "		low	1.547	5,466.8	60	619.2	2.19	85
12. "	no till	high	1.395	4,931.8	59	323.8	1.14	74
13. Soybeans after corn	reduced	high	1.299	4,592.9	57	238.9	.84	68
14. "		low	1.395	4,931.8	59	635.3	2.24	85
15. "	no till	high	1.208	4,270.9	55	155.2	.55	58
16. Oats after corn	no till	high	1.347	4,760.3	58	335.3	1.18	75
17. "	reduced	high	1.395	4,931.8	62	553.9	1.96	83
18. Hay	-		1.299	4,592.9	57	80.3	.28	42
19. Fallow	-	high	1.660	5,869.1	62	1,023.9	3.62	89
20. "	-	low	1.806	6,382.2	65	1,367.5	4.83	92

1. Only the 283 hectares in crop production are considered. All calculations are based on crop cover of 50 to 75 percent.
2. Land use alternatives were developed from four basic rotations: CCC, CCCOMM, CBCOMM, and COMM where C = corn, B = soybeans, O = oats, and M = hay.
3. Tillages include reduced or spring chisel plowing and disking and no till spray-plant operations.
4. Residue levels are high at 5,044 and low at 2,242 kilograms per hectare.
5. These activities produce the same hydrologic results but are considered different activities due to different economic costs.

reduced till soybean activities produce 68 percent of all soil loss. With all
cropland in hay, runoff falls to 57 percent of the watershed total. The hay
condition produces only 42 percent of all watershed soil loss.

A comparison of the impacts of all cropping activities (nonstructural
alternatives) on runoff at the structure site is presented in Table 2. The
best control is with no till soybeans in high residue (run 15). This is
closely followed by both reduced tillage soybeans in high residue (run 13) and
hay (run 18) which provide the same control. The hay-reservoir alternative,
combining the runoff control of hay and the temporary flow retarding structure,
provides the greatest downstream flow control.

Economic Impacts

Economic returns to cropping activities are considered on short run
(annual) and long run (years in a given rotation) planning periods. Annual
crop budgets and net returns are reported by tillage and residue level in
Table 3. Based on costs and prices used in the budgets, all three soybean
activities (runs 13, 14, and 15) have greater annual returns than other
cropping alternatives. The reduced tillage, high residue condition is the most
profitable of the three. Corn after hay and before soybeans (runs 4, 5, and 6)
is the next most profitable set of activities on an annual basis. When rotation profits are considered, the three continuous corn activities (runs 7, 8,
and 9) provide the greatest returns by a wide margin (Table 4). The no till
continuous corn activity is the overall leader. Profits of all other rotations
are impacted by negative returns due to establishment costs of hay and
associated oats.

Flood protection is not needed within the modeled watershed. However,
some consideration can be given to possible impacts on agricultural production
and returns if the watershed is to be managed for downstream flood protection.
The government of Iowa City might desire protection in addition to that
offered by the dry reservoir structure. Yet, it might not be willing to re-
strict agricultural activities to the exclusion of all crops except hay, which
is the most effective long run activity for controlling both runoff and soil
loss (run 18, Table 2).

The hydrologic data presented thus far, suggest that a cropping activity
involving runoff control from no till soybeans, the soil loss control from
no till corn after hay, and the overall control provided by hay and associated
oats might be quite effective. Such an alternative could be considered from
the regulatory standpoint. The cost of such regulation would include the
opportunity costs of foregoing any more profitable activities such as continuous
corn, the most profitable rotation, as well as significant administrative and
enforcement costs. If conditions warranted even greater control of agricultural
runoff and sediment, perhaps new rotations could be developed which emphasize
soybeans while decreasing the hay influence. This might provide some economic
advantage to farmers due to increased production of higher valued soybeans
while decreasing production of lower valued hay and oats on a rotation basis.
Resultant benefits might approach the economic benefits of continuous corn
production. Additional control over watershed runoff could also be achieved
with numerous other smaller temporary flow retarding structures throughout
the watershed. The analysis of such alternatives is beyond the scope of this
application.

Multiple Objective Linear Programming

The multiple objective programming model in WATERS uses hydrologic and
economic coefficients from the watershed model and agricultural budget and
returns generator as inputs to simultaneous optimizations of four generally

Table 2. North Branch Ralston Creek Watershed hydrologic data at proposed structure site (node 114): land use (nonstructural) and structural alternatives (for a selected storm).[1]

Land Use[2]	Conservation Management		In Channel (cms)	Peak Flood Elevation (Meters)	Channel Width (Meters)
	Tillage[3]	Residue[4]			
1. Corn after hay	reduced	high	199.9	208.2	52.2
2. "	"	low	217.3	208.2	53.2
3. "	no till	high	200.0	208.2	52.2
4. Corn before soybeans	reduced	high	199.9	208.2	52.2
5. "	"	low	217.3	208.2	53.2
6. "	no till	high	200.0	208.2	52.2
7. Corn, continuous	reduced	high	199.9	208.2	52.2
8. "	"	low	217.3	208.2	53.2
9. "	no till	high	200.0	208.2	52.2
10. Corn after soybeans	reduced	high	199.9	208.2	52.2
11. "	"	low	217.3	208.2	53.2
12. "	no till	high	200.0	208.2	52.2
13. Soybeans after corn	reduced	high	159.8	208.0	49.8
14. "	"	low	180.2	208.1	51.1
15. "	no till	high	155.3	208.0	49.5
16. Oats after corn	reduced	high	167.3	208.1	50.3
17. "	"	low	180.2	208.1	51.1
18. Hay	-	-	159.8	208.0	49.8
19. Hay, with reservoir	-	-	22.1	206.5	8.3

1. Only the 283 hectares in crop production are considered. All computations are based on crop cover of 50 to 75 percent.
2. Land use alternatives were developed from four basic rotations: CCC, CCCOMMM, CBCOMMM, and COMMM, where C = corn, B = soybeans, O = oats, and M = hay.
3. Tillages include reduced or spring chisel plowing and disking and no till spray-plant operations.
4. Residue levels are high at 5,044 and low at 2,242 kilograms per hectare.

241

Table 3. North Branch Ralston Creek Watershed crop budgets and annual returns.

Land Use[1]	Management Tillage[2]	Residue[3]	Chemicals	Fuel	Preplant	Operations	Total Variable Costs	Net returns Per Hectare
					(dollars)			
1. Corn after hay	reduced	high	35.58	13.79	141.27	187.94	329.21	315.57
2. "	"	low	34.59	13.79	139.06	187.94	327.00	317.79
3. "	no till	high	59.15	9.56	167.63	169.43	337.06	307.73
4. Corn before soybeans	reduced	high	36.82	13.79	142.57	187.94	330.51	314.26
5. "	"	low	35.82	13.79	140.38	169.43	328.32	316.48
6. "	no till	high	69.41	9.34	178.53	186.82	347.96	296.84
7. Corn, continuous	reduced	high	35.58	13.54	169.42	186.82	356.24	288.53
8. "	"	low	34.59	13.54	167.21	166.69	354.03	290.73
9. "	no till	high	39.26	9.19	174.99	186.82	341.68	303.09
10. Corn after soybeans	reduced	high	35.58	13.54	169.42	186.82	356.24	288.53
11. "	"	low	34.59	13.54	167.21	166.69	354.03	290.73
12. "	no till	high	39.26	9.19	174.99	166.69	341.68	303.09
13. Soybeans after corn	reduced	high	19.77	11.19	61.57	108.75	170.32	349.20
14. "	"	low	39.54	8.72	81.86	108.75	190.61	328.91
15. "	no till	high	49.42	6.84	95.97	88.61	184.58	334.94
16. Oats after corn	reduced	high	-	11.46	157.05	102.78	259.83	-33.70
17. "	"	low	-	11.46	154.43	102.77	257.20	-31.08
18. Hay	-	-	-	0.37	43.46	2.42	45.88	127.72

1. Land use alternatives were developed from four basic rotations: CCC, CCCOMMM, CBCOMMM, and COMMM where C = corn, O = oats, B = soybeans, and M = alfalfa hay.
2. Tillages include reduced or spring chisel plowing and disking and no till spray-plant operations.
3. Residue levels are high at 5,044 and low at 2,242 kilograms per hectare.

Table 4. North Branch Ralston Creek Watershed annual profits to selected crop rotations.

Land Use	Management Tillage[1]	Residue	Annualized Profits per Hectare (dollars)
1. Continuous corn	reduced " no till	high low high	288.53 290.73 303.09
2. 3 yr. corn, 1 yr. oats, 3 yr. hay	reduced " no till	high low high	147.52 148.90 150.56
3. 1 yr. corn, 1 yr. soybeans, 1 yr. corn, 1 yr. oats, 3 yr. hay	reduced " no till	high low high	155.99 154.08 153.55
4. 1 yr. corn, 1 yr. oats, 3 yr. hay	reduced " no till	high low high	91.10 92.07 89.52

1. Tillage management does not apply to oat and hay activities.

competing objectives. These are: 1) maximize annual crop profits, 2) maximize rotation profits, 3) minimize soil loss, and 4) minimize runoff.

Land use and conservation management alternatives and levels of crop and rotation profits, runoff, and soil loss achieved are summarized for selected optimizations. To the extent possible, both "within" and "between" solution results are summarized. Within any given single or multiple objective optimization, the levels of goal achievement portray relationships between objectives given resource constraints, resource use alternatives, and relative weightings on objectives. Solution specific tradeoff information is available within any optimization. Comparisons of results between given optimizations, allow the study of impacts of alternative choices (weightings) available to decision makers in their task of setting priorities for one or more planning objectives. Watershed level tradeoff information is determined by comparing optimal solutions. Results of all optimizations are displayed in respective rows of Table 5. For each optimization, the proportion of each objective maximum (minimum) achieved is given in appropriate units and percent.

Single objective analysis

Results of four single objective optimizations are presented in rows 1 through 4 of Table 5. When crop profit is maximized (row 1), rotation profit is just over five-tenths of its maximum, runoff is less than one-tenth above minimum, and soil loss is more than three times its minimum. The alternative is all land in reduced tillage soybeans after corn with high residue (Table 6). Rotation profit maximization (Table 5, row 2) provides an equal level of crop profit, produces runoff at substantially above one-tenth its minimum, and allows soil loss over twice its minimum. In this solution all land is in no till continuous corn with high residue (Table 6). Runoff minimization (Table 5, row 3) allows achievement of nearly maximum crop profit but only five-tenths of rotation profit maximum. In this analysis soil loss is twice its minimum. These results are achieved with no till soybeans after corn with high residue (Table 6). Soil loss minimization (Table 5, row 4) allows only two-tenths of crop profit and one-half of rotation profit maximums. Runoff is less than one-tenth above minimum. The minimum soil loss solution is all cropland in meadow in a CCCOMMM rotation (Table 6).

Unit weighted multiple objective analysis

An optimization with four equally weighted economic and hydrologic objectives provides the baseline conditions from which the constrained multiple objective optimizations begin (Table 5, row 5). In this LP, about nine-tenths of both crop and rotation profit maximums are achieved. Runoff is relatively high, over one-tenth above minimum, while soil loss is about six-tenths above minimum. This solution requires more than eight-tenths of all cropland in no till continuous corn with high residue, over one-tenth in no till soybeans after corn with high residue, and just under one-tenth in no till corn after meadow (Table 6). Unit weighting moderates achievement of objective levels from the extreme values.

Constrained multiple objective analysis

Four series of constrained optimizations are obtained totaling 77 individual solutions. Only four of 77 constrained solutions are discussed, namely those in which the highest level for each parametrically weighted objective was achieved.

In the crop profit LP, over nine-tenths of its respective maximum is achieved (Table 5, row 29). Rotation profit falls sharply from baseline level to less than six-tenths of its maximum and runoff falls to about one-tenth

Table 5. Linear programming results depicting relationships between economic and hydrologic objectives: North Branch Ralston Creek Watershed cropland.

Optimizations	Crop Profit (dollars)	(percent)	(weight)	Rotation Profit (dollars)	(percent)	(weight)	Runoff (cubic meters)	(percent)	(weight)	Soil Loss (metric tons)	(percent)	(weight)
1. Crop profit max. LP	99,283[1]	100.0	1.000	44,349	51.4	0	120,726	107.5	0	238.9	309.8	0
2. Rotation profit max. LP	86,174	86.7	0	86,174	100.0	1.000	129,634	115.4	0	163.5	212.0	0
3. Runoff min. LP	95,230	95.9	0	43,654	50.6	0	112,261[1]	100.0	1.000	155.2[1]	201.3	0
4. Soil loss min. LP	21,114	21.2	0	42,853	49.7	0	120,136	107.0	0	77.1[1]	100.0	1.000
5. Equal weighted objectives LP	87,267	87.8	1.000	77,929	90.4	1.000	127,569	113.6	1.000	144.4	159.2	1.000
6. Crop profit constrained LP	87,390[1]	88.0	2.182	77,537	89.9		127,497	113.5		122.4	158.6	
7. "	87,423	88.0	2.613	76,459	88.7		127,412	113.5		117.9	152.8	
8. "	87,554	88.1	2.962	76,043	88.2		126,991	113.4		117.9	152.8	
9. "	88,161	88.7	2.989	74,049	85.9		126,979	113.1		118.0	153.0	
10. "	88,179	88.8	3.010	74,103	85.9		126,936	113.1		118.0	153.0	
11. "	88,245	88.8	3.036	73,837	85.6		126,907	113.0		118.0	153.0	
12. "	88,289	88.9	3.072	73,696	85.5		126,617	112.7		118.2	153.4	
13. "	88,708	89.3	3.076	72,361	83.9		126,506	112.6		118.4	153.5	
14. "	88,878	89.3	3.086	71,816	83.3		126,454	112.6		118.4	153.5	
15. "	88,954	89.5	3.097	71,577	83.0		126,164	112.3		118.7	153.9	
16. "	89,398	90.0	3.110	70,161	81.4		126,017	112.2		118.9	154.1	
17. "	89,622	90.2	3.116	69,445	80.5		125,649	111.9		119.3	154.7	
18. "	90,186	90.8	3.139	67,645	78.4		124,672	111.0		120.6	156.4	
19. "	91,681	92.3	3.146	62,875	72.9		125,000	111.3		121.9	158.2	
20. "	91,807	92.4	3.148	62,897	72.9		124,885	111.2		122.2	158.4	
21. "	91,984	92.6	3.169	62,331	72.3		124,357	110.7		123.2	159.7	
22. "	92,760	93.4	3.243	59,857	69.4		124,170	110.6		123.6	160.3	
23. "	93,039	93.7	3.313	58,965	68.4		123,643	110.1		125.0	162.1	
24. "	93,802	94.4	3.350	56,533	65.6		123,085	109.5		126.6	164.2	
25. "	94,657	95.3	3.365	53,806	62.4		122,794	109.3		127.4	165.2	
26. "	95,077	95.7	3.388	52,464	60.8		122,532	109.1		128.5	166.6	
27. "	95,477	96.1	3.524	51,188	59.4		122,436	109.1		128.8	167.1	
28. "	95,618	96.3	3.541	50,740	58.8		122,108	109.0		133.3	172.9	
29. "	95,790	96.4	3.732	50,191	58.2			108.7				
30. Constrained rotation profit	86,734	87.3	1.000	80,430	93.3	1.062	128,578	114.5		130.8	169.7	1.000
31. "	86,453	87.0		81,752	94.8	1.069	129,237	115.1		134.9	174.9	
32. "	86,284	86.9		82,547	95.7	1.175	129,634	115.4		137.7	178.6	
33. "	86,263	86.8		83,218	96.5	1.381	129,634	115.4		141.8	183.9	
34. "	86,173	86.7		86,173	100.0	1.642	129,634	115.4		163.5	212.0	
35. Constrained runoff	87,352	87.9	1.000	77,530	89.9	1.000	127,432	113.5	3.617	122.2	158.4	1.000
36. "	87,772	88.4		75,559	87.6		126,624	112.7	4.106	121.1	157.0	
37. "	87,864	88.4		75,126	87.1		126,408	112.6	4.774	121.2	157.2	
38. "	88,155	88.7		73,762	85.5		125,849	112.1	5.349	121.1	157.1	
39. "	88,207	88.8		73,519	85.3		125,727	111.9	5.442	121.1	157.1	
40. "	88,218	88.8		73,463	85.2		125,120	111.4	5.452	121.2	157.1	
41. "	88,508	89.1		72,105	83.6		124,156	110.5	5.464	121.2	157.1	
42. "	89,034	89.6		69,633	80.8		123,938	110.4	5.513	121.3	157.3	
43. "	89,152	89.7		69,079	80.1		122,918	109.4	5.545	121.3	157.2	
44. "	89,688	90.3		66,564	77.2		122,733	109.3	5.545	121.3	157.3	
45. "	89,785	90.4		66,107	76.7		122,565	109.1	5.595	121.4	157.4	
46. "	89,875	90.5		65,684	76.2		122,201	108.8	5.599	121.4	157.3	
47. "	90,068	90.7		64,777	75.1		122,116	108.7	5.611	121.4	157.4	
48. "	90,114	90.7		64,561	74.9			108.7	5.650	121.4	157.4	
49. "	91,147	91.8		59,712	69.2		120,198	107.0	5.651	121.4	157.4	
50. "	91,177	91.8		59,569	69.1		120,142	107.0	5.659	121.4	157.5	

245

Table 5. (Cont.)

Optimizations	Crop Profit (dollars)	(percent)	(weight)	Rotation Profit (dollars)	(percent)	(weight)	Runoff (cubic meters)	(percent)	(weight)	Soil Loss (metric tons)	(percent)	(weight)
51. *	91,484	92.1	·	58,129	67.4	·	119,601	106.5	5.660	121.5	157.5	·
52. *	91,874	92.5	·	56,300	65.3	·	118,849	105.8	5.667	121.5	157.5	·
53. *	91,996	92.6	·	55,725	64.6	·	118,622	105.8	5.678	121.5	157.6	·
54. *	92,151	92.8	·	54,998	63.8	·	118,334	105.4	5.679	121.5	157.6	·
55. *	92,742	93.4	·	52,225	60.6	·	117,237	104.4	5.679	121.6	157.7	·
56. *	93,018	93.6	·	50,928	59.0	·	116,724	103.9	5.746	124.3	157.8	·
57. *	93,137	93.8	·	50,370	58.4	·	116,503	103.7	5.802	125.8	163.1	·
58. *	87,300	87.9	·	76,852	89.1	1.000	127,569	113.6	1.000	118.3	153.4	·
59. Constrained soil loss	82,459	83.0	1.000	76,957	89.3	·	126,955	113.0	·	104.4	135.4	1.053
60.	82,544	83.1	·	76,558	88.8	1.000	126,818	112.8	·	103.8	134.6	1.478
61.	81,445	82.0	·	76,582	88.8	·	126,685	112.8	·	101.5	131.6	1.906
62.	81,500	82.0	·	74,768	86.7	·	126,685	112.6	·	98.4	127.5	2.033
63.	79,735	80.3	·	74,807	86.8	·	126,420	112.6	·	95.7	124.1	2.538
64.	79,760	80.3	·	73,966	85.8	·	126,420	112.6	·	94.8	123.0	2.780
65.	80,180	80.7	·	71,994	83.5	·	125,612	111.8	·	93.8	121.6	4.294
66.	80,277	80.8	·	68,803	79.8	·	125,612	111.8	·	91.7	118.9	5.352
67.	80,291	80.8	·	68,362	79.3	·	125,612	111.8	·	91.5	118.6	6.900
68.	77,320	77.8	·	68,426	79.4	·	125,249	111.5	·	89.7	116.4	7.003
69.	77,338	77.8	·	67,857	78.7	·	125,249	111.5	·	89.4	116.0	7.352
70.	77,378	77.9	·	66,534	77.2	·	125,249	111.5	·	88.9	115.3	8.165
71.	77,392	77.9	·	66,069	76.6	·	125,249	111.5	·	88.7	115.0	10.960
72.	77,434	77.9	·	64,677	75.0	·	125,249	111.5	·	88.2	114.4	11.197
73.	72,801	73.3	·	64,718	75.1	·	125,734	112.0	·	86.5	112.2	11.748
74.	71,424	71.9	·	61,921	71.8	·	125,568	111.8	·	86.1	111.6	12.224
75.	71,510	72.0	·	59,398	68.9	·	125,568	111.8	·	85.1	110.4	12.788
76.	71,586	72.1	·	58,473	67.8	·	125,568	111.8	·	84.4	109.4	13.220
77.	71,615	72.1	·	55,907	64.8	·	125,568	111.8	·	84.2	109.1	14.748
78.	71,693	72.2	·	55,923	64.8	·	125,568	111.7	·	83.6	108.3	16.619
79. *	70,970	71.4	·	56,037	65.0	·	125,459	111.1	·	83.4	108.1	19.527
80. *	65,743	66.2	·	58,603	68.0	·	124,829	106.7	·	82.3	106.7	20.065
81. *	64,810	65.2	·			·	124,610	111.0	·	82.1	106.5	20.128

1. Single objective maximums (minimums).

246

Table 6. North Branch Ralston Creek Watershed agricultural land use and conservation management allocations.

Table 6. North Branch Ralston Creek Watershed agricultural land use and conservation management allocations.

Land Use	Single Objective LP				Multiple Objective LP				
	Crop profit maximization	Rotation profit maximization	Runoff minimization	Soil loss minimization	Unit weight optimization	Crop profit optimization	Rotation profit optimization	Runoff optimization	Soil loss optimization
				-------hectares-------					
Corn, high residue									
No till, after hay, CCCOMHH					23.78 (8.4%)	30.85 (10.8%)		23.78 (8.4%)	82.57 (29.0%)
No till, continuous	284.46 (100.0%)				229.78 (80.8%)	44.97 (15.8%)	284.46 (100.0%)	49.14 (17.2%)	83.04 (29.2%)
Soybeans, high residue									
Reduced till, after corn, CBCOMHH	284.46 (100.0%)[1]					186.59 (65.5%)			
No till, after corn, CBCOMHH		284.46 (100.0%)			30.89 (10.8%)	22.05 (7.8%)		211.53 (74.4%)	30.01 (10.6%)
Hay									
CCCOMHH				267.63 (94.1%)					
CBCOMHH									88.82 (31.2%)

1. Percent of cropland in a given land use.

247

above minimum. Soil loss increases moderately over baseline to about seven-tenths above minimum. Reduced till soybeans after corn with high residue, an activity not in the baseline solution, successfully competes for more than six-tenths of all cropland (Table 6). Most of this land use shift involved land in no till continuous corn with high residue in the baseline solution. Continuous corn, representing eight-tenths of all baseline crops, occupies less than two-tenths in the constrained crop profit solution. No till corn after meadow with high residue increases slightly to just over one-tenth of all cropland. Land in no till soybeans after corn with high residue decreases slightly to less than one-tenth of all cropland.

The constrained rotation profit LP (Table 5, row 34) duplicated the single objective rotation profit maximization (Table 5, row 2). This solution requires all cropland in no till continuous corn with high residue (Table 6), an increase from the baseline solution. The two-tenths of cropland in no till corn after meadow with high residue and no till soybeans after corn with high residue in the baseline solution shifts to no till continuous corn with high residue.

The constrained runoff LP (Table 5, row 57) has crop profit increasing from baseline level to over nine-tenths of maximum while rotation profit falls to under six-tenths maximum. Runoff nears minimum and soil loss increases to six-tenths over minimum. These changes reflect an increase in no till soybeans after corn with high residue from one-tenth of all cropland under baseline conditions to over seven-tenths of cropland acreage (Table 6). All of this shift occurs on land in no till continuous corn with high residue in the baseline solution. Land in no till corn after meadow with high residue, amounting to less than one-tenth of available cropland, remains in the same use.

The constrained soil loss LP (Table 5, row 81) requires crop and rotation profit decreases from baseline levels to under seven-tenths of their maximums. Runoff decreases slightly to about one-tenth of minimum while soil loss decreases to well under one-tenth of minimum. From baseline conditions, over three-tenths of all cropland shifts to meadow in a CBCOMMM rotation (Table 6). Land in no till soybeans after corn with high residue remains at about one-tenth of all cropland although some locational shifts occur.

Tradeoff and Land Use Analysis

Multiple objective programming solutions contain information concerning shadow prices or opportunity costs of marginal changes from solution conditions. This information allows analysis of tradeoffs between objectives without common units of measurement. However, economic and hydrologic effects of marginal changes must be interpreted in accordance with equation (2) which can be simply stated as:

objective function = (weight x crop profit) + (weight x rotation profit) +
    (weight x runoff) + (weight x soil loss)   (2)

The objective function value, objective levels, and weights are provided for each solution. A composite economic and hydrologic opportunity cost for an added activity at a site in the watershed is also provided in the data output. However, this shows only the magnitude and direction of the change in the value of the composite objective function. Effects of a marginal change from a given solution must be disaggregated. This is done by using data input files to determine crop and rotation profit, runoff, and soil loss effects of changing land use at a given location. These marginal effects, which must be combined with objective levels of the solution, are the tradeoffs between objectives for the proposed change from solution conditions. Figures 3 through 6 display tradeoff functions between objectives in this study.

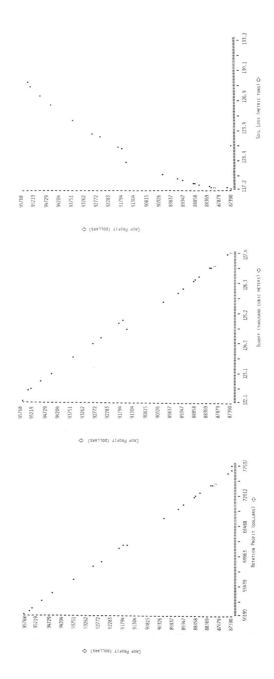

Figure 3. Tradeoff curves for North Branch Ralston Creek Watershed land use allocation objectives: emphasizing crop profit (arrows indicate relationships between objectives).

249

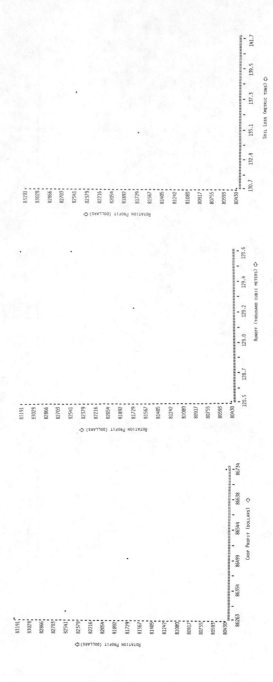

Figure 4. Tradeoff curves for North Branch Ralston Creek Watershed land use allocation objectives: emphasizing rotation profit (arrows indicate relationships between objectives).

250

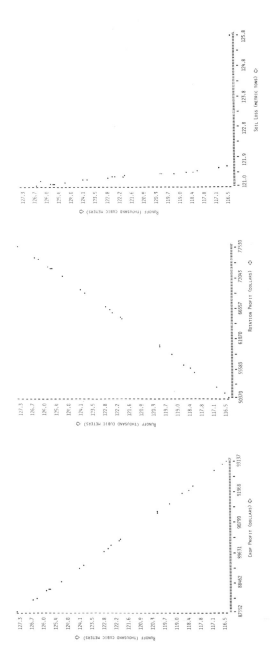

Figure 5. Tradeoff curves for North Branch Ralston Creek Watershed land
use allocation objectives: emphasizing runoff reduction (arrows
indicate relationships between objectives).

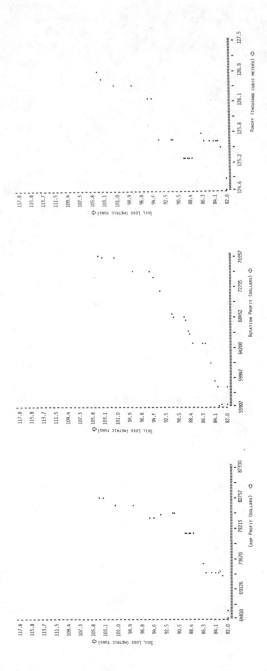

Figure 6. Tradeoff curves for North Branch Ralston Creek Watershed land use allocation objectives: emphasizing soil loss reduction (arrows indicate relationships between objectives).

In evaluating consequences of watershed resource use, those seeking to improve watershed conditions must know not only what but where adjustments are needed to achieve a desired mix of objectives. The multiple objective approach of WATERS specifies optimal types and locations of adjustments in accordance with objectives. As an example, baseline land use allocation when all objectives are considered equally important, is shown in Map 2. Changes in this allocation as runoff reduction is optimized, are determined by comparing Map 2 and 3. The primary shift is from no till corn to reduced till soybeans. Such information can be extremely important in planning and decision making. This is particularly true for farm operators faced with the choice concerning participation in resource conservation management programs.

Case Study Conclusions

A primary strength of WATERS is the nature of the hydrologic watershed model. Detailed description of watershed characteristics with a distributed parameter approach allows analyses of marginal changes of resource use and management and associated economic and environmental impacts. Such analyses are possible at the watershed level and for given streamtubes within a watershed. This model keys the physical and biologic nature of a watershed to impacts of rainfall events. In so doing, it also keys the economic use of watershed resources to hydrologic impacts and the identification of resource problems related to runoff, soil loss, and flooding. This model provides the basis for analyzing and evaluating resource capabilities and problems.

Another major strength of WATERS is the linking of hydrologic and economic models. This provides a relevant interdisciplinary framework for the study of related economic and environmental problems and evaluation of alternative solutions. Important hydrologic and economic linkages occur in the multiple objective LP model in the form of hydrologic and economic data. These data reflect economic resource uses and respective hydrologic consequences given one or more precipitation events. Such data describing crop and rotation profits, runoff, and soil loss relationships to land use and conservation management alternatives are noncommensurable, that is they do not have a common measure. An important characteristic of the multiple objective LP approach is that it allows the analysis of hydrologic and economic data in the study of multiple, noncommensurable, and often competing objectives. In essence, multiple objective linear programming allows an identification and expression of relationships between such objectives not usually considered jointly in the same analysis.

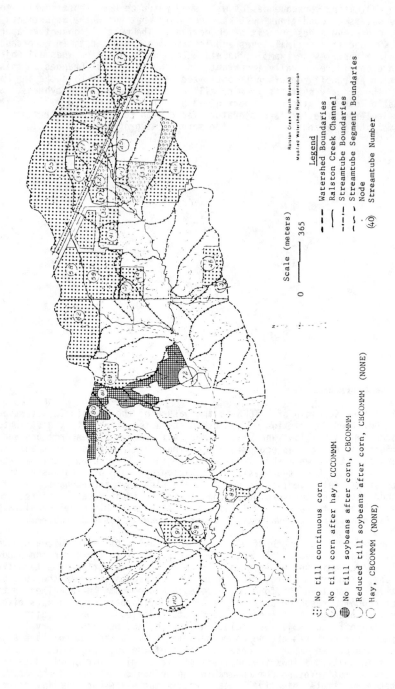

Scale (meters)

0   365

Ralston Creek (North Branch)

Modified Watershed Representation

Legend

- – · – Watershed Boundaries
- ——— Ralston Creek Channel
- - · · - Streamtube Boundaries
- - · - · Streamtube Segment Boundaries
- ◠ Node
- (40) Streamtube Number

⠿ No till continuous corn

◯ No till corn after hay, CCCOMM

⊕ No till soybeans after corn, CBCOMM

◯ Reduced till soybeans after corn, CBCOMM (NONE)

◯ Hay, CBCOMM (NONE)

Map 2. Land use with equal emphases on crop profit, rotation profit, runoff, and soil loss objectives.

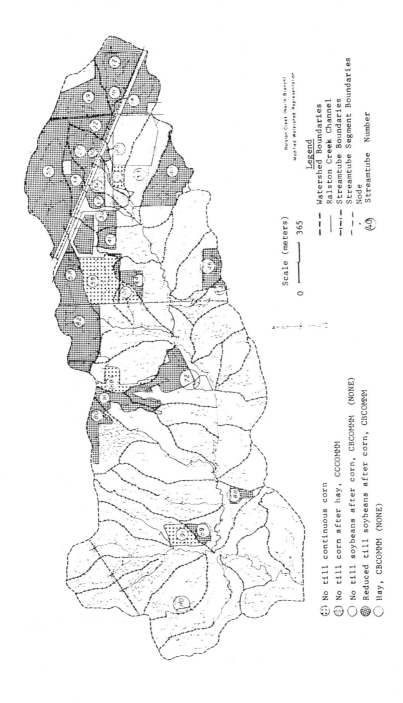

Map 3. Land use emphasizing runoff reduction with crop profit, rotation
profit, and soil loss objectives constrained with unit weights.

255

WATERS assists in developing relevant information pertaining to resource problems and related resource uses, thereby providing a basis for planning land use and conservation management activities given multiple, competing objectives. Information may be developed at the aggregate watershed and at the location specific streamtube and streamtube segment levels. WATERS accommodates local input in the form of participation in developing objectives for study, determining relative weights for objectives, and searching for that particular mix of economic activities which best meets desired objectives. Sensitivity analyses encompassing a wide range of economic activities and related hydrologic or environmental consequences are possible. Such marginal analyses include either aggregate or location specific tradeoff and land use analyses. Various baseline or land use and management conditions can be developed for purposes of impact analyses of alternative watershed management strategies. Such studies include traditional "with vs without" conditions useful for analyzing impacts of resource management programs. Studies of structural vs nonstructural" alternatives for reducing or controlling flooding problems and related damages are also possible with WATERS. This system is also useful for evaluating program and project effectiveness and would assist in identifying problem source areas and in making adjustments to increase program effectiveness through enrollment targeting.

A basic aspect of WATERS is the development of information which facilitates understanding relationships between resource allocation alternatives and objectives of those concerned. Such information allows comparison of consequences of a range of alternatives and assists decision makers in firming their judgements and deciding which tradeoffs are acceptable in order to best meet desired objectives. WATERS provides a useful interdisciplinary framework for economic and environmental information development and analysis within watersheds. It, therefore, provides a useful basis from which to move toward comprehensive watershed evaluation and research.

A major limitation of this application of WATERS is the emphasis on agricultural land use and management. The use of this type of system in a suburban fringe watershed sets the stage for additional analyses encompassing traditional urban pressures on agricultural land. This could include analysis of consequences of expanding residential development and perhaps the study of potential flooding problems in the watershed.

Another limitation is the use of a single storm event. Especially for agricultural studies, the use of an expected pattern of precipitation events over the course of cropland preparation and the growing season would be helpful in understanding the full extent of expected consequences of land use and management alternatives. This lack of multiple event analysis over time is related to the problem of analyzing present values of expected streams of benefits and costs related to production and conservation management activities over time. This includes discounted benefits and costs of flood reduction measured as reduced flood damages. Such an approach is necessary in a temporal analysis of watershed management alternatives to properly assess relationships between marginal costs and marginal benefits for a range of alternatives.

A third limitation concerns the treatment of the watershed as a single management unit. Although land use activities entered and left solutions according to optimality guidelines established by weighting schemes for the objectives analyzed, consideration was not given to natural productivity of soils as a control variable. The fairly uniform silt loam cropland soils were always treated equally with chemicals and fertilizers in accordance with the cropping alternatives.

Advancing the study and understanding of varied physical, biological, chemical, and socioeconomic implications of resource use and management decisions requires an appropriate analytical framework. Necessary considerations include the interdisciplinary identification of relevant processes and linkages and development of a good data base. WATERS incorporates many of these processes and linkages and provides comprehensive watershed description. This system can accommodate improved and added components needed to study implications of resource allocation problems and related policies and programs. In summary, WATERS provides a basis for integrating interdisciplinary needs and capabilities in search of improved information for guiding public and private resource allocation decisions.

REFERENCES

Barnes, H.H. 1967. Roughness Characteristics of Natural Channels. Geological Survey Water Supply Paper 1849. Washington, D.C.

Brill Jr., E.D. 1979. "The Use of Optimization Models in Public-Sector Planning." Management Science, Vol. 25, No. 5, pp. 415-422.

Carvey, D.G. and Robb, J.G. 1980. Hydrologic Economic Systems in Watershed Research. in: Watershed Management, Vol. II, pp. 1036-1047, American Society of Civil Engineers, New York.

Cohon, J.L. 1978. Multiobjective Programming and Planning. Academic Press, New York, pp. 99-162.

Chow, J.T. ed. 1964. Handbook of Applied Hydrology. McGraw-Hill, New York. pp. 12-1 to 12-30 and 20-1 to 20-45.

Croley II, T.E. 1980. A Microhydrology Computation Ordering Algorithm. Journal of Hydrology. Vol. 48, pp. 221-236.

_____, Jain, S.C., and Kumar, S. (to be published, 1981) A Distributed Watershed Model. Iowa Institute for Hydraulic Research Technical Report.

_____, _____, and Whelan, G. 1979. Overland and Channel Kinematic Cascades of Water and Sediment Flows. CANCAM 1979 Seventh Canadian Congress of Applied Mechanics, Paper B-18, University de Sherbrooke, Quebec, Canada.

_____, _____, _____, and Witinok, P. 1979. Overland and Channel Kinematic Cascades of Water and Sediment Flows. Symposium of Urban Runoff, University of Kentucky, Lexington, Kentucky, pp. 297-305.

Economics and Statistics Service. 1980 edition. Multiple Objective Resource Evaluation Budget Generator. U.S. Department of Agriculture [mimeo paper].

U.S. Department of Agriculture, Soil Conservation Service. Hydrology Guide for Use in Watershed Planning, National Engineering Handbook, Section 4, Hydrology, Supplement A.

EVALUATION OF RAINFALL-RUNOFF MODELS FOR THE HYDROLOGIC CHARACTERIZATION OF SURFACE-MINED LANDS

Philip B. Curwick, Hydrologist, U.S. Geological Survey,
Gulf Coast Hydroscience Center, NSTL Station, Mississippi 39529

Marshall E. Jennings, Hydrologist, U.S. Geological Survey,
Gulf Coast Hydroscience Center, NSTL Station, Mississippi 39529

ABSTRACT

The U.S. Geological Survey is presently testing and evaluating "state-of-the-art" rainfall-runoff watershed models that have potential use for characterizing the hydrology of surface-mined lands. The procedure used for testing and evaluating the models is presented. The procedure involves hypothetical catchment testing to select models to be tested with real catchment data. The hypothetical catchment is a simple geometric system of seven overland-flow planes and two cascading channels. This test was designed to identify applicable operational models and will demonstrate inherent differences between the models. Real catchment data will be derived from both historic and current records from U.S. Geological Survey field investigations in Indiana, Tennessee, and Pennsylvania. Real catchment testing will provide comparison information for a systematic evaluation of the accuracy, performance and predictive capabilities of each model when used to characterize the hydrology of surface-mined lands.

Test results using hypothetical catchment data are reported for three storm-event and two continuous rainfall-runoff models. Results obtained from hypothetical testing of the three storm-event models--ANSWERS, FESHM, and KINEROS--show all three models are applicable and can produce similar results. All three models will be tested further using real catchment data. Hypothetical testing of two continuous models--CREAMS and HYSIM--show a significant difference in model response. The CREAMS model is not believed to be applicable for characterizing the hydrology of surface-mined lands and will be excluded from further testing in the present study. However, additional testing of the HYSIM model will be performed using real catchment data.

Strengths, limitations and unique characteristics of the tested models that pertain to the hydrologic characterization of surface-mined lands, are also discussed.

INTRODUCTION

Surface-mining and reclamation operations can affect the quantity and quality of surface runoff from mined watersheds in a number of ways. The watershed surface is disturbed, causing infiltration and runoff characteristics to change. Soil and rock layers are excavated, thereby exposing previously undisturbed soils, rocks, and minerals. Subsequently, these soils, rocks, and minerals can be transported downslope during rainfall events possibly causing degradation of the environment.

Public Law 95-87, the Surface Mining Control and Reclamation Act of 1977, established regulatory procedures to protect the environment from the adverse effects of surface-mining operations. The regulatory procedures require characterization of surface-water hydrology at proposed mine sites and the prediction of the probable hydrologic consequences of the proposed surface-mining activities. Components of the hydrologic characterization may include information on base flow, daily flow, and storm flow as well as sediment loads and water quality conditions for all the afore mentioned flow regimes. Section 779.13 in the Regulations states that surface-water modeling techniques may be used to supplement data collection programs to furnish this information for both pre-mining and post-mining conditions. Deterministic watershed models may offer the best approach for characterizing surface-mining hydrology and predicting the probable hydrologic consequences of surface-mining activities.

The U.S. Geological Survey at the Gulf Coast Hydroscience Center is undertaking a research study to test and evaluate "state-of-the-art" watershed models that can be used to characterize surface-mining hydrology as required by Public Law 95-87. The study was limited to an evaluation of the most recently developed deterministic rainfall-runoff models. The models chosen to be evaluated were either specifically developed for use in surface-mining hydrologic assessment studies or were sophisticated enough to simulate the upland-watershed processes that may be affected by surface-mining activities. Ideally, three upland-watershed processes need to be simulated: (1) rainfall-runoff, (2) sediment erosion-transport, and (3) mineral and biological water quality.

To further serve the purposes of Public Law 95-87, the models must consider the spatial nonuniformity of both the upland-watershed processes and the model parameters associated in describing these processes. The models should be capable of simulating the movement of water and sediment in time and space as a direct result of rainstorms of different intensities and durations that can occur in different seasons of the year. The models should be physically based so that: (1) reliable results can be achieved in minimum calibration-data applications, (2) the model can be readily transferred to mining sites throughout the United States, and (3) input data are obtainable from the literature and supplemental site-specific field data.

A comprehensive review of research being performed at a number of universities and Federal agencies resulted in the selection of 12 models to be evaluated by the Survey (Table 1). The models are provisionally believed to be capable of simulating the watershed processes being affected by surface mining and are termed surface-mining hydrologic assessment models. The selected models are deterministic and characterize the watershed as either a lumped or a distributed system. Six models simulate discrete storm events while six are continuous daily simulation models. In addition, four of the continuous models are also capable of simulating storm-event hydrographs. An overview of each model's characteristics can be found in Jennings, Carey, and Blevins (1980).

In selecting these models, a minimum requirement was the capability to simulate flow and sediment processes both overland and in channels. Models which rely heavily on mathematical formulations of processes which do not allow model input parameters to be determined readily from the literature and/or from topographic maps, land-use maps, and soil surveys were not tested, nor were models whose oversimplifications restricts their use unnecessarily, considering the present "state-of-the-art" of hydrologic modeling.

The purpose of this paper is to describe the model evaluation procedure and to present selected model test results using a common hypothetical catchment data set. Test results are reported for the following models:

1. ANSWERS
2. FESHM
3. KINEROS
4. CREAMS
5. HYSIM

The ANSWERS, FESHM, and KINEROS models are only capable of simulating discrete events and therefore, will be tested with short-interval rainfall-runoff data. The CREAMS and HYSIM models are continuous simulation models and will be tested with daily data. The HYSIM model is also capable of simulating storm hydrographs, however, this capability was not tested in this study. The scope of the paper is limited to the rainfall-runoff process of the above models. However, in future testing of available models (table 1), other processes will be evaluated.

MODEL EVALUATION PROCEDURE

The procedure that will be used to evaluate the surface-mining hydrologic assessment models (table 1) is shown in figure 1. As shown in the flow chart (fig. 1), the evaluation will be based on:

1. Accessibility of the model,
2. Hypothetical catchment data tests, and
3. Real catchment data tests.

The accessibility of the model will be based on published information available to potential users and communication with model developers. Areas of evaluation will include availability of the model, adequate documentation, and technical support from model developers.

The hypothetical catchment tests will identify the applicable and operational models and will also serve as a screen to identify models that will be used in real catchment testing. The hypothetical testing will determine if the models, containing somewhat different mathematical formulations for the same watershed processes, can produce similar results with the same input data set. The hypothetical catchment tests are not intended to portray accuracy but instead to show the magnitude of implicit differences between models. From these tests, model deficiencies can be identified. These deficiencies may include difficulties in application, limitations, numerical computation problems, unreasonable computation costs, impractical input-data requirements, and technical inadequacies in modeling watershed processes. Also, only one of a group of generic models will be selected to be tested with real catchment data.

The real catchment tests, based on 8 sites in Tennessee, 10 sites in Indiana, and 8 sites in Pennsylvania, will provide comparison

Table 1.--Surface-mining hydrologic assessment models
selected to be evaluated by the Survey

Model	Developer	Simulation Type
ANSWERS	Purdue University (Beasley and Huggins, 1980)	Event
FESHM	Virginia Polytechnic Institute (Ross et al., 1978)	Event
KINEROS	U.S. Department of Agriculture (Smith, 1979)	Event
CREAMS	U.S. Department of Agriculture (Knisel, 1980)	Continuous
HYSIM	Tennessee Valley Authority (Betson et al., 1980)	Continuous/ Event
PRMS	U.S. Geological Survey, WRD (Leavesley, 1980)	Continuous/ Event
TENN-I	University of Tennessee (Overton, 1980)	Event
SEDMT	Colorado State University (Shiao, 1978)	Event
WATER	Universtiy of Arizona (Berkas, 1978)	Continuous
UTAH	Utah State University (Jeppson, 1980)	Continuous/ Event
SEDLAB	U.S. Department of Agriculture (Alonso et al., 1978)	Event
WATMOD	University of Kentucky (Barfield et al., 1979)	Continuous/ Event

information on model accuracy and performance. The tests will be
conducted similar to the procedure followed by Doyle and Miller
(1980). The procedure involved a series of calibration, sensitivity,
and verification analyses that were used to test the predictive
capabilities of a model.

The final results of the study can be used as guidance as to
what "state-of-the-art" watershed models can be used to simulate
surface-water characteristics required by Public Law 95-87.

HYPOTHETICAL CATCHMENT DATA

The hypothetical catchment is a simple geometric system patterned
after a watershed described by Smith (1979). Hypothetical land uses
of surface mining, hardwood and conifer forest, and pasture were
assigned to the catchment. Typical soil-water characteristics of a

262

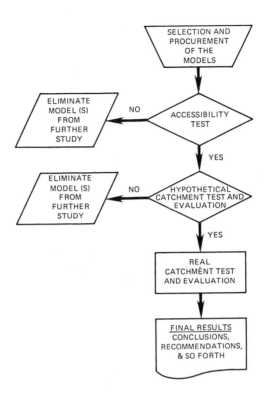

Figure 1.--Flow chart showing the model evaluation procedure

loam topsoil were assumed to occur uniformily throughout the catchment. Rainfall events were also assumed to occur uniformly across the catchment. Discrete rainstorms were used to test storm-event models, whereas a series of daily rainfalls were used to test the continuous models.

The physical and topographic characteristics of the hypothetical catchment were superimposed on a 7-plane 2-channel segmented drainage system, 22.0 hectares in size (fig. 2). Geometric shapes were selected since most models require this approximation for the overland flow routing on irregular shaped catchments. Assumptions restricting the representation of a natural channel cross section in some models dictated the need to use hydraulically equivalent triangular and rectangular channel cross sections. Each segment was assigned a uniform slope. Overland slopes were moderately steep and were either 0.12 or 0.20 m/m. Channel slopes were less steep and were 0.015 m/m.

Computational time steps for infiltration and flow routing were input to the models whenever possible. The time steps were chosen on the basis of rainfall characteristics and catchment response. A time step of 5 minutes was chosen for infiltration computations and 1 minute was chosen for surface-water flow routing for use in the storm-event models. Daily time steps were used in the continuous simulation models.

Roughness coefficients were also input to the models whenever

EXPLANATION
⑦ - PLANES
② - CHANNELS

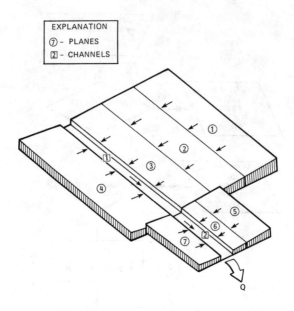

Figure 2.--Conceptual representation of hypothetical catchment.

possible. The Manning formula was often used to define the friction slope, therefore, Manning n values were assigned to each overland-flow plane and channel segment of the catchment. The n values ranged from 0.400 to 0.050 for overland-flow planes and were 0.025 for the channels.

Land uses were arbitrarily assigned to each plane of the hypothetical catchment. The land uses, Manning n values, and slopes assigned to the overland-flow planes (fig. 2) are given in table 2. The spatial variability in land use served to test the model's applicability for these conditions. It further tested the interactions of runoff computations for different land uses and will eventually permit testing of the sediment and water-quality components of the models.

Figure 3 shows the hypothetical soil-water characteristics as described by the soil-moisture variables of field capacity, wilting point, gravity water, available and unavailable plant water, and total porosity. These characteristics were assigned uniformly across the catchment and to a depth of 30.5 centimeters. Three antecedent moisture conditions of: (1) 100-percent field capacity, (2) 50-percent field capacity, and (3) 1-percent field capacity were used in the hypothetical testing of the storm-event models. For the continuous models, an antecedent moisture condition of 50-percent field capacity was used.

Hypothetical rainstorms were used to test each model's response to different intensities and durations of rainfall. Storm-event models were tested with six different rainstorms. Three rainstorms had uniform distributions with intensities and durations varying between rainstorms. The remaining three rainstorms represented hypothetical design rainstorms typical of those found in the Appalachian coal region of north-central Tennessee (Herschfield, 1963). The design rainstorms had 2-, 10-, and 25-year recurrence

Table 2.--Land use, roughness coefficient, and slope for each
overland-flow plane of the hypothetical catchment

Overland-flow plane	Land use	Manning n	Slope (m/m)
1	50% Hardwood/50% Conifer Forest	0.400	.12
2	100% Surface Mine	0.050	.12
3	100% Hardwood Forest	0.400	.12
4	100% Hardwood Forest	0.400	.12
5	50% Hardwood/50% Conifer Forest	0.400	.20
6	100% Pasture	0.150	.20
7	100% Pasture	0.150	.20

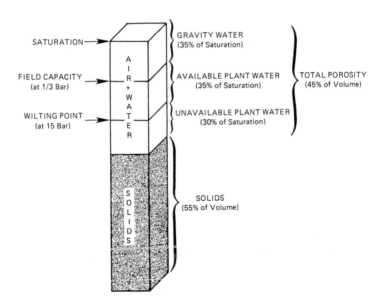

Figure 3.--Soil-block diagram showing hypothetical catchment
soil-water relationships

intervals and 3-hour durations. The six variations in rainstorm characteristics and three different antecedent moisture conditions resulted in 18 different hypothetical-test combinations to be used in testing storm-event models. Continuous models were tested with 365 days of daily rainfall data shown in figure 4. The daily rainfall data were derived from actual records at a rain gage in north-central Tennessee.

The hypothetical catchment was chosen so that the data set could be universally applied to each model so that a valid comparison could be made. It was further chosen to include spatial variability of some physical and hydrologic parameters, e.g., topography, land use, and so forth, in order to provide a rigorous test of the distributed parameter watershed models. Lumped models almost always employ some weighting function to account for spatial variability. Weighting of watershed parameters (such as flow resistance) was made in those instances and applied in the authors' best judgment. Deviations from the hypothetical data set presented in this section were sometimes required to meet specific model-input-data requirements.

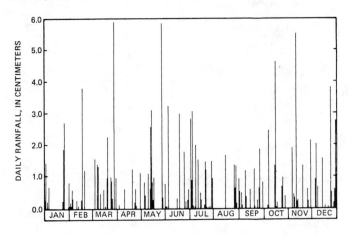

Figure 4.--Hypothetical daily rainfall data used to test continuous models

TEST RESULTS

Selected results of testing five surface-mining hydrologic assessment models using hypothetical catchment data are presented. Results are reported for the following models:
1. ANSWERS
2. FESHM
3. KINEROS
4. CREAMS
5. HYSIM

A deterministic-mathematical approach is used to compute the hydrologic response in all five models. The first three models--ANSWERS, FESHM, and KINEROS--utilize a distributed approach to assign model input parameters and to characterize the hydrologic response of a watershed. The last two models--CREAMS and HYSIM--use a lumped approach to assign model input parameters and to characterize the response of a watershed.

Storm-event models

The results of testing three storm-event models--ANSWERS, FESHM, and KINEROS--using four hypothetical-test combinations are summarized in table 3. Antecedent moisture conditions, rainfall, runoff, peak rates of runoff, and times to peak are reported. The computed runoff-hydrograph plots using the same four hypothetical-test combinations are shown in figure 5a-d. The results reported are representative of test results obtained using a total of 18 different hypothetical-test combinations.

The three storm-event models can produce similar results when tested with the hypothetical catchment data. Runoff volumes computed by the ANSWERS and FESHM models were generally in close agreement. The KINEROS model usually predicted lower runoff volumes for the same input rainstorms than did ANSWERS or FESHM. The shape and timing of the computed runoff hydrographs of the FESHM and KINEROS models usually bore a close resemblance. The ANSWERS model hydrograph always began before the hydrographs of the other two models and exhibited a trend of smaller peak discharges for the larger rainfall events. The hydrographs computed by FESHM and KINEROS, in all

Table 3.--Summary of selected storm-event simulations using hypothetical catchment data

Model	Antecedent moisture (% of field capacity)	Rainfall volume (cm)	Runoff volume (cm)	Peak discharge (m^3/s)	Time to peak discharge from beginning of runoff (h)
Test Run No. 4					
ANSWERS	100	7.62	2.62	0.93	2.50
KINEROS	100	7.62	1.19	0.67	1.83
FESHM	100	7.62	2.84	0.84	1.83
Test Run No. 8					
ANSWERS	50	15.24	7.95	2.20	2.70
KINEROS	50	15.24	5.61	1.97	2.40
FESHM	50	15.24	8.10	2.14	2.42
Test Run No. 10					
ANSWERS	50	30.48	20.09	2.57	5.70
KINEROS	50	30.48	16.36	2.29	5.40
FESHM	50	30.48	21.82	2.25	5.42
Test Run No. 15					
ANSWERS	50	7.75	2.51	2.94	0.50
KINEROS	50	7.75	1.37	1.76	0.50
FESHM	50	7.75	1.91	1.74	0.50

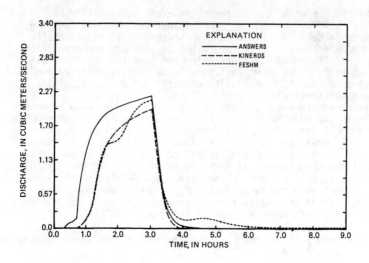

a. Results of test run no. 4

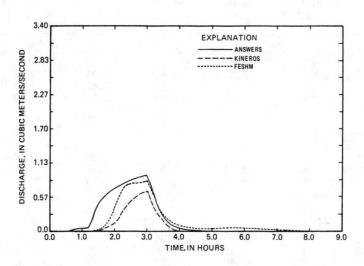

b. Results of test run no. 8

Figure 5.--Hypothetical catchment runoff hydrographs for the storm-event models

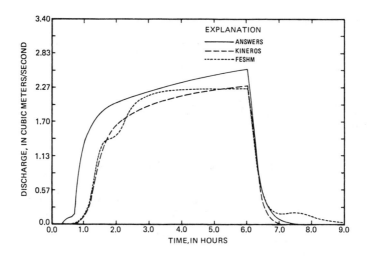

c. Results of test run no. 10

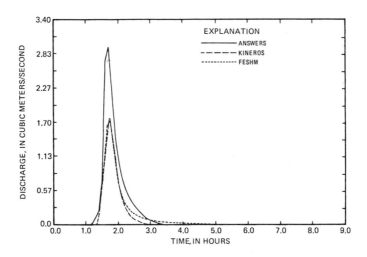

d. Results of test run no. 15

Figure 5.--Continued

cases, began at the same time. The recession limb of the FESHM model most always extended beyond the other two.

Continuous Models

The results of testing two continuous-simulation models-- CREAMS and HYSIM--using hypothetical catchment data are summarized in table 4. Rainfall, runoff, and evapotranspiration are reported in terms of depth of water on the watershed. The computed daily hydrograph for each model is shown in figure 6.
The two continuous models produced different results when tested with hypothetical catchment data. The annual runoff volumes differ by approximately six times even though the computed evapo- transpirations are somewhat similar for the two models.

The hydrology component of the CREAMS model contains two options to compute surface runoff. When daily rainfall data are available to the user, Option 1 is used to compute surface runoff. Option 1 employs the Soil Conservation Service curve number method for simulating daily runoff. Option 1 was used for hypothetical catchment testing

Table 4.--Summary of continuous simulations using hypothetical catchment data

| Month | Rainfall | Runoff | | Evapotranspiration | |
| | | CREAMS | HYSIM | CREAMS | HYSIM |
	(cm)	(cm)	(cm)	(cm)	(cm)
January	7.49	0.28	5.64	1.98	0.71
February	7.72	0.64	5.38	2.59	0.71
March	17.78	2.03	8.97	4.44	1.50
April	2.97	0.00	4.04	8.51	6.96
May	18.03	0.48	3.25	16.31	16.31
June	10.31	0.05	3.00	17.40	15.01
July	16.61	0.00	2.64	18.95	14.96
August	5.23	0.00	1.83	7.45	10.26
September	8.33	0.00	1.32	6.96	7.16
October	11.76	0.61	1.09	4.62	4.70
November	13.08	1.58	2.26	2.84	1.35
December	13.23	0.99	4.37	2.03	0.69
Total	132.54	6.66	43.79	94.08	80.32

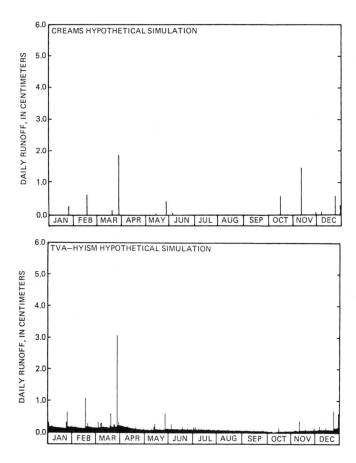

Figure 6.--Hypothetical catchment daily hydrographs for the
continuous models

since daily rainfall was used as input. If hourly or breakpoint (time-
intensity) rainfall data are available, Option 2 is used to compute
surface runoff. Option 2 uses an infiltration-based method (Smith and
Parlange, 1978) to simulate daily runoff.
     The HYSIM model was slightly modified for the hypothetical
catchment testing. The HYSIM model is normally driven by a stochastic
rainfall component. The rainfall component generates a daily rainfall
distribution from statistical measures of long-term rainfall records.
The daily rainfall distribution is subsequently used for simulating
daily streamflow. This component was altered so that a real-time
daily rainfall distribution could be input directly into the
daily streamflow simulation model. Furthermore, the HYSIM model is
also capable of simulating storm hydrographs. This capability was not
tested in this study.

DISCUSSION

     Model testing with hypothetical catchment data showed only
slight differences in model response for three storm-event models--
ANSWERS, FESHM, and KINEROS--and large differences in model response

271

for two continuous models--CREAMS and HYSIM. The results presented
in the previous section show the magnitude of the implicit differences
in model response for the same hypothetical input conditions. In
addition, the hypothetical testing identified strengths, limitations,
and unique characteristics of the models that pertain to surface-
mining applications and are discussed below.

One strength common to all three storm-event models is that they
disaggregate the overall hydrologic and hydraulic phenomena into their
respective physical components (such as infiltration or flow routing.)
By simulating the selected phenomena into separate components, each
individual process can be analyzed or altered to meet the needs of the
user. Use of component process models also results in a nonlinear
mathematical formulation of natural catchment behavior and can prove
to be an improvement over a linear formulation (such as the unit
hydrograph method.) Model parameters tend to be physically significant
or at least tend to take on clearer physical interpretations. This
results in improved capability to model in minimum data situations.

All three storm-event models are distributed-parameter models.
This attribute has both inherent strengths and liabilities. An obvious
strength of the distributed approach is that any degree of spatial
variability of topography, land use, soils and rainfall within the
watershed boundaries can be represented in the model. The distributed
approach also facilitates incorporation of relationships developed
from laboratory and small "plot" studies to yield predictions on a
watershed scale. However, the inclusion of spatial variation can
generate large and excessive data sets. The preparation of the
spatially variable data sets can be very time consuming thereby limiting the general attractiveness of a distributed-parameter modeling
approach. Further, the level of data discretization necessary
to achieve a given level of model predictability is basically unknown.
Unavailability of criteria to relate the level of discretization that
is necessary to provide a given level of predictability detracts from
the use of the models.

The ANSWERS and FESHM models utilize a modified Holtan equation
to compute infiltration. Both models provide documentation on user-
input infiltration parameters required by the models. However, each
model's documentation reports different interpretations and
recommendations for the same user-input infiltration parameters.
Subsequently, large differences in computed runoff volumes between
the models can result.

The KINEROS model utilizes an infiltration equation derived
from the mathematical description of the physics of soil-water
movement (Smith and Parlange, 1978). This approach is physically
based in contrast to an empirical-based approach such as the Holtan
equation utilized by ANSWERS and FESHM.

Values for only two infiltration parameters must be assigned
by the user to apply KINEROS. One of the parameters, the saturated
hydraulic conductivity is commonly required by many watershed models.
The other infiltration parameter, a soil infiltration constant, is
specific only to the KINEROS model. Both infiltration parameters
are best obtained from field infiltrometer data and, in fact, may
be the only way to assign the soil infiltration constant because of a
lack of documentation of the parameter. These field data requirements
may limit the use of the KINEROS model in surface-mining hydrologic
assessment studies where little or no infiltrometer data are available.
Real catchment testing can provide information to offset the liability
encountered when assigning model-infiltration parameters.

A strength and unique characteristic of the ANSWERS model is that
it was the only storm-event model tested that considers runoff
transmission losses. ANSWERS continuously simulates infiltration as

272

overland flow moves downslope. Therefore, any rainfall excess
routed from an upslope area has the opportunity to become infiltration
at some point downslope, if a soil with a greater infiltration
capacity is encountered. In the other storm-event models, a rainfall
excess distribution is computed for each segment of the watershed
and subsequently is routed downslope without considering transmission
losses. This feature is advantageous on flat slopes where runoff
is slow and when overland flow distances are extremely long.

The CREAMS model had the most comprehensive documentation of
all five models tested. However, the model was specifically designed
as an agricultural "field" scale model in contrast to a "basin" or
watershed scale model and as a result has limited routing capabilities.
It is applicable to a rectangular field-plot sized area defined as
having: (1) a single land-use, (2) relatively homogeneous soils,
(3) spatially uniform rainfall, and (4) a single agricultural
management practice. Although most agricultural fields fit these
criteria, watersheds such as those encountered in surface-mining
hydrologic assessment studies most often do not. Therefore, the
use of a model like CREAMS, that has been designed for such specific
conditions, is limited, and will be eliminated from further testing.

The HYSIM model contains regionalized lumped model components
that are roughly applicable only in the Tennessee Valley or in
areas with similar hydrologic characteristics as the Tennessee
Valley. The model must be used cautiously under conditions differing
from those used in regionalizing the model and its applicability
should be verified. In northern climates where snow and snowmelt
runoff are significant considerations, special provisions in the
HYSIM model will have to be made to handle these conditions.
Identifiable strengths of the HYSIM model are that the model is
adequately documented and the model is extremely easy to apply.
Furthermore, changes to the regional relationships are possible,
which will make the model attractive for use in areas outside the
Tennessee Valley. Additional testing of HYSIM is in progress.

SUMMARY

A model evaluation procedure for testing "state-of-the-art"
rainfall-runoff models for use in hydrologic characterization of
surface-mined lands has been presented. The procedure involves
the use of hypothetical and real catchment data.

Selected test results using hypothetical data only five models--
three storm-event models and two continuous models--are given in
this report.

The results of the hypothetical tests indicate that the storm-
event models--ANSWERS, FESHM, and KINEROS--produce similar results
for six rainstorms using a common hypothetical catchment data set.
Therefore, real catchment data will be used for further testing of
these models.

Tests of two continuous models--CREAMS and HYSIM--showed a
significant difference in model response. The CREAMS model computed
low runoff volumes relative to results computed by HYSIM. The
scale of applicability will exclude the CREAMS model from further
testing in this study, however, the HYSIM model will be tested further
using real catchment data.

REFERENCES

Alonso, C. V., DeCoursey, D. G., Prasad, S. N., and Bowie, A. J. 1978.
    Field Test of a Distributed Sediment Yield Model. in: Proceedings
    of 26th Annual Hydraulics Division Specialty Conference of Am. Soc.

of Civil Engineers, College Park, Maryland, pp. 671-678.

Barfield, B. J., Moore, I. D., and Williams, R. G. 1979. Prediction of Sediment Yield from Surface-Mined Watersheds. in: Proceedings of Symposium on Surface-Mining Hydrology, Sedimentology, and Reclamation, University of Kentucky, Lexington, Kentucky, UKY BU119, pp. 83-91.

Beasley, D. B., and Huggins, L. F. 1980. ANSWERS (Areal, Nonpoint Source Watershed Environment Response Simulation) User's Manual. Agricultural Engineering Department, Purdue University, West LaFayette, Indiana, 55 pp.

Berkas, W. R. 1978. Deterministic watershed model for evaluating the effects of surface mining on hydrology. M.S. Thesis, School of Renewable Natural Resources, University of Arizona, Tucson, Arizona, 122 pp.

Betson, R. P., Bales, J., and Pratt, H. E. 1980. User's Guide to TVA-HYSIM, A Hydrologic Program for Quantifying Land-Use Change Effects. Tennessee Valley Authority for the U.S. Environmental Protection Agency, EPA-600/70-80-048, 107 pp.

Doyle, W. H., Jr., and Miller, J. E. 1980. Calibration of a distributed routing rainfall-runoff model at four urban sites near Miami, Florida. U.S. Geological Survey Water-Resources Investigations 80-1, 95 pp.

Hershfield, D. M. 1963. Rainfall Frequency Atlas of the United States for Duration from 30 Minutes to 24 Hours and Return Periods from 1 to 100 years. U.S. Weather Bureau, Technical Paper No. 40, 61 pp.

Jennings, M. E., Carey, W. P, and Blevins, D. W. 1980. Field studies for verification of surface mining hydrologic models. in: Proceedings of Symposium on Surface Mining Hydrology, Sedimentology, and Reclamation, University of Kentucky, Lexington, Kentucky, UKY BU123, pp. 47-53.

Jeppson, R. W. 1980. Computer Solution of Total Water Movement On and Within a Watershed. Paper presented at Am. Water Resources Association Meeting, March 20, 1980, 33 pp.

Knisel, W. G., editor. 1980. CREAMS: A Field Scale Model For Chemicals, Runoff, and Erosion From Agricultural Management Systems. U.S. Department of Agriculture, Conservation Research Report No. 26, 640 pp.

Leavesley, G. H., Lichty, R. W., Troutman, B. M., and Saindon, L. G. 1980. Precipitation-Runoff Modeling System: User's Manual, (DRAFT), U.S. Geological Survey Water-Resources Investigations 80-XX, 245 pp.

Overton, D. E. 1980. Computer Program TENN II, A Nonlinear Model for Simulating Stormwater Runoff Volumes, Hydrographs, and Pollutant Yields for Small Watersheds: User's Manual. D. E. Overton and Associates, Consulting Engineers, Knoxville, Tennessee, 21 pp.

Ross, B. B., Shanholtz, V. O., Contractor, D. N., and Carr, J. C. 1978. A Model for Evaluating the Effect of Land Uses on Flood Flows. Virginia Water Resources Research Center, Bulletin 85, 137 pp.

Shiao, L. Y. 1978. Water and Sediment Yield from Small Watersheds. Ph.D. dissertation, Department of Civil Engineering, Colorado State University, Fort Collins, Colorado, 227 pp.

Smith, R. E. 1979. A Kinematic Model for Surface Mine Sediment Yield. ASAE Paper #79-253, Presented at the Winter Meeting of ASAE, New Orleans, Louisiana, Dec. 11-14, 1979.

Smith, R. E., and Parlange, J. Y. 1978. A parameter-efficient hydrologic infiltration model. American Geophysical Union Water Resources Research Journal, Volume 14, No. 3, pp. 533-538.

EVALUATION OF THE USDAHL HYDROLOGY MODEL FOR SIMULATING RUNOFF FROM NATIVE GRASS WATERSHEDS

Frank R. Crow, Professor
Agricultural Engineering Department
Oklahoma State University
Stillwater, OK 74078

ABSTRACT

The USDAHL-74 Model of Watershed Hydrology was evaluated on four widely separated native grass watersheds having areas of 6.3 ha to 83 ha. Emphasis was on evaluation of land use parameters and criteria for calibrating the model to the watersheds. In preliminary tests on a 37 ha watershed, the model performance was improved by using a variable grazing parameter to describe the change in hydrologic cover condition due to grazing. During a 4-yr test period, total simulated runoff was 16% greater than observed runoff. Further tests on the 37 ha watershed with an expanded data base consisting of 6-yr calibration and 18-year test periods resulted in total simulated runoff being underestimated by 11%, indicating that simulation accuracy was affected by the choice and length of the calibration period. When applied to a nearby 83 ha watershed, without prior calibration, the model overestimated total runoff during a 20-yr test period by 26%. The third phase emphasized calibration criteria and evaluation of the groundwater recharge parameter. The model was calibrated to a 6.3 ha watershed by two different methods and tested on a 7.8 ha watershed located 90 km away. Total simulation error for the type II calibration was 1.7%.

INTRODUCTION

The livestock industry of Oklahoma and the Southwest is highly dependent on native grass pastures as a source of food and water for grazing beef animals. Runoff from native grass watersheds constitutes an essential resource not only for the rancher but also for urban dwellers and industries who depend on downstream reservoirs for their water supplies. In times of water shortage a possible conflict may develop between these two groups of users. It is highly likely that rainfall-runoff models may be used to develop guidelines to be used for water resource management plans.

In order to learn more about the application of hydrologic models to such problems, a research program was established by the Agricultural Engineering Department of Oklahoma State University to assess the applicability and to evaluate the accuracy of one or more hydrologic models for simulating monthly and annual runoff from small native grass watersheds.

The purpose of this paper is to briefly discuss the progress of that research and to present some of the major findings in the process of identifying appropriate values of land use parameters during each phase of the investigation. More complete details of each investigation can be found in papers by Crow et al. (1977, 1980), and Bengtson et al. (1980), and Ph.D. dissertations by Ghermazien (1978) and Bengtson (1980).

The first model selected for study was the USDAHL Model of Watershed Hydrology, which was developed in 1970 and revised in 1974 by a USDA team headed by Holtan (1971, 1975). There has been a considerable amount of research interest in the USDAHL model with applications having been made in Ohio by Langford and McGuiness (1976); in Maryland by Fisher et al. (1979); in Idaho by Molnau and Yoo (1977) and Hanson (1979); and in Oklahoma by Nicks et al. (1977). Most of the above-cited work was concurrent with the research by Oklahoma State University described in this paper. In general, encouraging results were reported by most research workers.

MODEL CONCEPTS

The USDAHL model was selected for study because it appeared to be well suited to the small grassland watersheds for which a considerable amount of hydrologic data were already available for testing. The model uses input parameters that can be developed from existing records of soil surveys, soil infiltration data, land-use patterns, topographic and climatic data in terms that are familiar to agricultural hydrologists. The hydrologic response zones are closely associated with land capability classifications and provide input data on land slope, overland flow length, infiltration rate, soil depth, and soil moisture holding characteristics. Required meteorological input data include break-point rainfall, daily mean temperature, and daily class A pan evaporation. The precipitation is apportioned to surface runoff, infiltration, transpiration, lateral flow and ground water recharge; and the soil moisture balance is maintained on a continuous basis. Specific details of the hydrologic concepts are described by Holtan et al. (1975) and Holtan and Yaramonoglu (1977).

The USDAHL model places considerable importance on the role of land use parameters in describing the uniqueness of each watershed. The pertinent parameters and their definitions by Holtan et al. (1975) are listed below.

A   Vegetative Parameter - an index of surface-connected porosity

VD   Depression Storage Parameter - volume of depressions that would store rainfall until it infiltrated.

ET/EP   Evapotranspiration Parameter - the ratio of maximum evapotranspiration to maximum pan evaporation for a year.

TU   Upper Cardinal Temperature - temperature above which crop's ET is impaired.

TL   Lower Cardinal Temperature - temperature below which
      crop's ET does not function.

GRAZ Grazing Parameter - percent reduction in evapotranspiration
      attributed to grazing.

ROOT Root Depth Parameter - depth to which plants will extract
DEPTH moisture.

GR   Deep Groundwater Recharge Parameter - deep percolation
      which does not show up in the recession curve.

OBJECTIVES

The principal objective of this research project was to evaluate
the performance of the USDAHL-74 Model of Watershed Hydrology in simulation of monthly and annual runoff from native grass watersheds. Specific
objectives were:

1. To evaluate the effect of selected land use parameters on
    simulated runoff.

2. To evaluate the ability of the model to simulate runoff either
    with or without prior calibration on test watersheds.

3. To evaluate several methods of calibrating the model through
    the optimization of hydrologic parameters.

RESEARCH APPROACH

The work proceeded in several phases. A brief overview of each
phase is presented in this section, which is followed by more complete
discussion in the following sections.

The initial phase was essentially exploratory and involved relatively short calibration and test periods on a single watershed. The
model was calibrated to the watershed by evaluating the land use parameters over a 3-yr period. The parameters were then held constant and
the model was tested by a single computer run on the same watershed, but
over a different 4-yr period.

The second phase used the same watershed but with an expanded data
base. The model was calibrated over a 6-yr period and tested over a
different 18-yr period to determine the effect that length of calibration period might have on the model performance. This phase also
included a test over a 21-yr period on a nearby watershed to determine
the simulation accuracy when the model was used on a watershed for which
it had not been previously calibrated.

The third phase was devoted to formulating and testing criteria for
model calibration. A description of the two calibration types is provided in the section of this paper entitled Discussion - Phase III. To
avoid bias due to previous model experience on the watersheds used in
Phases I and II, the simulation runs used for the calibration and testing in Phase III used data from two different native grass watersheds,
neither of which had been previously used for model testing.

DISCUSSION - PHASES I AND II

Watershed Descriptions

Phases I and II of the USDAHL model evaluation program involved two watersheds established for cooperative hydrology research by Oklahoma State University and the USDA-SEA-AR Water Conservation Structures Laboratory. Watersheds W-3 and W-4, with areas of 37 ha and 83 ha respectively, are located 25 km north of Stillwater, Oklahoma in the Central Rolling Red Prairies resource area. The topography is gently rolling with slopes ranging from three to eight percent. Normally there is no base flow from the watersheds except after prolonged periods of rainfall. Vegetative cover consists of tall and mid-length native prairie grasses. The soils on W-3 are clay and clay loam with relatively shallow top soil and very slowly permeable subsoil. On W-4 the soils are somewhat deeper and have better internal drainage. The topographic maps and hydrologic response zones for both watersheds are shown in Figure 1. The data for the hydrologic zone and soil parameters are listed in Table 1. These data are required inputs to the model which were left as indicated throughout the testing and evaluation.

TABLE 1. HYDROLOGIC ZONE AND SOIL PARAMETERS FOR STILLWATER WATERSHEDS

	W-3 (37 ha)			W-4 (83 ha)		
	Zone 1	Zone 2	Zone 3	Zone 1	Zone 2	Zone 3
Zone area %	36	51	13	36	54	10
Average slope %	3.7	5.1	3.7	4.7	6.3	4.7
Overland flow length m	50	53	23	65	68	18
Soil texture	cl lm	cl lm	clay	loam	cl lm	si lm
Final infil. rate mm/h	0.5	0.5	0.5	0.5	0.5	0.5
Depth, upper layer cm	25	15	38	25	15	51
Depth, lower layer cm	90	100	115	114	100	152
G*, upper layer %	10	13	7	12	15	8
G*, lower layer %	7	7	7	11	11	12
AWC†, upper layer %	17	13	12	18	13	15
AWC†, lower layer %	12	12	12	11	12	16
Cracks‡, upper layer %	8	12	5	6	10	5
Cracks‡, lower layer %	12	8	5	9	7	5
Root depth cm	128	65	140	128	65	140

*Percent of soil depth drained by gravity
†Percent of soil depth drained by plants
‡Percent of soil depth subject to cracking

Calibration and Testing - Phase I

Phase I was devoted to preliminary calibration of the model on W-3 for the period 1970-1972. Monthly runoff was simulated using initial values of land-use parameters suggested by Holtan et al. (1975). The simulation results were compared with the observed runoff, and each of the parameter values was varied, one at a time, as necessary to increase or decrease simulated runoff to match the observed runoff. Both monthly runoff and cumulative runoff for the entire period were used in the calibration process. A satisfactory calibration was considered to be the set of land-use parameter values that minimized the difference between the cumulative simulated and observed runoff, and at the same time yielded a least squares regression equation of simulated versus observed monthly runoff that most nearly coincided with the equal value line.

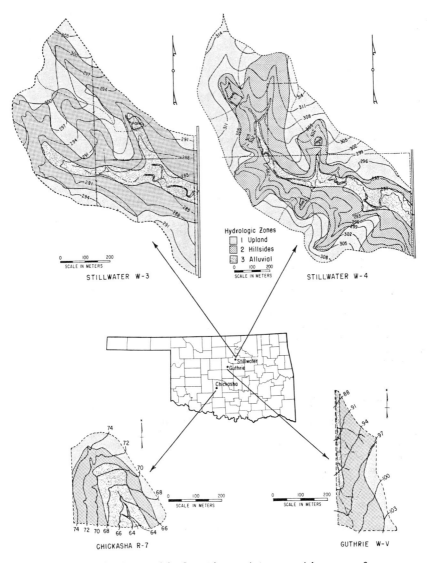

Figure 1   Geographic location and topographic maps of
watersheds used to evaluate the USDAHL model.

During this phase, it was discovered that the hydrologic cover condition of the watershed had considerable impact on the simulated monthly
runoff.   Preliminary simulation results were inconsistent from year to
year, depending on the condition of the vegetative cover.   More consistent results were obtained after a qualitative assessment of the hydrologic cover condition was transformed into a quantitative value that
could be used as a variable land use parameter.   The adjectives "good",
"fair", and "poor", as used by range management specialists to describe
the hydrologic cover condition, were represented by numerical values of
45, 55, and 65 respectively, to provide an appropriate value of the graz-

ing parameter in the model. This relationship proved to be very useful in further tests of the model.

An independent test of the model was made for the period 1956-1959. These years were deemed satisfactory for testing the suitability of the model and establishing the land use parameter values because they encompassed a wide range of moisture conditions for evaluating sensitive land-use parameters. Rainfall variations during this test period were extreme, varying from 46% less than to 54% more than the long term average.

The test was made by making a single computer run for each year, with only the grazing parameter, GRAZ, being varied to correspond to the grazing intensity. The test results were encouraging. Total runoff for the 4-yr period was overestimated by only 16%. The regression for the simulated (Q_s) and observed (Q_o) monthly runoff was

$$Q_s = 0.07 + 1.05 \ Q_o$$

with correlation coefficient (r) = 0.98. Small discrepancies, always on the side of overestimation, occurred in months of little or no observed runoff. The model also tended to overestimate during the cool season months, suggesting that the ET component was not functioning properly, or that a groundwater component was needed.

Calibration and Testing - Phase II

The main thrust of Phase II was to evaluate the model over longer calibration and test periods to determine how the greater variety of situations encountered would influence the simulated runoff. The calibration period, 1952-1957, included one exceptionally dry year and one year with high rainfall and runoff. The test period, 1958-1976, included two years each with exceptionally high and low runoff.

The base watershed for calibration again was W-3. Following the same procedure as in Phase I, the land-use parameters were adjusted during a series of computer runs until a set of parameter values was obtained that gave the best correlation of simulated and observed runoff. The final values of the land-use parameters are listed in Table 2. The

TABLE 2. LAND USE PARAMETERS FOR STILLWATER WATERSHEDS

Vegetative parameter, A	0.1
Depression storage, VD, cm	0.51
ET/EP ratio	1.02
Upper cardinal temperature, $^{\circ}C$	27
Lower cardinal temperature, $^{\circ}C$	0
Groundwater recharge, GR, mm/hr	0
Reduction in ET due to grazing, %	
Good hydrologic cover	45
Fair " "	55
Poor " "	65

relationship between the simulated and observed monthly runoff is shown in Figure 2. The regression equation was

$$Q_s = 1.5 + 0.95 \ Q_o$$

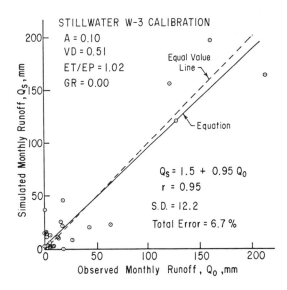

STILLWATER W-3 CALIBRATION

A = 0.10
VD = 0.51
ET/EP = 1.02
GR = 0.00

Equal Value Line

Equation

$$Q_S = 1.5 + 0.95 \, Q_0$$
$$r = 0.95$$
$$S.D. = 12.2$$
$$\text{Total Error} = 6.7\,\%$$

Figure 2   Monthly runoff for USDAHL calibration on
Stillwater W-3 for 1952-1957.

with r = 0.95 and standard deviation (S.D.) = 12.2 mm. The relatively
high standard deviation is an indication of the variability of simulation
accuracy of monthly events. For the total calibration period the cumulative simulated runoff was 6.5% greater than the cumulative observed runoff, which is indicated as total error on Figure 2 and all other figures
in the paper.

The model was tested by making a single computer run for 1958-1976,
using the parameter values listed in Table 2. The only variations permitted were changes in the average annual grazing parameter to account
for hydrologic cover condition. The result of this test was an 11.8%
underestimate of the cumulative runoff. In absolute terms this was an
improvement over the 16% error obtained in the Phase I test, but the
error was opposite in sign. Thus the choice of calibration period, and
its length, did make a difference in the runoff simulated by the model.
Given the stochastic nature of hydrologic events, this outcome probably
should not be too surprising. The equation for computed versus observed
monthly runoff was changed also, as may be seen in Figure 3. The
regression equation was

$$Q_s = 0.6 + 0.83 \, Q_0$$

with r = 0.93 and S.D. = 8.3 mm.

Another objective of Phase II was to determine the simulation error
that would result if the model were applied to an uncalibrated watershed
using land-use parameters determined by calibration on a nearby watershed. This test was made on watershed W-4 which lies 2 km north of W-3.
The practical reason for this test was that data collection on W-4 had
been discontinued in 1972, and we wanted to know how accurately the missing data could be simulated. The test run was for the period 1952-1972
for which observed runoff data were available. The land-use parameter
values were those listed in Table 2. The result of this test was that

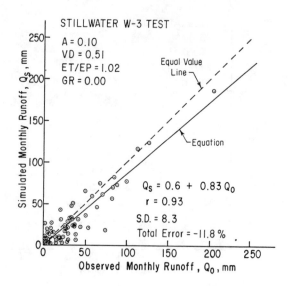

Figure 3 Monthly runoff for USDAHL test on Stillwater
W-3 for 1958-1976.

total runoff for the period was overestimated by 24.7%. The magnitude of
this overestimate would indicate that the hydrologic differences between
W-3 and W-4 were greater than previously believed. Some of the hydrologic inputs may have been incorrectly estimated or additional parameters
may be needed to describe the hydrology of watersheds more accurately.
The monthly simulated and observed data points are shown in Figure 4.
The regression equation was

$$Q_s = 1.3 + 1.10 \ Q_o$$

with r = 0.95 and S.D. = 8.0 mm.

DISCUSSION - PHASE III

Watershed Descriptions

The objective of Phase III was to evaluate several critieria for
model calibration. To make a fresh, unbiased start and avoid any watersheds that had been used in previous work with the model, W-3 and W-4
were eliminated; and two small research watersheds in central Oklahoma
were selected. Their locations in relation to the Stillwater watersheds
are shown on the vicinity map in Figure 1.

The calibration watershed, Guthrie W-V, is 7 km south of Guthrie,
Oklahoma. The distance between this watershed and W-3 is about 75 km.
Watershed W-V is currently inactive; but during the period 1941-1956, it
was operated by USDA-SEA-AR as part of the Rolling Red Plains Watershed
Research Center. The drainage area is 6.8 ha. The soil is relatively
shallow with good internal drainage. During the active period of the
watershed, the vegetative cover was native grass, which was mowed
annually and was moderately grazed. The previous land use had been clean-
tilled row crops.

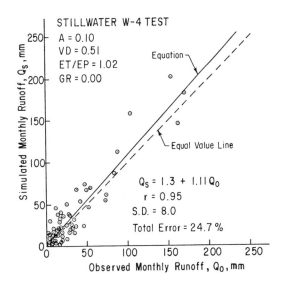

Figure 4   Monthly runoff for USDAHL test on Stillwater
W-4 for 1952-1972.

The test watershed, Chickasha R-7, is located 14 km southeast of
Chickasha, Oklahoma, approximately 90 km southwest of Guthrie W-V. The
drainage area is 7.8 ha. The topography and slopes are comparable to
the Guthrie watershed, but the soils are somewhat deeper and with slower
infiltration rates. During the period of the hydrologic record, the
vegetative cover was native grass which was in poor hydrologic cover
condition because of heavy grazing. This land had also been previously
farmed with clean-tilled row crops. Details of the soil and hydrologic
zone parameters for both watersheds are listed in Table 3.

Calibration and Testing - Phase III

Emphasis during Phase III was on procedures for calibration and
criteria for determining when a satisfactory calibration had been
achieved. Two different approaches are discussed in this paper.

Type I calibration depended on monthly runoff only. A satisfactory
calibration was the one for which the regression line between simulated
and observed monthly runoff for the period tested most nearly coincided
with the equal-value line. Type II calibration required that the total
simulated runoff must equal the total measured runoff for the entire
calibration period with the additional requirement that the regression
line for monthly runoff must be brought as close to the equal-value line
as possible within the equal runoff constraint.

A significant improvement over calibrations in the previous two
phases was made possible by making concurrent changes in the parameter A
from Holtan's infiltration equation and groundwater recharge, GR. To
change the slope of the regression line, one parameter had to be adjusted
to increase runoff while the other was adjusted to decrease runoff. By
using this approach, the regression line could be rotated to make it more
nearly coincide with the equal-value line.

TABLE 3. HYDROLOGIC ZONE AND SOIL PARAMETERS FOR GUTHRIE AND CHICKASHA
WATERSHEDS

	Guthrie W-V (6.3 ha)			Chickasha R-7 (7.8 ha)		
	Zone 1	Zone 2	Zone 3	Zone 1	Zone 2	Zone 3
Zone area, %	20	70	10	22	45	33
Average slope, %	2.6	4.5	2.5	2.8	4.5	3.6
Overland flow length m	37	35	9	44	28	16
Soil texture	loam	loam	loam	si lm	si lm	si lm
Final infil. rate mm/h	5.7	5.7	5.7	1.6	2.5	5.4
Depth, upper layer cm	30	23	23	23	25	33
Depth, lower layer cm	53	53	53	104	88	54
G*, upper layer %	14	14	14	11	11	11
G*, lower layer %	10	10	10	10	10	10
AWC$_+^{\dagger}$, upper layer %	16	16	16	20	20	20
AWC†, lower layer %	17	17	17	14	15	18
Cracks‡, upper layer %	6	6	6	8	8	8
Cracks‡, lower layer %	12	12	12	12	16	16
Root depth cm	83	76	76	128	114	88

*Percent of soil depth drained by gravity
†Percent of soil depth drained by plants
‡Percent of soil depth subject to cracking

    Calibration types I and II were evaluated on Guthrie watershed W-V.
The final values of the most important parameters are shown in Table 4.
Compared with the values used on W-3 and W-4 (Table 2), the values for
parameters A and VD were increased, and the value for ET/EP was decreased.
The change in ET/EP is especially noteworthy because of the relative
sensitivity of this parameter. The value of 0.88 reached by the calibration process compares well with values determined by lysimeter studies
by Doorenbos and Pruitt (1974) and Blad and Rosenberg (1974). It may be
noted that the type II calibration was obtained by reducing A and
increasing GR, resulting in the rotation of the regression line as
described above.

TABLE 4. LAND USE PARAMETERS FOR GUTHRIE AND CHICKASHA
WATERSHEDS

	Calibration Type	
	I	II
Vegetative parameter, A	0.80	0.60
Depression storage, VD, cm	1.27	1.27
ET/EP ratio	0.88	0.88
Upper cardinal temperature, $^{\circ}$C	27	27
Lower cardinal temperature, $^{\circ}$C	0	0
Groundwater recharge, GR, mm/h	0.0183	0.0229
Reduction in ET due to grazing, %		
Excellent hydrologic cover	35	35
Fair to Poor " "	60	60

Figures 5 and 6 show the plots of monthly simulated and observed runoff for the types I and II calibrations on watershed W-V. Each calibration was achieved after eight trials. The regression equation for type I was

$$Q_s = -0.51 + 1.01 \ Q_o$$

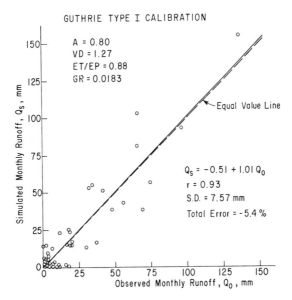

Figure 5   Monthly runoff for USDAHL calibration (type I) on Guthrie W-V for 1942-1953.

with $r = 0.93$ and S.D. = 7.57 mm. The total simulated runoff was 5.4% less than total observed runoff. For the type II calibration, the regression equation was

$$Q_s = -0.23 + 1.03 \ Q_o$$

with $r = 0.92$ and S.D. = 8.35 mm. The difference between the total simulated and total observed runoff was negligible.

While both calibration methods produced satisfactory results, the choice between them appears to depend on what the model is to be used for. Type I gave slightly better monthly predictions according to the statistical parameters of correlation coefficient and standard deviation, but the type II calibration gave the best estimate of total cumulative runoff.

Evaluation of simulated monthly runoff obtained in the calibrations of the Guthrie watershed showed that the model underestimated runoff in April and September, but overestimated in June. Generally the first storms after dry periods were underestimated while storms occurring

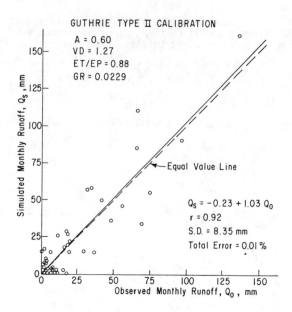

GUTHRIE TYPE II CALIBRATION

A = 0.60
VD = 1.27
ET/EP = 0.88
GR = 0.0229

Equal Value Line

$Q_S = -0.23 + 1.03\ Q_0$
$r = 0.92$
S.D. = 8.35 mm
Total Error = 0.01 %

Simulated Monthly Runoff, Q_S, mm

Observed Monthly Runoff, Q_0, mm

Figure 6   Monthly runoff for USDAHL calibration (type II)
on Guthrie W-V for 1942-1953.

within two weeks after wet periods were overestimated. These discrepancies suggest that the model may not be correctly simulating the rate
of soil moisture depletion, and may indicate an area of possible needed
improvement. Over the long term, however, the model gave good predictions on the calibration watershed.

To make an independent test of the model, it was applied to
Chickasha watershed R-7 for the period 1967-1974. This watershed had
poor overgrazed grass cover and hydrologic group D soils compared with
good grass cover and group B soils at Guthrie. This was a severe test
of the model because it involved differences in time, soils, cover condition, and distance between the calibration and test watersheds.

The results of the tests at Chickasha were good. For the type I
calibration, the total simulated runoff was 5.7% less than observed. The
regression equation of monthly runoff was

$$Q_S = -0.14 + 0.95\ Q_0$$

with r = 0.93 and S.D. = 7.12 mm.

The most accurate simulation was with the parameters developed by
the type II calibration. The total error was 1.73%. Figure 7 shows the
plot of monthly simulated and observed runoff. Good simulation was
obtained for small and large events. The regression equation was

$$Q_S = 0.13 + 1.01\ Q_0$$

with r = 0.93 and S.D. = 7.76 mm.

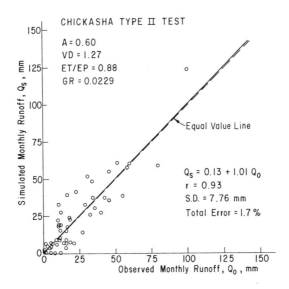

Figure 7   Monthly runoff for USDAHL test (type II) on
            Chickasha R-7 for 1967-1974.

SUMMARY

The USDAHL hydrology model was evaluated in an extended research
program, first on a single watershed, and later on three other native
grass watersheds in Oklahoma. The model had a tendency to underestimate
runoff from the first storms following dry periods and overestimate runoff following wet periods. In general, the model performed adequately
for simulating monthly and annual runoff from watersheds for which land
use parameters had been determined by previous calibration. Results were
less satisfactory when the model was transferred to a nearby watershed
with no prior calibration. The accuracy of simulation was affected by
the length and choice of calibration periods. The ratio of evapotranspiration to pan evaporation, ET/EP, and groundwater recharge, GR, were
found to be important parameters. By accounting for groundwater recharge
and by using an improved calibration technique, the model was calibrated
on a small watershed and tested on another watershed, located 90 km away,
with a total simulation error of 1.7%.

ACKNOWLEDGEMENTS

This paper was approved as Journal Article J-3941 of the Agricultural
Experiment Station, Oklahoma State University, Stillwater, OK, for
presentation at the International Symposium on Rainfall-Runoff Modeling,
Mississippi State University, May 18-21, 1981.

The contributions of Dr. Tesfai Ghermazien, Oklahoma Water Resources
Board, and Dr. Richard L. Bengtson, Agricultural Engineering Department,
Louisiana State University, are gratefully achnowledged for their part in
this research program.

REFERENCES

Bengtson, R. L. 1980. Predicting storm runoff from small grassland watersheds with the USDAHL hydrologic model. Unpublished Ph.D. dissertation, Oklahoma State University, Stillwater, Oklahoma.

Bengtson, R. L., Crow, F. R. and Nicks, A. D. 1980. Calibrating the USDAHL hydrologic model on grassland watersheds. TRANSACTIONS of the ASAE, Vol. 23(6), pp. 1473-1480.

Blad, B. L. and Rosenberg, N. J. 1974. Evapotranspiration by alfalfa and pasture in the east central great plains. Agronomy Journal, Vol. 77, pp. 248-252, March-April 1974.

Crow, F. R., Ree, W. O., Loesch, S. B. and Paine, M.D. 1977. Evaluating components of the USDAHL hydrology model applied to grassland watersheds. TRANSACTIONS of the ASAE, Vol. 20(4), pp. 692-296.

Crow, F. R., Ghermazien, T. and Bengtson, R. L. 1980. Application of the USDAHL hydrology model to grassland watersheds. TRANSACTIONS of the ASAE, Vol. 23(2), pp. 373-378.

Doorenbos, J. and Pruitt, W. O. 1977. Guidelines for the prediction of crop water requirements. Irrigation and Drainage Paper No. 24. Food and Agriculture Organization of the United Nations, Rome.

Fisher, G. T., Ayars, J. E., Holtan, H. N. and Nelson, D. L. 1979. USDAHL-74 model as a planning tool. TRANSACTIONS of the ASAE, Vol. 22(6), pp. 1347-1352.

Ghermazien, T. 1978. Transferability and capability of the USDAHL-74 model to simulate runoff from grassland watersheds. Unpublished Ph.D. dissertation, Oklahoma State University, Stillwater, Oklahoma.

Hanson, C. L. 1979. Simulation of arid rangeland watershed hydrology with the USDAHL-74 model. TRANSACTIONS of the ASAE, Vol. 22(2), pp. 304-309.

Holtan, H. N. and Lopez, N. C. 1971. USDAHL-70 Model of Watershed Hydrology. USDA-ARS Technical Bulletin No. 1435.

Holtan, H. N. Stiltner, G. J., Henson, W. H. and Lopez, N. C. 1975. USDAHL-74 Revised Model of Watershed Hydrology. USDA-ARS Technical Bulletin No. 1518.

Holtan, H. N. and Yaramanoglu, M. 1977. A user's manual for the University of Maryland version of the USDAHL model of watershed hydrology. University of Maryland, Baltimore, Maryland, MP 918.

Langford, K. J. and McGuinness, J. L. 1976. A comparison of modeling and statistical evaluation of hydrologic change. Water Resources Research. Vol. 12(6), pp. 1322-1324.

Molnau, M. and Yoo, K. H. 1977. Application of runoff models to a palouse watershed. ASAE Paper 77-2048. ASAE, St. Joseph, MI 49085.

Nicks, A. D., Gander, G. A. and Frere, M. H. 1977. Evaluation of the USDAHL hydrologic model on watersheds in the southern great plains ASAE Paper 77-2049. ASAE, St. Joseph, MI 49085.

APPLICATION OF CATCHMENT MODELS

PRECIPITATION-RUNOFF MODELING: FUTURE DIRECTIONS

L. Douglas James, Professor, Utah Water Research Laboratory, Utah
State University, Logan, Utah, 84322 and Stephen J. Burges, Professor,
Department of Civil Engineering, University of Washington, Seattle,
Washington 98195.

ABSTRACT

The development of hydrology as an applied science has been
shaped by needs for reliable quantitative estimation for project design
and facility operation. The future will bring new needs and produce
new capabilities. As to needs, trends are moving water resources
management 1) from flood control and water supply to water quality
control and instream environmental protection purposes, 2) from structural to nonstructural measures, 3) from hydrologic information for use
by professionals to information distributed to the public, and 4) from
deterministic to stochastic estimation. New capabilities are coming in
computational capability, understanding the spatial distribution of
hydrologic processes, and in model structure to take advantage of both
of these capabilities. To capitalize on these scientific advances in
producing more reliable hydrologic estimates for needed management
applications, rainfall-runoff modeling needs to 1) develop improved
methods for estimating the effect of watershed management on rural and
urban runoff quantity and quality, 2) estimate the hydrologic and water
quality effects of upstream water use on others downstream, 3) replicate
the temporal, spatial, and water quality variability in urban stormwater
systems, 4) establish a more scientific basis for pollutograph modeling,
5) establish a hydrologic base for estimating the environmental impacts
of instream flow changes, 6) improve replication of snow accumulation
and melt processes, 7) improve estimation of the magnitude of rare
events, 8) orient more toward information needs for reservoir operation,
9) take better advantage of new data management capabilities, 10) take
advantage of hydrologic modeling as a learning tool, and 11) deliver
results in a form more understandable to nontechnical audiences. All
these needs must face basic limitations of reasonable data availability.
Central issues exist in what we can afford to measure and how to design
measurement systems for maximum usefulness. Finally, we need to recognize that fundamental data and practicality limitations make certain
modeling directions inappropriate and provide signals to try new methods
when old ones prove unproductive.

291

INTRODUCTION

Hydrology is a science emerging from an art. Water project design, water use decision making, and water use regulation require quantitative estimates of hydrologic quantities. Hydrologic art began by formulating expressions to provide the needed estimates. Measurements of the presence of water (as it falls, is stored, or flows) and of catchment characteristics believed to be related were used to derive empirical expressions. Over the years, the benefits of more reliable estimation have motivated research that has transformed techniques for estimating hydrologic quantities from gross empiricisms to cause-effect relationships and moved hydrology from an art toward a science.

Hydrologic understanding has grown through interaction between needed applications and scientific advances. New opportunities to use or control water inspired new hydrologic tools, and new tools opened the door to previously unattainable water management benefits. As symbolized in Figure 1, the marriage of application with science in hydrology produced an understanding which enabled mankind to control more floods, irrigate more fields, and generate more hydroelectric power.

As the body of hydrologic information and the number of people using it expand, greater conscious effort is needed to match research to develop improved estimating methods with needs for better estimates. Researchers need to make an extra effort to understand in order to deal with user needs, and users should seek out and apply the best hydrologic methods.

In many aspects, the process of scientific advance is analogous to that of economic growth. Both processes occur through interaction between demand (applications) and supply (available science). Both processes are most efficient with a certain amount of tension. According to the unbalanced growth strategy of economic development (Hirschman 1958), economic growth depends on the "continuing out cropping of profitable opportunities." Development draws strength as supply grows in response to the tension created by excess demand and demand is acquired as marketing techniques dispose of excess supply. The analogy for hydrology is that even though we need to review the relevancy of our model building, we should expect science on a given topic to forge ahead of applications, and we should prioritize our research in areas where science trails the needs for application. This focus on estimating flows for design and sets of flows representing comparable conditions for water planning make hydrology an applied science.

THE DEMAND-USE TRENDS IDENTIFIED AND PROJECTED

Overview

From the days of the ancient hydraulic societies of China and the Middle East, water system builders and more recently design engineers have needed estimates of streamflow for water supply design. In some situations, information was available from direct measurements at a desired site. In other cases, gaging could be initiated to gather needed data. Most often, however, design engineers could not wait for relatively rare events of design magnitude to occur and be measured and have had to develop methods for estimating needed flows indirectly.

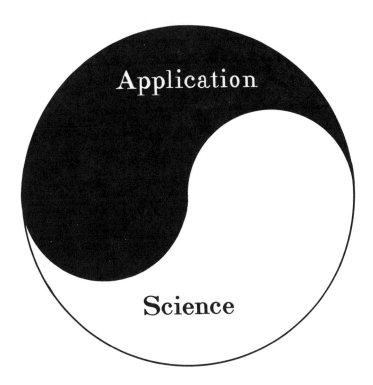

Figure 1. The practicality and theory of hydrology.

The historical advance in rainfall-runoff modeling has been shaped by the desired applications of the past; and, just as surely, one can expect future advances to be shaped by future information needs. These needs will be determined by trends in:

1. Water resources development purposes, use issues, and regulatory goals.

2. Water resources development and management methods.

3. The technical and nontechnical audiences who will use hydrologic information for designing and planning the methods used to achieve the chosen purposes, the information these groups consider important, and the methods of presentation which communicate to them best.

4. Design and planning philosophy with respect to risk and uncertainty.

Management Purposes

Historically, water resources management purposes in the United States have evolved through serial emphases on navigation, water supply and flood control, water quality protection, and preservation of natural environments (James and Rogers 1979). Each purpose introduced its own needs for hydrologic information as defined by requirements

as to type, spatial detail, and precision. Each need inspired the
development of new techniques as represented schematically in Figure 2.

Hydrologic information was needed in the design of reservoirs to
provide water for navigation canals and of facilities to pass flood
peaks through stream crossings and discharge groundwater intercepted
by cuts. Precise estimation, however, was not necessary because water
supply, stream crossing, and groundwater interception facilities

Management Purpose

Hydrologic Information	Navigation	Flood Damage Mitigation and Water Supply	Water Quality	Instream Flow Needs
Stream Flow Rates	X	O	O	O
Incremental Streamflow Rates		X	O	X
Streamflow Patterns		X	O	O
Water Quality			X	O
Quality Patterns				X

X Initial modeling
O Extension of previous modeling techniques

Figure 2. Hydrologic information requirements by management purpose.

could be made larger than needed for little extra cost. Design
mistakes could be corrected at an affordable expense. In other words,
hydrologic considerations were not major factors in system cost. The
rough estimating methods represented in Figure 2 for estimating flows
for navigation can only be called models in the most general sense.

The following management-purpose emphasis on water supply and
flood control stimulated improvement of hydrologic methodology to
estimate flow volumes and peaks, respectively. The earliest design
need was for annual yield or flood peak information at locations
selected for projects. Rainfall-runoff modeling began with graphs,
nomographs, and simple formulae for hand calculations. More complex
models were required to estimate streamflow volumes and rates as
storage was used to hold water over increasingly longer periods of
time, as the opportunities to increase output from a given facility
through more efficient operation became recognized, and as water and
land management needed to be coordinated.

As indicated by the row headings on Figure 2, with increasing
demands the needed hydrologic information progressed from the relatively
easily estimated reservoir size to provide a fairly small yield to
increasingly more difficult to estimate size for fuller development
of catchment yield. The economic analysis required to optimize size
required estimation of the incremental increase in yield with an
incremental increase in size (James and Lee 1971), a requirement
adding one more dimension of complexity to the needed hydrologic
sophistication.

A second added dimension was the need to estimate flow patterns

over time (flood hydrographs or stream yield by month) for design and
operation of storage facilities, flow patterns over space for design
and operation of storage and conveyance systems, and of both simultaneously for conjunctive land and water management.

Reflection on these increasing degrees of desired sophistication
also reveals a subtle progression from use of hydrologic information
to design functional engineering systems to more abstract applications
in proving economic justification and optimizing project design.
These abstract applications are seldom verified by post-construction
analysis. The resultant situation is one in which the most difficult
hydrologic quantities to estimate are those being estimated by models
which are least well validated.

The management-purpose emphasis on water quality introduced a
demand for quantitative estimates of sediment movements; movements of
inorganic, organic, and biological substances carried by the sediment
or otherwise with the flow; and interactions among these in transit.
This requires further refinement of flow estimation (as shown on
Figure 2) because of the ways water quality varies with flow magnitude
and temporal and spatial flow patterns. We are just beginning to
advance this estimation capability into feasibility and optimality
applications (Narayanan and Padungchai 1979) in response to a demand
being made increasingly clear by evidence that we are not effectively
targeting our large national expenditures on water quality control.

Finally, we are entering an era of increased concern that flow
diversions or quantity and quality modifications of flows in streams
are destroying environmental and esthetic values. Modeling to develop
hydrologic information for resolving these issues requires new capabilities in estimating incremental flows and flow quality patterns.
Instream flows are essentially residuals whose temporal, spatial, and
quality patterns and incremental changes in pattern are determined by
water manager use choices rather than natural events. The bottom row
on Figure 2 is included to show that the impacts of variations in
instream flow depends on the patterns in water quality and flow
required by environmental uses. Environmental needs must be analyzed
to identify critical parameters; hydrologic models should be structured
to estimate those parameters.

The hydrologist also needs to recognize that the management
purposes of navigation, flood control and water supply, water quality
control, and instream flow management are not ends in themselves but
rather means to more basic objectives. Present national policy
recognizes the four objectives of economic development, environmental
quality, regional development, and social well-being (U.S. Water
Resources Council 1980). National objectives, however, change with
time; each change alters optimal project designs and operation procedures. For example, a greater emphasis on environmental quality, or
even readjustment of priorities among quality parameters, alters the
hydrologic information needed to estimate impacts in priority domains
(perhaps flow variability during the spawning season for a given
species of fish). New areas of hydrologic art must be devised until
hydrologic science can be developed.

Management Methods

Until the 1930s, public programs in water resources were implemented almost entirely through project construction. Water resources
development meant a navigation channel, a water supply dam, a flood

levee or similar construction. Hydrologic information was used to
devise a functional design and later to formulate projects for optimal
performance.

The soil and water conservation movement introduced land treatment alternatives. Efforts were made to control floods, increase
catchment yield, and reduce erosion by such techniques as vegetation
management, land grading, and building many small rather than a few
large dams. Integration of these methods into project planning
required more advanced hydrologic techniques to estimate the effects
of land treatment on the various hydrologic information items shown on
Figure 2. Considerable effort has gone into developing the needed
quantitative relationships.

Another management change has been to supplement project construction with such nonstructural measures as floodplain management, conservation in water use, and water reuse to reduce waste disposal.
These measures are implemented as individual actions by large numbers
of people, each of whom requires hydrologic information for his or
her particular situation on such topics as flood hazard, water availability probabilities, and runoff conditions that would carry the
wastes away.

Special thought needs to be given to the changes in hydrologic
information requirements associated with growth in water demand to
the point where nature can no longer satisfy all needs. At that
point, supplying water to those with new needs requires shifts from
prior users. The hydrologic information needed is no longer that
required to increase basin yield but rather that required by the
market to effect efficient water transfers and by regulatory agencies
to protect the public interest during the transfer process. Each
water right has a value that depends on its use, time period, quality,
firmness, and other properties. These characteristics that determine
value need to be estimated and reliable information on them provided
to buyers, sellers, and regulators.

One more factor adding complexity to the need for hydrologic
information is that information dissemination on flood hazard, for
example, is not sufficient to ensure optimal floodplain use. The
disparity between optimal floodplain use (or water use or waste
disposal methods) from the public viewpoint and from the perceived
self interest of individual decision makers requires that public
nonstructural programs be implemented by regulatory or economic
incentives (James 1973). Implementation adds a new dimension to the
needed hydrologic information.

Regulatory programs require an order of magnitude of greater
hydrologic detail. As an example, estimates of average flood hazard
are satisfactory for structural project planning because overestimates
of damage in one area balance out with underestimates in another. For
regulation, each property is its own situation. It is not a reasonable
regulatory stance to excuse being too stringent with one party by not
being stringent enough with another. Consequently, accurate flood
hazard mapping has proved to be a major cost of the floodplain management program in the United States.

Varied Audiences

The hydrologic information for structural design is transferred from the hydrologist to the design engineer, individuals of similar disciplinary background. For structural feasibility studies, the transfer crosses discipline boundaries to economist and other social scientists untrained in hydrology but professionally obligated to using the information. For nonstructural measures, the audience expands to large numbers of nontechnical people, many of whom see no particular personal advantage in using the data. For implementing regulatory or other nonstructural programs, hydrologic information must be given to individuals who vary greatly in what they find useful (Larsen et al. 1979).

As water resources planning has moved into the modern era requiring integration of water with nonwater systems, hydrologic information has had to be communicated to more disciplines, to people at more levels of government, and to more people lacking technical training who are responsible for land and environmental resources (Weber 1979). The added complexity to water resources planning that has evolved (Schad 1979) has necessitated a search for economical planning methodology through efficient use of computer capability. The need becomes much greater for nonstructural programs where the hydrologic information must be communicated to nontechnical users wanting site-specific estimates at timing matching their convenience (James 1973).

Uncertainty

Rainfall-runoff models have historically been refined through better representation of physical processes occurring in catchments. Hydrologic quantities computed from mathematical expressions of physical laws tend to convey certainty. Many users overlook that unrepresentative sampling and imperfect measurement are inherent in any data collection process and particularly so for recording processes such as precipitation which are highly variable in space and time and for which one is forced to use a few square centimeters to represent many square kilometers. Furthermore, modeling economy requires approximating simplifications omitting theoretical higher order interactions over space and time and expressing a large amount of spatial variability in watershed characteristics as lumped parameters; calibration is handicapped by short data series and process and criteria issues. From the scientific viewpoint, the honest presentation is to convey hydrologic estimates stochastically as probability distributions rather than single quantities (Linsley 1979a).

The inherent uncertainties in hydrologic estimation are in direct contrast with the thinking of large segments of the public including many decision makers. Uncertain information is particularly difficult to reconcile with regulatory processes. An example issue is how to draw floodplain regulations flexible enough to be adjusted as estimates of hazards prove incorrect or as conditions change.

Hydrologists face major challenges in developing improved techniques for determining the uncertainty associated with their estimates, conveying probabilistic information to users, and dealing with situations where more recent information requires changing an estimate. Considerable progress is being made in applying probability and statistics to hydrologic data (see, e.g., Hann 1977) but progress in conveying the uncertainty associated with alternatives to decision makers is much more limited. Innovative techniques for doing so are urgently needed.

THE SUPPLY-CAPABILITY TRENDS IDENTIFIED AND PROJECTED

Biswas (1969) has traced how procedures for estimating flows for hydraulic design have evolved from the trial construction and performance observations of ancient times to methods based on a qualitative understanding of the hydrologic cycle. Linsley (1967) traced how qualitative understanding has improved quantitative estimation through the stages of empirical formulas based on drainage basin characteristics (1850-1932), correlations of streamflows with one another or storm characteristics (1932-1964), and digital computer modeling of the rainfall-runoff process (1964-present).

Recently Beard, in a thoughtful analysis of progress "toward scientific hydrology," concluded:

> If we are ever to integrate the meager knowledge that we have on the flood and drought potential of each individual stream into a body of information that can be applied to all stream systems and their components, it must be through an understanding of the cause-effect relationships that exist between precipitation and streamflow. Empirical determination or description of these relationships may help, but a complete understanding of the physical conditions and processes will eventually be necessary. (Beard 1981, p. 69)

Improvements in rainfall-runoff modeling are essential to providing better hydrologic information for meeting present and future needs.

Improvements in rainfall-runoff modeling can be expected to come through:

1. Expanded computational capability

2. More thorough understanding of hydrologic processes

3. Quantum jumps in learning how to take advantage of expansions in computational capability and more thorough hydrologic understanding to develop and apply new modeling structures or philosophies

Computational Capability

The origins of rainfall-runoff modeling predate the digital computer by decades. Morgan used rudimentary techniques in 1915 in his important design work for the Miami, Ohio, Conservancy District (Morgan 1971). By 1940, a multitude of methods from runoff coefficients to coaxial graphical correlation were developed for estimating rainfall excess; the unit hydrograph was developed for estimating the time pattern of basin runoff; and routing methods were developed for combining catchment hydrographs within a larger watershed. Since then, hydrologic models have moved from the computationally simple but unreliable formulas of the 1930s to the computationally more demanding but more useful continuous digital simulation models of the 1960s.

The digital computer (and related input-output devices) has become the dominant data management, computational, and display tool for hydrologic information. It provides hydrologists new power in collecting and organizing large volumes of information, making orders of magnitude more computations, and graphically displaying results in forms that users can quickly comprehend.

In practice, however, we see a persistence of the older methodologies (Linsley 1979b). For example, despite continued improvements in the speed and capacity of digital computers for data manipulation, most modelers continue to build modeling structures that use streamflow, evapotranspiration, and precipitation data collected with a network designed long before modern computer capabilities became possible.

Little has been done to take advantage of the possibilities that sophisticated remote sensing and automatic recording devices open for measuring and using spatial and temporal distributions of quantities important to soil moisture and precipitation-runoff phenomena. Greater effort is needed to update hydrologic practice in applying the best available methodology and in making better use of available computational capabilities in improving the methodology.

Technical Understanding

In almost every section of the United States, the amount of water available in a basin as a whole in an average year exceeds the demand (U.S. Water Resources Council 1978). A water supply project captures water at the times and in the portions of the basin where it is in excess, treats it to achieve quality standards, and delivers it when and where it is needed. Hydrologic information is used to size and operate the storage, treatment, and delivery systems.

For structural water supply design, the ideal information to have available is streamflow records of annual volumes, their distribution over the year, and the amounts at various sites over a long historical period for reliable statistical analysis for flow frequency. The data needs are similar (with different emphases on volumes, high flows, and low flows) for design for flood damage mitigation, hydroelectric power, low flow augmentation for salinity control, etc.

A long statistically stationary streamflow record is seldom found situated at a project location. Rainfall-runoff modeling provides a substitute estimating method but one with reduced reliability. In contrast with streamflows, precipitation records are available at many more sites and over longer periods. They can be used via precipitation-runoff models to estimate streamflow with particular reliability for design situations in which one needs flows on a frequency basis rather than for an historical storm. They also provide a basis for estimating how the flows would be changed by implementation of water resources development and management programs.

A dominant factor in formulating more reliable precipitation-runoff models is improved understanding of the hydrologic processes controlling soil wetting and drying. Hydraulic theory (Hillel 1980) and empirical observations need to be used to develop mathemathical relationships for estimating soil moisture and the rates over time of such processes as infiltration, evaporation, overland flow, interflow, and groundwater recharge and discharge as well as the primary desired end product, the streamflow itself. It is also important to recognize that the benefits from such scientific advances are far broader than those from improved flow estimation. For example, Krishna (1981) illustrated use of hydrologic modeling of soil moisture availability in planning management practices for greater crop production.

Catchment runoff is nature's integration of how the land surface responds to precipitation. Horton (1933) suggested that floods are generated from overland flow that occurs whenever the rainfall rate exceeds the precipitation rate. Betson (1964) hypothesized that small portions of a basin become saturated quickly during a storm and provide the bulk of the overland flow. Freeze (1972) found from field data in vegetated basins in a humid climate that overland flow is a rare occurrence and that precipitation on transient near-channel wetlands dominates flood hydrographs. Zaslavsky and Sinai (1981) hypothesized that interfaces between soil layers cause lateral flow leading to local moisture concentrations in areas with convex slopes at hillside bottoms and creating the partial source areas that generate most runoff. Overall, the literature shows general conceptual agreement about sources of streamflow; the concept of variable source areas as a way to estimate runoff from precipitation is included one way or another in many models. The concept of spatial heterogeneity of soil hydraulic conductivity and topography may be equally important (Freeze 1980). Obviously, a great deal remains to be done to improve runoff physics representations in rainfall-runoff models.

Urban catchments may ultimately be the easiest to model success-fully. In this application, Delleur (1981) notes trends away from use of the rational formula and toward continuous simulation models calibrated by using selected objective functions in parmeter optimiza-tion. He expects continuous simulation models to have a dominant role because of their capability to predict runoff quantity and quality with a reliability and speed that will eventually permit automated operation.of urban storm drainage systems. The most significant advances in real-time forecasting have utilized predictor-corrector methods. These methods make use of state equations that are updated formally (usually via an extended Kalman filter algorithm). Applica-tions have been reported by Kitanidis and Bras (1980a,b); similar techniques have been used in river quality modeling. While fuller analysis of these technical advances is not within the scope of this paper, the above summary indicates approaches where advances are likely.

The output requirements from hydrologic modeling have expanded as water resources have become more intensively used and new needs must be fulfilled by transferring water from lower to more highly valued uses. The hydrologic effects of the transfers are not simply to shift the use of a given water volume but also to change the flow patterns and water quality available to others. One modeling impli-cation of these trends is to require capability to estimate flows continuously over longer periods and simultaneously at many locations and to route them through complicated manmade water supply or flood water control systems. Hydrologists have responded with multisite flow generating models for water supply and flood control (Lumb and James 1976).

Modeling Philosophy

Computational and technological advances generate more sophisticated hydrologic models. When no better techniques were available, simple models had to suffice. Our advances make greater modeling sophistication possible, but it is important to remember that more complex models are not necessarily always better. More equations and parameters can improve replication of recorded flows without necessarily improving flow estimation outside the calibration period.

Furthermore, no matter how complicated a representation of the system physics is used, we will never be able to predict the exact hydrograph resulting from a particular storm. Multivariate time series of measured or estimated inputs contain temporal and spatial misrepresentations of actual inputs. There are errors in the model structure and errors in estimates of model coefficients and parameters. The principal outputs from precipitation-runoff models are estimated streamflow hydrographs, states of various moisture zones, groundwater levels and fluxes, etc. None of these can be measured with certainty, hence matching measurement does not necessarily mean matching catchment response.

Until the last decade, most hydrologic models used to convert precipitation to streamflow were deterministic. The assumption made by the modeler was that inputs and recorded streamflow data were reasonably accurate. If the model structure was a reasonably good approximation to catchment physics, it was assumed that the model parameters could be adjusted until recorded and estimated hydrographs were in reasonable agreement. This has often led to overfitting (i.e., any one can fit an nth order polynomial to n + 1 observations; embarrassment results when some inconsiderate individual provides more observations). There are two obvious ways to accommodate this shortcoming. The first is to follow standard modeling techniques using a split record, one part to calibrate, the other to verify and obtain prediction error statistics. The second approach is to recognize formally that all inputs and measured quantities are not known with certainty and to propagate these through the model. This latter approach will provide a probability distribution of outputs at each location and time when outputs are estimated. Such estimates can be made using first-order uncertainty propagation, e.g. Garen and Burges (1981), or more formally, state estimation techniques, e.g. Chiu (1978), Wood (1980). Such models are only useful for interpolating within the realm of observed phenomena. No existing model can be used to estimate with assurance extremes beyond what have been recorded at a particular site.

Future precipitation-runoff modeling must improve model structure and reduce model input and parameter uncertainty. Efforts to improve should be pursued in a systems framework that ensures that overmodeling is not attempted. We must recognize that our models are based on limited knowledge and on data restricted by limitations to our ability to observe and record states of the physical system and its inputs and outputs in space and time. We must balance improving model structure with improving information gathering, transmission, and use.

MODELING DIRECTIONS FOR THE FUTURE

The trends in needs for hydrologic information and in capabilities to develop and deliver it suggest 11 priority directions for rainfall-runoff modeling as follows:

1. Watershed Management: Land treatment in rural watersheds can be used to increase runoff volumes, reduce flood peaks and associated sediment production, and absorb pollution loadings. Water harvesting techniques are also receiving increased attention. In urban areas, other land treatment methods are available for reducing flood runoff (Lumb et al. 1974) and improving water quality.

Better rainfall-runoff modeling methods are needed to 1) estimate the effects of both rural and urban land treatment on flows and on water quality, 2) to provide hydrologic information on the comparative effectiveness of various land treatment methods, and 3) identify locations within a catchment where land treatment is most effective.

Hydrologic simulation by digital computer is the most promising tool for this sort of analysis (Crawford 1969). Present results suggest that the effects are greater for small than for rare events, greater in small than in large catchments, and greater on sediment production than on runoff volumes. It is precisely these relatively more frequent events in smaller watersheds, their associated sediment loads, and the pollutants carried with the sediment that need the most attention in integrating nonstructural methods with structural design. The nonstructural approach (whether floodplain management, water conservation, or waste confinement) addresses ordinary events rather than extreme ones. For dealing with smaller events and local situations,

> a. Urban design can feature land use patterns, pervious pavements, street cleaning, and other methods of reducing flood runoff and its associated nonpoint source pollution loading.

> b. Use controls for hillside areas can reduce flood runoff, and floodplain land use controls can reduce flood damage.

> c. Vegetative buffers near streams, measures to prevent pollutants from washing into the stream from newly exposed soil at mining sites, and general sheet erosion control can reduce storm pollutographs.

All of these watershed management methods are known to produce desired results, but rainfall-runoff modeling needs to be developed further for improved quantitative estimates of responses from catchments treated in the above ways. The challenge is greatest for water quality modeling. Given the poor success with attempts to model nonpoint source pollutant movement (even for hydrologically well defined problems such as pollutant removal from highway pavements), we are not optimistic that models will be capable of distinguishing signals from noise until more reliable pollutograph data (it can be enormously difficult to collect) are obtained.

2. Water Management: We are moving from an era in which water for each use can be planned independently, or even conjunctively from virgin sources, to one in which downstream water uses are supplied, not from hydrologic events, but rather from return flows from upstream use. For example, in basins developed for irrigated agriculture, upstream farm water use practices determine downstream water availability timing and quality. For further improvement in water use efficiency, water management techniques need to be formulated, hydrologically evaluated, and implemented to permit as many reuses as possible before the quality degrades to unacceptably low levels. Water utilization objective functions need to be formulated in terms of the maximum sum of values in use received from the series of uses and reuses that occurs.

The hydrologic key to development of water management methods that maximize reuse potential is modeling to determine how water quality is affected by use practices. For example, geochemical relationships need to be incorporated into future models.

A recent example of substantial change in water quality resulted from replacing forest with pasture in catchments near Perth, Western Australia (Bennett 1981). The changed evapotranspiration regime caused the water table to rise, bringing highly saline water to the surface and damaging the surface water supply. Investigation of this change beforehand with a rainfall-runoff-groundwater geochemical model might have provided timely information that would have substantially influenced the decision to change the vegetal cover.

3. Urban Stormwater Systems: Structural systems control storm runoff from urban areas most effectively as they integrate patterns of pervious and impervious areas, channels, sump storage, and gates directing flows through areas where rainfall has been less intense. Good hydrologic estimates are essential to evaluation of the effectiveness of alternative designs. Urban stormwater system designs are helped little by point estimates of flows and volumes. They need complete hydrographs, not just for one storm but for the series of rainfall bursts that occur from the time the rain begins until the system is entirely drained, not just at one location but reflecting the spatially variable precipitation that occurs over an urban area. Furthermore, the reservoirs and channels often become clogged with sediment, and the sediment often carries and stores considerable amounts of pollution.

4. Pollutant Transport: Exposed pollutants accumulate on a catchment surface over time. Pollutant washoff occurs with overland flow from impervious areas. Erosion from areas of bare earth adds sediment to the total stream load, and the moving sediment carries physical, chemical, biological, and radiological pollutants. Urban hydrographs rise and fall rapidly, often only long enough to carry sediment and pollutant loads a short way through the system. Other pollutants accumulate in the soil profile, are dissolved by percolating water, and discharged into streams. The various pollutants interact with one another chemically and are affected in various degrees by biological activity.

This complicated interactive process is usually modeled by linear pollution accumulation and exponential decay equations (Delleur 1981) which no more represents the total process than does the constant runoff coefficient in the rational formula. More accurate pollutant transport modeling requires careful flowpath identification, water movement quantification by flow path, and estimation of the associated sediment and pollutant movement.

Attempts at physically oriented modeling, however, are frustrated with the extremely difficult problem of trying to represent phenomena for which few data exist, spatial distribution is crucial, and even the ability to collect data that accurately reflect necessary system states is small. Without a reliable data base, attempts to model may actually obfuscate rather than clarify hydrologic understanding and its application in quantitative estimation.

5. Instream Flows: Interest is growing in promoting the environmental and recreational values of instream flows. Stalnaker (1979) explained how stream habitat productivity is largely determined by substrate distribution, channel configuration, and riparian cover. All these factors change with fluctuating stream water levels, and Stalnaker concluded that "the dynamic aspects of the stream habitat condition over seasons and years under changing water and nutrient supply must be understood and treated as stochastic processes."

Whereas structural flood control designs need flow peaks and hydrographs and water supply designs need flow volumes, management to
enhance instream flow values must examine many more parameters (velocity,
depth, temperature, and other flow conditions around the periphery and
longitudinal along the stream) describing flow patterns over various
durations and in various seasons. Estimation of these parameters
requires much more sophisticated modeling and is complicated by the
fact that the low flows which are most critical in determining these
sorts of flow properties are much more sensitive to changing upstream
conditions and channel seepage losses and gains than are either peaks
or volumes. Greater attention will need to be given to low flow
modeling.

6. Snowmelt Runoff: Modeling precipitation-runoff relationships
when snow or ice is involved poses additional problems. The four
principal ones are determining whether the precipitation is snow or
rain, melt rates, how the soil-water interaction is changed when one
is dealing with frozen ground or soil under a snow cover, and the
effects of snow cover on sediment and pollutant transport. There are
few data currently collected in a useful form to model snowmelt runoff
satisfactorily. Satellite data that indicate areal extent of snow
cover with time may find significant future use in seasonal snowmelt
runoff volume forecasts.

7. Extraordinary Events: Water resource development projects or
systems of projects control normal floods (those up to the magnitude
of the design flood) or supply water during normal droughts. As
people become dependent upon them, they tend to intensify the catas-
trophic losses during severe flood or drought situations. Water
resource systems are brittle in the sense that they behave well in
situations up to the magnitude of their design events; but beyond that
level, they fail and inflict major losses. Planning for dealing with
these losses requires estimation of magnitudes by frequency of rare
events. Estimation is complicated as the rare events result from
storms of atypical meteorological characteristics (e.g. rare, tropical
storms), flow paths that leave the channel and travel across urban
areas, sudden loadings of toxic pollutants, etc. Beard (1981) illustrates the difficulty by citing as an example the fact that occasional
extreme events are totally inconsistent with any rationale. He notes
that "from 1930 to 1940, virtually every annual flow of the Red River
of the North at Grand Forks, North Dakota, was smaller than flows in
any of the preceding 48 years of record (Beard 1981, p. 68). We need
to enhance our capability in replicating rare hydrologic events
(perhaps through global meteorological modeling), not just defined as
a rare flood peak or drought but as a spatial pattern of high or low
flows over a flood or drought period.

8. Reservoir Operation: As water resources development
approaches the point where additional structural measures are not
feasible, greater emphasis will have to be placed on more efficient
operation of existing facilities or shifting operation from one
purpose to another (e.g., reservoirs built for flood control may
need to be reanalyzed and reauthorized for water supply operation).
Reservoir operating rules need to be structured more effectively
for determining a) whether, as a flood occurs, flows should be stored
to reduce current damage or released to provide space for possibly
even larger flows to follow, b) whether, between flood peaks, storage
space should be filled to store water for beneficial use or emptied
to contain potential floods, c) whether stored water should be
released for present use or retained for use during possible future

droughts, and d) how much of the water to be released for beneficial
use should come from each reservoir in which water is stored (James
and Lee 1971). Examination of the hydrologic information required
for resolving these issues shows that rainfall-runoff modeling can
contribute by providing improved models of flow variations over time,
by season, and among proximate locations. Rapid reliable estimation
is particularly important for real time operation during storm events.

9. Data Network Design: Modeling activities are necessarily
conditioned by data availability. For design situations where flow
frequency information is required, existing precipitation networks
provide the major input data source. For situations where real-time
forecasts or forecasts for future streamflow volumes (say over
several months) are desired, the possibility of obtaining data for
an entire catchment via use of sophisticated remote sensing
instruments should influence model building, particularly when the
spatial distribution of soil moisture and precipitation are important
to predicting response. Clearly, the application that should benefit
most from this type of information is flood forecasting. The
possibilities for anticipating precipitation via use of satellite
information are described by the U.S. National Science Foundation
(1980):

> Satellites have unique capabilities to observe wind,
> water vapor, cloud pattern, temperature, and soil moisture
> conditions needed to predict the probability that flood
> producing storms will develop.... Radar is useful only
> after the storms form, rain gauges after the rain begins
> to accumulate, and stream gauges after the streams begin
> to rise. (p. 68)

Modeling activity is currently constrained by limited data.
Evaluations of hydrologic data networks have been attempted, and
conferences have been held for the specific purpose of examining
issues in hydrologic data network design (e.g. Chapman Conference,
Tucson, Arizona, December, 1978). The limitations of network design
as articulated by Moss (1979), Dawdy (1979), and Langbein (1979),
make it unlikely that a ground based network can be designed to meet
the needs of all potential users.

The availability of water quality data is much more limited.
Collected data are largely point measurements in space and time without
any real plan on how the data can be integrated to estimate total
annual pollutant loadings or the environmental impacts of those
planned. The quality of work done in laboratory analyses needs to
be carefully watched. Much more thought needs to be given to data
collection and data validity checking for integrating water quality
components into hydrologic models.

10. Learning Models: It is useful to distinguish modeling
purposes between learning and decision making. Much of the recent
literature on precipitation-runoff modeling describes learning
experiences. Many able workers have applied themselves diligently
to obtaining better descriptions of the physics of flow over the land
surface, through the soil, and in stream channels. Others have gone
to considerable effort to learn more about the physics of snowmelt.
Much of the current work is on point processes or descriptions of
flow physics for small land surface areas. Incorporation of these
works into models that represent entire catchments may or may not
lead immediately to improved model output. Process uncertainty
and inability to integrate in space and time the numerous necessary

inputs might dictate use of less complete and simpler models for quantitative flow prediction. It may be advisable, however, to continue building the more complicated models because of their greater potential in the long run.

11. Nontechnical Audience: The actual implementation of nonstructural measures is performed by members of the general public. Floodplain land use and floodproofing decisions are made by floodplain residents who may ask for hydrologic risk information at any time (but most often when a land sale or building construction is under consideration) and for precise sites (James 1973). Water rights sales occur between water users who need information or purchasing or sales opportunities and descriptive data on the time variability of water availability for a given right and its associated quality. Nonstructural efforts to control pollution require motivating people to reduce amounts of pollutants deposited on the land surface. Such programs are not built from the relatively simple one-time delivery of hydrologic information for a single location to a structural design engineer. Instead, the hydrologic studies need to produce estimates at many locations (floodplain, stream, or potential pollution source sites) and respond quickly at the demand of the user. Communication of the estimates must be comprehensible to the property manager. These estimation and communication needs create some major new rainfall-runoff modeling opportunities in terms of locations covered, speed of reply, and visual display. The presentation needs to be convincing to people who have often had years of experience with the locality.

IMPLICATIONS FOR THE FUTURE

Modeling Goals

The above eleven needed directions suggest four major themes for the way in which precipitation-runoff modeling should be headed:

1. Major advances are going to be needed to replicate sediment transport and the associated movement of pollutants. Processes requiring better coverage include washing of these substances from the ground surface or leaching from the soil, transport overland or in stream channels, deposition as hydrographs recede, picking materials up again when flows rise, and chemical and biological interactions in transport.

2. The rainfall-runoff modeling art is going to have to make major advances in quantitative representation of the pattern in which flows vary over time and space within a total management framework such as is found in urban areas for storm runoff; in integrated water supply systems combining wells, streams, and reservoirs; and in salinity and other pollutant movement through hundreds of miles of channels and reservoirs.

3. Model building requires reliable field data for checking proposed estimating methods. Available data are very limited for many important applications. Additional thought and effort need to be put into the design of data collection networks and the quality control needed to ensure usable results. We should pay attention to the thoughtful observation of Langbein (1979) that systematic analysis is needed so assure that proliferating data networks achieve their scientific objectives. Kemp and Burges (1978) addressed the hydrologic data requirements for urban streamflow modeling and found that much

could be gained by using relatively intense networks of precipitation gages and stream gages in a catchment for a period of two to three years to gather suitable information for calibrating a continuous streamflow simulation model. We are less optimistic about the possibilities of modeling sediment and pollutant movement in urban catchments.

4. Hydrologists are not making effective use of recent advances in displaying the results of digital computer modeling. Potential users of such information are not aware of hydrologic facts that they could put to good use. Much effort is needed to close this gap.

Modeling Concepts

How might future models differ from existing models? More accurate estimates of complete streamflow hydrographs at multiple locations throughout a catchment will require reduction in errors in model inputs as well as improvements in model structure, probably through quantum conceptual changes rather than marginal refinements. Greatest improvements are likely to result if the actual spatial distribution of precipitation over the catchment can be represented for time increments as small as a few minutes. This implies either a large number of rain gages or another way of recording the spatial and temporal distribution of precipitation. An excellent discussion of errors in estimating average precipitation over a catchment, how this influences rainfall-runoff modeling, and four significant attempts to solve the problem is given by Bras (1979). We will always be faced with trading off information collection and use with model output accuracy.

Most of the precipitation runoff models in current use were developed by professionals whose dominant training was in hydraulic engineering. Hence we see detailed representation of channel flow for average channel sections, lengths, etc. and usually a less detailed representation of the spatial distribution of soil moisture movement in the catchment. Some investigators have tied geomorphological information to unit hydrograph structure, e.g. Rodriguez-Iturbe and Valdes (1979) and Gupta et al. (1980). It seems that future models might beneficially exploit geomorphological relationships that can be obtained readily via remote sensing; modern information processing techniques permit handling of any degree of desired detail. The cost of data collection is the primary limitation.

SUMMARY REMARKS

The first portion of our paper reviewed how the information needed for water resources management provides the motivation for advances in hydrologic modeling capability. Reviews of needs and capabilities were used to suggest priority directions for rainfall-runoff model improvement. The assessment is only so many words unless theoretical and empirical work can be combined to make the needed modeling advances realities.

Model building needs to be kept in balance with what can be validated with reliable data. Precipitation patterns, catchment characteristics, and hydrologic responses all vary in complicated two or three dimensional spatial patterns and with time. Rainfall-runoff models need to be used systematically in the design of information collection systems.

Since complete measurement is impossible, we must consider the measurements we have as indices and seek to define the indices to achieve reliable model-estimated quantities. This must be done in an interactive mode between data collection and model refinement. The information collected should encompass precipitation and other weather patterns, physical properties of the watershed and how measurements of them relate to model parameters, and states of water storage within or movement through the watershed that can be used in model validation.

One possibility of this sort is to model parts of catchments using geomorphologically based models and to link these catchments with appropriate flow routing models. A Muskingum-Cunge scheme (see e.g. Cunge 1969; Flood Studies Report 1975) may be suitable. Observations at key locations could be coupled with a state-space type of model to estimate state variables of interest throughout the drainage network.

We must recognize that signal-to-noise issues continue to render many hydrologic quantities unsuitable to estimation by modeling. Many modeling assumptions limit the issues which a model can address. Since some issues remain outside these limits, there will always be a need for the back-of-the-envelope model; we might, however, need a few large envelopes.

Perhaps most important, our efforts to build and refine models must avoid the pitfalls of vertical as opposed to lateral thinking. de Bono (1967) articulates this case well.

> Logic is the tool that is used to dig holes deeper
> and bigger, to make them altogether better holes. But
> if the hole is in the wrong place, then no amount of
> improvement is going to put it in the right place. No
> matter how obvious this may seem to every digger, it is
> still easier to go on digging in the same place than to
> start all over again in a new place. Vertical thinking
> is digging the same hole deeper; lateral thinking is
> trying again elsewhere.

We have all been guilty at one time or another of digging the same hole deeper. Improved hydrologic estimation will not be achieved by more vertical thinking.

REFERENCES

Beard, L. R. 1981. Toward scientific hydrology. In Fluid Mechanics Research in Water Resources Engineering. (Eds. R.E.A. Arndt and M. H. Marsh). St. Anthony Falls Hydraulics Laboratory, Univ. of Minnesota, Minneapolis. pp. 63-70.

Bennett, D. 1981. Personal communication to S. J. Burges.

Betson, R. P. 1964. What is watershed runoff? Journal of Geographical Research. Vol. 69(8):1541-1551.

Biswas, A. K. 1969. A short history of hydrology. Progress in hydrology, Vol. 2. Dept. of Civil Engineering, University of Illinois, Urbana. pp. 914-936.

Bras, R. L. 1979. Sampling of interrelated random fields: The rainfall-runoff case. Water Resources Research, Vol. 14(1):84-96.

Chiu, C-L. (Ed.). 1978. Applications of Kalman filter to hydrology, hydraulics, and water resources. Proceedings of AGU Chapman Conference, University of Pittsburgh. May 22-24.

Crawford, N. H. 1969. Analysis of watershed changes. In Effects of Watershed Changes on Streamflow. (Eds. W. L. Moore and C. W. Morgan). University of Texas Press, Austin, pp. 27-34.

Cunge, J. A. 1969. On the subject of a flood propagation method. Journal of Hydraulics Research, IAHR, Vol. 7:205-230.

Dawdy, D. R. 1979. The worth of hydrologic data. Water Resources Research, Vol. 15(6):1726-1732.

de Bono, E. 1967. New think. Basic Books. New York.

Delleur, J. W. 1981. Mathematical modeling in urban hydrology. Proceedings of International Symposium on Rainfall-Runoff Modeling, Mississippi State University, Starkville.

Freeze, R. A. 1972. Role of subsurface flow in generating surface runoff 2: Upstream source areas. Water Resources Research, Vol. 8(5):1272-1283.

Freeze, R. A. 1980. A stochastic-conceptual analysis of rainfall-runoff processes on a hillslope. Water Resources Research, Vol. 16(2):391-408.

Flood Studies Report. 1975. Vol. III, Flood routing studies. Natural Environment Research Council., London. 76 p.

Garen, D. G., and S. J. Burges. 1981. Approximate error bounds for simulated hydrographs. In Press. Journal of the Hydraulics Division, Am. Soc. of Civil Engineers.

Gupta, V. K., E. Waymire, and C. T. Wang. 1980. A representative of an instantaneous unit hydrograph from geomorphology. Water Resources Research, Vol. 16(5):855-862.

Haan, C. T. 1977. Statistical methods in hydrology. The Iowa State University Press, Ames. 378 p.

Hillel, Daniel. 1980. Fundamentals of soil physics. Academic Press, New York. 413 p.

Hirschman, A. O. 1958. The strategy of economic development. Yale University Press, New Haven, Connecticut. 217 p.

Horton, R. E. 1933. The role of infiltration in the hydrologic cycle. Transactions AGU, Vol. 14:446-460.

James, L. D. 1973. Surveys required to design nonstructural measures. Proceedings of the American Society of Civil Engineers, Vol. 99(HY10):1823-1836.

James, L. D., and R. R. Lee. 1971. Economics of water resources planning. McGraw-Hill, New York. 615 p.

James, L. D., and J. R. Rogers. 1979. Economics and water resources planning in America. Proceedings of the American Society of Civil Engineers, Vol. 105(WR1):47-64.

Kemp, G. J., and S. J. Burges. 1978. Hydrologic modeling and data requirements for analysis of urban streamflow management alternatives. Technical Report No. 57, C. W. Harris Hydraulics Lab, Dept. of Civil Engineering, University of Washington, Seattle, Washington. 124 p.

Kitanidis, P. K., and R. L. Bras. 1980a. Real-time forecasting
with a conceptual hydrologic model 1. Analysis of uncertainty.
Water Resources Research, Vol. 16(6):1025-1033.

Kitanidis, P. K., and R. L. Bras. 1980b. Real-time forecasting
with a conceptual hydrologic model 2. Applications and results.
Water Resources Research, Vol. 16(6):1034-1044.

Krishna, J. H. 1981. Computer modelling of land and water management
systems. Indian Journal of Agricultural Engineering, Vol.
18(1):1-8.

Langbein, W. B. 1979. Overview of conference on hydrologic data
networks. Water Resources Research, Vol. 15(6):1867-1871.

Larson, D. T., L. D. James, and K. R. Kimball. 1979. Levels of
analysis in comprehensive river basin planning. Utah Water
Research Laboratory (UWRL/P-79/05), Utah State University,
Logan, Utah. 110 p.

Linsley, R. K. 1967. The relation between rainfall and runoff.
Journal of Hydrology, Vol. 5(4):297-311.

Linsley, R. K. 1979a. Two centuries of water planning methodology.
Proceedings of the American Society of Civil Engineers, Vol.
105(WR1):39-46.

Linsley, R. K. 1979b. Hydrology and water resources planning.
Proceedings of the American Society of Civil Engineers, Vol.
105(WR1):113-120.

Lumb, A. M., and L. D. James. 1976. Runoff files for flood hydrograph
simulation. Proceedings of the ASCE, Vol. 102(HY10):1515-1531.

Lumb, A. M., J. R. Wallace, and L. D. James. 1974. Analysis of
urban land treatment measures for flood peak reduction. Environmental Resources Center (ERC-0574), Georgia Institute of Technology,
Atlanta. 146 p.

Morgan, A. E. 1971. Dams and other disasters: A century of the
Army Corps of Engineers in civil works. Porter Sargent, Boston.
422 p.

Moss, M. E. 1979. Some basic considerations in the design of hydrologic data networks. Water Resources Research, Vol. 15(6):1673-1676.

Narayanan, R. and S. Padungchai. 1979. Effects of energy development in the Upper Colorado Basin on irrigated agriculture and salinity. Western Journal of Agricultural Econmics, Vol. 4(2):73-82.

National Water Commission. 1973. New directions in U.S. water policy. U.S. Government Printing Office, Washington, D.C. 197 p.

Rodriguez-Iturbe, I., and J. B. Valdez. 1979. The geomorphologic structure of hydrologic response. Water Resources Research, Vol. 15(6):1409-1420.

Schad, T. M. 1979. Water resources planning-historical development. Proceedings of the American Society of Civil Engineers, Vol. 105(WR1):9-26.

Stalnaker, C. B. 1979. Stream habitat and its physical attributes as influenced by water and land management. In Index Construction for Use in High Mountain Watershed Management (Ed. L. D. James). Utah Water Research Laboratory, Utah State University, Logan, Utah. pp. 145-170.

U.S. National Science Foundation. 1980. A report on flood hazard mitigation. Washington, D.C. 253 p.

U.S. Water Reources Council. 1978. The nation's water resources 1975-2000: Volume 1, Summary. U.S. Government Printing Office, Washington, D.C. 86 p.

U.S. Water Resources Council. 1980. Principles and standards for water and related land resources planning--level C; final rule. Part II. U.S. Government Printing, Washington, D.C. 35 p.

Weber, E. W. 1979. Water planning--overview. Proceedings of the American Society of Civil Engineers, Vol. 105(WR1):3-8.

Wood, E. F. 1980. Real-time forecasting/control of water resource systems. IIASA Proceedings Series, Vol. 8. Pergamon, New York. 330 p.

Zaslavsky, D., and G. Sinai. 1981. Surface hydrology: I--explanation of phenomena. Proceedings of the American Society of Civil Engineers, Vol. 107(HY1):1-16.

APPLICATION OF THE CONTINUOUS API CATCHMENT MODEL IN THE INDUS RIVER FORECASTING SYSTEM IN PAKISTAN

Jaromir Nemec
Director, Hydrology and Water Resources Department
World Meteorological Organization
Case Postale No. 5 - CH-1211 - Geneva 20, Switzerland

Walter T. Sittner
Private Consultant
719 Thayer Avenue,
Silver Spring, Maryland, 20910, U.S.A.

ABSTRACT

The paper describes the rationale which led the governments of most Asian countries to establish flood forecasting systems which operate on the basis of modern telemetry data collection and hydrologic modelling techniques in order to alleviate the curse of floods but to retain their blessings. It describes the different parts of the flood forecasting system installed in a joint UNDP/WMO/Netherlands/Pakistan project in the basin of the Indus River in Pakistan. These are: first, a rehabilitated hydrometeorological and stream gaging network; second, a radio operated automatic telemetry system of 49 stations; third, a 5.6 cm. radar used for QPF and areal assessment of rainfall; and finally, computerized rainfall-runoff and routing models. Of the two rainfall-runoff models used, the Antecedent Precipitation Index Continuous (APIC) Model is discussed in detail.

The paper describes the manner in which a number of problems of a somewhat unique nature were handled. These are, the treatment of snowmelt--a year-round component of flow in the Himalayas, the manner in which model parameters were estimated for catchments in which classical calibration methods could not be used, and the treatment of orographic effect on precipitation patterns. Also described is a method of automatic optimization of the rainfall-runoff relation. This was, to the best of the authors' knowledge, the first use of a fully automatic procedure to fit an API rainfall-runoff relation to individual storm data.

Two modifications were made to the mathematical formulations used in the model, as expressed in the original reference. These are presented in detail.

The flood forecasting system is now fully operational and was taken over by trained Pakistani specialists in October of 1980. An extension of the system to provide seasonal forecasts on the basis of snow observations is contemplated by the Pakistan Water and Power Development Agency (WAPDA).

INTRODUCTION

Floods have been, and are, both the blessing and curse of the countries in Asia. The water brought by the flood is a blessing since without it the fertile flood plains would not exist and could not produce the food needed so badly by the population. At the same time, when arriving suddenly and in quantities exceeding the absorption and storage capacities of the plains and channels, it becomes a curse destroying homes, crops, and lives. The conservation of the blessing, and the elimination of the curse, have been in the foreground of endeavors of governments of all countries, with results depending not only on technical means used, but also on social and political conditions prevailing. Nevertheless, adequate technical means have been, and will remain, the basic condition for the successful mitigation of the flood hazards. Among the non-structural methods of mitigation of the flood hazard, which are very cost effective as compared with the structural ones (dams and dykes), modern flood forecasting and associated real-time data collection systems for operation of water projects throughout the year have increasingly found favor with the governments of the region. In most cases, they sought advice and funding for the establishment of such systems with international agencies, in particular with the United Nations Development Program (UNDP) and the World Meteorological Organization (WMO). Bilateral cooperation between countries of this region and industrialized countries of the world has also resulted in installation of such systems. Among the first systems installed were those in the Mekong River Basin, in the Phillipines, in South Korea, and in some other countries. At present, almost all countries of this region are involved in the establishment of modern flood forecasting systems, and the one in the Indus River Basin in Pakistan is a good example of the modern instruments and techniques used.

RIVER FLOW FORECASTING AND FLOOD WARNING SYSTEM FOR THE INDUS RIVER BASIN IN PAKISTAN

The modern concept of this system was initiated in 1974 by a technical mission of hydrological experts of the World Meteorological Organization in which one of the authors of this paper participated. The establishment of the system was made possible by a UNDP/WMO/Pakistan and a Netherlands/Pakistan technical assistance project. The projects provided firstly for the rehabilitation and improvement of the hydrometeorological and hydrological networks in those parts of the basin which are accessible both geographically and politically. Secondly, selected precipitation and stream gaging stations, a total of 49, were equipped with automatic radio-telemetry, reporting to the flood forecasting center but also, when relevant, to the flow management offices at the two large dams in the Indus River Basin, the Mangla and Tarbela Dams. Thirdly, the data coming from the telemetry stations is supplemented by a quantitative precipitation forecast and areal assessment of rainfall, based on data provided by a 5.6 cm. radar with digital video-integrating processor, looking in those parts of the basin which are inaccessible. Finally, all the real-time information is used by two types of models—rainfall-runoff models for the upper part of the basin, and flood routing models for the lower part (the flood plain) reaching the Arabian Sea near Karachi.

In order to provide all the flexibility necessary for the proper functioning of the system in different circumstances, both types of models were developed at different levels of complexity and, in particular, with the use of different computing devices, starting with small desk and pocket calculators, or mini-computers with power supply from batteries, and ending with an IBM 360 computer with 196K of CPU memory.

Four different models were used: the Antecedent Precipitation Index Continuous (APIC) described by Sittner, Schauss, and Monro (1969); the Constrained Linear System (CLS) described by Natale and Todini (1971) as rainfall-runoff models and implicit solution of the differential equations of motion and continuity and a kinematic cascade of conceptual reservoirs as flood routing models.

APPLICATION OF THE APIC MODEL

The part of the project being reported on in detail in this paper involves the development of forecast procedure using the APIC Model for the five constituents of the Indus River. These are, the Indus, Jhelum, Chenab, Ravi, and Sutlej Rivers. The procedure was required to produce forecasts at five designated "rim stations" above the point where the five tributaries join to form the main Indus River. These are, Mandori on the Indus, Mangla Reservoir outflow on the Jhelum, Marala on the Chenab, Jassar Bridge on the Ravi, and Ferozepur on the Sutlej (map, Fig. 1). The following discussion deals with some of the problems encountered which were unique to the area, or at least unusual, and the methods used to solve them.

Snow and Snowmelt

The combined drainage area of the five tributaries, above the rim stations, is a little over 250,000 square kilometers. Two of these have their origins in the upper reaches of the Himalaya Mountains at elevations above 7,500 meters. While the entire basin lies between 32 and 37 degrees latitude and is subtropical, snow and snowmelt are prevalent the entire year in the headwaters of these two tributaries. Operational meteorological data from such areas is, of course, non-existent, and no attempt was made to model the accumulation or ablation of the snow mantle. Rather, advantage was taken of the fact that the portion of the channel discharge originating from snow is, during the flood season, a small part of the total flow and tends to change slowly from day to day. The planned operational solution to the problem was to simply assume each day that the snowmelt component would be the same on the subsequent days of the forecast period, and then adjust as necessary on those days.

Transposition of Model Parameters

In the calibration of any large scale river system for operational forecasting purposes, areas are usually encountered where direct calibration by classical methods is not feasible. There are three

315

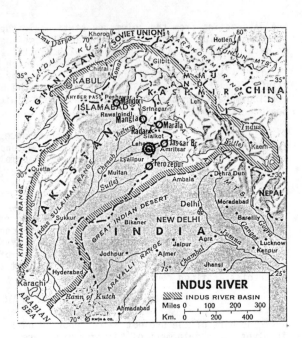

LEGEND

◉ Flood Forecast
    Center

↲ Radar

○ Rim Station

Figure 1................................Map of the Indus River Basin

reasons for this.

   First, some catchments which must be modelled are not headwater catchments. That is, their outflow is caused by two inputs, the precipitation over the area in question and channel inflow which is measured at an upstream gaging station. The "observed" runoff from such an area can only be determined as the difference between the measured channel inflow and channel outflow. If the contribution of the subject catchment is small in comparison to the upstream inflow, even small observational errors in the measured inflow or outflow will result in much larger errors in the computed residual runoff. This may well preclude any valid correlation with the causative meteorological variables.

   Second, the river system may involve catchments for which historical precipitation data is excessively sparse, non-existent, or unavailable.

   Third, while the APIC Model is a continuous model, some of its parameters must be evaluated on the basis of identifiable runoff events. Due to climatological conditions, some catchments may exhibit few such events among the historical data. This may result in the need for at least partial synthesis of model parameters for such areas.

   In the Indus Basin, all three of the above conditions were encountered, sometimes in the extreme. Thus, the work involved rather large scale synthesis and transposition of model parameters.

   This problem was recognized before the start of the actual calibration, and it was hoped that it would be possible to utilize one "master" rainfall-runoff relation for the entire area. A preliminary

316

calibration of two catchments, one with a mean elevation of 500 meters m.s.l., and the other with an elevation of 2,850 meters, disclosed marked differences in runoff characteristics and precluded any hopes for the applicability of a single relationship. A subsequent "on-the-ground" reconnaissance of the area readily disclosed the reason for the discrepancy and also suggested a solution. It was observed that certain physical characteristics of the catchments, type and quantity of vegetation, soil type and depth, slopes, and of course temperature, were strongly related to elevation. It seemed then that it might be possible to utilize elevation as an indirect parameter in the relationship. Subsequent calibration of a number of catchments, along with sensitivity analyses, resulted in a set of predictor equations by which two of the model parameters were expressed as functions of elevation. One of these parameters expressed the "width" of the season quadrant, and the predictor equation indicates a lesser seasonal variation in runoff conditions at lower elevations. The other parameter expressed the manner in which runoff conditions change from high to low during the growing season. Its predictor equation indicates that this progression takes place more rapidly at lower elevations. All other parameters were the same for all catchments, and the variation was accounted for with just these two.

The use of catchment elevation as an indicator of runoff characteristics is not unprecedented. Hopkins and Hackett (1960, 1961) reported on the solution of a similar problem in New England although their method of introducing elevation into the relationship was quite different.

Orographic Effect on Precipitation Patterns

Orographic effect is pronounced in the Indus Basin. Most of the runoff-producing storms occur during the monsoon season, roughly early July to early October. The moisture-carrying air masses sweep in from the Arabian Sea to the south and travel in a northwesterly direction across the plains of Pakistan and India for about 1,000 km. before encountering the foothills of the Himalayas. As they cross the Himalayas, the lifting creates increased precipitation to about 900 meters. Above this level, the moisture in the air mass is depleted and higher elevations receive lesser amounts of rainfall. This effect is well known, qualitatively by local meteorologists. For this project, however, it was necessary to quantify it, and this was done by analyzing rainfall records. Precipitation records in Pakistan are of good quality but tend to be sparse and poorly distributed areally. For this reason, some extrapolation was necessary to produce the relationship shown in Table 1. This relationship was used in two ways in the project.

First, in the determination of catchment mean precipitation volumes for the historical events used in calibration, the number of rain gages was usually insufficient to delineate the areal variation of rainfall over the catchments. The precipitation vs. elevation relationship was used to derive weights to apply to the station amounts to convert them to catchment means. These weights were functions of the station elevations and the distribution of elevation within the catchment.

Second, in the use of elevation as an indicator of certain rainfall-runoff parameters, as described in the preceding section, what was used was not the true catchment mean elevation, but an "effective" mean elevation, with each segment of the catchment weighted according to the amount of rainfall it typically receives. This procedure was felt

to be more logical than the use of a true mean elevation and also produced a better correlation with rainfall-runoff parameters.

A secondary effect of the orographic variation was somewhat more
subtle. As in most areas of the world, the rainfall patterns in
Pakistan consist of a systematic variation (orographic) with a random
variation (convective) superimposed upon it. At higher elevations, the
systematic variation appears to be the predominant component, resulting
in a reasonable degree of uniformity and consistency in precipitation
patterns. At lower elevations the systematic component is small and the
random component is dominant, resulting in runoff patterns which vary
greatly among events. The utlimate effect is probably a lower level of
accuracy in model calibration in the lower catchments and the need for a
more dense operational network of rainfall observing stations.

Table 1--Precipitation vs. Elevation Relationship--Indus River Basin

Elevation (meters)	Precipitation (*)	Elevation (meters)	Precipitation (*)
0	55	2,400	39
300	77	2,700	29
600	96	3,000	21
900	100	3,300	17
1,200	98	3,600	13
1,500	91	3,900	9
1,800	72	4,200	7
2,100	52	4,500	6

(*)--Expressed as a percentage of the 900 meter precipitation

Automatic Optimization of the APIC Model

The calibration of the APIC Model was described by Sittner,
Schauss, and Monro (1969). As noted there, the fitting of the incremental rainfall-runoff relationship begins with the calibration of a
conventional total storm relationship. The techniques for accomplish-
ing this are well known and involve a large amount of manual, graphical
correlation (Linsley, Kohler, and Paulhus, 1975, pp. 267-270 and
Appendix A). In this project, the calibration was done automatically
using the "Pattern Search" technique (Hooke and Jeeves, 1961). A
subroutine "OPT", published by Monro (1971) and originally written for
use with the Stanford Watershed Model was used for the APIC.

A problem which always exists with automatic optimization procedures, and one which is often given insufficient consideration, is the
devising of a suitable objective function. It is very difficult,
perhaps impossible, to compute a single error function which will
properly reflect all aspects of the manner in which the model fits the
data. In fitting API type relationships manually, it has been customary
to compute several error functions at each iteration and subjectively
consider all of them as a group. Such experience has not yielded
knowledge of a single objective function suitable for controlling a
fully automatic procedure.

318

In the work being reported on, a similar procedure was followed. For each new set of parameters, four error functions were computed:

OF1 - The sum of squared errors
OF2 - The average of absolute values of errors
OF3 - The average of algebraic values of errors
OF4 - The maximum error

The program, called "APIMO," was written so that any one of the four could be specified as the one to control the optimizing strategy. A sensitivity analysis indicated that OF2, the arithmetic average of the errors, should be used for this purpose. Optimization runs began using OF2 as the control and involving 11 parameters, seven for the season quadrant and four for the precipitation quadrant. When the pattern of change in OF2 indicated that iteration should stop, a twelfth parameter was introduced. It is called "FIDJ" (FI adjustment). It is an arithmetic adjustment to be applied to the Antecedent Index before entering the Retention Index (or Duration) quadrant. The use of such an adjustment, while not described in any of the references on API, is a commonly employed technique to accomplish the final fitting of a relationship, or to adapt a relationship to a catchment having similar but not identical runoff characteristics.

At this point in the run, ten more iterations were made, using the final values of all parameters other than "FIDJ," and varying "FIDJ" linearly from zero to a value 20 times OF3. The senstivity analyses had shown that the best overall combination of error functions would be found somewhere in this range. The value of "FIDJ" producing that combination was then determined by inspection.

This was, to the best of the authors' knowledge, the first use of a fully automatic procedure to fit an API rainfall-runoff relationship to individual storm data. The results were very satisfactory.

Modification to APIC Model

In projects of this type which are intended to make modern technology available to developing countries, one of the usual constraints is that the methods used must be well established through use elsewhere. In practice, it is seldom possible to accomplish this completely since problems usually arise which simply have not been encountered elsewhere. In the Indus project, the APIC Model differed in two ways from that presented by Sittner, Schauss, and Monro in 1969. For the benefit of those who may wish to use this model in a similar application, the two modifications are presented in detail below.

The season quadrant, as shown in the reference, is defined by 28 parameters. Such a formulation was quite adequate when the curves were produced graphically, and the equations then fit to those curves. But, the use of the automatic optimization procedure described earlier demanded a lesser number of parameters, and an alternative formulation involving only 7 parameters was adopted. This representation was devised at the Fort Worth, Texas River Forecast Center of the U.S. Weather Bureau in approximately 1960. It is as follows:

The two boundary curves of the season quadrant are defined by Eqs. (1) and (2) and four internal parameters of the model, ASES, BSES, AX, and BX.

$$AIW = (ASES)(AX)^{API} \qquad (1)$$

$$AID = (BSES)(BX)^{API} \qquad (2)$$

319

Two more parameters, AWK and BWK, are the week numbers corresponding to these two curves. That is, AWK is the week number corresponding to the right-hand curve, Eq. (1), representing the time of year when vegetative conditions are most conducive to runoff. BWK is the week number corresponding to the left-hand curve, Eq. (2), and represents the time when conditions are least conducive to runoff.

Having determined, with Eqs. (1) and (2), the values of Antecedent Index corresponding to the two extreme seasonal conditions, the Antecedent Index, AI, corresponding to the actual week number of the event, is then computed. The quantity, FRAC, is defined as the ratio of the number of weeks between the week in question and AWK to the total number of weeks between the boundary curves. For instance, if AWK = 10 (Mar. 5-11), and BWK = 34 (Aug. 20-26), and a storm occurs in week 17 (Apr. 23-29), then the number of weeks between the event and AWK is 17-10, or 7. The total number of weeks between the boundary curves is 34-10, or 24. FRAC is then 7/24, or 0.292. If the event occurs on say Nov. 15, however, this is week 46. The number of weeks between 46 and AWK, Nov. 15 to Mar. 8, is 16. The total number of weeks from week 34, through the end of the year and up to week 10, is 28. FRAC is then 16/28, or 0.571.

Since the change in runoff characteristics through the year is not linear with respect to time, a transformation is made, using Eq. (3) which involves the seventh parameter, EX.

$$FR = \left[\frac{1+Sin \ [\pi(1.5-FRAC)]}{2} \right]^{EX} \qquad (3)$$

Finally, the Antecedent Index, AI, is given by:

$$AI = AIW + FR \ (AID-AIW)$$

The other modification involved the use of a mathematical formulation to define the relationship between discharge and the groundwater recession factor, rather than the table lookup procedure used in the model reported on by Sittner, Schauss, and Monro (1969). The formulation involves two parameters, CQ and CK. It is:

$$Kg = 1 - \left[\frac{1-SIN \ [\pi(\frac{Q}{CQ} + 0.5)]}{2} \right]^{0.7} (1-CK)$$

If Q > CQ, then Kg = CK

where Q is the channel discharge and Kg is the groundwater recession coefficient as described by Sittner, Schauss, and Monro (1969).

Also described was a seasonal parameter used in the above relationship. This was not used in the Indus application since its effect is relatively small and the characteristics of the Pakistan rivers are such as to make it very difficult to accurately evaluate.

PRESENT STATUS OF THE PROJECT

The forecasting system, as described in the first part of this paper, has been made fully operational and successfully tested in the flood (monsoon) season 1980. The government agencies in charge of operation of both the flood forecasting center and the dams, the Pakistan Meteorological Department (PMD) and Water and Power Development Agency (WAPDA), respectively, have taken over the system as of October 31, 1980. Pakistani specialists trained both abroad and in Pakistan, and fully familiarized with the different parts of the system, will prepare it for the next flood season with some assistance from the UNDP/WMO, consisting mainly in a maintenance activity by consultants and equipment supplying companies. Further development of the system to provide seasonal forecasts from snowmelt and to tune up the calibration of the models is foreseen, but its implementation depends on several factors which are not of a technical nature. There is, however, no doubt that during the flood season 1980 the system has already proven its usefulness and was highly praised by the government authorities of Pakistan.

ACKNOWLEDGEMENTS

The authors acknowledge with thanks the permission of the Secretary General of WMO to use materials prepared under WMO auspices. The views expressed in this paper are those of the authors and do not necessarily represent the views of WMO.

REFERENCES

Hooke, R. and Jeeves, T.A. 1961. Direct Search Solution of Numerical and Statistical Problems. Journal of the Association for Computing Machines, Vol. 8, No. 2.

Hopkins, C.D., Jr. 1960. A Method of Estimating Basin Temperatures in New England and New York. Journal of Geophysical Research, Vol. 65, No. 11.

Hopkins, C.D., Jr. and Hackett, D.O. 1961. Average Antecedent Temperatures as a Factor in Predicting Runoff from Storm Rainfall. Journal of Geophysical Research, Vol. 66, No. 10.

Linsley, R.K., Kohler, M.A., and Paulhus, J.L.H. 1975. Hydrology for Engineers, Second Edition, McGraw-Hill Book Co., New York, pp. 267-270 and Appendix A.

Monro, J.C. 1971. Direct Search Optimization in Mathematical Modelling and a Watershed Model Application. NOAA Technical Memorandum NWS HYDRO-12, U.S. Department of Commerce, National Weather Service.

Natale, P. and Todini, E. 1977. A Constrained Parameter Estimation Technique for Linear Models in Hydrology. in: T.A. Ciriani, U. Maione, and J.R. Wallis (Editors), Mathematical Models for Surface Water Hydrology, J. Wiley & Sons, London.

Sittner, W.T., Schauss, C.E., and Monro, J.C. 1969. Continuous Hydrograph Synthesis with an API-Type Hydrologic Model. Water Resources Research, Vol. 5, No. 5.

APPLICATION OF A RAINFALL-RUNOFF MATHEMATICAL MODEL TO COMPUTE KARSTIC AQUIFERS OUTFLOWS. RESEARCH OF THE MODEL PARAMETERS AND FUNCTIONS SIGNIFICANCE

Claude Drogue, Professeur
Laboratoire d'Hydrogéologie, Université des Sciences et Techniques,
Place E. Bataillon - 34060 Montpellier Cedex - FRANCE

Lisbeth Gdalia, Chercheur
Laboratoire d'Hydrogéologie, Université des Sciences et Techniques,
Place E. Bataillon - 34060 Montpellier Cedex - FRANCE

Moumtaz Razack, Professeur Adjoint
Laboratoire d'Hydrogéologie, Université des Sciences et Techniques,
Place E. Bataillon - 34060 Montpellier Cedex - FRANCE

Alain Guilbot, Ingénieur
Laboratoire d'Hydrologie Mathématique, Université des Sciences et Techniques,
Place E. Bataillon - 34060 Montpellier Cedex - FRANCE

ABSTRACT

Conceptual models which allow computation of a basin discharges, taking into account rainfall, simulate principal hydrological processes by means of parameters and mathematical functions. One of the questions which then arises is related to the significance that could be attached to the model's parameters regarding the real phenomena and features they represent in theory. Such an investigation has been carried out in this paper dealing with, in particular, modeling of karstic basins' functioning in which runoffs are entirely underground. Our objective, throughout this research, was to place emphasis on the following two points :

(1) Could the application of such a model be generalized to modeling of karstic aquifers ?

(2) Could some correspondance be brought out when comparing physical characteristics of the basins with the parameters and mathematical functions of the model ?

The data used in this study is related to nine karstic basins located in Southern France. Their areas range between 0.5 and 200 km2. Geological formations are only of carbonaceous nature : limestones and dolomites.

The model used here, called CREC, includes a production function, with two parameters, (X3, X4), providing the efficient rainfall and a transfer function, four parameters (X1, X2, X5, X6), for computing the different forms of runoffs towards the basin's exit. A seventh parameter, (X7), is related to the recession of potentiel evapotranspiration.

Very satisfactory reconstitutions of hydrographs were achieved on the nine basins. The model sensitivity was also tested with regard to the variations of its parameters.

Investigation about the relationships between the model's parameters and the physical characteristics of the basins showed that two of the transfer parameters are sensitive (1) to the basins'dimensions; (2) to the lithology (more or less important of the dolomitical formations) ; and (3) to some features of geological structure. Besides, form of the production function varies according to the geological nature of the basins, (calcareous or dolomitical).

The conclusion drawn from this study is that a certain correspondance between the model's parameters and functions on the one hand, and some physical and geological features of the karstic aquifers on the other, could be brought out. Therefore it looks as if a more efficient application of such a conceptual model, (built essentially for a predictive use) could be envisaged, passing beyond the unique framework of rainfall-runoff modeling, as a tool for deterministic investigation of karstic aquifers. To achieve this goal, however it would be necessary to go beyond the study presented herein, which remains nevertheless a qualitative approach because of the few karstic aquifers involved.

INTRODUCTION

In so-called deterministic mathematical models, which allow the reconstitution of a basin's discharges taking into account rainfall, the principal processes of the hydrological functioning are represantated by means of mathematical equations. Thus real known physical phenomena are at best represented by numerical functions in conjunction with certain symplifying assumptions. But the schematizing of the concepts utilized in this representation lower importantly its truthfulness. So it is generally useless to search after an exact expression, in the elementary functions, of the real occuring mechanisms.

However the problem of these models' significance remains : is there any relation between the basins'physical nature and the model's parameters after adjustment. Such a research has been carried out concerning the modeling of karstic basins'functioning in which runoffs are entirely underground.

Previous work within this framework showed, as a matter of fact that a rainfall-runoff model, initially built for use in surface hydrology, could be extended to perfectly compute outflows at karstic basins'outlets (Guilbot, 1975 ; Drogue and Guilbot, 1976). It was also pointed out that such models could be well adapted for a global simulation of karstic aquifers'functioning. Keeping this in mind, our purpose in this paper is to examine, by modeling several karstic basins, the degree of correspondance between some features of the basins and the parameters of the model being fitted.

PRESENTATION OF THE MODEL AND OF THE KARSTIC BASINS

The model CREC

The model used here, called Crec, was proposed by Cormary and Guilbot in 1969. It is a conceptual model built with reservoirs and takes into account, using a relatively simple structure, the diffe-

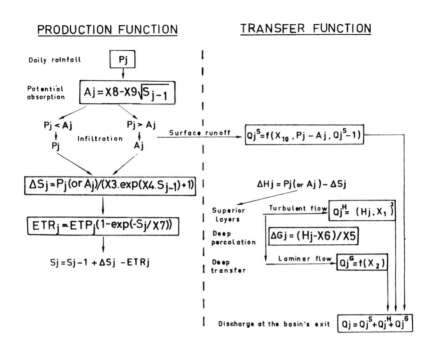

Figure 1. Mathematical Rainfall-Runoff Model CREC

P_j : daily rainfall
A_j : potential absorption
S_j : humidity of the soil on the day j
S_{j-1} : humidity of the soil on the day $j-1$
$X8, X9$: parameters of the potential absorption
ΔS_J : recharge of the soil humidity deficit on the day j
$X3, X4$: parameters of ΔS_j
ETR_j : evapotranspiration on the day j
ETP_j : évapotranspiration capacity
$X7$: parameter of ETR_j
Q_j^S : surface runoff
$X10$: parameter of Q_j^S
H_j : transfer towards the saturated zone
$X1$: parameter of the turbulent flow
ΔG_j : deep percolation
$X5, X6$: parameters of the deep percolation
Q_j^G : deep transfer
$X2$: parameter of the laminar flow
Q_j : basin runoff

rent terms of a basin's runoff. A summary description of this model follows.

Between the soil that receives the rainfall and the basin's outlet, two functions are defined : a production function and a transfer function (fig. 1). The production function will provide the rainfall fraction that is converted to runoff. This function takes into account the humidity of the soil, represented by the water depth of a reservoir S, that supplies evapotranspiration. The transfer function expresses the runoff's forms towards the basin's exit. It includes a quick flow in a reservoir H, with a non-linear depletion, and a slow flow in a reservoir G, with a linear depletion. These two reservoirs communicate when a threshold of the storage capacity, Ho, of the reservoir H, is reached. Finally, part of the basin's discharge can be attributed to the rainfall rate that streams for showers whose intensities exceed a critical threshold. This occurs however very seldomly in our karstic basins.

The parameters of the analytical expressions, which amount to seven, are determined numerically through non-linear optimization methods. These parameters are : X1 and X2 representing the depletion of reservoirs H and G respectively ; X3 and X4 for the production function ; X5 and X6 define the supply characteristics for the reservoirs H and G, respectively, and X7 for evapotranspiration reduction. In fact the model includes three more parameters (X8, X9, X10), which don't affect in anyway the model's adjustment, in so far as they are related to surface runoff (see fig. 1) and evaluated empirically.

The Karstic Basins

This study deals with nine carbonated basins located in Southern France (fig. 2). The available data (rainfall and runoff) extend over quite a long period. The areas of the basins range between 0.5 and 200 km2 and their climatic conditions are almost similar. The precipitation regime is marked with high rainfall intensities occuring over short periods. Annual rainfall ranges between 650 and 1500 mm. Runoffs, entirely underground, are controlled by means of gaging stations on the springs. Geological formations are only of carbonated nature : limestones and dolomites (table 1).

Figure 2. Location of the Study Basins in Southern France

Table 1 - Main characteristics of the Studied Basins

Basin	area (km2)	% of dolomites	Number of gaged outlets	observation's period	mean annual rainfall(mm)	mean annual runoff (mm)	runoff/ rainfall (%)
Saugras (a)	0.5	0	3	1966 1973	882	314	35.6
Lez (a)	170	0	2	1965 1973	1084	573	52.9
Hortus (a)	50	0	1	1966 1974	1044	345	33.1
Séranne (a)	70	0	2 groups	1967 1974	1498	754	50.3
Gapeau (b)	90	76	5	1967 1977	925	239	25.8
Gaou (b)	27	74	2	1967 1977	833	249	29.9
Rouvière (b)	6	99	1	1967 1977	659	247	37.5
Saint Clément (b)	16	46	1	1967 1977	891	209	23.5
Veyans (c)	200	20	3 groups	1970 1976	1114	558	50.1

Administrating body of the basins
(a) : Laboratoire d'Hydrogéologie de Montpellier
(b) : C.T.G.R.E.F. d'Aix-en-Provence
(c) : S.R.A.E. - E.D.F. d'Aix-en-Provence

Among the basin features, those which take a part in the runoff's modulation and could affect the model parameters, the following ones have been selected : they are in addition to the basin's area which obviously is an essential one, two morphological indexes and a lithological index. Each basin is then characterized by :

(1) its area : A

(2) an index of compactness : K (Roche, 1963, p. 144)

$$K = 0.28 \ P/A^{(1/2)} \qquad (1)$$

with P : perimeter of the basin

(3) an index of slope : T (Roche, 1963, p. 150)

$$T = (A1 - A2)/L \qquad (2)$$

with A1 : average altitude
A2 : minimum altitude
L : length of the aquivalent rectangle

(4) an index of lithology : D

$$D = (\text{outcrop area of dolomites'surface} \times 100)/A \qquad (3)$$

The two morphological indices are of course very formal ones. They express very unexplicitly the basin's physical reality. Nevertheless they are evaluated quite easily and as this study is just a first approach of the problem raised above, they seem sufficient.

ADJUSTMENT OF THE MODEL
AND IDENTIFICATION OF ITS SIGNIFICANT PARAMETERS

The adjustment of the model was accomplished on a daily basis over a period of four years for each basin. The goodness of fit was then controlled by the remaining years of data, amounting two to six years according to the basins. Potential evapotranspiration was computed according to Thornthwaite method (1954) based on the average monthly temperatures, and rainfall according to Thiessen method of polygones (Roche, 1963, p. 81). For all the studied basins, very satisfactory reconstitutions of the discharges were achieved (fig. 3)

After optimizing all the results, we have at our disposal, for each basin, the values of the model's parameters (table 2). But one should keep in mind that all the parameters don't affect the model's adjustment to the same degree. So it becomes necessary to search for those which are the most significant. This is performed by analyzing the model's sensitivity with regards to the variations of its parameters. The model's sensitivity is tested through a Criterion Function F, defined hereunder :

$$F = \sum_{i = 1}^{i = n} \{(Q_o - Q_c)/Q_o\} \{(Q_o - Q_m)/Q_m\} \qquad (4)$$

with Q_o : observed discharges

Q_c : computed discharges

Q_m : average discharges

The Criterion Function F is computed for time intervals from 1 to n days.

It was pointed out (L. Gdalia, 1979) that the only parameters which have a real affect upon the model's sensitivity were : X1 and X2 (transfer parameters) and X4 (production parameter). These three parameters, being most significant, will be utilized, thus, in the rest of his study.

COMPARATIVE ANALYSIS BETWEEN THE MODEL'S SIGNIFICANT
PARAMETERS AND THE BASINS' CHARACTERISTIC INDEXES

Parameters X1 and X2

It may be recalled that X1 determines the outflow from reservoir H and X2 outflow from reservoir G.

Comparison of X1 with the values of the compactness index K, shows that X1 decreases as K increases. The two are related hyperbolically as demonstrated in figure 4. Relation with the lithological index D is on the other hand not so explicit, even though for seven among the nine basins the importance of dolomites increases as well as the values of X1 (fig. 5). As for the slope index T, it doesn't show any meaningful correspondance with X1 (fig. 6). Thus, the parameter X1 might be linked to the basins'lithology and compactness.

As for the parameter X2, a similar analysis as above was not

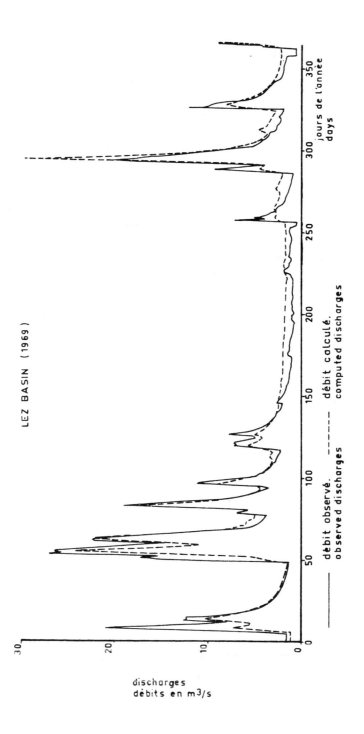

Figure 3. Example of the Runoff's Reconstitution on a Basin (Lez, 1969)

Table 2 - Values of the Model's Parameters after Adjustment

Basin	X1	X2	X3	X4	X5	X6	X7
Saugras	0.073	0.967	0.036	0.038	26	0.0039	0.16
Lez	0.0234	0.997	0.036	0.038	27	17	32.6
Hortus	0.074	0.5	0.012	0.033	22	1.2	21
Séranne	0.035	0.996	0.026	0.033	50	0.783	45
Gapeau	0.075	0.943	0.055	0.005	12.6	0.009	83
Gaou	0.106	0.8098	0.1585	0.008	14	0.05	5
Rouvière	0.084	0.818	0.003	0.08	50	0.05	7.5
Saint Clément	0.0569	0.986	0.1648	0.007	17.3	10	2.8
Veyans	0.036	0.9994	0.0172	0.16	47	0.0019	0.04

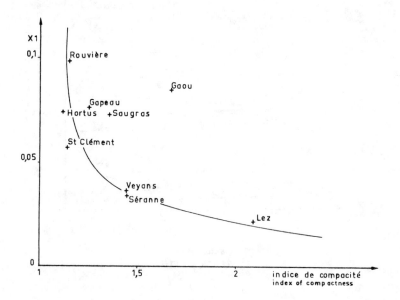

Figure 4. Parameter X1 and the Index of Compactness, K

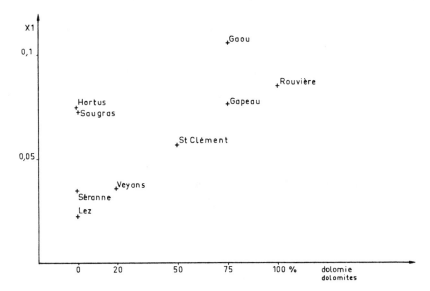

Figure 5. Parameter X1 and the Dolomitical Formation's Importance (% of the Dolomites'Area with respect to the Basins'whole Area)

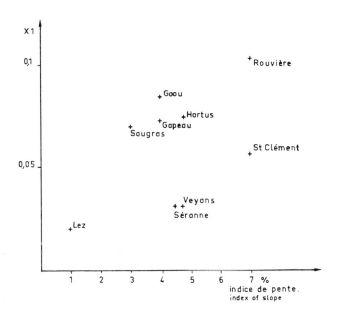

Figure 6. Parameter X1 and the Index of Slope, J

conclusive. So this led us to study a more global function of the depletion of the reservoir G, in which parameter X2 is inclued.

The discharge of this reservoir G, on any day j, is obtained by (transfer function) :

$$Q_j^G = X2 \, \Delta G_j + (1 - X2) \, Q_{j-1}^G \tag{5}$$

with Q_j^G : discharge of the reservoir G on day j

Q_{j-1}: discharge of the reservoir G on day j-1

ΔG_j : supply of the reservoir G
(deep percolation)

The discharge, Q_t^G as a function of time, for this reservoir, represents depletion of exponential type, and is given by :

$$Q_t^G = Q_0 \exp (-a.t) \tag{6}$$

with Q_t^G : discharge of the reservoir G with respect to time

$a = \text{Log}_e (1/X2)$

$Q_0 = 10 \text{ m3/day (arbitrary value)}$

Figure 7 shows the depletion characteristics of the basins, and can be categorized into two distinct groups : the first group represents slow depletion while in contrast the second group reflects quick depletion. Among the four basins with slow depletion, three have a large area (more than 60 km2 : Lez, Veyan and Séranne), while the fourth have a relatively small area (Saint-Clément : 16 km2). Basins with quick depletion are generally associated with small surfaces, except the Gapeau basin (A = 90 km2).

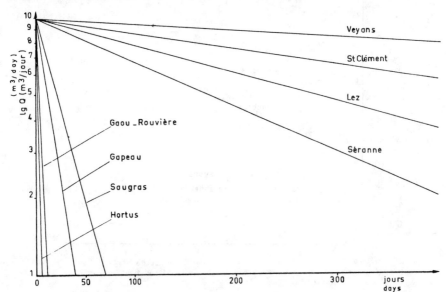

Figure 7. Depletion's Discharges of the Reservoir G for the Different Basins

So, with the exception of two basins (Saint-Clément and Gapeau), the influence of area on the transfer function of the reservoir G seems unquestionable. The larger the basin's area, the slower this reservoir's depletion. This is quite logical and agrees with the fact that large basins have more sustained recession's discharges than small ones. For the two basins which do not fit this classification, we cannot propose any interpretation. However it might be due to an imprecise evaluation of their areas.

Parameter X4 and the Production Function

Examining the relationship between the values of X4 and the basins' physical indexes ended in a failure. So, instead of studying only X4, we were led to consider the water depth variation of the reservoir S, which may be expression as :

$$\Delta S = P_{eff}/\{X3 \exp (X4.S) + 1\} \qquad (7)$$

with ΔS : water depth variation of the reservoir S

P_{eff} : infiltrated rainfall

S : water depth of the reservoir S

The fraction, expressed in percent, of the infiltrated rainfall which joins in the runoff is given by :

$$P_{ec} = \{P_{eff} - \Delta S)/P_{eff}\} \ 100 \quad \text{or} \qquad (8)$$

$$P_{ec} = \{(X3 \exp (X4.S))/(1 + X3. \exp (X4.S))\} \ 100 \quad (9)$$

For each basin, theoretical curves representing P_{ec} at various water depths (S) of this reservoir, ranging from 0 to 2000 mm, have been derived as shown in figure 8. As the aim, here, is just a study of the theoretical behaviour of the parameters X3 and X4, using high values for S is not objectionable.

Two groups of curves can be determined based on their slope :

- curves with a light slope represent three basins of principally dolomitical nature (more than 40 % of dolomites). These are Gaou, Saint-Clément and Gapeau.

- curves with a sharp slope, represent basins with only calcareous formations (Lez, Séranne, Hortus, Saugras) or mostly (Veyans : 80 %).

However, basin Rouvière which, though being entirely dolomitical, has its curve situated among the ones of calcareous basins. It is worthy of notice that for this basin the correspondance between its lithology and the parameter X1 is quite good. (see fig. 5).

Thus, the water depth of the reservoir S, representing the soil humidity, has a different effect depending on the basins' geological nature, dolomitical or calcareous, with just one exception. In dolomitical basins, this reservoir's water depth must be greater than 300 mm for underground runoff. On the other hand, for calcareous basins, the reservoir S takes part in the underground runoff for much smaller values of its water depths, i.e. about 100 mm.

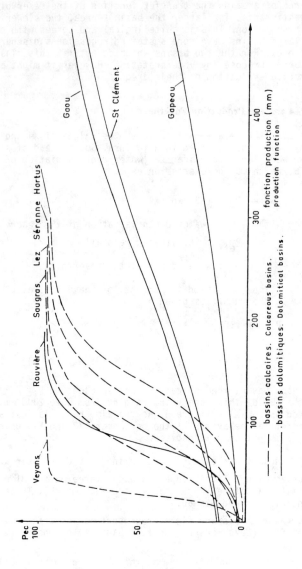

Figure 8. Efficient Rainfall P_{ec} taking part in the Runoff of the
Reservoir S and Production Function of this Reservoir

This is quite in agreement with the knowledge now prevailing concerning underground runoff within two carbonaceous formations. As a matter of fact, in dolomite formations water retention and slow circulations are facilitated by its porosity and by the accumulation of alteration sand. While in frank limestones porosity and permeability are only due to a network of fissures and karstic cavities, as in the calcareous aquifers here involved. Water retention is then weak and circulations very rapid.

Thus the relation between the production function of the reservoir S and the basins'geological nature is extremly clear and noteworthy.

About the basin Rouvière, its particular behaviour might be attributed to the existence of an open and developed karstification which is hydraulically very active. This karstification, brought out through underground works, is specific only to this system. There has been no such observation about the other basins of dolomite formations.

CONCLUSIONS

The major conclusions that can be drawn from this study, although a few karstic basins were examined are :

1) this model, built initially for surface hydrology, has been applied succesfully to underground runoff modeling in several karstic basins having different characteristics ;

2) some significant relations have been brought out when comparing physical and geological features of the basins and the model's parameters and functions.

Thus this model's application can be generalized for global rainfall-runoff simulation of karstic aquifers. Information concerning the basins'physical and geological nature and its hydrogeological functioning have been inferred from the parameters and functions of the adjusted model.

Therefore it looks as if a more efficient application of such a conceptual model, built essentially for a predictive use, could be envisaged, passing beyond the unique framework of rainfall-runoff modeling as a tool for an effective deterministic investigation of karstic aquifers. To this goal, it is necessary to go beyond the study presented herein, which nevertheless remains a qualitative approach because of the few karstic basins involved.

REFERENCES

Cormary, Y. and Guilbot, A. 1969. Relations pluie-débit sur le bassin de la Sioule. Rapport D.G.R.S.T., n° 30, 35 pp., Universite des Sciences et Techniques, Montpellier, France.

Drogue, C. and Guilbot, A. 1976. Représentativité d'un bassin témoin
en Hydrogéologie karstique : Application à la modélisation des
écoulements souterrains d'un aquifère de grande extension.
Journal of Hydrology, vol. 32, pp. 57-70.

Gdalia, L. 1979. Application d'un modèle mathématique conceptuel à
plusieurs aquifères karstiques de la bordure méditerranéenne
française. Thèse de 3ème cycle, Université des Sciences et
Techniques, Montpellier, France.

Guilbot, A. 1975. Modélisation des écoulements d'un aquifère karstique (Liaison pluie-débit) : Application aux bassins de Saugras
et du Lez. Thèse d'Université, Université des Sciences et
Techniques, Montpellier, France.

Roche, M. 1963. Hydrologie de surface, Gauthier-Villars Editions,
Paris, pp. 430.

Thornwaïte, C.W. 1954. The Measurement of Potential Evapotranspiration, J.P. Mather Seabrook, New Jersey, pp. 225.

EFFECT OF DRAINAGE SYSTEM DESIGN ON SURFACE AND SUBSURFACE RUNOFF FROM ARTIFICIALLY DRAINED LANDS

Richard Wayne Skaggs
Professor
Department of Biological and Agricultural Engineering
North Carolina State University
Raleigh, North Carolina 27650

and

Abdolhossien Nassehzadeh-Tabrizi
Research Associate
Department of Biological and Agricultural Engineering
North Carolina State University
Raleigh, North Carolina 27650

ABSTRACT

A computer simulation model, DRAINMOD, is used to analyze the effects of drain spacing and surface depression storage on total and the temporal distribution of surface and subsurface runoff. Effects on annual surface runoff and subsurface drainage as well as for single storm events, are presented and discussed. Simulations were conducted for ten combinations of surface and subsurface drainage (two levels of surface depressional storage and five subsurface drain spacings) for a sandy loam soil. Corn was grown on a continuous basis. A 25-year climatological record was used in the simulations to show variations from year to year.

The results showed that subsurface drainage had a very significant effect on both annual surface runoff and runoff from single storms. With good surface drainage, increasing the drain spacing from 15 to 100 m caused more than a 3-fold increase (from 19 cm to 57 cm) in the total surface runoff on a 5-year recurrence interval basis. With poor surface drainage, surface runoff was only 8 cm for the 15 m spacing as compared to 41 cm for the 100 m spacing. The effect of drain spacing on runoff and subsurface drainage from single storm events was similar to that observed for annual amounts. For high antecedent rainfall, the surface runoff from a 10 YRI -- 24 hour (17 cm) storm was 9.5 cm for a drain spacing of 15 m as compared to 14 cm for a 60 m drain spacing. The runoff from a 10 YRI -- 24 hour (17 cm) storm with no rainfall during 18 days prior to the storm was 3 cm for a 15 m spacing versus 6 cm for a 60 m spacing.

The results obtained in this study show clearly that the drain spacing and surface drainage treatment significantly affect the amount and rate of surface runoff. This implies an effect of drainage system design on erosion and pollutant movement through surface and subsurface drainage waters from flat or mildly sloping lands.

INTRODUCTION

Artificial drainage is needed for some of the world's most productive soils. Drainage may be accomplished by land forming to provide good surface drainage; installing drain tubes or ditches for subsurface drainage or a combination of surface and subsurface drainage practices. Artificial drainage is needed to provide trafficable conditions for planting and harvest and to remove excessive soil water from the profile during the growing season. The intensity of subsurface drainage depends on the drain depth and spacing as well as soil properties that govern the rate water moves to the drain. Surface drainage intensity depends on the amount of surface depressional storage which is directly related to smoothness and continuity of slope.

In most cases several alternative combinations of surface and subsurface drainage will satisfy the drainage needs. Providing these alternatives can be identified, the combination that would normally be chosen would be the one with least cost. A particular choice, however, may inflict other costs that are not as readily identified. For example, intensive surface drainage with poor subsurface drainage may greatly increase surface runoff rates and result in a need for outlet ditches and canals with larger capacities than would be required with good subsurface drainage. Improved surface drainage may increase erosion. On the other hand, more intensive subsurface drainage may increase the outflow of nitrates and other potential pollutants in drainage waters. Thus, it is important to be able to characterize the effect of surface and subsurface drainage design on surface runoff and subsurface drain outflow.

The purpose of this paper is to use the water management model, DRAINMOD, to analyze the effect of drainage system design on total and temporal distribution of surface and subsurface runoff.

THE MODEL

The simulation model, DRAINMOD, used in this study, was described in detail by Skaggs (1978, 1980). It was developed for design and evaluation of multicomponent water management systems which may include facilities for subsurface drainage, surface drainage, subirrigation and sprinkler irrigation. Long-term simulations are conducted to evaluate the performance of a given water management system design over a long period of climatological record. The rates of infiltration, drainage, evapotranspiration and the water content distribution in the profile can be computed by obtaining numerical solutions to nonlinear differential equations (Freeze, 1971). However these methods would require prohibitive amounts of computer time for long term simulations; they also require soil property inputs that are not available for many soils. Thus they are not used in the present model. In order to simplify the required inputs and to make them consistent with available data, approximate methods are used for each component in the model.

The water management system considered is shown schematically in Figure 1. The model is based on a water balance for a thin section of soil of unit surface area which extends from the impermeable layer to the surface and is located midway between adjacent drains. The water balance for a time increment of Δt may be expressed as,

$$\Delta V_a = D + ET + DS - F \tag{1}$$

where, ΔV_a is the change in the water free pore space or air volume

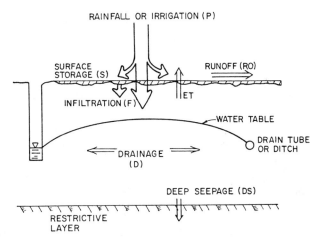

RAINFALL OR IRRIGATION (P)

SURFACE STORAGE (S)

RUNOFF (RO)

INFILTRATION (F)

ET

WATER TABLE

DRAIN TUBE OR DITCH

DRAINAGE (D)

DEEP SEEPAGE (DS)

RESTRICTIVE LAYER

Figure 1. Schematic of the water management system considered. The model treats subsurface drainage to either ditches or drain tubes, surface drainage, subirrigation and sprinkler irrigation.

(cm) in the section. D is drainage from (or subirrigation into) the section (cm), ET is evapotranspiration (cm), DS is deep seepage (cm), and F is infiltration (cm) entering the section in time Δt.

A water balance is also computed at the soil surface for each time increment and may be written as,

$$P = F + \Delta S + RO \qquad (2)$$

where P is the precipitation (cm), F is infiltration (cm), ΔS is the change in volume of water stored on the surface (cm), and RO is runoff (cm) during time Δt. The basic time increment used in equations 1 and 2 is 1 hour. However when rainfall does not occur and drainage and ET rates are slow such that the water table position moves slowly with time, equation 1 is based on a Δt of 1 day. Conversely, time increments of 3 min. are used to compute F when rainfall rates exceed the infiltration capacity.

The guiding principle in the model development was to assemble the linkage between various components of the water balance, allowing the specifics to be incorporated as subroutines, so that they can be readily modified as better methods are developed. Methods used to evaluate components of equations 1 and 2 are briefly discussed below.

Precipitation

Precipitation records are read into DRAINMOD as hourly values. Hourly records are stored in the computer based HISARS (Wiser, 1975) for North Carolina and are automatically accessed as inputs to the model. Precipitation records on more frequent time intervals could be used, but such data are not normally available so the model was programmed and tested for hourly records.

Infiltration

Infiltration is predicted by an approximate equation of the type presented by Green and Ampt (1911). The equation may be written as,

$$f = K_s + K_s M_d S_f/F \tag{3}$$

where f is the infiltration rate, F is accumulative infiltration, K_s is the hydraulic conductivity of the transmission zone, M_d is the difference between final and initial volumetric water contents and S_f is the effective suction at the wetting front.

In addition to uniform profiles for which it was originally derived, the Green-Ampt equation has been used for profiles that become denser with depth (Childs and Bybordi, 1969), for soils with partially sealed surfaces (Hillel and Gardner, 1969) and for nonuniform initial water contents (Bouwer, 1969). Resistance to air movement may be significant for shallow water tables where air may be entrapped between the water table and the advancing wetting front (McWhorter, 1971, 1976). Morel-Seytoux and Khanji (1974) showed that the Green-Ampt equation retained its original form when the effects of air movement were considered. The equation parameters were simply modified to include effects of air movement.

For a specific soil with a given initial condition, equation 3 may be written as,

$$f = A/F + B \tag{4}$$

where A and B are parameters that depend on the soil properties, plant factors such as extent of cover and depth of root zone; and rainfall factors such as intensity, duration, and time distribution. The model requires input for infiltration in the form of a table of A and B versus water table depth. Although it is normally assumed that the A and B matrix is constant, it is possible to allow it to depend on events that affect surface cover, compaction, etc.

When the rainfall rate exceeds the infiltration capacity as given by equation 4, equation 2 is applied for time increments of 3 min. Rainfall in excess of infiltration is accumulated as surface storage. When the storage depth exceeds the maximum storage depth, the excess is allocated to surface runoff. These values are accumulated so that, at the end of the hour, infiltration and runoff as well as the depth of surface storage are predicted. Infiltration is accumulated from hour to hour and used in equation 4 until rainfall terminates and all water stored on the surface has infiltrated. The same A and B values are used as long as the rainfall event continues. An exception is when the water table rises to the surface, at which point A is set to A = 0 and B is set equal to the sum of the drainage, ET and deep seepage rates.

Surface Drainage

Surface drainage is characterized in the model by the average depth of depression storage that must be satisfied before runoff can begin. Depression storage has a micro-component in small depressions created by surface structure and cover and a macro-component due to larger surface depressions and which may be altered by land forming, grading, etc. A field study was conducted by Gayle and Skaggs (1978) to measure the micro and macro storage components for artificially drained soils in eastern North Carolina.

Another storage component that should be considered is the surface

340

water that is accumulated, in addition to the depression storage, during the runoff process. This detention storage depends on the rate of runoff, slope, and hydraulic roughness of the surface. It is neglected in the present version of the model which assumes that runoff moves immediately from the surface to the outlet.

Subsurface Drainage and Subirrigation

The methods used in DRAINMOD to calculate drainage and subirrigation rates assume that lateral water movement occurs mainly in the saturated region. The lateral flux is evaluated in terms of the water table elevation midway between the drains and the water level or hydraulic head in the drains. There are several methods that can be used to calculate drainage rates (e.g. van Beers, 1976, Kirkham, 1958). DRAINMOD employs the Hooghoudt steady state equation, as used by Bouwer and van Schilfgaarde (1963). The equation may be written as,

$$q = \frac{8 \ K \ d_e m + 4 \ K \ m^2}{L^2} \tag{5}$$

where q is the flux in cm/hr, d_e is the equivalent depth of the impermeable layer below the drain, m is the midpoint water table height above the drain, K is the effective lateral hydraulic conductivity and L is the distance between drains. Equation 5 may be modified for layered soils by the methods presented by van Beers (1976). Although this method was derived for steady state conditions, it compares well with transient methods for predicting drainage flux when applied sequentially for short time increments or for small changes in water table position. The model uses a modification of equation 5 presented by Ernst (1975) to calculate subirrigation rates. This method, along with Moody's (1968) procedures for determining d_e are discussed in more detail by Skaggs (1980).

Evapotranspiration

The determination of evapotranspiration (ET) is a two-step process in the model. First the daily potential evapotranspiration (PET) is calculated from atmospheric data and distributed at a uniform rate over the 12 hours between 6:00 AM and 6:00 PM. In case of rainfall, hourly PET is set to zero for any hour in which rainfall occurs. After PET is calculated, calculations are made to determine if ET is limited by soil water conditions. If soil water conditions are not limiting, ET is set equal to PET. When PET exceeds the amount of water that can be supplied from the soil system, ET is set equal to the smaller amount.

There are several excellent methods for predicting PET based on climatological variables (Jensen, 1973; McGuinness and Borden, 1972). In application, however, the choice is limited by the availability of meteorological data. PET is routinely calculated in the model by the method presented by Thornthwaite (1948) which requires only daily maximum and minimum temperature and geographic location as input information. Other methods can be used when the input data are available.

Calculations to determine if ET is limited by soil water conditions consider both the water content in the root zone and upward flux from the water table. Methods for making these determinations are discussed in detail by Skaggs (1980).

The validity of DRAINMOD has been tested using data from field experiments in North Carolina (Skaggs, 1978); Ohio (Skaggs et al., 1980) and Florida. A total of more than 35 site-years of data have been used to test the model; it has performed well for a wide range of soils and climatological conditions.

PROCEDURES

Analyses were conducted for a Rains sandy loam soil (fine-loamy, siliceous, thermic Typic Paleaqualts). The Rains soil series consists of deep, poorly drained, moderately permeable soils that formed in loamy sediments on marine terraces. The surface is nearly level and the soil requires drainage for trafficability and crop protection. The soil is underlain at a depth of about 2 m by a heavy subsoil that restricts vertical water movement. Relationships for drainage volume and steady upward flux from the water table are plotted as functions of water table depth in Figure 2. Other soil properties and drainage system parameter inputs are summarized in Table 1.

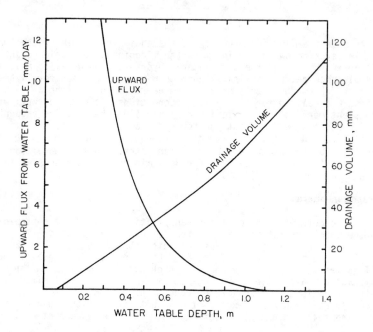

Figure 2. Drainage volume and steady upward flux as affected by water table depth for Rains sand loam (from Skaggs, 1978).

Simulations were conducted for a unit area using climatological data from Wilmington, N.C. The crop considered was corn, grown on a continuous basis. It was planted on April 15 and harvested after August 15 each year. The effective root depth as a function of time after planting was obtained from data presented by Mengal and Barber (1974) and used as input to the model. Simulations were conducted for drain spacings of 15, 30, 60, 100 and 300 m for both good surface drainage (depression storage, S = 2.5 mm) and poor surface drainage (depression storage, S = 25 mm). The results were analyzed to determine the effect of drainage system design on surface and subsurface drainage from both single storms and on an annual basis.

Table 1. Summary of soil property and other input data used in simulations for Rains sandy loam.

Property or parameter	Input Value		
1. Soil properties:			
Depth to restricting layer:	2.0 m		
Saturated hydraulic conductivity:			
Depth < 1.1 m	4.3 cm/hr		
1.1 m < depth < 2 m	1.0 cm/hr		
Water content at lower limit available to plants:	0.09 cm^3/cm^3		
Planting date	April 15		
Length of growing season	120 days		
Saturated water content in root zone:	0.37 cm^3/cm^3		
Green-Ampt Infiltration Parameters:			
	Effective water table depth (cm)	A cm^2/hr	B cm/hr
	0	0.0	0.0
	50	1.2	1.0
	100	3.3	1.0
	150	6.0	1.0
	200	9.2	1.0
	500	25.0	1.0
2. Drainage system parameters:			
Drain depth	90 cm		
Drain diameter	10 cm		
Effective drain radius	5.1 mm		
Surface depressional storage	2.5 mm, 25 mm		
Drain spacing	15, 30, 60, 100, 300 m		

RESULTS AND DISCUSSION

The effect of drain spacing on annual surface runoff and subsurface drainage is shown for a 5-year recurrence interval (5 YRI) in Figure 3. Relationships are given for surface drainage treatments (land leveling and smoothing) corresponding to depressional storage depths of S = 2.5 mm and S = 25 mm. For both surface drainage treatments, annual surface runoff is significantly reduced by greater subsurface drainage intensity. For example, reducing the drain spacing from 100 m to 15 m decreased the 5 YRI annual surface runoff by a factor of 3 (from 57 to 19 cm) for good

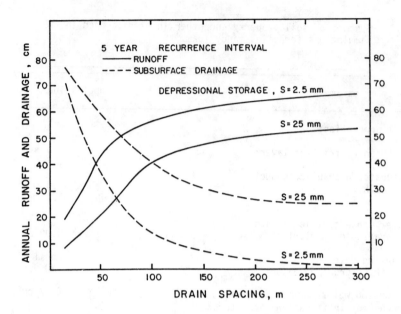

Figure 3. Effect of subsurface drain spacing and surface depressional storage on 5 year recurrence interval annual runoff and subsurface drainage from a Rains sandy loam soil.

surface drainage (S = 2.5 mm) and by more than a factor of 4 for poor surface drainage (S = 25 mm). The increased drain spacing caused subsurface drainage to be reduced as expected, with a larger reduction for good than for poor surface drainage treatments.

The effect of surface drainage intensity on both annual runoff and subsurface outflow is also indicated by the results in Figure 3. Improving surface drainage by land forming and smoothing to reduce the effective depth of depressional storage increases surface runoff and reduces subsurface drainage for all drain spacings. The effect of surface drainage on both runoff and subsurface flow is least for very intensive subsurface drainage (small drain spacing) but has a relatively constant effect for drain spacings in excess of about 50 m.

In most cases several combinations of surface and subsurface drainage intensities will satisfy trafficability and plant protection requirements for crop production. While actual drainage requirements depend on the individual situation, a common choice is either good subsurface drainage and poor surface drainage or poor subsurface and good surface drainage. The effect of these extremes on surface and subsurface outflows may be examined by considering results for the following combinations: (a) drain spacing, L = 15 m and S = 25 mm, and (b) L = 60 m and S = 2.5 mm. Although the drainage benefits are not the same for these two combinations, they can be used to demonstrate the effect of alternative drainage system designs on the hydrology. Notice the large differences in the predicted surface runoff (Figure 3). For good subsurface drainage (L = 15 m, S = 25 mm) the predicted 5 YRI surface runoff is 8 cm as compared to 48 cm for L = 60 m, S = 2.5 mm. A similar comparison shows that the 5 YRI subsurface drainage is 77 cm for combination (a) as compared to 31 cm for combination (b). The total water leaving the field, i.e. runoff plus subsurface drainage, decreases somewhat with the intensity of subsurface drainage (Figure 4). For example, increasing the drain spacing from 15 m to 60 m would decrease the total 5 YRI outflow from 80 to 70 cm, about 12 percent, regardless of the surface drainage treatment. However, the drainage system design has a tremendous effect on the partitioning of outflow between surface runoff and subsurface drainage (Figure 3).

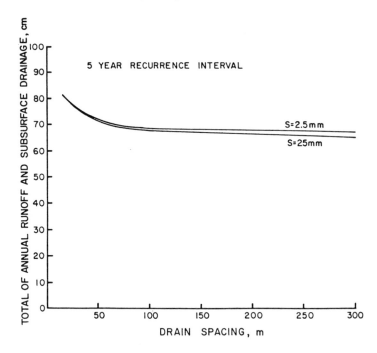

Figure 4. Effect of drain spacing on the 5 year recurrence interval total outflow - surface runoff plus subsurface drainage - for two values of surface depression storage.

The effect of drainage system design on annual surface runoff and subsurface drainage is not the same in all years. This is shown clearly in Figure 5 where the year to year variation in predicted surface runoff is plotted for the two combinations discussed above. Similar results for predicted annual subsurface drainage are plotted in Figure 6.

The effects of drainage system design on runoff and drainage from single 24 hour storms are shown in Figures 7 and 8. Both storms had a total rainfall of 166 mm and represent 10-year recurrence interval events at Wilmington, N.C. Results in Figure 7 show that, even though the amount of water removed by subsurface drainage during the 24-hour period was small (less than 0.6 cm), drain spacing had a very important effect on the total surface runoff. For example, runoff for L = 15 m and S = 2.5 mm was 2.8 cm as compared to 6.5 cm for L = 60 m and S = 2.5 mm. The effect of subsurface drainage is to reduce antecedent soil water conditions as reflected by the initial water table depths, plotted versus drain spacing in Figure 9. Closer drain spacing increased the water table depth at the beginning of the storm, providing more storage for infiltrating water and decreasing surface runoff.

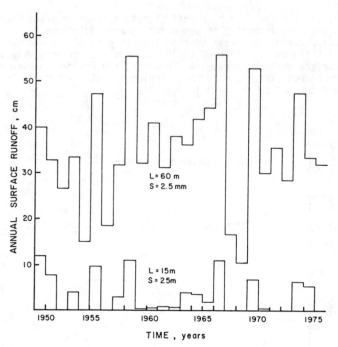

Figure 5. Year-to-year variation in annual surface runoff for two
drainage system designs.

    The importance of antecedent conditions on surface runoff is clear-
ly demonstrated by comparing the results in Figure 7 with those in
Figure 8 for a 10 YRI storm with an initially shallow water table (Figure
9). Rainfall prior to the storm in Figure 8 caused the water table to be
higher, and the storage volume for infiltrating water to be lower, than
for the equal size storm of Figure 7. Thus, surface runoff was greater
for the storm with higher antecedent rainfall for all combinations of
drain spacing and surface depressional storage. Again, intensive subsurface drainage lowered the initial water table depth and reduced the
surface runoff by as much as a factor 2 for the wet initial conditions.
The volume of water removed by subsurface drainage was also greater for
the wetter initial conditions because the water table was higher and
drainage occurred over a longer period of the day. It is interesting
that the drain spacing can have as much or more influence on runoff as
the surface drainage intensity regardless of antecedent rainfall
(Figures 7 and 8).

    Results presented in Figures 3-8 show clearly that the partition-
ing of the outflow between surface and subsurface drainage can be significantly affected by the design of the drainage system. More directly,
the volume of surface runoff, on either an annual or storm-by-storm
basis, can be reduced by increasing the intensity of subsurface drainage.
The total outflow may be slightly increased (Figure 4) but surface runoff would be reduced. Although the effect of subsurface drainage on
surface runoff may be very important from flood and erosion control
standpoints, few experimental studies document such effects. Schwab

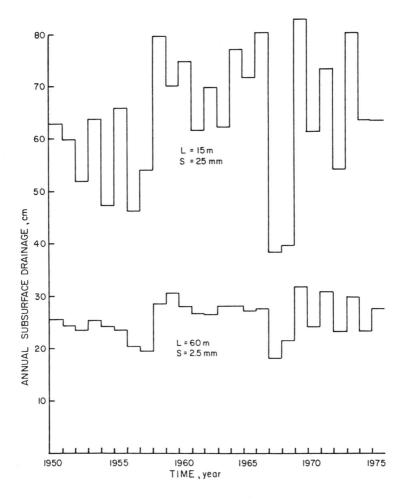

Figure 6.   Year-to-year variations in annual subsurface
drainage for two drainage system designs.

et al. (1975) presented the results of an 11-year field study which
included replicated plots for surface drainage only, subsurface drainage
only, and combination surface and subsurface drainage. Analysis of the
experimental results showed that the average surface runoff during the
growing season from the combination plots was only 45 percent of that
measured for the plots with surface drainage alone. These results are
in qualitative agreement with the simulated results presented herein.
The data of Schwab et al. (1975) were used in a previous study (Skaggs
et al., 1980) to test the validity of DRAINMOD for the North Central
Ohio conditions. Results of the study showed that the model gave
reliable predictions for the frequency and magnitude of surface and subsurface runoff for all three drainage treatments.

Because water is removed slowly over long periods of time by subsurface drainage, increasing the subsurface drainage intensity would

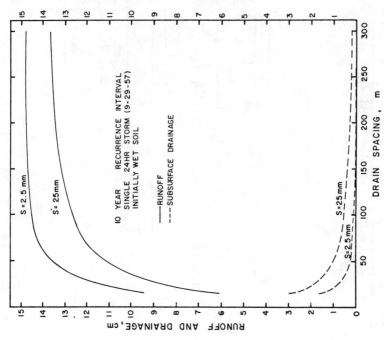
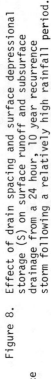

Figure 8. Effect of drain spacing and surface depressional storage (S) on surface runoff and subsurface drainage from a 24 hour, 10 year recurrence storm following a relatively high rainfall period.

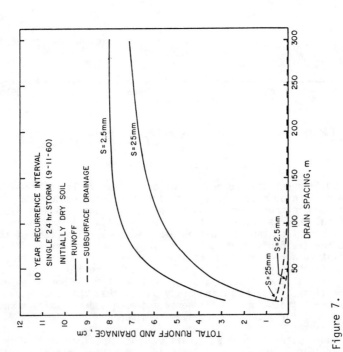

Figure 7.

Effect of drain spacing and surface depressional storage (S) on surface runoff and subsurface drainage from a 24 hour duration, 10 year recurrence interval storm following a low rainfall period.

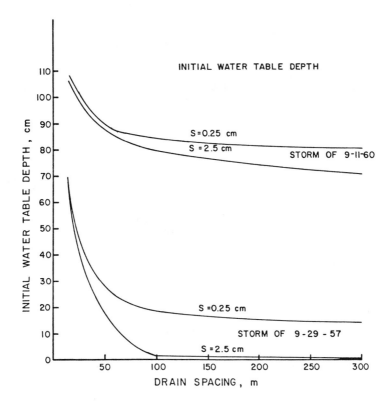

Figure 9.   The effect of drain spacing and surface depressional
             storage on the initial water table depths for the
             10 year recurrence interval storms of Figures 7 and 8.

have an important effect on the time distribution of the total outflow
(runoff plus drainage). Cumulative rainfall, surface runoff and subsurface drainage are plotted in Figure 10 for the 10 YRI storm of
9-29-57. Simulation results are given for drain spacings of 15 and 60 m,
both in combination with good (S = 2.5 mm) surface drainage. As
expected, runoff begins earlier for the 60 m spacing because of a higher
initial water table (Figure 9) and lower storage volume, as discussed
previously. The initial water table was deeper for the 15 m spacing
and the beginning of runoff was delayed by three hours over that of
the 60 m spacing. Simulations (not plotted) for the 15 m spacing with
poor surface drainage (S = 25 mm) showed an additional 5 hour delay in
the initiation of runoff for this storm because of the time required to
satisfy depressional storage.

     Once runoff starts, the rate of subsurface drainage, as indicated
by the slope of the curves given in Figure 10, is much slower than
surface runoff rates during the storm period. However, surface runoff
ceases at, or shortly after, the cessation of rainfall, while subsurface drainage continues over a much longer period of time. Thus
improving subsurface drainage causes surface runoff, which occurs over
a relatively short period of time, to be replaced by subsurface outflow

349

over a much longer period of time. This is demonstrated on an annual
basis in Figure 11 for the year 1960. Most of the outflow for a drain
spacing of 60 m occurs as surface runoff in many storm events over the
year. While there was greater total outflow during the year for the
15 m spacing, there were only six events for which surface runoff was
predicted. Subsurface flow occurred more slowly over longer periods
of time and was nearly continuous for both drain spacings.

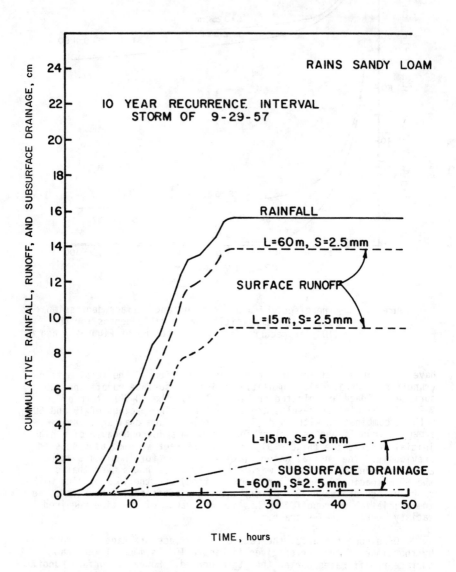

Figure 10. Simulated cumulative surface runoff and subsurface drainage
for two drainage system designs during and following a 24 hour --
10 year recurrence interval storm.

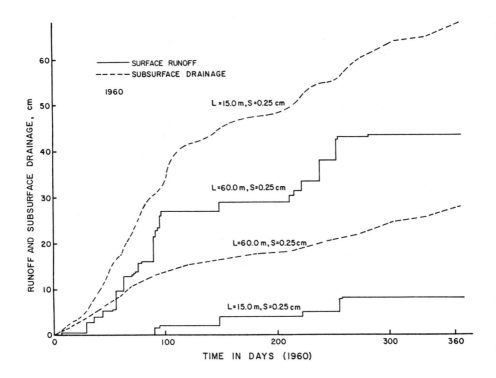

Figure 11. Annual (1975) simulated cumulative surface runoff and sub-
surface drainage for a system with good subsurface drainage
(L = 15 m, S = 2.5 mm) and for poor subsurface drainage
(L = 60 m, S = 2.5 mm).

The results of Figures 10 and 11 show that soils with good subsurface drainage buffer the effects of storm events, reducing surface
runoff and distributing the outflow over a longer period of time.
Since surface runoff is reduced, it seems logical to assume that sediment, pesticide and nutrient movement from the soil surface would also
be reduced by more intensive subsurface drainage. However, subsurface
flow is increased so the movement of nutrients such as nitrates or
other potential pollutants that are transported via subsurface drainage
water would also be increased by more intensive subsurface drainage.

SUMMARY AND CONCLUSIONS

The computer simulation model, DRAINMOD, was used to analyze the
effect of drainage system design on surface runoff and subsurface drainage from a Rains sandy loam soil. Simulations were conducted for ten

combinations of surface and subsurface drainage (two levels of surface drainage and five subsurface drainage spacings). A 25-year period of climatological record was used in the simulations to show the effect of drainage system design on total annual runoff and its variation from year to year.

The results showed that the intensity of subsurface drainage had a very significant effect on both surface annual runoff and surface runoff from single storms. Increasing the drain spacing from 15 to 100 m for good surface drainage caused more than a 3-fold increase (from 19 cm to 57 cm) in the total surface runoff on a 5-year recurrence interval basis (5 YRI). For poor surface drainage, surface runoff was only 8 cm for the 15 m spacing as compared to 41 cm for the 100 m spacing. In both cases the wider drain spacing decreased annual subsurface drainage by an amount similar to the increase in surface runoff. The effect of drainage system design on runoff and subsurface drainage from single storm events was similar to that observed for annual amounts. For high antecedent rainfall, the surface runoff from a 10 YRI -- 24 hour storm was 9.5 cm for a drain spacing of 15 m as compared to 14 cm for a 60 m drain spacing. The runoff was smaller for drier antecedent conditions. The runoff from a 10 YRI -- 24 hour storm with no rainfall during 18 days prior to the storm was 3 cm for a 15 m spacing versus 6 cm for a 60 m spacing.

The results obtained in this study show clearly that the drainage system design significantly affects the amount and rate of surface runoff -- on both an annual basis as well as single storm events. This implies an effect of drainage system design on erosion and pollutant movement through surface and subsurface drainage waters from flat or mildly sloping lands.

REFERENCES

Bouwer, H. 1969. Infiltration of water into nonuniform soil. J. Irrigation and Drainage Division, ASCE. 95 (IR4):451-462.

Bouwer, H. and J. van Schilfgaarde. 1963. Simplified method of predicting fall of water table in drained land. Transactions of the ASAE 6(4):288-291.

Childs, E. C. and M. Bybordi. 1969. The vertical movement of water in stratified porous material - 1. Infiltration. Water Resour. Res. 5(2):446-459.

Ernst, L. F. 1975. Formulae for groundwater flow in areas with subirrigation by means of open conduits with a raised water level. Misc. Reprint 178, Institute for Land and Water Management Research, Wageningen, The Netherlands. 32 pp.

Freeze, R. A. 1971. Three dimensional transient saturated - unsaturated flow in a groundwater basin. Water Resources Research, 7:347-366.

Gayle, G. A. and R. W. Skaggs. 1978. Surface storage on bedded cultivated lands. Transactions of the ASAE, Vol. 21(1): 102-104, 109.

Green, W. H. and G. Ampt. 1911. Studies of soil physics, part I - the flow of air and water through soils. J. Agricultural Science, 4:1-24.

Hillel, D. and W. R. Gardner. 1969. Steady infiltration into crust topped profiles. Soil Science, 108:137-142.

Jensen, M. E. (Editor). 1973. Consumptive use of water and irrigation water requirements. Report by Technical Committee on Irrigation Water Requirements, Irrigation and Drainage Division, American Society of Civil Engineers. 215 pp.

Kirkham, D. 1958. Seepage of steady rainfall through soil into drains. American Geophysical Union Transactions, 39:892-908.

McGuinness, J. L. and E. F. Borden. 1972. A comparison of lysimeter-derived potential evaportranspiration with computer values. USDA Technical Bulletin 1452:71 pp.

McWhorter, D. B. 1971. Infiltration affected by flow of air. Hydrology Paper No. 49. Colorado State Univ., Fort Collins.

McWhorter, D. B. 1976. Vertical flow of air and water with a flux boundary condition. Transactions of the ASAE, 19(2):259-261, 265.

Moody, W. T. 1966. Nonlinear differential equation of drain spacing. Journal of the Irrigation and Drainage Division, ASCE, 92(IR2):1-9.

Morel-Seytoux, H. J. 1973. Two phase flows in porous media. Advances in Hydroscience, 9:119-202.

Schwab, G. O, N. R. Fausey and C. R. Weaver. 1975. Tile and surface drainage of clay soils. Research Bulletin 1081, Ohio Agricultural Research and Development Center, Wooster.

Skaggs, R. W. 1978. A water management model for shallow water table soils. Technical Report No. 134 of the Water Resources Research Institute of the University of North Carolina, N. C. State Univ., 124 Riddick Building, Raleigh, NC 27650.

Skaggs, R. W. 1980. A water management model for artificially drained soils. Technical Bulletin No. 267. North Carolina Agricultural Research Service, North Carolina State Univ., Raleigh, 54 pp.

Skaggs, R. W., N. R. Fausey and B. H. Nolte. 1980. Water management model evaluation for North Central Ohio. Transactions of ASAE. In press.

Thornthwaite, C. W. 1948. An approach toward a rational classification of climate. Geog. Rev., 38:55-94.

van Beers, W. F. J. 1976. Computing drain spacings. Bulletin 15. International Institute for Land Reclamation and Improvement/ ILRI, P. O. Box 45, Wageningen, The Netherlands.

Wiser, E. H. 1975. HISARS - Hydrologic information storage and retrieval system, Reference Manual, North Carolina Agricultural Experiment Station Tech. Bull. No. 215, 218 pp.

WATER RESOURCES SIMULATION ON MICRO-COMPUTER

John Elgy
University of Aston in Birmingham
Gosta Green, Birmingham B4 7ET.

David Elgy
Late of the North West Water Authority,
Great Sankey, Lancashire

INTRODUCTION

One of the most popular means of designing Water Resources
Systems is by simulating the behaviour of a real system on a
digital computer. A model of a system comprising reservoirs,
rivers, aquifers etc is stored in the digital computer, the
designer then tests how this model will behave under various
sets of operating procedure until he arrives at an optimum.
There are therefore two important criteria for good simulation;
the first is an accurate and realistic model, the second, that
the simulation shall be as quick as possible to enable an
optimum to be reached in a reasonable time. The second can be
accomplished when the designer can perform a large number of
different tests and interpret the results of such tests well.

To enable a large number of systems and operating rules to be
tested the designer needs to have ready access to a computer,
preferably working interactively with it. In the past
computers have been expensive and gaining the interactive
facility has been very difficult. However the price of very
small computers has dropped substantially in recent years and
it is now possible for the designer to have on his desk a
computer capable of simulating quite complex water resources
schemes.

In addition to the hardware it is also essential to have the necessary software, often written specifically for each system under consideration. It is impractical to produce a general purpose simulation program that can meet the needs of every situation, but it is possible to create a program which will be sufficiently general for most cases and yet still allow straightforward user modification.

A general purpose simulation model based upon the concept of activities within subcatchments is presented. Flow diagrams are given to aid in the explanation of the algorithm and to enable it's implimentation. An example comprising two river absractions, a reservoir and a ground water source is used to illustrate the algorithm.

THE MICRO COMPUTER

At present there are a large number of micro computers on the market with a vast range of facilities and prices. Most are equipped with: (i) an 8 bit micro processor consisting of the 8 bit central processor together with a number of internal registers (ii) up to 64 K bytes (1 byte = 8 bits) of Memory and (iii) a BASIC interpreter. The type of machine used in this study also included two 5 1/4 inch floppy disks each capable of storing up to 175 K bytes.

The two important aspects from the simulation point of view are the 8 bit micro processor and the BASIC interpreter. Firstly the 8 bit processor limits available memory to 64 K bytes (2^{16}), and means that eight digit precision decimal numbers can be stored in 5 bytes (in packed binary coded decimal). This means for example that when comparing decimal numbers, up to 5 bytes have to be compared. Secondly the use of a basic interpreter as opposed to the more usual compiler slows the system down further, but does allow for greater flexibiltiy in the development and running of a program. In Main frame and mini computers a program is converted from a high level language to a machine code by a compiler. On the other hand in a micro computer the program is actually stored in BASIC then interpreted line by line as the program is run. Every time a line is executed it is interpreted, thereby allowing a program to be interupted during execution and variables to be printed, extra calculations carried out and variables changed. The program may then be continued, though not in itself changed.

It was decided that a machine with a Zilog Z80A processor working at an instruction cycle rate of 4 Mhz was the minimum required for simulation purposes. A North Star Horizon micro computer with dual disk drive was therfore used which had 56 K bytes of random access memory. Since on initial trial runs program execution was unduly slow a North Star Floating point Arithmetic board was later added. This speeded up the execution time of programs with large amounts of arithmetic but only marginally improved programs with large number of decisions in them.

The computer was used with the following proprietory software: North Star DOS 5.0 and 5.2, North Star BASIC and Floating point BASIC versions 5.0 and 5.2 and A. Ashley inc. machine code assembler and hybrid BASIC. The sections to be coded only need to be the cores of loops etc which are repeated many times, this considerably improves the execution time of some programs.

In addition an Intertube V.D.U. was the principal terminal to the computer with the occasional use of a teletype paper tape reader for the input of data and a dot matrix printer for listings and occasional output. It is envisaged that in use the computer and V.D.U. will be all that are needed for the execution of the program.

SIMULATION

The increasing use of multi source, multi purpose water resource schemes has posed considerable problems for the designers of such schemes. Not only must the conflicting interest of, for instance, water supply and flood protection be satisfied but operating procedures must be stipulated and the cost of the scheme minimised. Though dynamic programming and classical optimisation techniques (see Kottagoda (1969) for a full discussion) are more elegant the only practical way to solve many such problems is by simulation.

Simulation techniques have been used for many years – some early examples being developed by Tocher (1963), Maass et al (1962) and Dorfman (1960). Later Wyatt (1973) considered planning, economic aspects and operation of water resources systems and proposed an integrated model of a total water system based upon a common data base. Simulation programs developed in the UK have in general been designed to simulate a specific water resources system, for example, the very practical work of Walsh (1971, 1973, 1977), Jamieson and Wilkinson (1972). These programs are, however, constructed to be as general as possible whilst still allowing the idiosyncrasies of each system to be modelled.

Of the general purpose simulation models, probably the most important is that written by Fisher (1974) called GENSIM. Though the program was used extensively by the Water Resources Board to study the resources of Wales and the Midlands it has not been universally adopted. This is because of its complexity and the difficulty experienced in modelling some control rules. The GENSIM program required the following hardware: ICL 1902A computer with 18240 words of storage, disk drives, paper tape reader line printer and 2 magnetic tape decks. Fisher (1974) reported that to simulate a system comprising two reservoirs and two regulation control points using 73 pentads (one pentad = 5 days) of data required a runtime of 52.5 seconds of which 3.4 secs were mill time. This clearly cannot be used interactively on a desk top.

THE ALGORITHM

To write a totally general simulation model is a daunting and perhaps unrealistic task – to quote Walsh (1977) "Unfortunately as every water resources system has its own idiosyncrasies the development of a completely general model which can cater for all is virtually impossible." These idiosyncrasies are often license restrictions which for example, permit the building of a reservoir provided that flush releases are made on the first Friday of each month April to October to ensure that the river remains free of weeds! It is impractical for a general purpose model to account for every possible such case. However in reality these cases do provide real and significant restrictions

to the system's optimum solution. The algorithm presented
therefore is as general as is practicable but the program is
written in such a way as to allow the user to tailor the program
to his own requirements. The algorithm is now described with an
explanation of how this is converted to a computer program. It
may assist in following the algorithm if the computer flow
diagram figures 3 to 8 are consulted.

The algorithm considers the system as a series of numbered subcatchments each having at its downstream boundary a control or
decision point. These control points are reservoirs, river
abstractions, regulation points and flood danger areas - in
fact any point where the designer wishes to control the system
or investigate the effect of his controls. Subcatchments may
not overlap but they need have no physical size and this use of
"dummy catchments" permits complex systems to be modelled.

The subcatchments abstracting from and contributing to each
subsequent subcatchment are stored in an array, with up to
three contributing catchments and one abstracting catchment
being permitted. The contributing catchments are those which
flow into the one under consideration. Abstracting catchments
are those linked by interbasin transfer or pumped storage
schemes. Table 1 gives the linking array related to the
example system shown in Figure 1.

Associated with each subcatchment is one of the following
activities:
>   1. river abstraction,
>   2. storage reservoir,
>   3. regulating reservoir,
>   4. flood danger area.

Minimum and maximum channel flows at the exit to a subcatchment
are also specified. The maximum channel flow applies in the
case of flood danger areas or gives the maximum release
possible from reservoirs. The minimum flow is either the
"hands-off" or regulating level at river abstractions, or the
compensation release from a reservoir. This can be seasonally
varied.

The river abstraction activity is able to supply water to
direct supply, pumped storage reservoirs or interbasin
transfers. Initially the natural flows at the abstraction
point are modified to account for inflows into reservoirs,
abstractions upstream and additions due to interbasin
transfers. The permitted abstraction is found from a look up
table which is input to the program and gives various
abstractions at different river levels. If this permitted
abstraction is less than demand then a routine is called which
modifies the compensation released from regulating reservoir
according to a proportional drawdown rule and hence increases
the flow at the abstraction point. If the river abstraction
has an aquifer associated with it to augment supply and the
demand cannot be met, then the aquifer abstraction routine is
called. The drought register and counter are increased if the
supply still cannot be met. Water abstraction from the river
for transfers and for pumped storage schemes is then added to
the natural inflow from these subcatchments. There may be a
number of river abstractions in one river basin.

TABLE 1 - Linkage and order arrays for example in figure 1

a) Contribution array

Catchment	1	2	3	4
Links	000	100	000	000

b) Abstraction array

Catchment	1	2	3	4
Links	0	0	0	0

c) Minimum and maximum channel capacities in 10^3 m^3/d

Catchment	Minimum		Maximum
	Winter	Summer	
1	14	18	28
2	400	550	10000
3	90	400	10000
4	0	0	0

d) Abstraction rates in 10^3 m^3/d

Catchment	ABOVE Control Curve	BELOW Control Curve
1	127	55
2	180	180
3	280	280
4	up to a total of 251	

e) Order of decisions

Catchment	1	2	3	4
Rank	1	3	2	4

The storage and regulating reservoir activities are handled by the same routine, a flag being set to show that regulation releases can be made from reservoirs. This section can be split up into five sub sections:

  1. Modify inflows into the reservoir to account for inflows into upstream reservoirs, diversions etc.

  2. Water for direct supply is abstracted from the reservoir. Two abstraction rates are permitted depending on whether the reservoir level is above or below a control curve. The abstraction rate can be varied to suit a seasonally varying

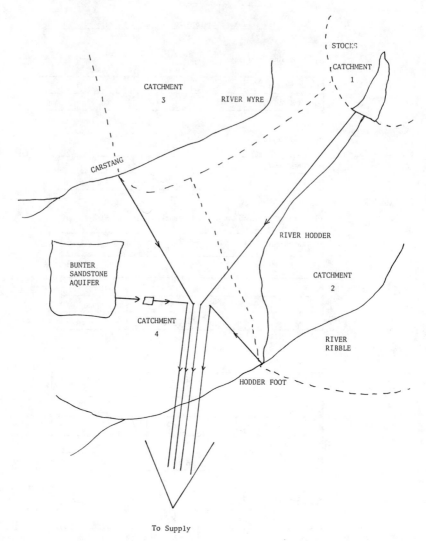

To Supply

FIGURE 1 Example of Water Resources system to be analysed by simulation

demand.

3. The new reservoir volume is calculated (note the control curve comparison is made on the previous periods reservoir volume) and if this is above the flood control curve then a release is made to either lower the volume to the control curve or equal to the maximum downstream channel capacity.

4. Spills are recorded and the downstream volume amended accordingly.

5. If the demand has not been met and an augmenting

aquifer is associated with the reservoir then that aquifer subroutine is called upon to provide the extra supply if this is possible.

Since the inflow into a pumped storage reservoir has been modified to account for transfers these are treated in the same way as other reservoirs.

Flood danger areas are points on the river where floods are likely to occur. In this activity the number of times the flow exceeds a threshold value is counted and the magnitudes of such floods recorded, the flow having been modified by reservoirs etc. upstream.

The length of time it takes a flood wave to travel downstream varies accordingly to the magnitude of the flood and to whether it is on the rising or falling limb. This makes the calculation of travel time of intercepted water or released water very difficult to quantify. The algorithm therefore allows for a number of possible travel times between the upper and subsequent subcatchments which are positive and less than 5 time intervals apart. Releases from reservoirs are assumed to take the same travel time as normal flows. Travel times are accounted for merely by staggering the downstream flow data at downstream points by the travel time to that point.

The modelling of aquifer behaviour is a complex subject which could easily dominate the simulation algorithm, however, for this model it is assumed that maximum abstraction rates will be given for the aquifer. These are the maximum daily rate weekly rate etc. If a more complex or realistic aquifer model is required then this can replace the aquifer subroutine.

It is necessary for decisions on the operation of a water resources scheme to be taken in a prescribed order, for example it is pointless to release water from a reservoir to meet an abstraction downstream prior to deciding whether it is necessary to have that water available. Therefore it is necessary to consider the subcatchments in a prescribed order which is not necessarily the order in which they are numbered. The catchment numbering describes their physical inter-relationship, a separate array holds the order in which decisions are made, this is shown in Table 1e for the example given in Figure 1.

THE SIMULATION PROGRAM

Figure 3 shows a flow diagram giving the outline of the simulation package. There are two main programs in the suite SYST-IN and SYST-SIM. SYST-IN allows for a system to be created onto disk for later simulation, specifying system linkages, constraints, operating rules etc. The program SYST-SIM is the actual simulation program, which takes as input any system previously created by SYST-IN. This system may be changed by the SYST-SIM program, though it is only expected that the operating rules will be modified in this way. These changes are not permanently stored.

Since it is envisaged that the user may wish to modify the program to model some idiosyncrasies of his system the program is

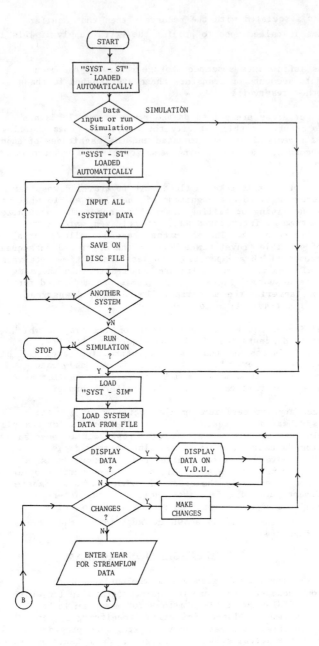

Figure 3. Overall flow diagram for simulation model

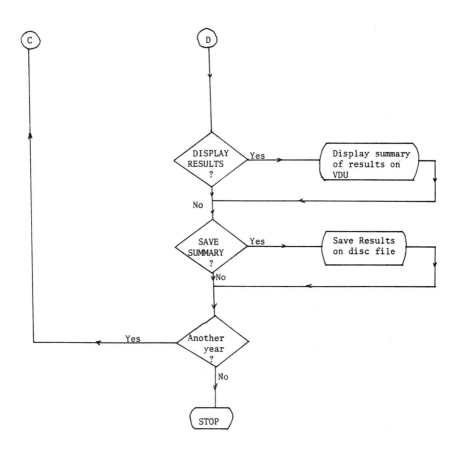

FIGURE 3 CONTINUED

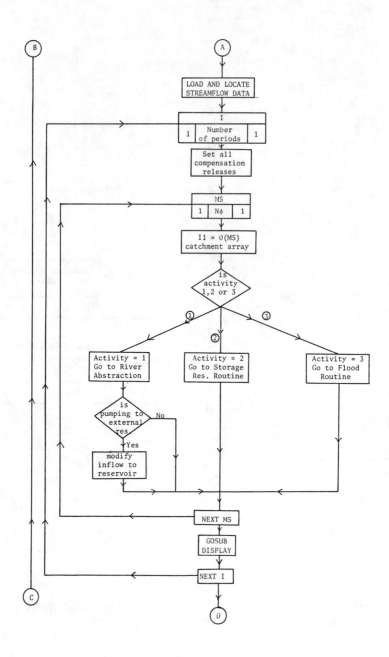

FIGURE 3 CONTINUED

written in structured programming BASIC [see Koffman and
Friedman (1979)] and provided with full documentation. By
splitting the program up into these building blocks the user may
discard a large portion of the simulation part of the program and
use only the input, output and data handling sections, these
often being the most time consuming parts of a program to write.

Figure 4 gives a flow diagram for the river abstraction section
of the program. Two points should be noted about this part of
the program.

   1. No releases are made to support a pumped storage
scheme.

   2. When releases are made the exact release from each
reservoir is given by the RELEASE subroutine. This subroutine
is shown in Figure 5. This routine adopts a proportional
release algorithm whereby releases are made in proportion to
the volume held in storage. It first finds which are
regulating reservoirs then modifies the release made from these
reservoirs according to their volume.

The correction of natural stream flows at the start of the
abstraction subroutine means that as the simulation passes
downstream any other abstraction points will have correct
releases made to them.

The reservoir management routine shown in Figure 6 can be split
into four sections. One section deals with releases and supply
when the reservoir is above the control curve, in this case it
is assumed that the reservoir will not fail having made its
contribution to supply and releases. The enhanced yield is
taken from the reservoir. If the reservoir is below the control
curve then the second section is executed where checks are made
on reservoir volume and the reservoir managed as it approaches
empty. If need be then the reservoir is lowered to the flood
control curve and finally checked for spills. The aquifer
subroutine is then called if a ground water source exists in
conjunction with that reservoir and the supply is still not
met, if this fails to supply extra water then an attempt is made
to abstract more from the reservoir.

The flood routine is a simple checking routine which is shown
in Figure 7. The crude model of aquifer behaviour is shown in
the flow diagram in Figure 8.

The algorithm presented offers a means by which any water
resources system can be simulated allowing an optimisation of the
capital works and of the operating rules. The program is written
in structured BASIC to enable any user modifications necessary to
corectly model the idiosyncrasies that the systems may possess
This program fitted within the 56K of storage available on the
North Star Horizon yet still left sufficient room for some
machine code subroutines.

THE DATA BASE

A large amount of programming time in any simulation study is
concerned with input and output of data, to ease this problem a
consistant data base was designed allowing access to data via

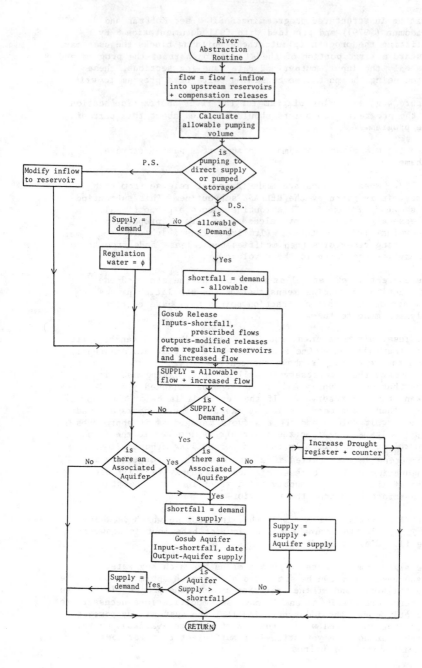

Figure 4. River abstraction routine

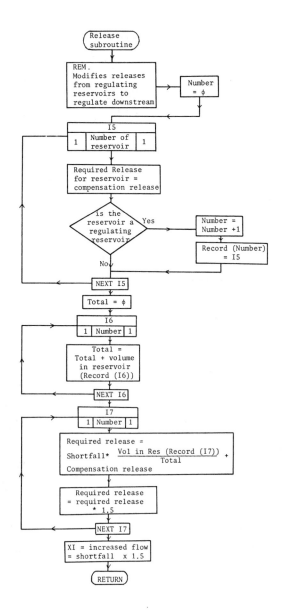

Figure 5. Reservoir release subroutine

367

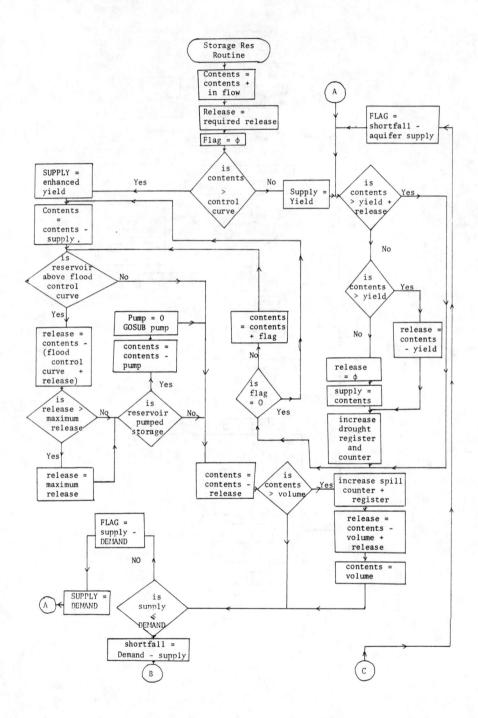

Figure 6. Reservoir management routine

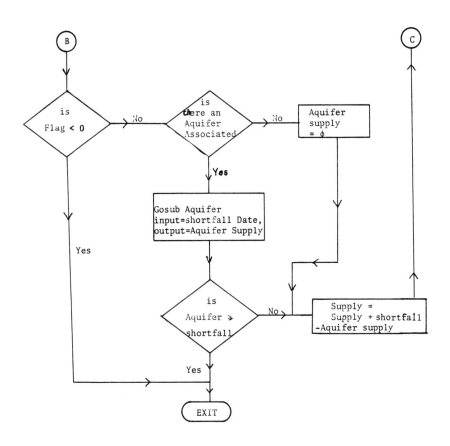

FIGURE 6 CONTINUED

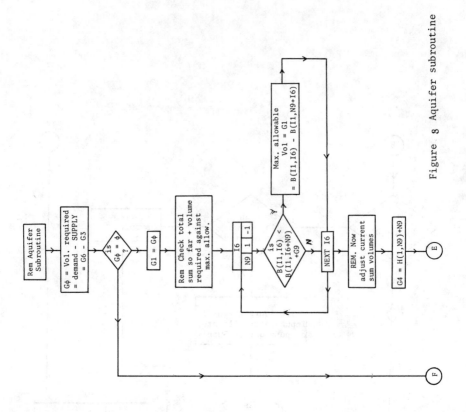

Figure 8 Aquifer subroutine

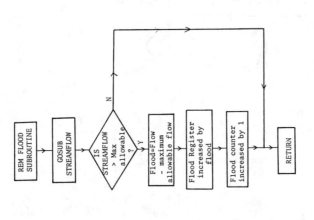

Figure 7 Flood Routine

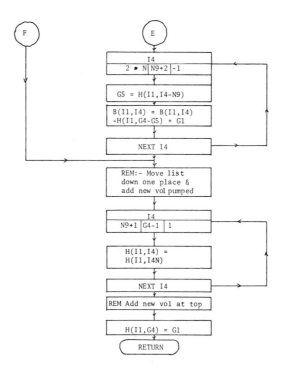

FIGURE 8. CONTINUED

one of the following: station key number, name, date, type of record.

The daily streamflows used in the simulation study were read into a work file from a paper tape reader. By using the utility program DATABASE segments of data from this work file are then displayed upon the V.D.U. to allow for updating and correcting. Once the user is satisfied that the data is correct it is then stored in the data base. Figure 2 shows the structure of the data base. The streamflow records are stored sequentially in the random access file DATA, the file HEADER contains a record relating to each set of data, each record containing the addresses of the data in the random access file. The file HEADER is 2K bytes and DATA is 171K bytes, which allows for a total of 95 years of daily data distributed between 30 stations. This makes possible storage of sufficient data for the simulation on one 5 1/4 inch disk.

The important aspect of each record in the file HEADER is the key number. This consists of 2 groups of 2 digits separated by a space, the first two digits representing the disk number, the second two the record number on that disk. The linking key holds the key number of the continuation, if any, of that streamflow record.

The utility program DATABASE allows for the editing or updating of streamflow records as more data becomes available. Separate

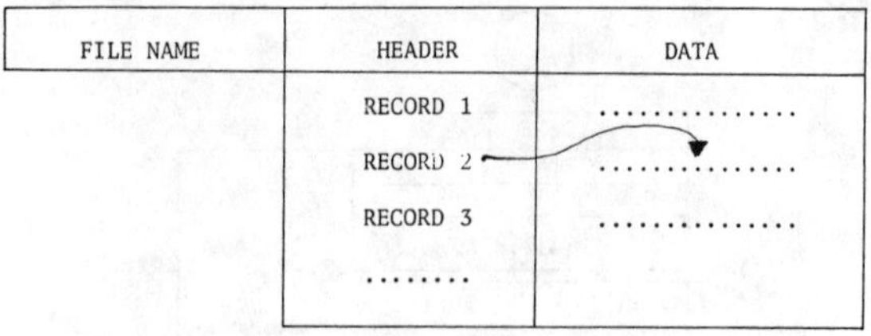

FILE NAME	HEADER	DATA
	RECORD 1	....
	RECORD 2	....
	RECORD 3	....
	....	

Record n in HEADER CONTAINS THE FOLLOWING BYTES

BYTES	TYPE OF VARIABLE	TITLE OF VARIABLE	EXAMPLE
1 - 6	string	Key number	∇ 1 ∇∇ 5
7 - 18	string	Data description	STOCKS ∇ INF
19 - 25	string	Data units	cum/s
26 - 30	B.C.D.	interval of record (in Hrs)	24
31 - 35	B.C.D.	Starting time of record (hrs after 00.00 JAN 1.1900)	315552
36 - 40	B.C.D.	Finishing time of record	675000
41 - 45	B.C.D.	Starting address of Data (in bytes from beginning of DATA	299541
46 - 50	B.C.D.	Finishing address	374426
51 - 56	string	Continuation key number	-
57 - 67	spare		

FIGURE 2 STRUCTURE OF DATA BASE

utility subroutines allow the database to be searched either by
key number or station name and a given year's data to be loaded
into memory, or outputted to the data base. By checking the
units and time interval and if necessary applying a conversion
factor the data returned to or received from the main program is
in units of 10^3 m^3/day

Though in this example the data was originally available in
paper tape form no difficulties were encountered when the data
came from a terminal line to a main frame computer.

AN EXAMPLE FROM THE LANCASHIRE CONJUNCTIVE USE SCHEME

The Lancashire conjunctive use scheme supplies central
Lancashire with water from a number of sources, these sources
include a reservoir, a groundwater source and a river
abstraction. These sources are used in conjunction, that is,

used in such a way as to minimise the total net cost of the system. This then is an ideal example to show the use of a simulation model and indicate how it may be used to find the optimum operating rules of the system.

There is little point in repeating the work already done on the system so the program was used to study the variation of the scheme shown in Figure 1. Water is abstracted from the river Ribble at Hodder foot just below its confluence with the river Hodder. Stocks is a reservoir of $28.6 \times 10^3 m^3$ and it supplies a safe yield of $55 \times 10^3 m^3$/day when the level is below the control curve and $127 \times 10^3 m^3$/day when above the control curve. The control curve is given in Table 2. Up to $280 \times 10^3 m^3$/day can be abstracted from the river Wyre at Garstang provided a prescribed flow of $90 \times 10^3 m^3$/day is left during October to April and $400 \times 10^3 m^3$/day is left during May to September. The Bunter Sandstone aquifer can supply water subject to the license limits given in Table 3. Walsh (1977) gives the relative costs of these sources as

> Stocks reservoir   1.0
> Wyre abstraction   1.48
> ground water   1.93 - 2.93

since the river Ribble has similar water quality to that of the Wyre its cost may be taken the same. An optimum operating policy would therefore seem to be

    1. Use non stored water, that is river water when it is available.

    2. Use Stocks water if it is above the control curve at the enhanced level, otherwise use its safe yield.

    3. Use groundwater as a last resort to top up supplies during periods of deficiencies.

The order of decisions shown in Table 1 was therefore decided upon, with the aquifer used to augment the supply from Stocks reservoir.

In use certain aspects of the program were noted:

    1. A major drought occured during 1976 and the simulation concentrated upon optimising the operating rules for this drought. Since this drought would not occur again in the future it would be necessary to generate synthetic data for a full appraisal of the reliability of any design arrived at.

    2. A small amount of program customisation had to take place to operate all 4 sources conjunctively.

    3. Once a failure was observed it proved easier after a run had been aborted to start the simulation from the beginning of the drought rather than continue with changed variables.

    4. The program took approximately 1.5 seconds to process one days simulation or 5 minutes to reach the critical part of the simulation in August 1976. Compare this with that reported by Fisher (1977) given earlier.

TABLE 2 - Stocks Control Curve

Month	Level in $10^3 m^3$
January	23100
February	25400
March	28300
April	27800
May	26300
June	24500
July	22200
August	20700
September	16750
October	12250
November	18500
December	22100

TABLE 3 - Permissible abstraction rates from

Bunter Aquifer

Time Period	Rate
Any 60 days	$177 \times 10^3 m^3/d$
Any 90 days	$157 \times 10^3 m^3/d$
Any 189 days	$128 \times 10^3 m^3/d$
1 calendar year	$79 \times 10^3 m^3/d$
3 calendar years	$51 \times 10^3 m^3/d$

No results are presented for this simulation because it is not felt that an optimum had been reached though around 5 runs per hour have been recorded. Sufficient, however, has been learnt about the program to be satisfied with its performance but disappointed with its response times.

To improve response time the cores of some frequently repeated loops have been converted to machine code. The parts chosen for machine coding are parts which either the user will never change or parts which can be thrown out totally and replaced. The first section for change was therefore the aquifer subroutine, which unfortunately is not called frequently enough for any noticeable increase in speed. By machine coding the seasonal look up table, the part of the program wich decides which season a particular day is in, a noticeable improvement in speed occurred.

In use the program has proved to be very successful, enabling up to 5 simulations an hour to be carried out and by observing the values at the various decision points a greater understanding of the processes involved has occurred. This has lead to more rapid convergence on an optimum than would be possible submitting background jobs to a mainframe computer.

CONCLUSIONS

The micro computer has proved to be popular with users, particularly the high degree of interaction achieved. This degree of interaction has proved to be an essential factor of the simulation program produced. The model has performed well on test simulations but the execution time has been slower than desired, though not excessively so. Execution time would ideally be three times faster and in an attempt to achieve this sections of the model have been machine coded, giving a slight improvement in speed. Overall the model has achieved its objectives in that it allows interactive simulation to take place on the desk of the designer.

REFERENCES

Dorfman R. (1960) "Operations Research" A.E.R.

Fisher R. (1974) "'GENSIM', A General Purpose Simulation Program" Water Resources Board, H.M.S.O.

Jamieson D.G. & Wilkinson J.C. (1972) "River Dee Research Program, A Short-Term Control Strategy for Multipurpose Reservoir Systems", Water Resources Research, August 1972.

Koffman E.B. & Friedman F.L. (1979) "Problem Solving and Structural Programming in BASIC" Addison-Wesley.

Kottagoda N.T. (1979) "Stochastic Water Resources Technology" Macmillan.

Maass et al (1962) "The Design of Water Resources Systems" Harvard University Press.

Tocher K.D. (1963) "The Art of Simulation" English University Press.

Walsh P.D. (1971) "Designing Control Rules for the Conjunctive Use of Impounding Reservoirs" Journ of Inst. of Water Engineers 25 No. 7.

Walsh P.D. (1973) "The Application of a Simulation Model in the Planning and Management of Water Resources in Lancashire" Water Research Association Conference on Computer Uses in Water Systems, Reading Sept. 1973.

Walsh P.D. (1977) "The Practical Analysis and Operation of Multi-Source Water Supply Systems with Particular Reference to the Lancashire Conjunctive Use Scheme" Ph.D. Thesis University of Birmingham 1977.

Wyatt T. (1973) "An Intergrated Model for the Planning and Operations of Water Systems" Water Research Association Conference on Computer Use in Water Systems, Reading Sept 1973.

The authors would like to acknowledge the considerable assistance given to them in the construction of this model by Mr. P.G. Gaskin and Miss A.E. Callcott.

WATER MANAGEMENT INVESTIGATIONS ON SMALL WATERSHEDS IN THE TROPICS

J. Hari Krishna
Agricultural Engineer
Farming Systems Research Program
ICRISAT Patancheru PO
A.P. 502 324   INDIA

and

Robert W. Hill
Associate Professor
Agricultural & Irrigation
Engineering Department
UMC 41, Utah State University
Logan, Utah 84322

ABSTRACT

Several small research watersheds have been developed at the International Crops Research Institute for the Semi-Arid Tropics (ICRISAT) near Hyderabad, India where different land and water management treatments have been laid out. In semi-arid climates, runoff can be an important component of the water balance because it can be collected and reused as supplemental irrigation to provide adequate moisture for crop growth. Hydrologic data from the research watersheds at ICRISAT were therefore used to develop a parametric runoff prediction model which computes daily soil water balance and storm runoff volumes and can be used to estimate rainfall productivities. The model performance has been quite satisfactory, particularly with regard to runoff prediction.

Results of field studies indicate that watershed based systems of farming which permit greater opportunity time for infiltration while also providing for runoff collection and reuse show promise in stabilizing rainfed agriculture in the Semi-Arid Tropics. Actual yield data show that rainfall productivities can be significantly increased through the adoption of a comprehensive crop production technology for these regions.

INTRODUCTION

The rainy season in most parts of India lasts on an average from 4 to 5 months in a year. During this period, the several high-intensity storms that occur cause large amounts of runoff and erosion on many soils, particularly under conditions of limited vegetative cover. Although radiation levels and temperatures permit year-round crop production, inadequate water availability frequently results in only one crop per year in rainfed areas; the erratic rainfall distribution often causes drought stress even to rainy-season crops. Improved estimates of runoff are an important prerequisite for the design of more effective land and water management systems. Runoff, if collected, stored, and used later for supplemental irrigation may also contribute to greater stability of rainy-season crops and expansion of the double cropped area.

Because of the limited understanding of the hydrology of agricultural watersheds in the semi-arid tropics (SAT), several small research watersheds were developed at the International Crops Research Institute for the Semi-Arid Tropics (ICRISAT) near Hyderabad, India, in 1973 and 1974. The data collected provide initial estimates for the design of waterways and structures for excess water disposal or storage of runoff. However, rainfall in the SAT varies greatly from year to year. A simulation model was therefore developed to estimate the occurrence of runoff and also to extrapolate the information gained to other areas. The available information in developing countries is often limited. Thus, a physically based model with minimal input data which would result in estimates of storm runoff volumes was considered appropriate.

WATERSHED TREATMENTS AND DATA COLLECTION

The watershed treatments ranged from traditional methods to improved systems of resource management applied to a deep Vertisol. The Vertisols have a low hydraulic conductivity and are often imperfectly drained (Kampen and Krishna 1978). In large areas, it is common to fallow the land in the rainy season and to grow a crop on residual soil moisture, and this practice was simulated on one watershed (BW4C). The improved land management methods consist of land smoothing and graded (0.6% slope) ridges or broadbeds and furrows in order to provide adequate surface drainage during wet periods and simultaneously allow sufficient opportunity time for infiltration. With this watershed treatment on BW1, it was feasible to grow a rainy-season crop as well as a second crop in the postrainy season. These two treatments represented by BW1 and BW4C were used to develop a simulation model for runoff prediction and water balance analysis (Krishna and Hill 1979).

The rainfall in 1975 was above normal and that in 1976 below normal. The hydrologic data of these two years were therefore used for calibration, and the model was tested with 1974 data when rainfall was about normal at the ICRISAT Center. There was no runoff in 1977 and additional testing of the model was possible with 1978 and 1979 data. The total rainfall on individual watersheds was determined using a Thiessen network. Runoff was monitored by means of Parshall flumes and continuous waterstage recorders located at the watershed outlets. Soil moisture measurements were made by the gravimetric method and also with a neutron probe. Daily pan evaporation data were used to compute the crop evapotranspiration demands.

STORM RUNOFF PREDICTION

A parametric simulation model (RUNMOD) was developed to predict storm runoff volumes and to compute other components of the water balance on a daily basis (Krishna 1979). The daily input data required are rainfall amount, storm duration or rainfall intensity, and evaporation. If pan data are not available, empirical or other methods of determining crop evapotranspiration may be used. Information on the soil moisture status at the beginning of the growing season and knowledge of the waterholding capacity of the soil are also required.

The model utilizes the concept of two soil moisture zones, an upper zone of 20 cms and a lower zone of 160 cms. The daily evapotranspiration loss is assumed to exclusively occur from the upper zone initially; only after the moisture there is depleted, will evaporation loss occur from the lower zone. The soil water budget is maintained on a daily basis and when rainfall is received, the upper zone is fully recharged before any moisture is added to the lower zone.

The evapotranspiration computation is similar to that used by Ligon et al. (1965) and by Haan (1972):

$$AE_{cr} = PE_{cr}* \ (AW/AWX), \ (P = 0, M_u = MUI) \qquad \ldots\ldots \ (1)$$

$$AE_{cr} = PE_{cr}, \ (P = 0, MUI < M_u \leq MUX) \qquad \ldots\ldots \ (2)$$

$$AE_{cr} = 0.5*PE_{cr} * (AW/AWX), \ (P> 0, M_u = MUI) \quad \ldots\ldots \ (3)$$

$$AE_{cr} = 0.5*PE_{cr} \ \ (P> 0, MUI< M_u \leq MUX) \qquad \ldots\ldots \ (4)$$

Where

AE_{cr} = Actual crop evapotranspiration, mm

PE_{cr} = Potential crop evapotranspiration, mm

AW = Available moisture in lower zone on any given day, mm

AWX = Maximum available moisture in lower zone at field capacity, mm

P = Daily precipitation, mm

M_u = Moisture in upper zone on any given day, mm

MUI = Initial moisture content in upper zone, mm

MUX = Maximum moisture content in upper zone, mm

An "infiltration rate" (F) is computed on the basis of two parameters RIH and RIL, which refer to the "high" and "low" rates of infiltration. The higher value is assumed when the soil surface is totally dry, which corresponds to an MUI of 60 mm in the montmorillonitic clay soils at ICRISAT. The parameter RIL is used when the upper zone moisture equals 90 mm (MUX). Depending upon the value of M_u on any given day, the infiltration index F (mm/hr) for a given storm is computed according to one of the following relationships:

$$F = RIH \ (M_u = MUI) \qquad \ldots\ldots \ (5)$$

$$F = RIL \ (M_u = MUX) \qquad \ldots\ldots \ (6)$$

$$F = RIH - [(RIH-RIL) * (M_u-MUI)/MUA] \ .. \ (7)$$

$$(MUX> M_u > MUI)$$

Where

MUA = Maximum available water in the upper zone (30 mm)
and the others are as defined earlier.
= (MUX - MUI)

The infiltration rate F obtained from the above equation is used
with the storm duration (in hours) to obtain the infiltrated amount INF:

INF = F * SD   ...... (8)

Where

INF = Infiltration depth, mm

F = Infiltration rate mm/hr

SD = Storm duration, hours

The computed storm runoff CRO (mm) is then computed as

CRO = P - INF   .... (9)

where the terms are as defined earlier.

The computed runoff and the measured runoff data are compared
through a simple univariate optimization procedure to select the values
of RIH and RIL. The values of RIH and RIL were identified from two
years data (1975, 1976) and the model was tested with data from 1974,
78 and 79. Even though the model uses a lumped approach, it nevertheless predicts surface runoff with a fair degree of accuracy. Further
refinement of the model is possible with additional hydrologic data.

Parameter calibration with data from BW1 collected in 1975 and
1976 resulted in values of 34 mm/hr and 27 mm/hr for RIH and RIL,
respectively. When these values were used with 1974, 78 and 79 data
for testing, satisfactory predictions of runoff were obtained. The
results of the model are summarized in Table 1. The comparison of
individual runoff events in BW1 for 1974 and 1978 are shown in Figures
1 and 2.

In BW4C, the calibrated parameter values were 21 mm/hr and 16 mm/hr.
These values were used to obtain estimates of runoff for the other years.
The comparisons of individual storm events for BW4C during 1974 and 78 are
illustrated in Figures 3 and 4. To test the model's capability to simulate
the soil moisture variation, the measured soil moisture data were compared
with those computed by the model and they were observed to match each other
well.

The model appears to perform better in situations where the rainfall
events are of a high intensity and short duration. Since most tropical
storms are of this nature, this model appears appropriate under such
conditions.

WATER BALANCE AND RAINFALL PRODUCTIVITY

Besides providing a simple method of estimating runoff from small
watersheds, the model was used to compute the water balance with traditional and improved systems of land and water management.

The analysis showed that for watershed BW4C with traditional
cropping in the postrainy season, 42% of the seasonal precipitation was
lost as evaporation from soil during the wet season; the computed runoff
was 28%, the profile moisture accretion amounted to 23% and the deep

380

Table 1 : Summary of RUNMOD performance (1974-79)

Watershed	Year	Rainfall** (mm)	Measured Runoff (mm)	Simulated Runoff (mm)
BW1	1974	776	114.1	112.3
"	1975*	963	156.7	155.4
"	1976*	650	71.6	70.2
"	1977	523	0.0	0.0
"	1978	1063	270.6	290.9
"	1979	610	72.9	78.6
BW4C	1974	774	210.3	204.1
"	1975*	966	249.7	280.6
"	1976*	666	209.3	190.6
"	1977	516	52.0	50.5
"	1978	1054	409.2	405.3
"	1979	600	178.0	165.7

* Calibration period
** June - October rainfall

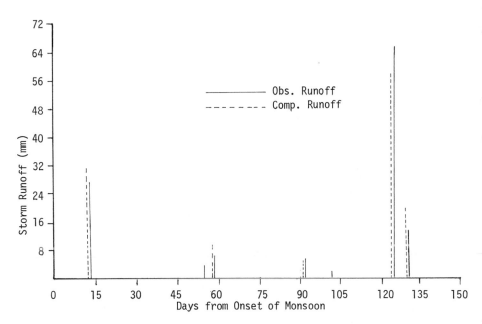

Figure 1 : Comparison of observed and Computed Runoff events for BW1 in 1974.

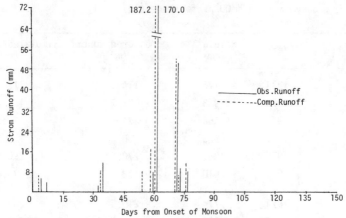

Figure 2 : Comparison of observed and computed runoff events for
BW-1 in 1978.

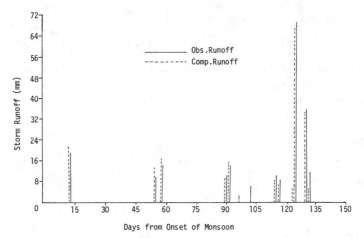

Figure 3 : Comparison of observed and Computed Runoff events
for BW4C in 1974.

percolation was 7%. In the graded broadbed and furrow system without
field bunds and with a well-defined grassed waterway (BW1), the corresponding figures for evapotranspiration, computed runoff, profile moisture
accretion, and deep percolation were 45%, 14%, 22% and 19%. Figures
of this order have also been reported in other studies at ICRISAT Center
(ICRISAT 1976, ICRISAT 1977, ICRISAT 1978).

No crop was grown in the rainy season on the traditionally managed
watershed (BW4C), yet the moisture loss from bare soil was quite high
due to frequent wetting of the soil surface. The deep percolation losses
on BW4C were less than those in BW1. This reflects the fact that a
substantial part of the rainfall on BW4C was lost as surface runoff
(twice as much as BW1). The capacity of the model to compute all components of the water balance facilitates an evaluation of hydrologic
response and rainfall utilization potential of traditional and improved
land and water management systems.

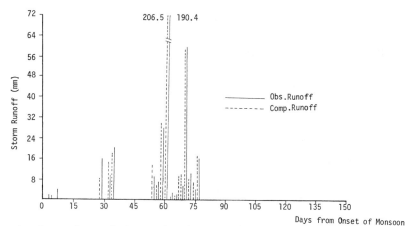

Figure 4 : Comparison of observed and computed runoff events for BW 4C in 1978.

Rainfall productivity (RP) can be defined as the agricultural production or its monetary equivalent in relation to the seasonal precipitation (Kampen and Krishna, 1978). The crop yields obtained on some of the watersheds at ICRISAT indicate the potential for vastly increased levels of RP under improved resource management. During 1976 when total rainfall was 71 cms, in BW4C with local post-monsoon sorghum, a yield of only 600 kg/ha was obtained whose value was Rs 660 (ICRISAT, 1977). This resulted in a RP of about 9 Rs/ha/cm of rainfall (660/71). When an improved variety of sorghum was used on the same watershed, a yield of 1200 kg/ha was measured whose monetary value was Rs 1320. The resulting RP was 18 Rs/ha/cm.

In BW1 which was graded and laid out in broadbeds and furrows along with a grassed waterway for improving drainage, a rainy season crop of maize was possible along with post monsoon chickpea. A yield of 3310 kg/ha of maize and 600 kg/ha of chickpea with a total gross value of Rs 3840 was obtained. This resulted in a computed RP of 54 Rs/ha/cm (3840/71). In BW3 which is very similar to BW1, runoff was collected in the rainy season and applied to the chickpea crop later during moisture stress. This resulted in a yield of 3350 kg/ha of maize and 1300 kg/ha of chickpea with a total gross monetary value of Rs 4511/ha, which in terms of RP is as high as 63 Rs/ha/cm (4511/71).

Similar results were obtained also from Alfisol watersheds at the ICRISAT center. In 1978 for instance even when rainfall was above average, an intercrop of sorghum and pigeonpea under improved management and with on site runoff water availability for supplemental irrigation, a RP of 54 Rs/ha/cm was obtained in contrast to a traditional intercrop system where RP was about 7 Rs/ha/cm (ICRISAT 1980). Thus the graded broadbed and furrow system of land management for in situ moisture conservation and for subsequent application of irrigation water along with a complete range of improved inputs* can substantially increase the rainfall productivity under semi-arid tropical conditions.

---

*
Improved varieties, adequate plant protection and fertilization at 22-57-0 at planting + 58 N side-dressed later.

REFERENCES

Haan, C.T. 1972. A water yield model for small watersheds. Water
Resources Research 8(1) : 58-69.

ICRISAT 1976. Annual Report 1975-76, International Crops Research
Institute for the Semi-Arid Tropics, Hyderabad, India.

ICRISAT 1977. Annual Report 1976-77, International Crops Research
Institute for the Semi-Arid Tropics, Hyderabad, India.

ICRISAT 1978. Annual Report 1977-78. International Crops Research
Institute for the Semi-Arid Tropics, Hyderabad, India.

ICRISAT 1980. Annual Report 1978-79. International Crops Research
Institute for the Semi-Arid Tropics, Hyderabad, India.

Kampen, J. and Krishna, J.H. 1978. Resource Conservation, management
and use in the Semi-Arid Tropics, ASAE Technical Paper No.78-
2072 presented at the American Society of Agricultural Engineers.
Summer Meeting, Utah State University, Logan, Utah.

Krishna, J.H. 1979. Runoff prediction and rainfall utilization in the
Semi-Arid Tropics. Ph.D. Dissertation, Department of Agricul-
tural and Irrigation Engineering, Utah State University, Logan,
Utah.

Krishna, J.H. and Hill R.W. 1979. Hydrological Investigations on small
watersheds at ICRISAT. Paper presented at the joint meeting of
American and Canadian Societies of Agricultural Engineering,
Winnipeg, Canada.

Ligon, J.T., Renoit G.R. and Elam A.B. Jr. 1965. Procedure for estimating occurrence of soil moisture deficiency and excess.
Transactions of the ASAE, 8 : 219-222.

SOME METHODS OF DISCHARGE FORECASTING AND THEIR APPLICATION IN SWITZERLAND

Manfred Spreafico
Head of Hydrology Department
Swiss National Hydrological Survey
Postfach 2742, CH-3001 Bern, Switzerland

ABSTRACT

Long-term and short-term discharge forecasts of Swiss rivers are applied to solve various water resources problems. On the one hand these forecasts are made by using rainfall-runoff or snowmelt-runoff models. On the other hand operational forecasts in connexion with flood warning are made mostly by discharge-discharge models. Because of the great differences of the geological, topographical, morphological, hydrological and meteorological parameters within small areas it is very difficult to forecast the runoff in a sophisticated way. Additional problems arise since the flow regime of the rivers is more and more influenced by human activities on the water cycle.

The first part of the paper explains a range of problems existing in Switzerland in setting up discharge forecasts. Then the application of rainfall-runoff and snowmelt-runoff models is shortly described. The last part presents two provisional procedures of discharge forecasts, which are in operational use for flood warning and lake regulation.

INTRODUCTION

In Switzerland long-term and short-term discharge forecasts are necessary for several application purposes; the long-term forecasts concern one month to six months, the short-term hours to days.

The long-term forecasts are mainly needed for the regulation of the reservoirs situated in the Swiss Alps, but also for discharge forecasts in the River Rhine. Long-term forecasts are especially needed to optimize the operating schedules of the hydroelectric power-plants and to plan the utilization of ships in view of a superannuated middle loaded draught. The domestic and industrial water supply, in particular in the Netherlands, needs long-term forecasts for the operation of pumping and treatment plants. The forecasted summer discharge of the River Rhine constitutes a decisive value in the regulation of the Isselmeer.

The short-term discharge forecasts serve mainly the following purposes:

- real-time operation of power-plants and lake regulation
- flood warning and initiation of short-term measures for flood-damage prevention
- calculation of the transport capacity for up-stream navigation

Besides their use in water resources management, forecasts based on rainfall-runoff models contribute mainly to scientific research in small catchments.

Discharge forecasts within the flood-warning systems of the Swiss National Hydrological Survey are made mainly by discharge-discharge models.

PARTICULAR PROBLEMS IN SETTING UP DISCHARGE FORECASTS IN SWITZERLAND

Switzerland, with regard to its geological, topographical and hydrological aspects as well as to its area of 41 000 sq. km, is a multiform country. Within a small space, the soil characteristics, the land use and the climate show great differences; thus considerable variations are resulting in the precipitation's quantity and distribution, furthermore part of it even falls as snow. Therefore, neighbouring hydrological basins can show very varying runoff characteristics:

River basin	Specific Discharge 1976/77 in $1/s/km^2$			
	Winter (Mean)	Summer (Mean)	Maximum	Minimum
Massa	8.8	101	303	1.2
Riale di Roggiasca	88.9	230	4986	8.5
Melera	94.8	124	1714	21.0
Mentue	25.3	20	395	4.5

The 94 reservoirs in Switzerland, with an exploitable storage capacity of about 3400 millions cubic meters, the dense urbanization of the country and the intensive cultivation of the soil have a great influence on the natural discharge in numerous rivers. Figure 1 shows the rivers with flow regimes considerably (more than 20 % of the mean annual discharge) affected by water resources management. The influence of the reservoir depends on the storage capacity, the size of the reservoir system, the water adduction and diversion and also on its utilization.

Figure 2 shows the discharge in the river Drance de Bagne, before and after the construction of the Mauvoisin dam. The discharge is influenced by storage and diversion of water. Therefore, it is often necessary that a forecast-model for discharge takes somehow into account the operation of the reservoirs.

Rivers in the lower Alps react very fast to intensive precipitation. Due to the high slope of the rivers and to their relatively short courses,

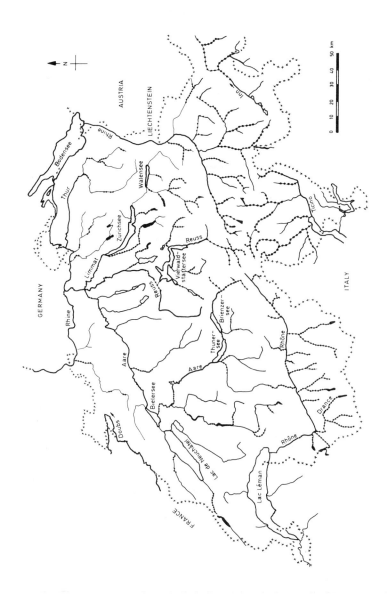

Figure 1   River reaches with flow regimes considerably affected by
water resources management

DRANCE DE BAGNES, LE CHABLE

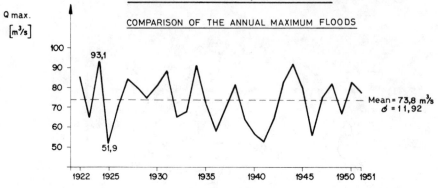

COMPARISON OF THE ANNUAL MAXIMUM FLOODS

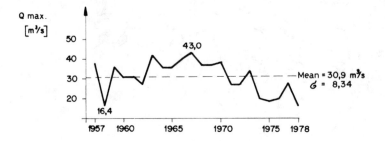

Figure 2   Maximum discharges of the river Drance de Bagne before and
after the commencement of operating the Mauvoisin Reservoir

the resulting running period of the flood wave in the channel is short-
lived. In the case of flood in the upper course, the time for taking the
necessary protection measures in the lower region is short, generally
just a few hours.

The said factors lead to difficulties in the application of rainfall-
runoff models. Although Switzerland possess' a very efficient meteorological network and its National Hydrological Survey has a dense hydro-
metrical observation network for surface and groundwater, the gauging
dates are rarely sufficient for calibrating a detailed deterministic model. Small, for hydrological analysis well equipped catchments are an exception. Insufficient observations as well as the inaccuracy of the measurements occuring in the case of extremly high floods, lead to errors in
the discharge forecasting, especially in the mountainous regions. Thus,
investigations have showed clearly that forecasts have been inaccurate
in those cases, where it has been impossible to take into consideration
the exact quantity of areal precipitation. The problem in determining
the precipitation is the great variation of rainfall from one area to
the other and the great altitude differences in the catchment, with more
rainfall in the higher altitude.

Comparative investigations (Naef, 1977) have been made with rainfall-
runoff models within three small catchments having a surface of 1,6 sq.

km, 10 sq. km, 120 sq. km. Complete conceptual models (Stanford Watershed Model IV, Sacramento River Forecast Center Hydrologic Model, Constrained Linear System) , but also partial models (effective rainfall-
runoff models, infiltration models, etc.), have been tested. The comparison between the three catchments proves that the results obtained with
complicated models are not substantially better than the one with a simple effective rainfall-runoff model for one linear channel and one linear reservoir. All tested models have satisfactorily simulated middle
floods, but in the case of extreme floods the errors were more important.
The main reason is believed to be the missing knowledge concerning the
details of the distribution of infiltration and rainfall, related to
space and time. It is also interesting to mention that the best model
has been for every catchment a different one. The extension and the
type of the catchment have an influence on the choise of the suitable
model. Therefore, it remains questionable if a model studied for a small
catchment is convenient to a larger one.

In the mountainous basins of Switzerland water is stored in the form of
snow and ice. The accumulation of snow during the winter months provides
an opportunity to forecast the resulting runoff. Therefore, various
snowmelt-runoff models have been developed for discharge forecasting.
Martinec (1975) e.g. has developed a simple model, which takes into
account the variability of the degree-day factor, the recession coefficient and the snow coverage. Therefore, only input data of practical
interest, as they can be obtained or at least extrapolated in hydrological catchments, are used. The model allows forecasting of the snow coverage's daily discharge.

Because the complicated conditions in larger catchments in Switzerland
make practically impossible the utilization of detailed deterministic
models, lumped system models are used for the long-term and for the
short-term discharge forcastings. Further, as it is described later, discharge-discharge models are used in the flood warning systems.

LONG-TERM AND SHORT-TERM OPERATIONAL DISCHARGE
FORECASTS OF THE RIVER RHINE

Since 1954 the Laboratory of Hydraulics, Hydrology and Glaciology at the
Federal Institute of Technology, Zurich, has established long-term forecasts for a period of one or several months; also short-term forecasts
about the Rhine discharge at Rheinfelden have been established for one
to three days (Vischer and Jensen, 1978). The basin area is 34000 km^2
and the mean annual discharge at Rheinfelden is about 1000 m^3/s.

The forecasts are based on regression analysis. The summer discharge at
Rheinfelden being mainly determined through snow and ice stored during
the winter, the water equivalent of the snow coverage appears to be an
important variable. As an example, the equation for the discharge forecast in June, which will be made on June 1st is as follows:

$$\delta^{(A_{June, 50\%})} = U_{50} = a_0 + a_1 J_{May\ 31st} + a_2 W_{(1)\ March\ 31st}$$

$$+ a_3 W_{(2)\ March\ 31st} + a_4 W_{(3)\ March\ 31st}$$

$$+ a_5 N_{\text{April-May}} + a_6 A_{\text{April - May}}$$

$$A_{\text{June, 50 \%}} = \delta^{-1} (U_{50}) \qquad A_{\text{June, 10 \%}} = \delta^{-1} (U_{50} + s_u t_{10})$$

$$A_{\text{June, 90 \%}} = \delta^{-1} (U_{50} - s_u t_{90})$$

δ	A transformation, introduced for the purpose of discharges; it leaves the standard deviation of a prediction error independent of forecasts' value (δ^{-1} means back transformation).
s_u	Standard deviation of error
t_{10}	10 % quantile of t (Student distribution)
$A_{\text{June, 50 \%}}$	Forecasted 50 % value of natural discharge in June (natural discharge = gauged discharge + volume variation of reservoirs)
$J_{\text{May 31st}}$	Total of storage capacity of the lakes in the hydrological basin, per May 31st
$W_{(1) \text{March 31st}}$	Water equivalent of the snow at the highest station (2540 m above sea-level) on March 31st
$W_{(2) \text{March 31st}}$	Water equivalent - average at 4 reprensentative stations
$W_{(3) \text{March 31st}}$	Water equivalent - average at 7 other stations
$N_{\text{April-May}}$	Precipitations at 32 stations in April and May
$A_{\text{April-May}}$	Natural discharge of the River Rhine at Rheinfelden in April and May
$a_0, \ldots, a_6$	Regression coefficients

For the short-time forecasting of discharges in various Swiss rivers exist efficient models. In spite of the difficulties mentioned these models supply in many cases good results. But if the rainfall-forecast, as an important input parameter, is inaccurate, there are sometimes important differences between the forecast and the real discharge.

FLOOD-WARNING SYSTEMS

Because of the difficulties mentioned, several water resources problems which under optimal conditions can be solved by means of rainfall-runoff models, must be settled nowadays with the help of other procedures.

Flood-control along Swiss rivers is settled on the one side through structural protection (water course corrections, reservoirs, etc.) and on the other side through sophisticated flood-warning systems. For flood-warning we frequently renounce on forecasting by means of rainfall-runoff models. As a matter of fact, the advantage of such a type of forecasting is cancelled by the technical and the psychological problems resulting by the uncertainty of the forecast. The warning system used by the Swiss national Hydrological Survey is based on gauging of water-level and discharge at selected points and on the use of flood-routing models. When the measured water-level exceeds a certain limit, flood-warning will be given by the help of automatic teletransmitters. The warning is transmitted on a strictly prescribed way to all people interested. The responsible authority has the possibility to watch the development of the discharge in the upper course of the river by means of additional teletransmitters for water-level control, to calculate the progression of the flood in the lower course of the river by means of flood-routing models and to observe the discharge of other rivers in the same basin by means of teletransmitters for water-level control.

The Swiss National Hydrological Survey has 6 teletransmitters for flood-warning and 46 teletransmitters for water-level control in operation. It further may use 11 teletransmitters belonging to independent organizations (see Figure 3).

A teletransmitter for water-level control is a passive instrument because it only gives information when asked for. Each teletransmitter is attached to a telephon extension, consequently it can be called up over telephon network. The delivered informations about the water-levels are coded in acoustic signals. The first series of signals, following the ringing tone, are delivering the number of meters, the second one the number of decimeters and the third one the number of centimeters. Each series is separated from the next one by a long, penetrating sound. When the third series is finished, a group of short sounds are resounding. Afterwards the transmitted water-levels are repeated twice. The figure 0 is coded with 10 signals.

A water-level of 6.43 m e.g. is given from the teletransmitter by 6 short signals, 1 long signal, 4 short signals, 1 long signal and 3 short signals.

There are two important reasons for having chosen such a code: it is easier for the technical receiver to treat impulses than written language and also Switzerland being a country with 4 different languages, a code is better and quicker understood by all interested persons.

It is possible to store the measures of the water-levels in the gauging station for a future use. Also a special instrument enables automatic questioning of the water-level informations.

On the contrary to the one for water-level, the teletransmitter for flood-warning is an active instrument. When the water-level reaches the previously fixed critical points important for the flood- and catastrophe-warning, the teletransmitter automatically phones up the responsible authority. Thus all the telephon-numbers of the persons which have to be warned must be known at the time of teletransmitter-programming. The teletransmitter for flood-warning calls the people interested in a given

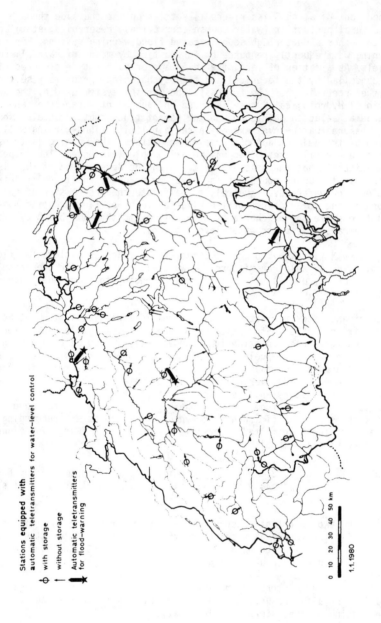

Figure 3. Automatic teletransmitters for water-level control and for
flood-warning in stations of the Swiss National Hydrological
Survey

order. When a connection is established, the teletransmitter sounds the
warning and the receiver must give a receipt for the warning. Then the
teletransmitter calls the next one.When a connection cannot be established at the first call, the transmitter calls the subscriber later
again, four times at most.

During continuous water-level rising, flood- or catastrophe-warning is
released once only. A second warning is released if the water-level
reaches the flood-level again, after having gone down to a lower water-
level, previously determinated.

Great attention is given to the reliability in operation of the flood-
warning system. The reliability is guaranteed through periodical servi-
cing and control, achieved by the observers or the specialized staff of
the Hydrological Survey.Test-warnings are also performed by collaborators
of the Hydrological Survey. On the other side, the reliability of the
system is guaranteed by automatic control of the installations. In case
of failure in the mechanism, a disturbance-warning is released to ·the
house of the observer. He takes immediately charge of the function of
the failing transmitter and calls if necessary the subscribers. He tries
to repair the failure, according to special instructions. If he doesn't
succeed, the repair is done by a specialist of the Hydrological Survey.

The answer given by the persons receiving the warning (flood-, catastro-
phe- or failure-warning) is registered, as well as the time, so that it
is possible to control quickly later the transmission of the warning.

In Switzerland, priority is given to flood-warning. The other extreme
event, the low flow, is e.g. important for navigation. Thus, the gauging
station Rheinfelden on the Rhine River releases flood-warning as well as
low flow-warning.

Due to the relatively steep bed slope, flood waves show in a lot of Swiss
rivers the same behaviour as kinematic waves. This means that the changes
in water depth($\partial h/\partial x$) and the terms of acceleration
$(\frac{v}{g} \frac{\partial v}{\partial x} , \frac{1}{g} \frac{\partial v}{\partial t})$ in the equations of Saint Venant

$$\frac{\partial F}{\partial t} + v \frac{\partial F}{\partial x} + F \frac{\partial v}{\partial x} = 0$$

$$\frac{\partial h}{\partial x} + \frac{1}{g} \frac{\partial v}{\partial t} + \frac{v}{g} \frac{\partial v}{\partial x} = Js - Jr$$

F : Cross-sectional area of the flow in the channel
t : Time
v : Mean velocity in the cross-sectional area
x : Distance along the channel
h : Elevation of water surface
g : Acceleration of gravity
J_s: Bed slope
J_r: Roughness width

are negligible compared with the bed slope Js. For that reason the equation of continuity is able to describe the routing of flood waves.

For many Swiss river reaches, good results are obtained in the determination of flood events by using flood-routing models, e.g. with simple

hydrological methods such as Kalinin-Miljukov or Muskingum (Spreafico, 1972). As an example, the measured and the calculated (by the Kalinin-Miljukov method) flood hydrograph of the River Thur between Halden and Andelfingen (47,6 km) are shown in Figure 4.

The routing of a flood wave can therefore be calculated easily downstream of the flood-warning station. Another advantage of that type of warning-system is that it also works by floods produced artificially. Such an event can be produced by the rupture of a dam or the outburst of an ice-dammed marginal glacier.

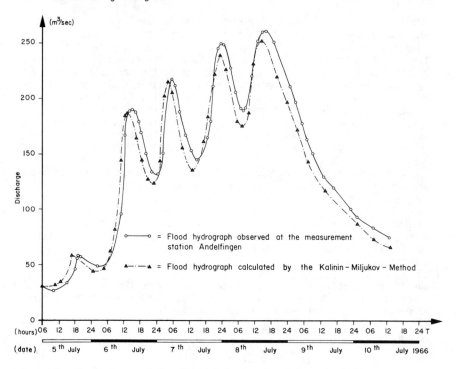

Figure 4   Comparison of the flood hydrographs observed at the measurement Station Andelfingen of the river Thur and calculated with the Kalinin-Miljukov-method

REGULATION OF LAKES

Two third of the greater lakes in Switzerland are regulated, some of them already for centuries. Their regulation is based upon a regulation plan, which fixes the outflow in relation to the water level of the lake. Until now the main object of the regulation was flood control, but actually the lake regulation plan is becoming more complex and the areas being investigated (water supply, power generation, navigation, recreation, fishery and environmental impacts) are growing. A regulation plan that fully satisfies all interests involved cannot be found. Therefore the task is to devise a regulation plan which offers the maximum benefit to the system of water resources as a whole, while ensuring that none of the concerned parties is subjected to unacceptable limitations.

The storage equation of a lake may be written as follows:

$$\Delta V = (N + QZ_o + QZ_u - VE - QA_o - QA_u) \, \Delta t$$

ΔV = Change in lake storage
N = Precipitation on the lake
QZ_o = Surface water contribution
QZ_u = Groundwater contribution
VE = Evaporation from the lake
QA_o = Outflow from the lake
QA_u = Groundwater outflow

When the lake's inflow is known or when long-term forecasts of the inflows have been estimated through rainfall-runoff models, the water-level
of the lake can be influenced by means of outflow regulation, accordingly to the requirements of the riparians. Unfortunately such accurate
long-term forecasts of inflows for our lakes are at the moment inopera-
ble. Therefore the following procedure is applied:

Considering results obtained from past events (change of water-level, inflow and outflow), a long-term regulation plan is calculated by using
deterministic simulation models, deterministic models with mathematical
programming algorithm, queuing models, stochastic models with mathematical programming algorithm or stochastic simulation models (Spreafico,
1977 and Spreafico,1979).

The long-term regulation plan is very often completed by a short-term
regulation (see Figure 5).

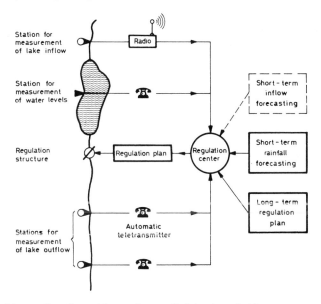

Figure 5   Operation scheme of lake regulation

The authority responsible for the regulation has the opportunity, in the case of an unusual meteorological and hydrological situation of deviating during a short period from the long-term regulation plan. For such a proceeding it uses short-term rainfall forecasts and the latest information about inflow, outflow and water-level of the lake. If the water-level of the lake is temporary high and if a larger inflow can be expected, the responsible person may preventively lower the water-level. Thus he creates an increased volume of retention.

Such measures require great professional training and much practical experience. Short-time forecasting of the expected inflow could improve the situation.

REFERENCES

Martinec, J. 1975. Snowmelt-runoff model for stream flow forecasts. Nordic Hydrology, Vol. 6, pp. 145-154

Naef, F. 1977. Ein Vergleich von mathematischen Niederschlag-Abfluss-Modellen. Mitteilungen der Versuchsanstalt für Wasserbau, Hydrologie und Glaziologie, Nr. 26, 114pp., ETH-Zentrum, Zürich

Spreafico, M. 1972. Anwendung von hydrologischen Verfahren zur Berechnung des Ablaufs von Hochwasserereignissen in der Thur. Mitteilungen des Institutes für Hydromechanik und Wasserwirtschaft, Nr. 4, 86 pp., ETH-Hönggerberg, Zürich

Spreafico, M. 1977. Verfahren zur optimalen Regulierung natürlicher Seen. Mitteilungen der Versuchsanstalt für Wasserbau, Hydrologie und Glaziologie, Nr. 25, 211 pp., ETH-Zentrum, Zürich

Spreafico, M. 1979. Methods of developing regulation plans for lakes. in: Proceedings of the III. World Congress on Water Resources of the IWRA in Mexico, Vol. 4, pp. 1574-1583

Vischer, D., Jensen, H. 1978. Long range forecast of the river Rhine discharges at Rheinfelden. Wasserwirtschaft, 68. Jahrgang, Heft 9, pp. 259-264

Section 4
APPLICATION IN URBAN ENVIRONMENT

MATHEMATICAL MODELING IN URBAN HYDROLOGY

J. W. Delleur
School of Civil Engineering
Purdue University
West Lafayette, Indiana 47907, U.S.A.

ABSTRACT

The Clean Water Act of 1977 recognized the need to quantify the pollution from nonpoint sources. As a result the U. S. Environmental Agency developed the Nationwide Urban Runoff Program for the period 1978-1983 to assess the nature, cause and severity of urban runoff problems and the possible means of controling them. The USGS is cooperating with EPA in the data collection, data storage and retrieval, model calibration and refinement. The program is expected to provide an enhanced understanding of the physical and chemical processes involved in urban drainage and to test various methods of urban runoff management.

For the modeling of urban runoff the use of the rational formula is rapidly decreasing. Hydrograph applications have received renewed attention due, in part, to the fact that hydrograph identification and convolution can be performed very simply on small programmable calculators. Several synthetic unit hydrographs and instantaneous unit hydrographs have recently been proposed.

The application of single rainfall event models usually requires the formulation of a design storm. This may be either in the form of a uniform intensity for a given duration and recurrent interval or the development of a synthetic hyetograph. The design storm method has some shortcomings, particularly for the determination of nonpoint source pollutant loads and concentrations in streams. Recent research shows that the design storm and antecedent soil moisture condition have important effects on the peak flow-frequency curve.

One of the simpler techniques for estimating the initial and infiltration losses is the Soil Conservation Service Curve Number Method. Better results could be obtained if the method was refined. One of the more sophisticated methods of soil moisture accounting and infiltration estimation is used in the parametric-deterministic ADS (Alley-Dawdy-Schaake) model of the USGS.

Recent improvements of some of the large scale models such as SWMM, STORM, and new versions of ILLUDAS are presented. The ADS model and SMADA (Stormwater Management and Design Aid), are discussed briefly. The trend is toward more attention to the selection of the objective functions which are used with the runoff models and to the parameter optimization.

Among the methods for obtaining the least cost drainage systems, one approach makes use of the discrete differential dynamic programming (DDDP) as the basic search technique. An alternative method consists of coupling a hydrologic model such as ILLUDAS with a dynamic programming subroutine.

Detention storage plays an important role in the strategies to control urban runoff and combined sewer overflows. Mathematical models have been developed for the movement, decay, storage, and treatment of stormwater runoff pollutants and dry weather wastewater flows through storage/treatment systems. These models may then be used in conjunction with models of the receiving water quality that simulate the effects of upstream sources in combination with dry and wet weather urban sources.

INTRODUCTION

Many mathematical models in urban hydrology were developed during 1967-1974, the period called "tool making" by McPherson (1980) and followed by the period of "tool wielding." The models developed during the initial period and their improvements through 1978 were reviewed by Delleur and Dendrou (1980). This paper is intended to update that review, emphasizing recent developments through 1980. Another extensive review of urban hydrology was prepared by McPherson (1979a) which summarized the progress made in the United States during the period of 1975-1978, and was part of the U. S. National Report to the seventeenth General Assembly of the International Union of Geodesy and Geophysics.

INTERNATIONAL PERSPECTIVE

Twelve national reports on urban hydrological modeling and catchment research in the U.S.A., Australia, Canada, the United Kingdom, the U.S.S.R., the Federal Republic of Germany, Sweden, France, Norway, the Netherlands, Poland and India, were prepared under the International Hydrological Programme (IHP) and were published in three UNESCO reports (UNESCO 1977, 1978).

These reports (with the exception of the U.S.S.R. and Poland) were updated in 1979 for the International Symposium on Urban Hydrology held as part of the Spring Annual Meeting of the American Geophysical Union. The papers were collected in a report edited by McPherson (1979b).

Nearly all nations reported an increased use of urban runoff mathematical models which are being employed by practioners at large and are no longer limited to a few researchers.

The recent developments in the United Kingdom are of significance. The U.K. Meteorological office has a program of automatic digitization of rainfall records which makes it possible to obtain rainfall depths at 1 minute intervals (Folland and Colgate, 1978). These rainfall data can be used as input in the storm drainage package recently developed by the Hydraulics Research Station and the Institute of Hydrology (Hydraulics Research Station, 1980).

UNITED STATES PERSPECTIVE

Many of the recent developments in urban hydrology in the United States were influenced by legislation of the U. S. Congress which focused attention on environmental problems. The Federal Water Pollution Control Act of 1972, Public Law 92-500, under Section 208, mandated area-wide planning for water pollution abatement and management in all metropolitan areas. Large grants were authorized by the U. S. Environmental Protection Agency for the construction of sewage treatment plants which were to attain secondary treatment level of "point sources" of pollution of receiving streams. These point-sources included municipal and industrial wastewaters. Further reduction and abatement of "nonpoint sources" were also mandated. These include runoff from agricultural areas, urban runoff discharged directly or as combined sewer overflows. As the pollution from point sources was gradually controlled the relative importance of nonpoint sources increased.

Public Law 95-217, known as the Clean Water Act of 1977, made extensive amendments to the previous legislation. It recognized the difficulty in quantifying pollution from nonpoint sources. Federal grants for stormwater projects were deferred until a better understanding of the nature of urban nonpoint pollution could be obtained.

Nationwide Urban Runoff Program

In response to the Clean Water Act, the U. S. Environmental Protection Agency has developed the Nationwide Urban Runoff Program. During the period 1978-1983 it will assess the nature, causes and severity of urban runoff problems and the means of controlling the nonpoint source of pollution from urban areas. Auxiliary objectives are: 1) the development of an information base to identify, assess, and implement effective controls; 2) the examination of current dry weather water quality standards for judging the significance of storm-dominated pollution problems. There are about 30 intensive projects nationwide which cover a wide range of climatic and hydrologic regimes. By mid 1980 several of the projects were operational, and by the end of the year most projects were in a position to collect and report field data.

The U. S. Geological Survey is cooperating in the data acquisition, data storage and analysis of many of these projects. A data management system was developed to store, retrieve, update and to interface data files with statistical procedures and rainfall-runoff-quality models. Meteorologic data include precipitation, pan evaporation, snowfall, air temperature, solar radiation and wind velocity measurements. Discharge or stage is recorded synchronously with rainfall. Automated sampling equipment is used for water quality monitoring. The urban hydrology monitoring system incorporates a system control based on a microprocessor to acquire and record data on site, and to control the automated water quality sampling device. Analysis procedures are specified by Friedman (1978). A cone splitter device was developed to obtain accurate and representative subsamples. The water quality constitutents include suspended sediments, organic and inorganic chemical constituents, and bacteriological parameters. One atmospheric deposition collector is recommended for each watershed. Basin characteristics, storm event characteristics, environmental practices are recorded for each site. The data are stored and updated in the USGS WATSTORE system. The data are periodically transferred to EPA's STORET system. An intermediate data system based on the Statistical Analysis System (SAS) is used to interface data from the daily values file, unit-values files, and water quality files of WATSTORE and to perform statistical analyses (Doyle and Lorens, 1981).

U. S. Geological Survey Urban Hydrology Studies

Recent urban hydrology studies have been performed by the U. S. Geological Survey in Miami, Florida, Portland, Oregon (Miller and McKenzie, 1978), and Denver, Colorado. In the Miami area, four sites were used with the following homogeneous land uses: a single-family residential, a multifamily residential, commercial and a high traffic highway site. Rainfall, runoff quantity and quality, and bulk precipitation quality data were collected (Hardee, 1979). Regression analysis showed that the storm loads from single family residential areas are strongly correlated to the peak discharge and the length of antecedent dry periods. The loads from the highway area correlated with rainfall and that from the commercial area correlated with peak discharge and rainfall (Miller et al., 1978).

ELEMENTARY RAINFALL-RUNOFF MODELS

Rational Formula

The rational formula is still used by some for small urban areas. Rossmiller (1980) recently reviewed the assumptions and misconceptions associated with the rational formula. He proposes a new formula to estimate the runoff coefficient in terms of the soil curve number (as defined by the U. S. Soil Conservation Service), the recurrence interval of the storm, the slope of the watershed, the rainfall intensity and the watershed imperviousness. He also reviews the several methods of estimating the time of concentration and recommends Kerby's formula or the SCS formulas for the estimation of the overland flow time. The use of the rational formula is rapidly decreasing because of its inability to yield the runoff hydrograph which is needed for the control of runoff quantity and quality. In addition Burke and Gray (1980) in a comparison of the rational method and of a more advanced model (ILLUDAS) show that although a drainage system designed according to the rational method for a N-year storm might be adequate for long duration storms, it would fail for shorter ones, thus not providing the degree of protection specified.

Unit Hydrographs

Hydrograph techniques are advantageous in those cases where the time distribution of the runoff is needed only at a small number of points in the urban watershed. The watershed characteristics are lumped in a system generally assumed be linear, and the mathematical description of detailed conveyance of the flow is avoided.

Hydrograph applications have recently received attention due, in part, to the fact that hydrograph identification and convolution can easily be performed on small programmable calculators and on small computers. Croley (1980) has published a series of programs for small programmable calculators including those for the Izzard overland flow hydrograph, the kinematic wave overland flow hydrograph, the linear reservoir cascade unit hydrograph, synthetic and dimensionless hydrographs, numerical convolution and hydrograph transformation by S-curve technique. These and other calculations on open channel flow such as routing and surface profiles techniques and pipe flow hydraulics have also been programmed in BASIC for small computers (Croley, 1981).

Conceptual Models

The system which transforms the effective rainfall into direct runoff is conveniently represented by conceptual models. The simpler conceptual models are the single linear reservoir and the cascade of identical linear reservoirs called the Nash model. The single linear reservoir has been used in the past by Rao et al. (1972) for urban watersheds and more recently by Pedersen et al. (1980). Diskin (1978) proposed the use of two parallel cascades of linear reservoirs, one representing the pervious area of the watershed, the other representing the impervious area which is directly connected to the drainage system. To obtain the rainfall excess, Diskin first substracted an initial abstraction D, and the infiltration losses were estimated using the phi-index method with ϕ_A and ϕ_B for the impervious and pervious areas, respectively. The other parameters of the model are α, the fraction of the watershed that is impervious, the number

of linear reservoirs, N_A and N_B with storage coefficients K_A and K_B representing the flow contributions from the pervious and impervious areas, res-
prectively. In the application of the model, Diskin used $\phi_A = 0$, ϕ_B was
adjusted to obtain the correct volume of runoff, the number of reservoirs
were set as $N_A = 2$ and $N_B = 3$ to simulate the faster response of the impervious areas while K_A, K_B and α were optimized. The model has the distinct
merit that increase in imperviousness due to urbanization will increase the
value of α, while construction and improvement of the sewer and conveyance
system will primarily affect the value of K_A.

The linear reservoir $S = KQ_o$ (where S = storage, Q_o = outflow rate,
and K is a constant) which is a building block of the conceptual models is
a particular case of the Muskingum storage equation

$$S(t) = K[rQ_i(t) + (1-r)Q_o(t)] \tag{1}$$

where Q_i is the inflow rate, r is a constant, t is time. The quantity
KQ_o is usually called the prism storage and $Kr(Q_i-Q_o)$ is the wedge
storage. When this equation is replaced in the discrete continuity equation (applied to a time increment Δt)

$$\frac{S(t+\Delta t) - S(t)}{\Delta t} = \frac{Q_i(t)+Q_i(t+\Delta t)}{2} - \frac{Q_o(t)+Q_o(t+\Delta t)}{2} \tag{2}$$

the usual Muskingum routing equation is obtained

$$\tag{3}$$
$$Q_o(t+\Delta t) = C_1 Q_i(t) + C_2 Q_i(t+\Delta t) + C_3 Q_o(t)$$

with $C_1 = \dfrac{2Kr+\Delta t}{2K(1-r)+\Delta t}$; $C_2 = \dfrac{\Delta t-2Kr}{2K(1-r)+\Delta t}$; $C_3 = \dfrac{2K(1-r)-\Delta t}{2K(1-r)+\Delta t}$;

$$\tag{4}$$

$$C_1 + C_2 + C_3 = 1$$

Obviously for the linear reservoir $S = KQ_o$ corresponds to the case $r = 0$
and the routing coefficient become

$$C_1 = C_2 = \frac{\Delta t}{2K+\Delta t} , \quad C_3 = \frac{2K-\Delta t}{2K+\Delta t} ; \quad C_1 + C_2 + C_3 = 1 \tag{5}$$

The Postulate of the Kalinin-Miljukov model $S(t) = KQ_i(t)$ corresponds to
the case $r = 1$, then the routing coefficient becomes

$$C_1 = \frac{2K+\Delta t}{\Delta t} , \quad C_2 = \frac{\Delta t-2K}{\Delta t} , \quad C_3 = -1 ; \quad C_1 + C_2 + C_3 = 1$$

As the Muskingum routing equation can be shown to be a particular
simplification of the complete St. Venant equations for free-surface
unsteady flow, the macroscopic conceptual models of urban hydrology may
be related to the microscopic general dynamic flow equations, and a framework for generalization of these models can be established.

Thiebault et al. (1980) consider a reach of length Δx in which the
geometric characteristics of the flow are constant and let

$$r(y,t) = \frac{A(y,t)-A_o(t)}{A_i(t)-A_o(t)} \quad , \quad A_i(t) \neq A_o(t)$$

$$A(y,t) = A_i(t) \quad , \quad A_i(t) = A_o(t)$$

(6)

where A is the flow cross section, y is the depth, the subscript i and o represent the inflow and outflow sections, respectively. They further consider the points k in Δx where the cross sectional area is the average area of the reach at any one time:

$$A(k,t) = \frac{1}{\Delta x} \int_i^o A(x,t)dx$$

(7)

For these points

$$r(k,t) = r \quad \text{and} \quad \frac{\partial r(k,t)}{\partial t} = 0$$

(8)

so that the differential continuity equation (the first of the two St. Venant equations) becomes, according to Thiebault et al.,:

$$r\frac{dA_i(t)}{dt} + (1-r)\frac{dA_o(t)}{dt} + \frac{A_o(t)V_o(t) - A_i(t)V_i(t)}{\Delta x} = 0$$

(9)

where V_i, V_o and A_i, A_o are the velocities and the flow areas at the inflow and outflow sections of the reach Δx. Given the upstream boundary conditions that fix V_i and A_i, and a downstream rating curve relating V_o and A_o, then (9) can be solved for $A_o(t)$. The second St. Venant equation (the momentum equation) is no longer needed as it is replaced by these boundary conditions. For the case when r = 0, and with the assumption

$$V_i(t) = V_o(t) = V$$

(10)

equation (9) is integrated as

$$Q_o(t) = \frac{V}{\Delta x} e^{-Vt/\Delta x} * Q_i(t)$$

$$= \frac{1}{K} e^{-t/K} * Q_i(t)$$

(11)

where * indicates a convolution. The term $\frac{1}{K} e^{-t/K}$ is recognized as the instantaneous hydrograph or impulsive response of linear reservoir. (See for example Chow 1964, p. 14-27). With a lesser restriction that $V_i(t) = V_o(t) = V(t)$, the numerical integration of (9) gives, according to Thiebault et al. the following coefficients in equation (3):

$$C_1 = 1 - \frac{e^{-\lambda}}{1-r}$$

$$C_2 = \frac{r}{r-1} \frac{V_i(t+\Delta t)}{V_i(t)} e^{-\lambda}$$

(12)

404

$$C_3 = \frac{V_i(t+\Delta t)}{V_i(t)} \, e^{-\lambda}$$

with

$$\lambda = \frac{V_i(t+\Delta t)}{\Delta x(1-r)\Delta t}$$

The conceptual models used in urban hydrology may thus be traced to the St. Venant flow equations through assumptions of spatial and temporal homogeneities (eq. 8 and 10) which appear reasonable for approximately steady uniform flow, that is when the slope of the energy grade line and that of the bottom surface are approximately equal. Further generalization of the conceptual models appear possible through equation (12).

Using the kinematic wave approximation to the St. Venant equation, Chan and Bras developed the frequency distribution of the volume of water above a given threshold discharge. They used the joint probability density function of rainfall intensity and duration to obtain a closed form expression for the annual exceedance series of flood volumes. This provides a fast and reliable aid for the analysis of flood control measures in urban areas.

Design Storm Concept

The application of the unit hydrograph and of single event rainfall-runoff models usually requires the formulation of a design storm. The design storm is typically chosen with a prescribed return period which is assumed to be equal to the desired return period for the runoff peak discharge used in sizing an urban drainage system.

There are two principal types of design storms: 1) the uniform intensity storm for a given duration and frequency obtained from the intensity-duration-frequency relationships. This type is commonly used in conjunction with the rational formula; 2) the synthetic storm profiles such as the quartile hyetographs developed by the Illinois State Water Survey, the Chicago hyetograph, the U. S. Soil Conservation type II hyetograph, the summer storm profiles from the U. K. Flood Studies Report, or the dimensionless triangular hyetograph recently proposed by Yen and Chow (1980).

The fundamental assumption that the recurrence interval of the design storm can be transferred to the runoff flowrate or volume has not been fully investigated in a broad scale, in spite of the fact that the establishment of the limits of validity of this concept of is practical engineering importance. The areal variability of rainfall is typically ignored in the application of design storms. Obviously the design storms do not yield the probability information such as flow duration curves that may be needed for planning purposes. Specifically for storm water management procedures which involve storage and treatment of the runoff, the probability distribution of the outflows and overflows become a function of the storage capacity and treatment rate. As a result the design storms are not applicable to the determination of nonpoint source pollutant loads and pollutant concentrations in receiving streams (Delleur, 1979).

The widespread use of the design storm is due to the fact that it greatly reduces the complexity of stormwater runoff analysis. It usually gives conservative results for small impervious basins and is usually considered an acceptable practice, occasionally mandated by some ordinances. Nevertheless, the temporal distribution is quite different from the rainstorms occurring in nature. The use of design storms may result in runoff peaks and of different frequency and magnitude, thus resulting in flood control facilities that may be over or under designed.

The main alternative to the use of the design storm is continuous simulation. The principal limitations of this approach are cost of long computer runs, the availability of long precipitation series for the study area and the availability of a continuous model which simulates the situation under study.

Marsalek (1978) compared discharge and volume probability relationships obtained from computer simulation with historic storms to those obtained with the Chicago and Illinois State Water Survey design storms. The two approaches produced significantly different results and the use of historic storms rather than a design storm was recommended.

Because in many cases the cost and complexity of continuous simulation cannot be justified many efforts are underway to improve or extend the design storm concept. Walesh et al. (1979) present a technique of historical storms in which the major storms from the historic precipitation record are screened, then processed through an event model to obtain simulated hydrographs of direct runoff. These hydrographs are then used in a discharge and volume probability analysis to obtain the discharge-probability and volume-probability relationship. This technique takes advantage of the low cost of event models while eliminating the need to select a design storm. The results of this type of analysis may then be compared with the results of several design storms to select an appropriate design storm.

Wenzel and Voorhees (1978) used a similar approach with the ILLUDAS model, modified to permit continuous simulation, applied to the Boneyard Creek in Champaign-Urbana, Illinois. The results were compared with those obtained with two design storms: a uniform distribution and the distribution developed by the Illinois State Water Survey. It was concluded that the design storm hyetograph and the antecedent soil moisture condition are very important parameters and have similar effects on the peak flow-frequency curve. The Illinois State Water Survey distribution design storm with dry antecedent soil moisture conditions produced a frequency curve which is close to the historical storm curve for the total catchment, but this was not the case for the subcatchments. This illustrates the difficulties encountered in making general statements about the design storm.

Wenzel and Voorhees (1979) also tested the Grey Haven Catchment in Baltimore, Maryland, subject to a 25 year rainfall series observed at Coshocton, Ohio. The model used was the USGS model of Alley et al. (1980). Three design storms were used: triangular, Illinois State Water Survey and uniform with dry and wet antecedent soil moisture conditions. The frequency curves obtained with Illinois State Water Survey design storm and the dry antecedent soil moisture conditions agreed well with the frequency curves obtained with the historical peaks. However, changes in the hyetograph and in the antecedent soil moisture can shift the design frequency curve over a wide range.

As part of a comprehensive study reported by Urbonas (1979), the Urban Drainage and Flood Control District serving metropolitan Denver, Colorado, has developed design storms for drainage areas with 35, 40, 45, and 97% imperviousness. The respective drainage system consists of streets and grass channel; streets and concrete channel; streets and pipes and large pipes. The determination of the design storm required

substantial rainfall-runoff data for the calibration of the models,
long term simulation of the runoff and statistical analysis of the simulated peaks and volumes. Because of the semi-arid climate in the Denver
region, the antecedent moisture condition was not a variable in the
determination of the design storm.

Packman and Kidd (1980) in the United Kingdom used a sensitivity
analysis to find the antecedent conditions and design storm that consistently give flows which match an observed flood frequency distribution. The rainfall duration of 15, 30, 60, or 120 min. is chosen as that
which gives the largest peak. The depth is defined from the local depth-
duration frequency relationships and the rainfall profile used is the
50% summer profile which can be expected to be exceeded in terms of peaked-
ness 50% of the time. For the English climate the optimal antecedent
moisture condition called the Urban Catchment Wetness Index was found
from a sensitivity analysis and was correlated to the standard average
annual rainfall. The T-year design storm so obtained is applied to the
Wallingford model to obtain the T-year flood. This model incorporates a
regression equation yielding the percentage runoff from the percentage
imperviousness, the soil index and the wetness index.

If design storms are to be used, it appears desirable to develop
them for the given region. It appears, however, that the amount of
work necessary to develop the design storm is such that the task should
be undertaken by the appropriate government agency. Once the design
storms are available, the analysis of drainage systems are greatly
simplified.

Initial and Infiltration Losses

Several methods of estimation of initial and infiltration losses
are used in urban hydrology. The runoff coefficient and the ϕ-index
methods are often used in conjunction with the rational formula and unit
hydrographs. The U. S. Soil Conservation Service Curve Number is often
used, in particular with the Soil Conservation Service (1975) method
for small urban watersheds. The Curve number is based on the hydrologic
soil classification and on land use, and is adjusted for antecedent precipitation and season of the year. An antecedent moisture condition
index is also used in the ILLUDAS model.

It must be noted that by definition the curve number is not dimensionless as

$$CN = \frac{1000}{10+S} \qquad\qquad (13)$$

where S is the water storage in the mantle in inches. As a result in
metric units, if CN is defined as

$$CN = \frac{25400}{254+S'} \qquad\qquad (14)$$

where S' is the storage in mm, it retains its scale of 100 to 0 corresponding to zero to infinite soil storage.

The accumulated rainfall excess Q is then given in terms of the
accumulated precipitation P, and the initial abstraction I_a, all in
inches by

$$Q = \frac{(P-I_a)^2}{(P-I_a)+S} \quad ; \quad P \geq I_a \tag{15}$$

Hjelmfelt (1980) has indicated some of the shortcomings of the CN method to estimate infiltration. In its differential form the rate of infiltration f_t at time t is

$$f_t = \frac{dF}{dt} = \frac{S^2}{(P-I_a)+S^2} \frac{dP}{dt} \tag{16}$$

The infiltration rate is seen to vary with the rainfall intensity dP/dt, and does not approach a terminal or equilibrium infiltration rate, which is contrary to the Hortonian concept of infiltration. Hjelmfelt shows that equation (13) is equivalent to the Holtan-Overton relation for infiltration when the terminal infiltration is zero. In that case

$$f_t = \left(\frac{S_t}{S}\right)^2 f_o \tag{17}$$

where S_t is the available soil moisture, and S and f_o are the initial soil storage and infiltration rate at t = 0.

Altman and Feldman (1980) have studied some of the limitations of the SCS method applied to urban watersheds. They state that difficulties can be encountered in accurately determining the curve number for an urban area due to the compaction of soil by heavy equipment, inability to estimate variable vegetation conditions, introduction of fill material and mixing of surface and subsurface soil. They also feel that further study is needed to improve the SCS relationships for hydrograph lag time in urban areas since these are based on limited data and analysis. The SCS method using the equation for the time lag adjusted for channel improvement and impervious areas result in the estimation of relatively low peak discharge estimates in urban areas. The frequency curves obtained by this method were relatively low in three out of four watersheds studied. The frequency curves were shown to depend considerably on the method of determination of the time lag.

The application of the curve number method to predict runoff is very sensitive to the antecedent moisture index. Burke and Gray (1979) cite 8 to 18 fold variations in runoff volume (from 23000 ft^3 to 186,000 ft^3 and 422,000 ft^3) with antecedent moisture conditions I, II, and III in an urban watershed which is 30% impervious and has SCS type B soil. Gray and Cogo (1980) obtained the probability distribution of antecedent moisture conditions in Tippecanoe County, Indiana. They used the 3 AMC classes and 2 seasons used in the SCS method, and added a fourth AMC class during the dormant season which corresponds to frozen soil or snow cover. They used 17 years of daily water equivalent precipitation values and daily soil temperature values. During the growing season AMC I is the dominant condition. It can be expected on 88% of the days and AMC II is occurring 7% of the time and is not an "average" condition of the soil. During the dormant season AMC IV occurs 28% of the time and AMC II is occurring 14% of the time. Again AMC II is not the average condition. AMC I occurs 49% of the time in the dormant season and again is the dominant condition. For runoff calculation in the dormant season it would seem prudent to consider the AMC III and IV together since they are associated with reduced infiltration capacity or melting of the snow cover during rain storm. This combination occurs 20% of the time.

As the parameters needed for the SCS model depend on land cover distributions, they can after some modification, be made compatible with the Landsat digital data. A study by Ragan and Jackson (1980) shows that the curve numbers for the modified land use categories obtained from the satellite imagery compared well with those obtained using conventional categories. They state that the improvement in the ability to extract topographic and soil information with Landsat should make the satellite based SCS method increasingly attractive for hydrologic analysis of ungaged watersheds. In a previous paper, Jackson et al. (1977) have discussed the capability of using Landsat imagery in urban hydrologic modeling, particularly in conjunction with the model STORM.

LARGE SCALE MODELS

Many of the large-scale models for urban storm drainage simulation were developed in the 70's. These include MITCAT, SWMM, STORM, HEC-1, ILLUDAS, the Penn State Model, CAREDAS and QQS which were previously reviewed by Delleur and Dendrou (1980). The purpose of this section is to summarize some of the recent improvements and extensions that have been made to these models. Brandstetter (1976) reviewed the objective, advantages and limitations of selected computer models.

The current trend is towards more exact models and towards automatic calibration of the models. Singh (1979) has suggested a systematic evaluation of the models which requires an objective function, goodness of fit criterion, sensitivity analysis, error analysis and comparison of models. Diskin and Simon (1977) have given an extensive analysis of the procedure for the selection of the objective function to be used in the parameter optimization. Han and Rao (1980) considered 17 objective functions and concluded that the sum of the squares of the differences between the observed and calculated hydrograph ordinates gives the best overall performance. Several search algorithms have been used in the parameter optimization of hydrology models. Soroosian (1980) compared Rosenbrock's (1960) search technique and the Nelder and Mead (1965) simplex algorithm. He concluded that Rosenbrock's method was preferable for the estimation of the more sensitive parameters while the simplex algorithm gave better results for the less sensitive parameters.

The single event ILLUDAS model (Illinois Urban Drainage Area Simulator) has been improved and extended. The Illinois State Water Survey expanded the flow routing algorithm in December 1978, and in October 1979 two routing options were added: a hydrograph time shift and a storage routing using the implicit method of solution.

Han and Delleur (1979) performed a sensitivity analysis of the parameters in ILLUDAS. They also developed an extension which includes a subroutine for the calculation of the total rainfall during the five days prior to a storm to determine the antecedent moisture condition at the beginning of each storm. With this modification, the program ILLUDAS may be used for continuous simulation over a long period. The paved and grassed area abstractions are calibrated so as to reproduce the volumes of runoff. The report includes a complete listing of the program.

Wenzel and Voorhees (1980) developed a continuous version which includes a daily soil moisture and initial abstraction scheme coupled with the Horton infiltration model. New routing options were added which include 1) a simple hydrograph time shift, and 2) a reservoir type routing using either one of two relationships between inflow, outflow and storage. A calibration procedure has been added which makes use of Rosenbrock's algorithm to optimize the model parameters in two steps. In the first step all parameters are optimized by matching the runoff volumes, and in a second step the initial abstractions on pervious and impervious areas are optimized by matching hydrograph peaks. A screening procedure is included which eliminates from simulation those events which are unimportant in the analysis. This last feature is of interest in the frequency analysis of flood peaks. Han and Rao (1980) also used Rosenbrock's method to optimize the parameters in the single event version of ILLUDAS. They showed that the optimization procedure clearly improves the regeneration of peakflows and of hydrograph volumes. Because Wenzel and Voorhees (1980) used their continuous simulation version, they optimized 11 parameters while Han and Rao (1980) optimized 6 parameters in the single event version.

An interactive version of Illudas was developed by Patry et al. (1979) for use on minicomputer while most rainfall-runoff models are of the batch process type. This version of ILLUDAS establishes a dialogue with the user. This reduces errors in data input, increases the user's confidence in the model, accelerates the training period and has excellent pedagogic capabilities. This version can handle mutliple rainfall hyetographs, as many as one per reach of the catchment. The infiltration parameters in Horton's equation are specified directly. The kinematic wave equation was introduced to calculate the pervious and impervious surfaces inlet times.

Han and Delleur (1979) have added a runoff quality module to the ILLUDAS model. This version called DRAINQUAL estimates suspended solids and BOD in a manner similar to that used in STORM, except that a shorter time increment of 5 minutes is used. With this small time interval the prediction ability for BOD and SS was superior to that of the model STORM The report includes a complete listing of the program. A version of ILLUDAS called QUAL-ILLUDAS has been developed by the Illinois State Water Survey (Terstriep, et al. 1978) but has not yet been released and is currently being improved.

The model STORM (Storage Treatment, Overflow, Runoff Model) computes runoff quantity by the coefficient method, by the SCS curve number technique or by a combination of the two (the coefficient method for the more impervious areas, and the CN method for the more pervious areas). The SCS curve number technique implemented in STORM includes a continuous soil moisture accounting and a two-parameter unit hydrograph which differ substantially from the SCS hand calculation procedure, as described by the Hydrologic Engineering Center (1980).

Dendrou et al. (1980 a,b) integrated STORM with an urban growth simulation model so as to explicitly evaluate the effectiveness of alternate growth scenarios. By partitioning the watershed into subbasins, the capacity of the drainage system, the placement and size of detention storage facilities and the size of a central treatment facility are optimized for minimum cost in a two level solution procedure.

An outgrowth of the program STORM is the new model SEM-STORM prepared by Shubinski et al. (1977) for the Southeast Michigan Council of Governments. Although the objectives of SEM-STORM are the same as those of STORM, it has been completely reprogrammed, and allows for the combination of many watersheds. A minicomputer version of STORM has been installed on the PDP 11/70 computer at the Environmental Research Laboratory in Athens, Georgia.

The new version of the Hydrologic Engineering Center computer program

HEC-1 contains options for the use of the kinematic wave theory to compute outflow hydrographs and to route hydrographs. DeVries and MacArthur (1979) give the details of the application of the program to urban runoff calculations.

In the SWMM model (Storm Water Management Model) the TRANSPORT model simulating the flows in the sewer system has been improved by WRE-CDM in the new EXTRAN version. The model is based on a complete solution of the Saint Venant equations for unsteady free surface flow. It is capable of computing the surcharges and backwater effects, it handles flow diversions by orifices, weirs, pumps, etc. The program accepts the hydrographs computed by the SWMM-RUNOFF routine or hydrographs input directly from cards. Improvements have been made on the surcharge computation, and improvements are being made on stability with orifices and weirs.

The Version III of SWMM is expected to be available in early 1981. It includes a continuous simulation of urban snowmelt/removal. The storage/treatment block is improved to include dry and wet weather units, storage units with sedimentation, particle size and specific gravity characterization, sludge handling, capital and operation and maintenance costs. The Runoff Block is improved to include linear and nonlinear pollutant accumulation, pollutant rating curves, porous pavement, natural channels and baseflow. The Extended Transport Block (EXTRAN) mentioned above is also included in Version III. Han and Rao (1980) introduced a parameter optimization procedure in the Runoff Block of SWMM using Rosenbrock's algorithm. Regeneration tests on two watersheds show that a considerable improvement in the fit between observed and optimized hydrographs is obtained as compared to the traditional trial and error parameter adjustment.

A research and service project for the implementation of storm water management modeling (IMPSWM) has been started at the University of Ottawa, Canada in 1979. The project assists in the implementation and application of practical modeling in urban drainage. The firms of CDM and IMPSWM are cooperating on improvements of EXTRAN.

Alley et al. (1980) presented a parametric-deterministic model for urban watersheds. Although the user's manual has been available from the USGS for about two years, the formal publication appeared in 1980. This model will be referred in the sequel as the ADS model using the initials of the three authors: Alley, Dawdy and Schaake. The model uses a parametric soil moisture accounting procedure to determine the antecedent moisture condition and the Philip equation to compute infiltration. A modified Rosenbrock optimization technique can be used to optimize the soil mosisture and infiltration parameters. The analysis of infiltration and soil moisture is much more detailed in this model than in those discussed previously. The kinematic wave method is used to route the flows over the contributing areas and through the pipe and channel system. A water-quality algorithm has been drafted.

The ADS model of USGS, SWMM II, the Penn State runoff model, HEC-1, and others use the kinematic wave method for routing the overland flow. Kibler and Aron (1980) verified the basic validity of the method by simulating the highly controlled overland flows over asphalt and other surfaces used in the classical experiments of Izzard. Although the agreement between observed and computed outflows was not perfect the reproduction of the patterns of the hydrographs were good. The study also showed that rainfall intensity patterns and the degree of imperviousness are the most sensitive parameters of the surface runoff model. Also

411

considered was the detail of the segmentation or the effect of scale, and it was shown that coarser-scale models tend to yield hydrographs with peaks that overshoot the observed peak.

Curran and Wanielista (1980) developed SMADA (Stormwater Management and Design Aid) which is written in BASIC in the interactive mode. Each subwatershed is divided into pervious, impervious, and impervious that is directly connected to the drainage system. For each time increment the available abstraction and infiltration are subtracted from the rainfall with the excess running off. This instantaneous runoff is subject to a time delay by the routing procedure of the Santa Barbara unit hydrograph method to deliver the outflow hydrograph. Retention for the purpose of pollution control is accounted for by setting runoff equal to zero until the necessary diversion volume is acquired. The diversion required is based upon the desired degree of pollutant removal. Detention for peak flow reduction basins are sized by noting the time increments in which the hydrograph values exceed a specific discharge.

The HSPF model is currently being updated. The software which was intended for application on the minicomputer Hewlett Packard HP-3000 is being converted so that it can be used on large systems such as CDC or UNIVAC. Version 7 is expected to be released in mid 1981.

Least Cost Drainage Systems

Important improvements have been made in the methods for obtaining least cost drainage systems. Rawls and McCuen (1978) have reviewed some of the important cost functions for urban storm sewer systems and have applied them in the context of the rational formula, using rainfall intensity-frequency-duration curves for Washington, D. C., Los Angeles, Cal., and Dallas, Tex. The effects of imperviousness, design return period and time of concentration on system costs is exhibited by means of graphs.

Froise and Burges (1978) consider the least cost of drainage systems which include conveyance and storage elements. They assumed that the layouts of the network alternatives were known. The optimization problem was solved by the technique of dynamic programming and unsteady flow routing. The model does not include a generation of inlet hydrographs which are specified separately. They concluded that cost reductions of the order of 30% can be achieved as compared to conventional methods. About 60% of these savings is achieved by optimal combination of conduit sizes and the use of advanced hydraulic simulation techniques and the remainder is attributed to the use of storage elements.

Wenzel et al. (1979) extended a previous least cost design model which makes use of discrete differential dynamic programming to search for the optimal design (Yen et al., 1976). The drainage system is divided into stages by isonodal lines. These are lines which pass through all manholes (nodes) that are separated from the system outlet by the same number of pipes (links). The optimization progresses stage by stage in the downstream direction. The new version utilizes the runoff computation similar to that from ILLUDAS, and a capability of utilizing detention storage has been added which includes two types of optional constraints. In the first type, the maximum allowable flow downstream of the reservoir is specified. In the second type the total volume can be limited to a maximum specified value. The hydrograph time shift is used for routing the flows. There is a version which incorporates the consideration of risk. The average annual expected damage cost, which can be discounted to a present worth value using an assumed economic life and discount rate, is added to the

412

pipe installation cost. The total cost of the pipe and expected damage is minimized using DDDP.

Han et al. (1980) have independently constructed a similar model which uses dynamic programming instead of DDDP. The inflow hydrographs are obtained from the runoff portion of ILLUDAS. The effect of uncertainties in the model inputs is investigated by an extensive sensitivity analysis of the results with respect to storm duration, return period, antecedent moisture condition and temporal rainfall distribution. For a given watershed there exists a storm duration which produces the highest peak runoff and hence the highest costs. The costs were found to vary more in the lower return periods than in the high ones. They also recommend the use of the median first quartile storm profile derived from the local rainfall data. The report includes a self-contained listing of the computer program.

Pollutant Washoff

The pollutant washoff is usually described by the exponential equation

$$P_o - P = P_o(1 - e^{kt}) \tag{18}$$

where P_o is the initial mass of pollutants on the land surface, P_o-P is the mass of pollutant removed in time t and k is the decay coefficient. In STORM and SWMM it is assumed that k is proportional to the runoff rate. It is usually set at a constant value of 4.6 inches^{-1} or 116.8 mm^{-1}; assuming that 1/2 inch of runoff would wash away 90% of the pollutants in 1 hour. Recent investigations by Smith and Jennings (1979) and Ellis and Sutherland (1979) reported that k varies for different constituents and for different watersheds. In SWMM and STORM an availability factor is used to account for the effects of runoff intensity on the constituent washoff. One approach consists in replacing these equations and equation (18) by a washoff scheme based on sediment transport theory. Alley, Ellis and Sutherland (1980) used the Particle Transport Model (PTM) to simulate the washoff of sediment from impervious areas based on the Erosion, Transport, and Deposition (ETD) system of Meyer and Wischmeier (1969). An alternate approach consists of obtaining the best values of the decay coefficient and of the availability factors from the measurements currently being made as part of the Nationwide Urban Runoff Program of EPA and USGS. This approach has been followed by Alley (1980) to optimize the k values which were found to vary between storms and between constituents for an urbanized basin in South Florida. The optimized k values for total nitrogen were affected by assumed wetfall contributions to runoff quality.

Retention Basins and Receiving Waters

Retention basins are a practical method of control of nonpoint source effects. The diversion of first flush collects most of the pollutants from urban areas. Wanielista (1979, p. 248-252) reports the efficiencies up to 99.9% for diversion of 1.25 inch in an onsite system in Orlando, Florida (on an annual basis and diversion of every storm is assumed), and has given regression equations for the prediction of retention basin volume. Regression equations developed by McCuen (1980) also suggest that the large trap efficiencies are evident for most water quality parameters, particularly for the more frequent storm events. The complete analysis of the behavior of retention basins requires the use of the one-dimensional version of the transient, convective-dispersion equation

$$\frac{\partial c}{\partial t} = \frac{\partial}{\partial x} [E \frac{\partial c}{\partial x} - Vc] \pm \sum S \qquad (19)$$

where c is the concentration of the pollutant, E is the longitudinal dispersion coefficient, V is the longitudinal velocity in the system, and S represents the sources and sinks of the pollutant. Medina et al. (1981) have used this fundamental approach to develop models of the movement decay, storage and treatment of stormwater runoff pollutants and dry weather wastewater flows through natural and manmade systems. The methodology has been successfully applied to Des Moines River receiving outflows from the city of Des Moines, Iowa, and to the Humboldt Avenue detention tank in Milwaukee, Wisconsin.

A receiving water quality model for urban stormwater management based on the same principles has been developed by Medina (1979) for EPA. The model may be interfaced with STORM and SWMM. The model is useful for the evaluation of many urban pollution control alternatives in terms of other subsequent impacts on receiving water quality.

Another comprehensive and versatile stream water quality model which has recently been improved is the QUAL-II model (Roesner, Giguere and Evenson). The model can be used to study the impact of waste loads on instream water quality and to identify the magnitude and quality characteristics of nonpoint source waste load.

CONCLUSIONS

During the last two years the EPA/USGS Nationwide Urban Runoff Program has served as a catalyst to advance the science of urban hydrology in the United States. It is expected that this will continue at least for the next three years. It is hoped that, with the help of this program, the effectiveness of stormwater management alternatives will be tested and that the magnitude and frequency of pollutants from urban runoff in receiving streams will be better known. It is expected that in the future rainfall forecasting coupled with appropriate hydrologic models will produce runoff quantity and quality predictions. Such information would perhaps make it possible to enhance the operation of urban drainage by using the available in-line storage capability. This would lead to automated operation of sewer systems, a solution which could be more economical than the addition of storage. McPherson (1981) foresees this possibility through the use of weather radar.

Due to the randomness of rainfall and the desirability of characterizing the pollution impacts of receiving waters in terms of magnitude and frequency of occurrence it is likely that continuous simulation models will play a dominant role while single events models will tend to be used primarily for the design of storm sewer systems.

Much research yet remains to be done, particularly regarding the response of natural receiving waters to time varying urban runoff loads; regarding the rates, constants and kinetics formulation for water quality modeling; the transport and disposition of suspended matter; the selection of the proper models, the resolution of conflicting requirements of drainage versus flood control and water quality objectives. M. B. Sonen

in a recent article entitled "Urban Runoff Quality: Information Needs",
(ASCE, Journal of the Technical Councils, Vol. 106 No. TC1, August 1980,
pp. 20-40) states that the current perceptions of runoff water quality
tend to be limited to immediate and pressing regulatory requirements. To
overcome these limitations he advocates more comprehessive theory and
sampling and filling the many information gaps through research.

REFERENCES

Alley, W. M. 1980. Determination of the Decay Coefficient in the Ex-
    ponential Washoff Equation, Proceedings, International Symposium
    on Urban Runoff, University of Kentucky, July 28-31, 1980, pp.
    307-311.

Alley, W. M., Dawdy, D. R. and Schaake, J. C. 1980. Parametric-
    Deterministic Urban Watershed Model, Journal of the Hydraulics
    Division, ASCE, Vol. 106, No. HY5, pp. 679-690.

Alley, W. M., Ellis, F. W. and Sutherland, R. C. 1980. Toward a More
    Deterministic Urban Runoff-Quality Model, Proceedings, International
    Symposium on Urban Storm Runoff, University of Kentucky, July 28-31,
    1980, pp. 171-182.

Altman, D. G. and Feldman, A. D. 1980. Investigation of Soil Conserva-
    tion Service Urban Hydrology Techniques, Paper presented at the
    1980 Spring Meeting of the American Geophysical Union, Toronto,
    Canada.

Brandstetter, A. 1976. Assessment of Mathematical Models for Storm and
    Combined Sewer Management, EPA-600/2-76-175.

Burke, C. B. and Gray D. D. 1979. A Comparative Application of Several
    Methods for the Design of Storm Sewers, Purdue University Water
    Resources Research Center, West Lafayette, Indiana, Technical
    Report 118, 134 pp.

Burke, C. B. and Gray, D. D. 1980. A Comparative Application of the
    Rational Method and the Illinois Urban Drainage Area Simulator to
    an Indiana Subdivision, Proceedings of the Indiana Academy of
    Science for 1979, Vol 89, pp. 199-203.

Chan, S. L. and Bras, R. L. 1979. Urban Storm Water Management: Dis-
    tribution of Flood Volumes, Water Resources Research, Vol. 15,
    No. 2, pp. 371-392.

Chow, V. T. (Ed.). 1964. Handbook of Applied Hydrology, McGraw-Hill
    Book Co., New York.

Croley, T. E. 1980. Synthetic Hydrograph Computations on Small Pro-
    grammable Calculators, Iowa Institute of Hydraulic Research, The
    University of Iowa, Iowa City, Iowa, 236 pp.

Croley, T. E. 1981. Hydrologic and Hydraulic Calculations in Basic for
    Small Computers, Iowa Institute of Hydraulic Research, The University
    of Iowa, Iowa City, Iowa.

Curran, T. M. and Wanielista, M. P. 1980. SMADA-Storm Management and
    Design Aid, ASCE Preprint 80-603.

Delleur, J. W. 1979. The Design Storm Concept: Is It a Sufficient Criterion to Determine the Reliability of Modern Storm Drainage Systems, in G. Patry and M. B. McPherson (Editors), The Design Storm Concept, Proceedings of a seminar held at Ecole Polytechnique de Montreal, Quebec, May 23, 1979, Report EP80-R-GREMU-79/02.

Delleur, J. W. and Dendrou, S. A. 1980. Modeling the Runoff Process in Urban Areas, CRC Critical Reviews in Environmental Control, pp. 1-64.

Dendrou, S. A., Delleur, J. W. and Talavage, J. J. 1978a. Storm Drainage Systems for Urban Growth, Journal of the Water Resources Planning and Management Division, ASCE, Vol. 104, No. WR1, pp. 1-16.

Dendrou, S. A., Talavage, J. J. and Delleur, J. W. 1978b. Optimal Planning for Urban Storm Drainage, Journal of the Water Resources Planning and Management Division, ASCE, Vol. 104, No. WR1, pp. 17-33.

DeVries, J. J. and MacArthur, R. C. 1979. Introduction and Application of Kinematic Wave Routing Techniques Using HEC-1. Hydrologic Engineering Center, Training Document No. 10.

Diskin, M. Ince, S. and Oben-Nyarko, K. 1978. Parallel Cascades Model for Urban Watersheds, Journal of the Hydraulics Division ASCE, Vol. 104, No. HY2, pp. 261-276.

Diskin, M. H. and Simon, E. 1977. A Procedure for the Selection of Objective Functions for Hydrologic Simulation Models, Journal of Hydrology, Vol. 34, pp. 129-149.

Doyle, W. H. and Lorens, J. A. 1981. Urban Stormwater Data Management System, Proceedings, International Symposium on Rainfall-Runoff Modeling, Mississippi State University, May 18-21, 1981.

Ellis, F. W., and Sutherland, R. C. 1979. An Approach to Urban Pollutant Modeling, Proceedings, International Symposium on Urban Storm Runoff, University of Kentucky, July 23-26, 1979, pp. 325-340.

Folland, C. K. and Colgate, M. G. 1978. Recent and Planned Rainfall Studies in the Meteorological Office with Application to Urban Drainage Design in P. R. Helliwell (Editor) Urban Storm Drainage, Proceedings of an International Conference on Urban Drainage, Southampton, 1978, Pentech, London.

Friedman, L. C. 1978, 1979. Water Quality Laboratory Services Catalog, U. S. Geological Survey, Open-File Report 78-842, 449 pp.

Froise, S. and Burges, S. J. 1978. Least Cost Urban Drainage Networks, Journal of the Water Resources Planning and Management Division, ASCE, Vol. 104, No. WR1, pp. 75-92.

Gray, D. D. and Cogo, N. P. 1980. Antecedent Moisture Condition Probabilities in Tippecanoe County, Indiana Proceedings, First Annual Indiana Water Resources Symposium, Indiana Water Resources Association.

Han, J. and Delleur, J. W. 1979. Development of an Extension of ILLUDAS Model for Continuous Simulation of Urban Runoff Quantity and Discrete Simulation of Runoff Quality, Purdue University Water Resources Research Center, Technical Report No. 109, 136 pp.

416

Han, J. and Rao, A. R. 1980. Optimal Parameter Estimation and Investigation of Objective Functions of Urban Runoff Models, Purdue University Water Resources Research Center, Tech. Rept. No. 136, 153 pp.

Han, J., Rao, A. R. and Houck, M. H. 1980. Least Cost Design of Urban Drainage Systems, Purdue University, Water Resources Research Center, Technical Report 138, 119 pp.

Hardee, J. 1979. Instrumentation of Urban Hydrology Sites in Southeast Florida, U. S. Geological Survey Water Resources Investigations, 79-37, 38 pp.

Hjelmfelt, A. T. 1980. Curve Number Procedure as Infiltration Model, Journal of the Hydraulics Division, ASCE, Vol. 106, No. 6, pp. 1107-1111.

Hydraulics Research Station 1980. The Design and Analysis of Urban Storm Drainage, Rept. DE50, Wallingford, England.

Hydrologic Engineering Center 1980. Guidelines for Calibration and Application of Storm, HEC Training Document, No. 8.

Jackson, T. J., Ragan, R. M. and Fitch, W. N. 1977. Test of Landsat-Based Urban Hydrologic Models, Journal of the Water Resources Planning and Management Division, ASCE, Vol. 103, No. WR1, pp. 141-158.

Kibbler, D. F. and Aron, G. 1980. Observations on Kinematic Response in Urban Runoff Models, Water Resources Bulletin, Vol. 16, No. 3, pp. 444-452.

Marsalek, J. 1978. Research on the Design Storm Concept, ASCE Urban Water Resources Research Projgram, Tech. Memo No. 33.

McPherson. 1979a. Urban Hydrology, Reviews of Geophysics and Space Physics, Vol. 17, No. 6, pp. 1289-1297.

McPherson (Editor). 1979b. International Symposium on Urban Hydrology ASCE Urban Water Resources Research Program, Technical Memo No. 38.

McPherson. 1980. Urban Water Management in the 1980's: Introduction, American Society of Civil Engineers, Preprint 80-511.

McPherson, 1981. Study of Integrated Control of Combined Sewer Regulators, Report of ASCE Urban Water Resources Research Council to EPA Storm and Combined Sewer Division, Wastewater Division.

McCuen, R. H. 1980. Water Quality Trap Efficiency of Storm Water Management Basins, Water Resources Bulletin, Vol. 16, No. 1, pp. 15-21.

Medina, M. A. 1979. Level III: Receiving Water Quality Modeling for Urban Stormwater Management, EPA Report EPA-600/2-79-100, 205 pp.

Medina, M. A., Huber, W. C. and Heany, J. P. 1981. Modeling Stormwater Storage/Treatment Transients, Journal of the Environmental Engineering Division ASCE, in press.

Meyer, L. D. and Wischmeier, W. H. 1969. Mathematical Simulation of the Process of Soil Erosion by Water, Transactions of the ASAE, 1969, pp. 754-762.

Miller, R. A., Mattraw, H. C. and Jennings, M. E. 1978. Statistical Modeling of Urban Stormwater Processes in Broward County, Florida, Proceedings, International Symposium on Urban Stormwater Management, University of Kentucky, July 24-27, 1978, pp. 264-274.

Miller, T. L. and McKenzie, S. W. 1978. Analysis of Urban Storm-Water Quality from Seven Basins near Portland, Oregon, U. S. Geological Survey Open File Report 78-662, 44 pp.

Nelder, J. A. and Mead, R. 1965. A Simplex Method for Function Minimization, The Computer Journal, Vol. 7, pp. 308-313.

Packman, J. C. and Kidd, C. H. R. 1980. A Logical Approach to the Design Storm Concept, Water Resources Research, Vol. 16, No. 6, pp. 994-1000.

Patry, G., Raymond, L. and Marchi, G. 1979. Description and Application of an Interactive Mini Computer Version of the ILLUDAS Model Proceedings, Stormwater Management Model (SWMM) Users Group Meeting, May 24-25, 1979, EPA-600/9-79-026, pp. 242-274.

Pedersen, J. T., Peters, J. C., and Helweg, O. J. 19 . Hydrographs by Single Linear Reservoir Models, Journal of the Hydraulics Division ASCE, Vol. 106, No. HY5, pp. 837-852.

Ragan, R. M. and Jackson, T. J. 1980. Runoff Synthesis Using Landsat and SCS Model, Journal of the Hydraulics Division, ASCE, Vol. 106, No. HY5, pp. 667-678.

Rao, A. R., Delleur, J. W. and Sarma, P. B. S. 1972. Conceptual Hydrologic Models for Urbanizing Areas, ASCE Journal, Hydraulic Division, 98, HY7, 1205.

Rawls, W. J. and McCuen, A. M. 1978. Economic Assessment of Storm Drainage Planning, Journal of the Water Resources Planning and Management Division, ASCE, Vol. 104, No. WR1, pp. 45-54.

Rosenbrock, H. H. 1960. An Automatic Method for Finding the Greatest or Least Value of a Function, The Computer Journal, Vol. 3, pp. 175-184.

Roesner, L. A., Giguere, P. R. and Evenson, D. E. 1977. User's Manual for Stream Quality Model (QUAL II), Computer Program Documentation for Stream Quality Model (QUAL II), Camp, Dresser & McKee, Annadale, VA.

Roesner, L. A., Kassem, A. M., and Wisner, P. E. 1980. Improvements in EXTRAN, Proceedings Stormwater Management Model (SWMM) Users Group Meeting, January 10-11, 1980, EPA 600/9-80-017, pp. 132-141.

Rossmiller, R. L. 1980. The Rational Formula Revisited, Proceedings, International Symposium on Urban Storm Runoff, University of Kentucky, July 28-31, 1980, pp. 1-12.

Shubinski, R. P., Knepp, A. J. and Pristol, C. R. 1977. User's Manual for the Continuous Storm Runoff Model SEM-STORM, Camp, Dresser & McKee (CDM), Annandale, Virginia.

Singh, V. P. 1979. A Note on a Systematic Evaluation of Urban Runoff Models, Proceedings, International Symposium on Urban Storm Runoff, University of Kentucky, July 23-26, 1979, pp. 37-46.

Smith, P. E. and Jennings, M. E. 1979. Accumulation and Washoff of

Pollutants on Urban Watersheds, abstract in EOS, Transactions of the American Geophysical Union, Vol. 16, No. 18, p. 259.

Soil Conservation Service. 1975. Urban Hydrology for Small Watersheds, Technical Release No. 55.

Soroosian, S. 1980. Comparison of Two Direct Search Algorithms Used in Calibration of Rainfall-Runoff Models, Preprints, IFAC Symposium on Water and Related Land Resources Systems, Cleveland, Ohio, 28-31 May 1980, Pergamon Press, pp. 441-449.

Terstriep, M. L., Bender, G. M. and Benoit, D. J. 1978. Buildup, Strength and Washoff of Urban Pollutants, Paper presented at the ASCE Convention and Exposition, Chicago, Ill., October 1978.

Thibault, S. Chocat, B. and Botta H. 1980. Relations Théoriques Entre Différents Modèles d'Ecoulement et de Ruissellement Utilisés en Hydrologie Urbaine, Journal of Hydrology, Vol. 48, pp. 313-326.

UNESCO. 1977. Research on Urban Hydrology, Vol. 1, State-of-the Art Reports from Australia, Canada, USSR, United Kingdom, USA, Technical Paper in Hydrology, 15, 185 pp.

UNESCO. 1978a. Research on Urban Hydrology, Vol. 2, State-of-the-Art Reports from Federal Republic of Germany, Sweden, France, Norway, Netherlands, Poland and India, Technical Paper in Hydrology, 16, 265 pp.

UNESCO. 1978b. Research in Urban Hydrology, International Summary, Technical Paper in Hydrology, 17, 48 pp.

Urbonas, B. 1979. Reliability of Design Storms in Modeling Proceedings, International Symposium on Urban Storm Runoff, University of Kentucky July 23-26, 1979, pp. 27-35.

Walesh, S. G., Lau, D. H. and Liebman, M. D. 1979. Statistically-Based Use of Event Models, Proceedings International Symposium on Urban Storm Runoff, Univ. of Kentucky, July 23-26, 1979, pp. 75-81.

Wanielista, M. P. 1979. Stormwater Management, Quantity and Quality, Ann Arbor Science, 384 pp.

Wenzel, H. G. and Vorhees, M. L. 1978. Evaluation of the Design Storm Concept, Paper Presented at the 1978 Fall Meeting of the American Geophysical Union in San Francisco.

Wenzel, H. G. and Voorhees, M. L. 1979. Sensitivity of Design Storm Frequency in G. Patry and M. B. McPherson (Editors), The Design Storm Concept, Proceedings of a Seminar at Ecole Polytechnique de Montreal, Quebec, 23 May 1979, and of a related Session of the American Geophysical Union Spring Meeting, Washington, D. C., 29 May 1979, Ecole Polytechnique de Montreal, Report EP80-R-GREMU-79/02.

Wenzel, H. G., Yen, B. C. and Tang, W. H. 1979. Advanced Methodology for Storm Sewer Design-Phase II, University of Illinois, Water Resources Research Center, Research Report UILU-WRC-79-0140, 78 pp.

Wenzel, H. G. and Voorhees, M. L. 1980. Adaptation of ILLUDAS for Continuous Simulation, Journal of the Hydraulics Divsion, ASCE, Vol. 106, No. HY11, pp. 1795-1812.

Yen, B. C., and Chow, V. T. 1980. Design Hyetographs for Small Drainage

Structures, Journal of the Hydraulics Division ASCE, Vol. 106, No. HY6, pp. 1055-1076.

Yen, B. C., Wenzel, H. G., Mays, L. W., Tang, W. H. 1976. Advanced Methodologies for Design of Storm Sewer Systems, University of Illinois Water Resources Center, Research Report UILU-WRC-76-0112, 224 pp.

RAINFALL-RUNOFF-QUALITY MODEL FOR URBAN WATERSHEDS

Peter E. Smith
Hydrologist, U.S. Geological Survey
Gulf Coast Hydroscience Center
NSTL Station, Mississippi

and

William M. Alley
Hydrologist, U.S. Geological Survey
Reston, Virginia

ABSTRACT

A rainfall-runoff-quality model for urban watersheds is described. The model consists of two separate programs: one for rainfall-runoff simulations and the other for runoff-quality simulations.

The rainfall-runoff part of the model provides detailed simulations of storm runoffs specified by the user and a daily soil-moisture accounting between storms. A drainage basin is represented as a set of overland-flow, channel, and reservoir segments which jointly describe the drainage features of the basin. Kinematic wave theory is used for routing flows over contributing overland-flow areas and through the channel network. The model provides three options for kinematic wave routing. These include a method of characteristics formulation and both implicit and explicit finite difference solutions. A means of avoiding kinematic shock problems in the method of characteristics formulation has been developed.

The water-quality component of the model considers three sources of water-quality constituents in the runoff: impervious-area runoff contributions, pervious-area runoff contributions, and precipitation contributions. A daily accounting of constituent accumulation on impervious areas is maintained between storms.

The water-quality component of the model can be applied on either a lumped-parameter or distributed-parameter basis. For distributed-parameter simulations, constituent transport through channel segments is modeled using a Lagrangian method. Constituent transport through reservoir segments is modeled using a plug-flow scheme.

INTRODUCTION

The U.S. Geological Survey has been involved in developing simulation models of rainfall-runoff processes since the late 1960's. Dawdy, Lichty, and Bergmann (1972) reported on the first simulation model from this research: a lumped-parameter rainfall-runoff model for small rural watersheds. Subsequent work by Dawdy, Schaake, and Alley (1978) produced a distributed routing rainfall-runoff model (DR$_3$M). This model was largely a synthesis and update of a version of the Massachusetts Institute of Technology catchment model (Leclerc and Schaake, 1973) and the original Geological Survey model. DR$_3$M was developed principally for application to urbanized watersheds.

The purpose of this paper is to describe developmental work on this urban rainfall-runoff model since its release in 1978. Much of this development work has centered on the addition of a water-quality component to the model. However, many additions to the runoff routing structure have also been made.

RAINFALL RUNOFF

DR$_3$M operates on two time intervals. The model provides detailed simulation of storm-runoff periods during days for which short-time-interval rainfall data are input to the program. These days are referred to as "unit days" and it is only during unit days that flow routing is performed. Between unit days, the model uses daily precipitation and daily evaporation data to provide a continuous accounting of soil moisture on a daily basis. Thus, the advantages of continuous simulation are combined with those of a single event model.

During simulation of a period of storm runoff, the generation of rainfall excess and flow routing are treated independently. The time series of rainfall excess is determined first, and then in a second step it is routed to the watershed outlet.

Rainfall Excess

Pervious and impervious areas are handled differently in DR$_3$M. For every period of storm runoff, separate volumes of pervious and impervious area rainfall excess are determined. These volumes are later combined for input into the flow-routing component of the model.

Pervious areas

Soil moisture is modeled using a dual storage system, one representing the antecedent base-moisture storage (BMS) and the other representing upper-zone storage caused by infiltration into a saturated moisture storage (SMS). During unit days, moisture is added to SMS based on the Green-Ampt infiltration equation (Green and Ampt, 1911). Between unit days, a specified proportion of daily rainfall is added to BMS.

Evapotranspiration takes place from SMS when available, otherwise from BMS, with the rate determined from pan evaporation multiplied by a pan coefficient. On unit days, moisture in SMS drains into BMS during periods of no rainfall. Storage in BMS has a maximum value (BMSN) equivalent to the field-capacity moisture storage of the active soil zone. Zero storage in BMS is assumed to correspond to wilting point conditions in the active soil zone. When storage in BMS exceeds BMSN, the excess is spilled to deeper storage. These spills could be the basis for routing interflow and baseflow components. However, this option is not included in the present version of the model.

Point-potential infiltration (FR) is computed by a variation of the Green-Ampt equation (Green and Ampt, 1911):

$$FR = KSAT (1 + PS/SMS) \qquad (1)$$

where KSAT is the effective saturated value of hydraulic conductivity and PS is defined as:

$$PS = P (m - m_0) \qquad (2)$$

where P is the average suction head across the wetting front, m is the moisture content of the soil after wetting, and m_0 is the antecedent soil moisture content.

The capillary potential at the wetting front is not a constant but varies according to the soil-moisture conditions. The model determines the effective value of PS as varying linearly between a value at plant wilting and value at field capacity. Point-potential infiltration computed by the Green and Ampt equation is converted to effective infiltration over the basin, assuming a linear variation of infiltration capacity with area percentage (Crawford and Linsley, 1965). A substantial part of the above scheme for determining pervious-area rainfall excess has been adopted from the model developed by Dawdy, Lichty, and Bergmann (1972).

Impervious areas

Two types of impervious surfaces are considered by the model in computing impervious-area rainfall excess. The first type, effective impervious surfaces, are those impervious areas that are directly connected to the channel drainage system. Streets, roofs that drain onto driveways, and paved parking lots that drain onto streets are examples of effective impervious surfaces. The second type, noneffective impervious surfaces, are those impervious surfaces that drain to pervious ground. An example of a noneffective impervious surface is a roof that drains onto a lawn.

The only abstraction from rainfall on effective impervious area is impervious retention. This retention, which is user specified, must be filled before runoff from effective impervious areas can occur.

Rain falling on noneffective impervious areas is assumed to run off onto the surrounding pervious area. The model assumes that this occurs instantaneously and that the volume of runoff is uniformly distributed over the pervious area. This volume, expressed as millimeters over the pervious area, is added to the rain falling on the pervious areas prior to computation of pervious-area rainfall excess.

Parameter optimization

An option is included in DR$_3$M to determine "optimum" parameter values using a technique devised by Rosenbrock (1960) that is a modified, improved version of the steepest descent method. In the 1978 version of the model only the soil-moisture-accounting and infiltration parameters could be calibrated using Rosenbrock's technique. However, effective impervious area can now be included in the optimization.

During an optimization run, storm-runoff volumes for a series of storms having measured rainfall and runoff data are simulated by the model. An objective function which is the sum of the squared deviations of the logarithms of computed and measured storm-runoff volumes is computed. One of the parameter magnitudes is then revised and a second simulation made. If the result is an improvement, the revised set is accepted; if not, the method returns to the previous best set of parameter values. This procedure is repeated for a user-specified number of iterations.

Rosenbrock's method of optimization proceeds by stages. During the first stage, each parameter represents one axis in an orthogonal set of search directions. Adjustments are made in these search directions until end-of-stage criteria are satisfied. At the end of each stage a new set of orthogonal directions is computed based on the experience of parameter movement during the preceding stage. The major feature of this procedure is that, after the first stage, one axis is aligned in a direction reflecting the net parameter movement experienced during the previous stage.

Flow Routing

A drainage basin is represented by DR_3M as a set of segments which jointly describe the drainage features of the basin. The segment types include overland-flow, channel, and reservoir segments.

Overland-flow segments receive uniformly distributed lateral inflow from rainfall excess. They represent a rectangular plane of a given length, slope, roughness, and percent imperviousness. The volume of rainfall excess distributed to an overland-flow segment is determined from the percentages of pervious and impervious areas within the segment.

Channel segments are used to represent natural or man-made conveyances such as gutters or storm-sewer pipes. Channel segments may receive up-stream inflow from as many as three other segments, including combinations of other channel segments and reservoir segments. They also can receive lateral inflow from overland-flow segments.

Reservoir segments can be used to describe an on-channel detention reservoir. Alternately, they can be used to simulate culverts which detain water due to limited capacity and for which outflows are uniquely described as a single-valued function of storage upstream of the culvert. Flows are routed through reservoir segments using the modified-Puls routing method.

A schematic illustrating the relationships between channel and overland-flow segments is shown in figure 1. There is wide flexibility to the approach one can take in dividing a basin into segments for runoff computations. A useful feature of the model is that it allows the same overland-flow segment to be used at different places in the watershed. The model routes the flow through a given overland-flow segment only once and uses the outflow in cubic feet per second per foot of channel length as lateral inflow to all appropriate channel segments. A simple use of this option is shown in figure 1 where ØF01 drains to both CH02 and CH03.

Rainfall excess is routed through channel and overland-flow segments by solving the unsteady continuity equation and the kinematic-wave form of the equation of motion. These can be expressed as

$$\frac{\partial Q}{\partial x} + \frac{\partial A}{\partial t} = q \tag{3}$$

$$S_0 = S_f \qquad (4)$$

in which Q is discharge, A is the flow cross-sectional area, q is lateral inflow per unit length, x and t are space and time coordinates, and S_0 and S_f are the bed slope and friction slope, respectively. A flow resistance formula, such as the Manning equation, is used to reduce the equation of motion to a power relationship of the form

$$Q = \alpha A^m \qquad (5)$$

where α and m are constants that must be determined for each overland-flow and channel segment, and are functions of the geometry, slope, and roughness of the segment.

The model has three different methods available for the solution of equations 3 and 5: a method of characteristics and implicit and

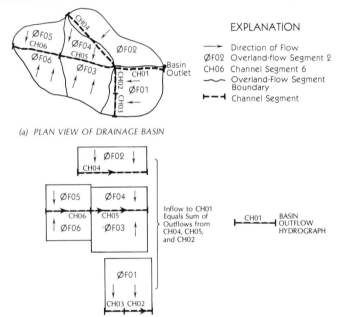

(a) PLAN VIEW OF DRAINAGE BASIN

EXPLANATION

→ Direction of Flow
ØF02 Overland-flow Segment 2
CH06 Channel Segment 6
～ Overland-Flow Segment Boundary
├─┤ Channel Segment

(b) SCHEMATIC REPRESENTATIONS OF MODEL SEGMENTS

Inflow to CH01 Equals Sum of Outflows from CH04, CH05, and CH02

CH01   BASIN OUTFLOW HYDROGRAPH

Segment	Inflow to Segment	
	Lateral Inflow	Upstream Inflow
ØF01	Rainfall Excess	—
ØF02	" "	—
ØF03	" "	—
ØF04	" "	—
ØF05	" "	—
CH01	—	CH02, CH04, CH05
CH02	ØF01	CH03
CH03	ØF01	—
CH04	ØF02	—
CH05	ØF03, ØF04	CH06
CH06	ØF05, ØF06	—

(c) SEGMENT INTERRELATIONSHIPS

Figure 1.--Segmentation of an urban watershed for DR₃M illustrating the relationship between channel and overland-flow segments

425

explicit finite-difference methods. The user has the option to select a particular method for each model segment.

Method of characteristics

Substituting equation 5 into equation 3 yields the following:

$$\frac{\partial A}{\partial t} + \alpha m A^{m-1} \frac{\partial A}{\partial x} = q \tag{6}$$

It is this partial differential equation that is solved in the model by the method of characteristics. The solution provides values of A that can be converted to discharge using equation 5.

Equation 6 can be represented by the following characteristic equations (Eagleson, 1970):

$$\frac{dx}{dt} = \alpha m A^{m-1} \tag{7}$$

$$\frac{dA}{dt} = q \tag{8}$$

Integration of equations 7 and 8 can be done explicitly if the lateral inflow, q, is assumed to be uniform in time and space. Since, for a model segment in DR3M, q is constant in space and piecewise constant in time, this assumption can be met by integrating over time steps where q remains constant. The result after integrating between two points on a characteristic path, (x,t) and $(x+\Delta x, t+\Delta t)$, has been given by Harley et al., (1970) and can be expressed as:

$$\Delta x = \frac{\alpha}{q} \left[(q\Delta t + A(x,t))^m - A(x,t)^m \right] \tag{9}$$

$$A(x+\Delta x, t+\Delta t) = A(x,t) + q\Delta t \tag{10}$$

for $q \neq 0$, and

$$\Delta x = \alpha m A(x,t)^{m-1} \Delta t \tag{11}$$

$$A(x+\Delta x, t+\Delta t) = A(x,t) \tag{12}$$

for $q = 0$.

The systematic application of equations 9-12 to route flows can be done by several different approaches. The method of solution in DR3M moves all characteristics within a segment through a time step before proceeding to the next time step. Since, in the general case, a characteristic does not intersect the downstream boundary of a segment exactly at the end of a time step, the flow area at that point is found by iteratively tracing the characteristic path backward from the downstream boundary to the beginning of the time step. At that point the flow area is interpolated from the two nearest characteristics and used in equation 10 or 12, whichever is applicable, to compute the area at the downstream boundary. This approach is different from previously published solutions (Li et al., 1975, 1978, and Borah et al., 1980) that route individual characteristics consecutively through all time steps. The solution deviates from a truly analytical solution because of the interpolation that is necessary during each time step.

426

In the past, use of the method of characteristics has been hampered by the presence of kinematic shocks. In the present method those model segments that are subject to kinematic shocks (channel segments with upstream inflows) are treated by a modification of the procedure outlined above. For each time step the nonlinear relationship expressed by equation 5 is approximated, over the range of flow-area occurring during the time step, by a linear relationship of the form

$$Q = \alpha^*A + b \tag{13}$$

where α^* and b are constants. The constant α^* (which varies with each time step) is used in place of α in equations 9 and 11 to route characteristics through a time step. Since equation 13 is linear, the characteristic paths defined by equations 9 and 11 are linear and also parallel over a time step. This prevents the crossing of characteristic paths and thus avoids the problem of kinematic shock. In the application of this method, one is effectively representing the curve of equation 5 by a series of straight line segments. This is a good approximation as long as the range of flow areas occurring during each time step is not large. Testing of the method thus far has shown that very little error is introduced by this approximation, and solutions are close to analytical solutions.

The method of characteristics is provided as an option in the model. Its use is recommended for the modeling of small watersheds where model segments are short and no attenuation of routed hydrographs is expected. It is possible that, for these situations, the method of characteristics solution may be superior to finite-difference solutions. The artificial smoothing and numerical dispersive effects that are unavoidable in finite-difference schemes may introduce some error in simulated watershed hydrographs. This error may be important if the peak discharge is of interest. Borah et al. (1980) offer some interesting comparisons between the method of characteristics and finite-difference solutions applied to a variety of unsteady-flow problems, ranging from simple cascades to complex natural watersheds. While the superiority of the method of characteristics in reproducing the observed watershed response at this point has not been proven, it can at least be said that the method offers the advantage of an efficient algorithm that does not suffer from problems with numerical stability. It also relieves the model user of the need to carefully select proper time and space increments that are necessary in finite-difference methods to achieve good precision.

Finite-difference methods

A second option is available in DR$_3$M for solving equations 3 and 5 by a nonlinear finite-difference method. The method is a four-point implicit formulation that requires an iterative procedure to solve for the unknown discharge. The linear explicit finite-difference method that was contained in the 1978 release of the model is now used to obtain the initial estimate of the unknown discharge for the nonlinear method. The user has the option of using the linear method by itself with no iterations, if it is desirable to save computer time and if the accuracy is acceptable. This provides the third flow routing option in the model. A space interval, Δx, and a time interval (time step), Δt, form the computational box for the finite-difference method. Four points of a computational box are represented in figure 2. The purpose of the finite-difference method is to solve for A and Q at point d, given values of A and Q at points a, b, and c.

427

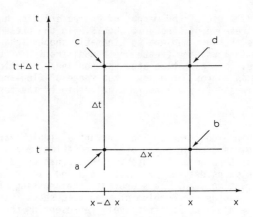

Figure 2.--Four-point finite-difference mesh

In the nonlinear method the continuity equation is represented by a finite-difference equation using quantities at all four corners of the box and a weighting factor for the space derivative. The equation can be written as

$$\frac{\gamma(Q_d-Q_c) + (1-\gamma)(Q_b-Q_c)}{\Delta x} + \frac{(A_d-A_b) + (A_c-A_a)}{2\Delta t} = q \qquad (14)$$

where γ is the weighting factor that is assigned by the user to a value between 0.5 and 1.0.

Equation 14 has two unknowns, Q_d and A_d, but they are related by equation 5. By rewriting equation 5 in the form

$$A_d = \left(\frac{Q_d}{\alpha}\right)^{1/m} \qquad (15)$$

and substituting for A_d in equation 14, the resulting equation is non-linear with one unknown, Q_d.

The solution for Q_d is done by an iterative procedure using Newton's second order method for finding the roots of an equation. How rapidly the iterative procedure converges to a correct solution depends on how good a first estimate is made for the unknown discharge. To speed convergence, DR3M obtains the first estimate using a linear explicit finite-difference method that was presented by Leclerc and Schaake (1973).

The linear method requires the use of two different finite-difference equations. The selection of the appropriate equation to use within each computational box depends on a stability parameter

$$\theta = m \frac{Q_b}{A_b} \frac{\Delta t}{\Delta x} \qquad (16)$$

If θ is greater than or equal to unity, finite differences are written using grid points a, c, and d. The continuity equation is represented as

$$\frac{Q_d - Q_c}{\Delta x} + \frac{A_c - A_a}{\Delta t} = q \qquad (17)$$

and can be solved directly for Q_d.

If Θ is less than unity, finite differences are written using grid points a, b, and d. The continuity equation is represented as

$$\frac{Q_b - Q_a}{\Delta x} + \frac{A_d - A_b}{\Delta t} = q \qquad (18)$$

and can be solved for A_d. Once A_d is known, Q_d can be determined by equation 5. The linear explicit method is computationally very fast and, if used alone, can save computer time.

RUNOFF QUALITY

The water-quality component of DR3M has been developed as a separate program and is referred to as DR3M-QUAL. DR3M-QUAL operates similarly to DR3M. The model provides detailed simulation of the quality of storm runoff during days for which short-time interval (1-minute to hourly) data are input to the program (unit days). Between unit days, daily precipitation data are used to provide a daily accounting of constituent accumulation on the effective impervious areas of the watershed.

DR3M-QUAL can be operated as either a lumped-parameter model or as a distributed-parameter model. As a lumped-parameter model, spatial variations in model parameters are not accounted for. This approach uses runoff values at the outlet from the watershed and assumes that the runoff load originates entirely from the constituent load washed off the effective impervious area of the watershed and the constituent load in precipitation. Pervious-area contributions to runoff loads are not accounted for. The effects of street sweeping on runoff quality can be simulated. The runoff values can be included in the input deck to DR3M-QUAL or can be read from flow files created during a run of DR3M.

As a distributed-parameter model, DR3M-QUAL can simulate impervious-area, pervious-area, and precipitation contributions to runoff quality. Impervious-area and pervious-area model parameters can have different values at different locations in the watershed. The effects of street sweeping and detention storage on runoff quality can also be simulated. Segment flow data needed by DR3M-QUAL for a distributed-parameter run are read from flow files generated by DR3M. Because DR3M and DR3M-QUAL are separate programs, many water-quality runs can be made without the expense of repeating the same flow routing for every run.

Impervious-Area Runoff Quality

The impervious-area runoff quality component simulates the processes of constituent accumulation and constituent washoff and accounts for the removal of constituents by street sweeping.

Constituent accumulation

Several studies have suggested that the rate of accumulation of water-quality constituents on urban surfaces is nonlinear and that there is a limit to the amount of constituents that can accumulate between storms, regardless of the length of dry period. Data collected by

Sartor and Boyd (1972) and Pitt (1979) suggest that the accumulation rate is largest for several days after a period of street cleaning or rainfall, and then the rate decreases and approaches zero. Apparently constituents are resuspended by wind and land-use activities such as vehicles moving along a highway. This phenomena can be modeled as:

$$\frac{dL}{dt} = K - K_2 L \qquad (19)$$

where L is the amount of constituent on effective impervious areas, in kilograms per hectare; K is a constant rate of constituent deposition, in kilograms per hectare per day; K_2 is a rate constant for constituent removal, in day^{-1}; and t is time, in days. Integration of equation 19 yields:

$$L = K_1[1 - \exp(-K_2 T)] \qquad (20)$$

where $K_1 = K/K_2$ is the maximum amount of constituent on effective impervious areas, in kilograms per hectare; and T is time since last period of street sweeping or storm runoff (accumulation time), in days.

Traditionally, equation 20 has been derived with the assumption that urban impervious surfaces were completely washed by the last cleaning, either mechanical (street sweepers) or by storm runoff. In order to eliminate this assumption, T in equation 20 can be redefined as the equivalent accumulation time:

$$T = t + t_e \qquad (21)$$

where t is the time since last cleaning, and t_e is the time required for a land-surface load to accumulate equal to that at the end of the last period of street sweeping or storm runoff, assuming initially clean urban impervious surfaces. The variable t_e is computed as:

$$t_e = -\frac{1}{K_2} \ln \left(1 - \frac{L_e}{K_1}\right) \qquad (22)$$

where L_e is the land-surface load in kilograms per hectare at the end of the last period of street sweeping or storm runoff. Equation 22 is derived from equation 20 by substituting L_e for L and solving for $T = t_e$.

Figure 3 illustrates the exponential accumulation equations. The upper curve accounts for a residual amount of constituent remaining on effective impervious surfaces at the end of the last period of street sweeping or storm runoff, whereas the lower curve assumes no residual.

Constituent washoff

Constituent washoff from effective impervious areas can be simulated using an exponential washoff equation:

$$WSHOFF = L_0[1 - \exp(-K_3 R \Delta t)] \qquad (23)$$

where WSHOFF is the amount of constituent removed from the effective impervious area during a time step, in kilograms; L_0 is the amount of constituent on the effective impervious area at the beginning of the time step, in kilograms; K_3 is the washoff coefficient; R is the runoff rate, in millimeters per hour[1]/; and Δt is the time step, in hours.

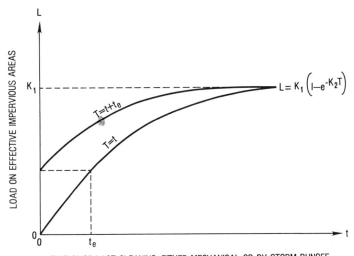

Figure 3.--Constituent accumulation

Constituent concentrations are determined by dividing WSHOFF by $AR\Delta t$ where A is the effective impervious area of the watershed. The primary assumption for use of equation 23 is that the rate of constituent washoff is proportional to the amount remaining on the land surface. Equation 23 is used to simulate constituent washoff in several well-known urban runoff models such as SWMM (Metcalf and Eddy, Inc., and others, 1971) and STORM (U.S. Army Corps of Engineers, 1976). Because washoff in equation 23 is a function of the product of runoff intensity and duration, the amount of constituents washed off during a storm is assumed to be a direct function of the total volume of storm runoff.

During simulation on a daily basis, the washoff of constituents is accounted for as well as dry-weather accumulation previously described. Because washoff simulated by the exponential washoff equation (equation 23) is a function of the volume of storm runoff, it can be rewritten for daily accounting as:

$$WSHOFF_d = L_{o,d}[1 - \exp(-K_3(PPT-IMP))] \qquad (24)$$

where $WSHOFF_d$ is the amount of constituent removed from the effective impervious area during a day, in kilograms; $L_{o,d}$ is the amount of constituent on the effective impervious area at the beginning of the day, in kilograms; PPT is the daily rainfall, in millimeters, and IMP is the impervious retention, in millimeters.

If the exponential washoff equation is applied on a distributed basis, the routing of constituents through a channel network and the mixing of contributions from different subbasins can result in many

---

1/ Millimeters of runoff refers to the volume of runoff in millimeters distributed over the effective impervious area of the watershed.

different distributions of simulated constituent concentrations over a storm hydrograph. However, applied on a lumped-parameter basis, the exponential washoff equation predicts decreasing concentrations of a given constituent with increasing time since start of storm.

Street sweeping

A schematic illustrating DR3M-QUAL's conceptualization of constituent accumulation and removal from effective impervious areas is shown in figure 4. Removal of constituents can be by rainfall-runoff or by street sweepers.

DR3M-QUAL simulates removal of constituents by street sweeping as:

$$L_s = L_0 - E(L_0 - L_r), \text{ if } L_0 > L_r \qquad (25a)$$

$$L_s = L_0, \text{ if } L_0 \leq L_r \qquad (25b)$$

where L_s is the amount of a constituent remaining after street sweeping, L_0 is the amount of a constituent before street sweeping, L_r is the base residual load which cannot be removed by street sweepers, and E is the efficiency (ranging from 0 to 1.0) of street sweepers at removing the load of a constituent in excess of the base residual load.

Only a part of the effective impervious area of a watershed (lumped-parameter simulation) or overland-flow segment (distributed-parameter simulation) may be swept by street sweepers. This is accounted for in the model as:

$$L_A = L_s \cdot SSAREA + L_0(1 - SSAREA) \qquad (26)$$

where L_A is the adjusted amount of the constituent on effective impervious surfaces remaining after street sweeping and SSAREA is the fraction of the effective impervious area that is swept.

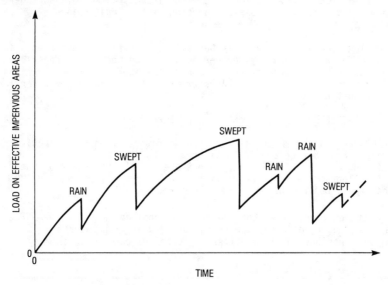

Figure 4.--Constituent accumulation and removal by rainfall and street sweeping

Street sweeping can be simulated according to a fixed schedule, or actual days during which street sweeping occurred can be input to the model. Street sweeping operations are suspended on unit days or on daily accounting days when rainfall exceeds impervious retention. If street sweeping is suspended because of rainfall, the model assumes that the streets are not swept until the next day on which street sweeping was scheduled. This assumption conforms to common street sweeping practices of public works departments.

Pervious-Area Runoff Quality

Little information is available on the contribution of pervious areas to urban runoff quality. Conceptually, constituents in the pervious-area runoff are classified according to two sources: (1) that part dissolved or desorbed from the soil matrix into the flowing water without soil erosion, and (2) that part associated with soil erosion. Mathematically,

$$CP(J) = K_6 + K_7 \cdot SED(J) \tag{27}$$

where $CP(J)$ is the concentration of a constituent in pervious-area runoff at time step J, K_6 is a constant representing the "non-erosion" contribution, K_7 is a constant representing the ratio of the constituent concentration to the sediment concentration, and $SED(J)$ is the concentration of sediment in the pervious-area runoff at time step J.

Williams (1975) has determined that the total sediment yield resulting from a period of storm runoff can be estimated by:

$$TMASS = PFAC1 \ (V_R Q_p)^{PFAC2} \tag{28}$$

where TMASS is the total sediment yield in metric tons, V_R is the storm runoff volume in cubic meters (m^3), Q_p is the peak flow rate in m^3/s, and PFAC1 and PFAC2 are constants.

Based on studies on 18 watersheds in Nebraska and Texas, Williams (1975) found that PFAC2 could be estimated as 0.56 and PFAC1 as a function of parameters in the Universal Soil Loss Equation (Wischmeier and Smith 1965):

$$PFAC1 = 11.8 \cdot K \cdot LS \cdot C \cdot P \tag{29}$$

where K is the soil erodibility factor, LS is the slope length gradient ratio, C is the cropping management factor, and P is the erosion-control practice factor. Equations 28 and 29 explained about 92 percent of the variation in sediment yield from the 16 Nebraska and two Texas watersheds (Williams, 1975). The transferability of the above relationships for PFAC1 and PFAC2 is undocumented and use of locally collected data is recommended for estimating these parameter values.

Instantaneous sediment concentrations can be estimated from TMASS by assuming that sediment concentrations are proportional to instantaneous flow rates. Based on studies by Rendon-Herrero (1974), Kuo (1975), and Williams (1978), this appears to be a reasonable assumption.

Precipitation Quality

Urban runoff can be viewed as chemically modified rainwater. Accounting for the original chemical composition of rainwater may be

433

important in terms of both model reliability and evaluation of management practices. For example, precipitation contributions to runoff loads would be unaffected by street sweeping practices.

DR$_3$M-QUAL accounts for precipitation contributions to runoff quality by adding a concentration representing precipitation quality to the concentrations predicted from constituent washoff. This adjustment can be a constant for all storms, can vary monthly, or can vary on a storm-by-storm basis. Although it is well known that the chemistry of precipitation can change during a storm, few data exist to quantify this effect. Thus assigning a constant concentration to precipitation on a storm basis is the maximum level of detail currently included in the model.

Constituent Transport

Application of DR$_3$M-QUAL as a lumped-parameter model requires runoff data at the watershed outlet. For applications of DR$_3$M-QUAL as a distributed-parameter model, a drainage basin is represented as a set of segments which jointly describe the drainage features of the basin. The same types and designations of segments apply for DR$_3$M-QUAL as for DR$_3$M. That is, segments are designated as channel, overland-flow, or reservoir segments. Constituent loads from overland-flow segments are determined using the discharge from these segments and the algorithms for impervious-area, pervious-area and precipitation quality previously described. These loads are contributed laterally to channel segments.

Channel segments

The transport of constituents through channels is performed using a Lagrangian method that is precise and conceptually simple. The method is designed to solve the simplest transport problem, the plug flow (dispersionless) transport of a conservative substance. A necessary feature of the method is that it accounts for pollutants that enter channel flows by lateral inflow.

In a Lagrangian method, one conceptually follows an individual fluid parcel while keeping track of all factors which tend to change its concentration. If dispersion (mixing between parcels) is assumed negligible, and the pollutant being modeled is considerd conservative, then the only factor that will tend to change the concentration of a parcel as it moves downstream is the addition of lateral flow of a different concentration.

The Lagrangian method in DR$_3$M-QUAL operates on an individual channel segment. Knowing the time-varying concentrations of the upstream and lateral flows entering a segment, the method is used to determine the time-varying concentrations of the flow leaving the segment. The method is really no more than a bookkeeping that keeps track of the volume and concentration of fluid parcels that are in a segment during a time step.

Considering a channel segment, the method operates on a time-step basis. During each time step, a parcel enters the segment and is assigned the volume and concentration of the flow entering from the upstream end. At the beginning of a time step there are anywhere from one to many parcels in the segment, and the number of parcels increases by one when the new parcel enters. The lateral inflow that enters the segment during a time step is distributed among all parcels in proportion to parcel volumes. The volume and concentration of each parcel is adjusted to account for the addition of the lateral inflow.

434

At the downstream end of the segment, part of one or more parcels leaves the segment depending on the volume of flow that leaves the segment during the time step. At the end of a time step the concentration of the parcel at the downstream end of the segment is saved in an array and the number of parcels is updated. After completing all time steps, this array of concentrations is used as input to the next downstream channel segment.

Reservoir segments

DR3M-QUAL transports water-quality constituents through reservoir segments (detention basins) using plug flow concepts. Plug flow assumes no mixing between plugs and routes the flow on a first-in, first-out basis.

Particles entering a reservoir are assumed to settle according to Stokes' Law and particles are considered trapped as soon as they reach the reservoir bed. Resuspension or saltation of particles is not accounted for. The model accounts for the variation of constituent concentrations with depth by subdividing each plug into as many as eight layers. Selective withdrawal at the reservoir outlet from these layers is provided for in the model. Because of the short detention times that generally occur in urban detention basins, chemical reactions are not considered by the model. Many of the concepts used by this model have been adapted from the DEPOSITS (DEposition Performance Of Sediment In Trap Structures) model documented by Ward and others (1979).

Stokes' Law has generally been applied to problems involving estimation of the removal of discrete particles in settling tanks for water treatment facilities (Fair and others, 1968). However, urban detention basins, unlike water treatment facilities, are generally not well defined hydraulic structures. Rather, these basins are often irregular in shape. Basin geometry in DR3M-QUAL is determined by the input of a stage-area curve. This curve is input as a set of points defining the surface area of the basin at selected stages. From this set of points a stage-capacity-average depth curve is defined.

The rate at which water-quality constituents settle in a detention basin depends to a large extent on the particle-size distribution of the constituent. For this reason, the input to the model includes a particle-size distribution curve for influent concentrations of each water-quality constituent to each detention basin. Although particle-size distributions of the influent should vary throughout a storm event, no data presently exist to accurately quantify this effect. Therefore, for a given constituent and detention basin, the particle-size distribution of the influent is assumed to remain constant.

DR3M-QUAL retrieves the inflow and outflow hydrographs to and from a reservoir segment from the segment flow files generated by DR3M. The outflow hydrograph is then subdivided into plugs of flow of equal time increment. The time increment used is the unit time interval specified in the model. Figure 5 shows the location of plug number 7 on typical inflow and outflow hydrographs for a unit time interval of 1 hour. The starting point location of the plug on the inflow hydrograph is determined by first ascertaining the point at which the accumulated inflow (point A on inflow hydrograph) is equal to the accumulated outflow (point C on outflow hydrograph). The ending point location of the plug on the inflow hydrograph (point B) is determined in a similar manner but based on point D. The detention time (DETTME) of the plug is then assumed equal to the time between the centroids of the inflow and out-

435

flow plugs as shown on figure 5. The constituent mass of the inflow plug is determined by linear interpolation at the plug starting and ending times (A and B) of a curve relating accumulated constituent mass of the inflow and time since start of storm.

Finally, the average depth of a plug (DEPTH) while in a detention basin is determined by first computing the area under the average depth versus time curve contained between the centroid of the plug on the inflow hydrograph and the centroid of the plug on the outflow hydrograph. This area is then divided by the detention time of the plug (DETTME). This computation is illustrated on figure 5. The average depth versus time curve is obtained by linear interpolation of the stage-average depth curves previously defined from the basin geometry.

The model assumes that constituents are evenly distributed through-out a plug at its entrance to the detention basin. Constituents then gradually settle out as the plug moves through the basin. Settling of discrete nonflocculating particles that settle without hindrance from other particles can be described by Stokes' Law:

$$V_s = \frac{g}{18} (SG - 1) \frac{D^2}{\nu} \tag{30}$$

where V_s is the settling velocity of a particle in cm/s, g is the gravitational constant in cm/s^2, SG is the specific gravity of the particle, D is the particle diameter in cm, and ν is the kinematic viscosity in cm^2/s.

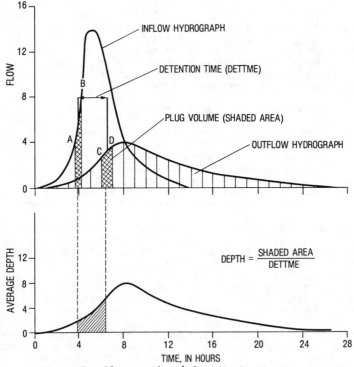

Figure 5.--Plug routing (after Ward and others, 1979)

All particles with a settling velocity greater than a critical velocity are assumed to be removed from suspension. The critical velocity (V_0) is defined as:

$$V_0 = DEPTH/DETTME \qquad (31)$$

where DEPTH and DETTME for a plug are determined as on figure 5.

If N (out of N_T) particles having a settling velocity $V_s \leq V_0$ compose a particle-size fraction, the proportion X of the particles to be removed is:

$$X = N/N_T = V_s/V_0 \qquad (32)$$

The size-weight composition of a particular suspension can be expressed by a curve showing the cumulative frequency distribution of settling velocities (see figure 6). Defining F_0 as the fraction of particles having settling velocities $V_s \leq V_0$, then r_1 of the particles have settling velocities $V_s \geq V_0$ and are thus totally removed, where

$$r_1 = 1 - F_0 \qquad (33)$$

The fraction of the remaining particles with settling velocities $V_s < V_0$ that are removed is:

$$r_2 = \int_0^{F_0} (V_s/V_0)dF \qquad (34)$$

The fraction of particles remaining in suspension (R) is:

$$R = 1 - (r_1 + r_2)$$
$$= F_0 - \frac{1}{V_0} \int_0^{F_0} V_s dF \qquad (35)$$

The integral in equation 35 can be evaluated by substituting Stokes' Law for V_s.

Each plug is subdivided into as many as eight layers of equal depth. The amount of constituent remaining in suspension within each layer is computed using equation 35.

The model provides for various outflow distributions based on the value of the model parameter JFLOW:

JFLOW = 1 -------- uniform withdrawal with depth.

JFLOW = 2 -------- withdrawal only from top layer.

JFLOW = 3 -------- withdrawal only from the bottom layer.

JFLOW = 4 -------- withdrawal is uniform, from the top layer or from the bottom layer depending on the relative positions of the stage of the riser crest and the reservoir stage.

If JFLOW = 4, the model compares the reservoir stage at the time step of interest with the stage of the crest of the riser. If the

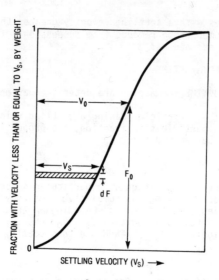

Figure 6.--Cumulative frequency distribution of settling velocity

stage of the riser crest is in the top layer, then the flow is assumed
to come completely from the top layer. Likewise, if the stage of the
riser crest is contained within the bottom layer, then the flow is
assumed to come completely from the bottom layer. If the stage of
the riser crest is contained in neither the top nor bottom layers,
then a uniform withdrawal with depth is assumed.

In an ideal settling pond, plug flow will occur and the flow will
pass through the pond on a first-in, first-out basis. However, flow
in all detention basins is characterized by mixing, turbulence,
short-circuiting, and resuspension. A well designed detention pond
will minimize these factors. A pond which is likely to exhibit a
high degree of short-circuiting, mixing, and turbulence will probably
have a low trap efficiency and will not be suitable for evaluation with
the detention storage algorithms included in this model, although
the model can account for dead storage (previously stored volumes of
flow which are bypassed during the flow event).

MODEL APPLICATIONS

DR3M and DR3M-QUAL can be used for a wide variety of applications.
Applications of DR3M will be discussed first followed by a discussion
of DR3M-QUAL's applications.

DR3M

DR3M can be applied to drainage basins ranging from tens of acres
to several square miles. It is not generally recommended for use on
drainage basins over 10 square miles unless sufficient rainfall data
are available to adequately define its spatial variation. The model
does not have a baseflow component. However, baseflow or upstream inflow
to the watershed can be input to the model through use of an input-
hydrograph or input-discharge point. The model is intended primarily
for application to urban or urbanizing watersheds. It may have limited
use for rural applications where baseflow and interflow contributions to
runoff are either negligible or can be estimated and input to the model.

438

The capability to use the same overland-flow segment repeatedly throughout the watershed can be used to define short distances of overland flow representative of many rural watersheds without use of an overwhelming number of segments.

DR$_3$M can be calibrated and verified using data collected over a short period of time. Long-term historical records of rainfall can then be input to the model to extend the records of storm-runoff. Rainfall and runoff data collected by the U.S. Geological Survey can be retrieved from WATSTORE (National Water Data Storage and Retrieval System) in the format required by the model. Long-term records of rainfall data at short-time intervals (usually 5 or 10 minutes) can be obtained from the National Weather Service for many cities in the U.S. The data are for anywhere from 3 to over 10 "major" storms per year for a period of record often exceeding 50 years. Much of these data are stored in the U.S. Geological Survey WATSTORE computer files and can be retrieved in the format required by the model. Records of daily precipitation and daily evaporation for the period of record spanned by the short-time interval data can also be retrieved from WATSTORE for most of these stations.

DR$_3$M can be used for urban-basin planning purposes by its determination of the hydrologic effects of different development configurations. A set of model segments can be arranged easily into a network that will represent many complex drainage basins. Certain assumptions would have to be made to determine the changes required in model parameters to represent various types of development. Examples of the above application of the model might include assessing the effects of increased impervious cover, detention ponds, or culverts on runoff volumes and peak flows.

The separation of rainfall excess computations and flow routing in the model results in several advantages of DR$_3$M over most other rainfall-runoff models. The first of these advantages is that the soil-moisture accounting and infiltration parameters as well as the effective impervious area can be calibrated through repeated application of the Rosenbrock algorithm without having the expense of routing at each iteration. Secondly, a long-term sequence of runoff volumes can be inexpensively simulated. This information could be useful for purposes such as design of detention storage facilities (Raasch, 1979) or for determination of runoff volumes for pollutant load computations.

The assumptions behind the kinematic wave equations for channel routing should be recognized by any potential user of the model. The kinematic wave solution is based on the assumption that disturbances are allowed to propagate only in the downstream direction and that the acceleration terms in the equation of motion are negligible. Therefore, the model does not account for backwater effects, flow reversal, or dynamic acceleration effects in channels of small slopes. In addition, the capacity of circular-pipe segments is limited to nonpressurized-flow capacity.

DR$_3$M-QUAL

DR$_3$M-QUAL can be used for a wide variety of applications. The model is continuous and, hence, an accounting of impervious-area constituent accumulation is made between storm events. Rather than operating on a fixed time step, the model provides short-time interval simulation of storm events specified by the user and a daily accounting of constituent accumulation between storm events. Therefore, many of the advantages of continuous simulation are combined with those of a single event model.

DR$_3$M-QUAL can be used to generate long-term records of storm-runoff loads. These can be used to estimate frequency distributions of loads or for comparison with concurrent flow rates and quality characteristics of receiving waters. Also, an option exists in the model to output impervious-area runoff loads simulated during the daily accounting phases of a model run.

The model can be run on one of three spatial modes:

1. Lumped-parameter

2. Distributed (no transport)

3. Distributed transport

As a lumped-parameter model, DR$_3$M-QUAL uses runoff data from the watershed outlet. These data can be included in the input to the model or can be read from flow files generated by DR$_3$M. Runoff loads are assumed to originate predominantly from the effective impervious areas of the watershed. The direct-search technique devised by Rosenbrock (1960) can be used to aid in determining "optimum" parameter values. Lumped-parameter simulations can be used to inexpensively estimate impervious-area model parameters for later, more detailed distributed-parameter runs.

When an application involves determining the effects of runoff quality on the quality of local receiving waters, the time interval of interest may be days or even weeks or months, while DR$_3$M-QUAL simulation intervals are on the order of minutes. Thus, the magnitude of storm-runoff loads may be much more important than within storm variations. For this reason DR$_3$M-QUAL contains an option for distributed simulations without constituent transport. Model simulations in this mode are equivalent to a distributed-parameter run with instantaneous transport in place of Lagrangian transport. Considerable savings in computer costs can be made by using this mode. Unlike lumped-parameter simulations, pervious-area contributions to runoff loads can be accounted for as well as spatial variations in impervious-area runoff-quality parameters. Like lumped-parameter simulations, distributed (no transport) runs can be used for initial calibration of model parameters prior to final calibration as a distributed transport model. Output from distributed (no transport) runs includes storm-runoff loads but no information about within-storm variations is given. Because of the importance of transport computations in reservoirs, distributed (no transport) runs cannot account for the effects of reservoir segments.

Both DR$_3$M and DR$_3$M-QUAL have been designed for ease of calibration. A user has the option of reading in measured runoff and runoff-quality data for numerical and graphical comparisons. Output from the models includes several different types of plots.

DR$_3$M is currently being calibrated and verified using rainfall-runoff data from drainage basins in Colorado, Florida, Georgia, Hawaii, Oregon, and Pennsylvania. Additional use of both DR$_3$M and DR$_3$M-QUAL is expected as part of the urban runoff program of the U.S. Geological Survey and U.S. Environmental Protection Agency. User's manuals for the current version of DR$_3$M and DR$_3$M-QUAL are currently being written.

REFERENCES

Borah, D.K., Prasad, S.N., and Alonso, C.V. 1980. Kinematic Wave Routing Incorporating Shock Fitting. Water Resources Research, Vol. 16, No. 3, pp. 529-541.

Crawford, N.H. and Linsley, R.K. 1966. Digital Simulation in Hydrology, Stanford Watershed Model IV, Technical Report No. 39, 210pp., Civil Engineering Dept., Stanford University.

Dawdy, D.R., Lichty, R.W. and Bergmann, J.M. 1972. A Rainfall-Runoff Simulation Model for Estimation of Flood Peaks for Small Drainage Basins. Professional Paper 506-B, 28pp., U.S. Geological Survey.

Dawdy, D.R., Schaake, J.C., Jr. and Alley, W.M. 1978. User's Guide for Distributed Routing Rainfall-Runoff Model. Water-Resources Investigations 78-90, 146 pp., U.S. Geological Survey.

Eagleson, P.S. 1970. Dynamic Hydrology, McGraw-Hill, New York.

Fair, G.M., Geyer, J.C. and Okun, D.A. 1968. Water and Waste Water Engineering, John Wiley and Sons, Inc., New York.

Green, W.H. and Ampt, G.A. 1911. Studies on Soil Physics; I, Flow of Air and Water Through Soils. Journal Agriculture Research, Vol. 4, pp. 1-24.

Harley, B.M., Perkins, F.E., and Eagleson, P.S. 1970. A Modular Distributed Model of Catchment Dynamics. Hydrodynamics Laboratory Report No. 133, Massachusetts Institute of Technology.

Kuo, C.Y. 1975. Evaluation of Sediment Yield due to Housing Construction--A Case Study, Department of Civil Engineering, Old Dominion University, Norfolk, Virginia.

Leclerc, G. and Schaake, J.C., Jr. 1973. Methodology for Assessing the Potential Impact of Urban Development on Urban Runoff and the Relative Efficiency of Runoff Control Alternatives. Report No. 167, 257pp., Ralph M. Parsons Laboratory, Massachusetts Institute of Technology.

Li, R.M., Simons, D.B., and Stevens, M.A. 1975. On Overland Flow Water Routing. in: National Symposium on Urban Hydrology and Sediment Control, pp. 237-244, Proceedings: University of Kentucky, Lexington.

Li, R.M., Eggert, K.G., and Simons, D.B. 1978. An Interactive Digital Watershed Simulation For Assessing Storm Water Runoff. in: International Symposium on Urban Storm Water Management, pp. 103-112, Proceedings: University of Kentucky, Lexington.

Metcalf and Eddy, Inc., University of Florida and Water Resources Engineers, Inc. 1971. Storm Water Management Model, EPA-11024 Document 07/71, 4 volumes, U.S. Environmental Protection Agency.

Pitt, R. 1979. Demonstration of Nonpoint Pollution Abatement through Improved Street Cleaning Practices. EPA-600/2-79-161, 269pp., U.S. Environmental Protection Agency.

Raasch, G.E. 1979. Urban Storm Water Detention Sizing Technique.

in: International Symposium on Urban Storm Runoff, July 23-26,
pp. 55-60, Proceedings: University of Kentucky, Lexington.

Rendon-Herrero, O. 1974. Estimation of Washload Produced on Certain
Small Watersheds, Vol. 100, No. HY7, pp. 835-848.

Rosenbrock, H.H. 1960. An Automatic Method of Finding the Greatest
or Least Value of a Function, Vol. 3, pp. 175-184, Computer Journal.

Sartor, J.D. and Boyd, G.B. 1972. Water Pollution Aspects of
Street Surface Contaminants, EPA-R2-72-081, 236pp., U.S.
Environmental Protection Agency.

U.S. Army Corps of Engineers. 1976. Storage, Treatment, Overflow,
Runoff Model (STORM), Hydrologic Engineering Center, Davis,
California.

Ward, A.J., Haan, C.T. and Tapp, J. 1979. The DEPOSITS Sedimentation Pond Design Manual, 190pp., Institute for Mining and
Minerals Research, University of Kentucky, Lexington.

Williams, J.R. 1975. Sediment-Yield Prediction with Universal
Equation Using Runoff Energy Factor. in: Present and
Prospective Technology for Predicting Sediment Yields and
Sources, ARS-5-40, pp. 244-252, Agricultural Research Service.

Williams, J.R. 1978. A Sediment Graph Model Based on an Instantaneous
Unit Sediment Graph, Vol. 14, No. 4, pp. 659-664, Water
Resources Research.

Wischmeier, W.H. and Smith, D.D. 1965. Predicting Rainfall-Erosion
Losses from Cropland East of the Rocky Mountains, Handbook No. 282,
Agricultural Research Service.

Section 5
APPLICATION IN FOREST ENVIRONMENT

SIMULATING HYDROLOGIC BEHAVIOR IN THE OUACHITA MOUNTAINS OF CENTRAL ARKANSAS

Thomas L. Rogerson
Research Forester
Southern Forest Experiment Station
Forest Service, USDA, Fayetteville, Arkansas 72701

ABSTRACT

A simple simulation model to predict the water balance and hydrologic changes from thinning small pine-hardwood drainages in central
Arkansas is described. The model simulates the water balance on a
daily time interval. Rainfall input to the model can be either observed
values or simulated by a submodel developed from 13 years of rainfall
records. Variables needed, besides rainfall, are initial soil water
deficits in the surface foot and profile and three parameters--stand
basal area and maximum soil water deficits in the surface foot and
profile. Model output includes a rainfall and runoff summary, a soil
water deficit graph, and a monthly and annual water balance.

Managing forests to optimize timber, water, wildlife, and recreational resources is becoming increasingly difficult as resource demands
become greater and more complex. Thinning and clearcutting, the most
common timber management practices, greatly affect other forest resources.
Determining the effects of these practices on the hydrology of an area
would require years of study at considerable expense. However, hydrologic simulation models can evaluate different thinning levels in a few
minutes of computer time. In addition, simulation models help determine
the most sensitive parameters or those that have the greatest influence
on the response of the model (Hill et al., 1972). These parameters should
be measured as precisely as possible when collecting field data.

Most hydrologic simulation models are complex and require many hydrologic variables and watershed parameters (Crawford and Linsley, 1966; Holtan
and Lopez, 1971). This investigation was designed to develop a simple simulation model that would predict hydrologic behavior before and after vegetative forest management practices on small pine-hardwood drainages.

DESCRIPTIONS OF DRAINAGES

Hydrologic and vegetation data from three small, 1.28- to 1.63-acre
(.52 to .66 ha), drainages were used for model development. The drainages
have a northeast aspect and a slope of 15 percent. Soils are Goldston,

Herndon, and Alamance stony loams. The soils are shallow, 2½ to 3 feet
(76 to 91 cm) deep, and moderately permeable. Water storage capacity is low
because of the thin solum. Parent material is predominately uplifted shale
with some sandstone interbedding.

Annual precipitation on the drainages averaged 55.0 inches (1397 mm)
for the 1961-1973 period and ranged from 39.9 to 82.8 inches (1013 to 2103
mm). Precipitation is uniformly distributed throughout the year. Almost
all precipitation is rain, but light snow may fall one or two times each
year. Runoff is ephemeral and almost always ends within 24 hours of the
rainfall event. Mean annual temperature is 62.1°F (16.7°C); the January
mean is 42.3°F (5.7°C) and the July mean is 81.6°F (27.6°C).

The initial vegetative cover consisted of a shortleaf pine (Pinus
echinata Mill.) overstory of 105 ft²/acre (24.1 m²/ha) and a mixed hardwood
understory of 35 ft²/acre (8.0 m²/ha). In the spring of 1970, the pine overstory was logged and the hardwood understory was injected with herbicide
on one drainage; on another, the understory was injected, and the overstory
was thinned to 60 ft²/acre (13.8 m²/ha); no thinning or clearing was done
on the third drainage.

THE MODEL

The model was developed with data from the three drainages for calendar
years 1968 and 1969. It was written in GPSS, an IBM discrete simulation
language (IBM 1971, 1973), which is available on IBM and Univac computers.
The language offers great flexibility and allows model modifications without extensive reprogramming. The model was programmed for the IBM 370-155
computer at the University of Arkansas. It was classified as stochastic-
empirical, according to Clarke's (1973) definitions: It is stochastic
in that the rainfall variable is random and has distributions in probability, and it is empirical in that it is based on observations and experimental data.

Hydrologic variables of rainfall, soil water in both the surface foot
and profile (including the surface foot), and runoff, along with the
watershed parameter of basal area, were used to estimate the equations,
functions, and other parameters in the model.

The generalized hydrologic model is:

$$PR = RO + ET + DS \pm \triangle SW$$

where

PR = precipitation
RO = runoff
ET = evapotranspiration
DS = deep seepage
$\triangle SW$ = change in soil water

Input

The model can be run with two input modes. The first mode uses observed
storms on a daily basis as input and the second mode uses simulated storms.
A storm was defined as a period of rain followed by at least 12 hours without
rain. The time unit selected for this model was one day, so only one storm
per day could occur. Because the model only allows one storm per day, if two
storms were observed on the same day, the storm nearest to another day was
said to occur on that day. Likewise, an observed storm lasting two or more
days was credited to the day on which most of the rain fell.

446

In the first mode, observed rainfall events for the period to be simulated are stored in a rainfall function by date (January 1 = 1 and December
31 = 365 or 366). Days with no rainfall are entered as zeros. This mode was
used in developing the model (Rogerson, 1976) and in testing it against actual
hydrologic data.

The second mode uses a rainfall submodel (Rogerson, 1980) to simulate
rainfall events from four functions that determine days to next storm and
amount of rainfall. The submodel was developed from 13 years of rainfall data
from the study area. The submodel assumes that whatever happens one day (storm
or no storm) is independent of the previous day.

Other inputs needed are soil water deficits in the surface foot and in the
profile—two variables which describe initial conditions in the model—and
three parameters—basal area and maximum soil water deficits in the surface
foot and in the profile.

Precipitation—Throughfall Phase

The simulation begins each day by determining if there is precipitation;
if not, the simulation proceeds to the evapotranspiration phase. if
precipitation occurs, storm size is either determined from the observed
values or simulated from parametric precipitation functions, depending
on the input mode selected. A throughfall value is then calculated from
a modified form of Rogerson's (1967) equation:

$$\text{THFAL} = (.98 \ PR - .00097 \ (PR \times BA) - .0184) \ (DGR)$$

where

THFAL = throughfall in inches
PR = precipitation in inches
BA = basal area in ft^2/acre
DGR = dormant-growing season function

The seasonal function (DGR) increases throughfall by 5 percent during the
November 16 - April 15 dormant season. This value was derived from
throughfall data (Lawson 1975) collected on the experimental drainages
being simulated.

Runoff Phase

Several conditions, developed from data from the study area, determine
whether runoff occurs. First, if throughfall is less than or equal to 0.35
inch (8.9 mm), there is no runoff. Second, if throughfall is greater than
0.35 (8.9 mm) but less than 1.00 inch (25.4 mm), runoff cannot occur unless
throughfall is greater than the soil water deficit in the surface foot. Third,
if throughfall is greater than or equal to 1.00 inch, (25.4 mm), there can
be runoff. An empirically derived, stepwise multiple regression equation
containing the variables throughfall and soil water deficits in both the
surface foot and profile determines the amount of runoff:

$$RO = .89498 \ (\text{THFAL}) - .15 \ (\text{THFAL}) \ (SWD_p) + (.25574/\text{THFAL})$$
$$+ .05597 \ (\text{THFAL} - SWD_l) - .03 \ (\text{THFAL} - SWD_p)$$
$$(SWD_p) - .919$$

where

RO = runoff in area inches
THFAL = throughfall in inches
SWD_l = soil water deficit in the surface foot in inches
SWD_p = soil water deficit in the profile in inches

This phase also reduces the soil water deficits for the surface foot

447

and profile by the throughfall amount if no runoff occurs or by the difference between throughfall and runoff if runoff has occurred.

Evapotranspiration Phase

Evapotranspiration for the surface foot and profile is simulated by multiplying factors derived from soil water deficit, vegetation, and maximum transpiration-surface evaporation functions, and then adding interception if precipitation has occurred.

The soil water deficit functions (figs. 1 and 2) were developed from soil water data collected on the drainages with a neutron probe system. Maximum soil water deficits were 2.56 inches (65 mm) for the surface foot and 5.60 inches (142 mm) for the profile. As a general rule, the surface

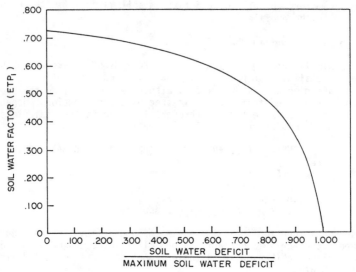

Figure 1. - Relationship between soil water deficit ratio and soil water factor for the surface foot.

foot has 2.5 inches (64 mm) of available water, and each additional foot has 1.5 inches (38 mm) in the root zone or to bedrock (whichever is shallower). These values could be used in the model if soil water values were unknown. Soil water deficit/maximum soil water deficit ratios were used instead of maximum deficits to make the model adaptable to other areas. When these ratios are zero, the soil water factors are 0.727 for the surface foot and 1.000 for the profile. Thus, when both the surface foot and the profile are saturated, nearly 73 percent of the transpiration-surface evaporation occurs in the surface foot. Both deficit factors remain near maximum until the ratios are greater than 0.5 and then they decrease rapidly.

The vegetation function (fig. 3) was derived from limited empirical findings from the three small watersheds with basal areas of 0, 60, and 140 ft^2/acre (0, 13.77, and 32.14 m^2/hectare). At 0 ft^2 the vegetation factor was 0.36 because of surface evaporation and transpiration by herbaceous species. Between 0 and 60 ft^2 (13.77 m^2) the factor increased only slightly because the trees occupied only a portion of the site. As the basal area increased to 140 ft^2 (32.14 m^2) and the site was more fully occupied, the vegetation factor rose rapidly; it was assumed to reach a

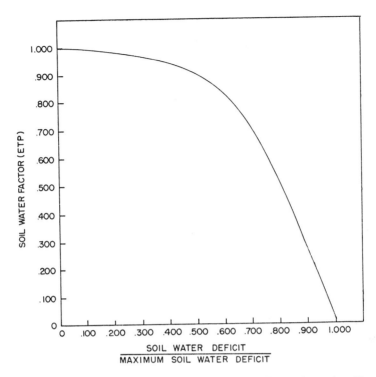

Figure 2. - Relationship between soil water deficit ratio and soil water factor for the profile.

maximum of 1.0 at 200 ft^2 (45.91 m^2). The function's form was similar to half of a normal curve. In a different situation and with more data, a different form might be more appropriate and could easily be substituted.

Maximum transpiration-surface evaporation was observed to range between 0.21 and 0.02 inch (5.3 and 0.5 mm) per day. Within this range, a function (fig. 4) was developed that is similar to average daily temperature and solar radiation curves. Further investigation is needed to determine if this function could be empirically derived for various geographic locations.

Interception was computed as the difference between precipitation and throughfall.

The general equations for transpiration-surface evaporation and evapotranspiration are:

$$TE = (ETP) (ETV) (ETM)$$
$$\text{and } ET = TE + I$$

where TE = transpiration-surface evaporation
ETP = soil water deficit function
ETV = vegetation function
ETM = transpiration-surface evaporation function
ET = evapotranspiration
I = interception

449

If interception is greater than or equal to 0.35 inch (8.9 mm), it was assumed that transpiration-evaporation from the profile would be zero because of the high moisture content in the atmosphere and on the vegetation. In this case, profile evapotranspiration would equal interception. If transpiration-surface evaporation (TE) plus interception (I) is greater than 0.35

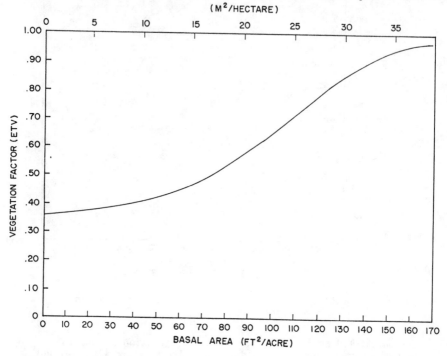

Figure 3. - Vegetation factor as a function of watershed basal area.

inch, a new transpiration-surface evaporation value is calculated to equal 0.35 inch minus interception. The same procedures are used for surface foot simulations, except 0.30 inch (7.6 mm) is used instead of 0.35 inch. The values 0.35 and 0.30 inch were assumed for this simulation.

In the final steps of the evapotranspiration phase, the soil water deficits for the surface foot and profile are increased by their simulated daily transpiration-evaporation rates.

Deep Seepage Phase

Daily seepage values for both surface foot and profile are determined from soil water deficit-seepage functions (figs. 5 and 6). These functions were derived empirically from soil water deficit data during periods of low transpiration and no rainfall. When a soil water deficit was negative (soil water was greater than the maximum base value), seepage rates for both the surface foot and profile were about equal to the absolute value of the negative deficit. Thus, at the end of a day when this phenomenon occurred, the soil water deficit was approximately zero. At zero deficits, both surface foot and profile seepage rates were 0.20 inch (5.1 mm) per day. Seepage rates were nearly zero (but never reaching zero) when the soil water deficits approached 40 percent of their maximum values.

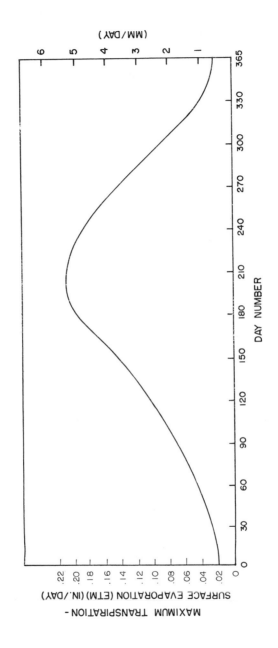

Figure 4. - Relationship between data and maximum transpiration-surface evaporation.

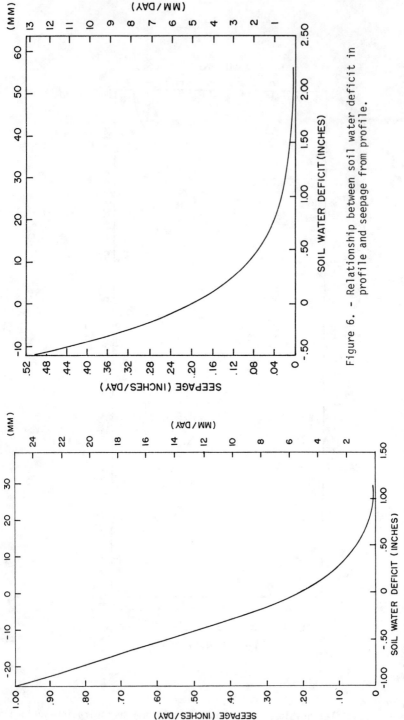

Figure 6. - Relationship between soil water deficit in profile and seepage from profile.

Figure 5. - Relationship between soil water deficit in surface foot and seepage from surface foot.

In the final steps of each day's simulation, the model updates the soil water deficit values for both the surface foot and profile by their respective seepage losses.

Output

Output from the model is easily varied and can be limited or expanded as necessary. Any variable in the simulation can be in the statistical output on a daily to annual basis. Variables may be tabulated in frequency distribution tables with the following output: sum of the variable, number of table entries, mean variable value, standard deviation, and the number of times the variable fell within each frequency class. Graphs can be plotted using any of the variables in the simulation. Additional output can be obtained by access to user-written FORTRAN subroutines.

Normal output for both input modes consists of (1) a graph of profile soil water deficits at biweekly intervals by 0.25-inch (6.4 mm) increments, (2) a daily rainfall and runoff summary which includes day number, rainfall, throughfall, runoff, and profile soil water deficits before the storm for all dates on which rainfall occurs, and (3) a water budget summary by months which includes monthly and annual values for rainfall runoff, evapotranspiration, deep seepage, and the change in soil water deficits.

If mode 2 — simulated precipitation — is used, hydrologic behavior can be simulated for any number of years; but with mode 1 only 1 year can be simulated in each computer run. In mode 2, different precipitation event sequences can be obtained by changing random number multipliers. Using the same multipliers in successive runs will produce the same sequence of precipitation events. This duplication is useful if hydrologic differences between various levels of basal area are being investigated. Caution must be exercised when thinning or clearcutting are simulated by modifying the vegetation factor of basal area because stand regrowth is not included in the model. After the vegetation factor has been modified, the simulation should be limited to no more than 3 years to avoid erroneous values.

Two simulation runs, in mode 2, were made for 10-year periods. The same random number multipliers were used in both runs to produce the same sequence of precipitation events. In one run the parameter of basal area was 140 ft^2/acre (32.1 m^2/hectare) for the entire simulation, and in the other run it was reduced to 60 ft^2/acre (13.8 m^2/hectare) at the beginning of the eighth year. Soil water deficit graphs (fig. 7) and water budget summaries (table 1) for the eighth year are shown as an example of model output. There is little difference between the graphs for stands of 60 and 140 ft^2/acre during the mid-December through early-May period when evapotranspiration is low. However, large differences occur during the late-May through early-December period when evapotranspiration is high and before soil water recharge is completed. Comparing the water budget summaries (table 1) reveals that runoff was 63 percent greater for the year and 102 percent greater for the June-December period on the 60 ft^2/acre stand than on the 140 ft^2/acre stand. Annual evapotranspiration was 38 percent less, and deep seepage was 50 percent more at the lower basal area.

MODEL ACCURACY

The goals at the onset of model development were to simulate annual runoff within 15 percent of the measured runoff 70 percent of the time and to simulate profile soil water deficits within 0.50 inch (12.7 mm) of the measured values 70 percent of the time. One can never exactly match simulated to recorded hydrologic values because recorded values contain measurement error (James, 1972). Both measured runoff and soil water de-

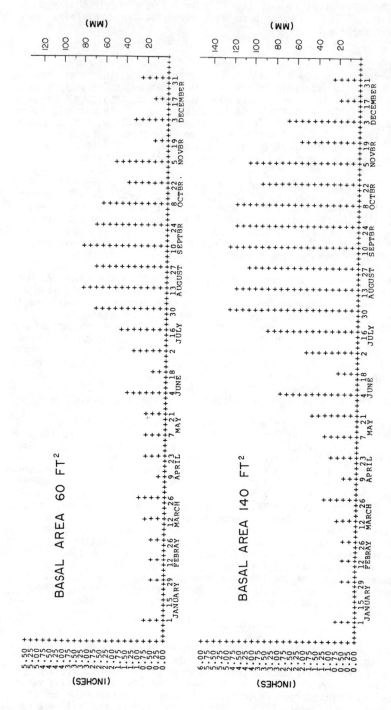

Figure 7. – Comparison of biweekly profile and water deficits for basal areas of 60 and 140 ft2/acre during the 8 years of simulation.

Table 1.--Comparison of simulated water budget summaries for basal areas of 13.8 and 32.1 m²/hectare during the eighth year of simulation

Period	Rainfall	Runoff		Evapotranspiration		Deep seepage		Change in soil water	
		13.8	32.1	13.8	32.1	13.8	32.1	13.8	32.1
				- - - - - millimeters - - - - -					
January	218.2	38.9	29.7	21.8	48.0	155.4	139.7	2.3	0.8
February	229.4	85.1	70.6	22.9	53.6	123.2	108.2	-1.8	-3.0
March	61.5	2.0	00.0	25.4	51.8	41.4	25.1	-7.4	-15.2
April	132.1	17.0	9.9	42.2	83.1	66.8	32.0	6.1	7.1
May	61.7	00.0	00.0	51.1	97.5	29.7	7.6	-18.8	-43.2
June	263.9	106.7	62.0	82.6	146.1	69.1	31.8	5.6	24.1
July	38.6	00.0	00.0	76.5	117.3	6.4	1.8	-43.9	-80.5
August	73.2	00.0	00.0	70.1	60.2	1.5	1.5	1.5	11.4
September	85.9	1.0	00.0	60.2	73.4	2.0	1.5	22.4	10.9
October	55.9	4.1	00.0	43.9	53.3	3.3	1.5	4.6	1.0
November	80.0	12.4	0.5	24.4	41.7	24.9	1.5	18.3	36.3
December	142.2	63.2	30.2	14.2	33.5	62.7	39.1	2.0	39.4
Annual	1442.6	330.4	202.9	535.3	859.5	586.4	391.3	-9.1	-10.9

455

Table 2.--Annual precipitation, measured runoff and
simulated runoff for the years tested[1]

Year	Vegetation	Precipi- tation	Measured runoff	Simulated runoff
		- - - - - - -millimeters- - - - - - -		
1967	Uncut	1668.8	189.5	215.1
1970	Uncut	1594.6	149.4	206.0
	Thinned		304.8	361.2
	Clearcut		495.6	420.6
1971	Uncut	1187.7	109.5	115.8
	Thinned		190.0	193.8
	Clearcut		267.0	234.4
1972	Uncut	1265.9	166.6	129.3
	Thinned		291.3	255.5
	Clearcut		303.5	313.9
1973	Uncut	2100.8	528.1	370.3
	Thinned		691.4	609.3
	Clearcut		702.6	713.5

[1]Data for 1968 and 1969 were not used to test the model as
these years were used to estimate parameters and develop
the model.

ficits probably contained errors of up to 10 percent of the actual values.

The model was tested by simulating the hydrologic system for 13 water-shed-years when runoff and soil water deficits were measured. These in-cluded 5 years for a pine-hardwood watershed with a basal area of 140 ft^2/acre (32.1 m^2/ hectare), 4 years for a pine watershed thinned to 60 ft^2/acre (13.8 m^2/hectare), and 4 years for a clearcut watershed. For 8 of the 13 watershed-years, the model simulated annual runoff within 15 percent of the measured values (table 2). This percentage was slightly less than the original goal but was considered acceptable because of the complexity of the system being modeled and the possibility of errors of up to 10 percent in the measured values.

The model predicted profile water deficits with slightly better accuracy than stated in my original goals. Thirty-eight percent of the simulated values were within 0.25 inch (6.4 mm) of the measured values and 73 percent were within 0.50 inch (12.7 mm). Figure 8 shows the differences between measured and simulated profile soil water deficits for the three vegetative conditions during 1971. Simulated deficits for the thinned and clearcut watersheds were 0.25 to 0.75 inch (6.4 to 19.0 mm) greater than measured deficits throughout most of 1971. Also, measured deficits on the clearcut watershed were less than zero several times during the year. Thus, the soil profiles were holding more water after treatment than before when the zero deficits were determined. This increased capacity was attributed to in-creased organic matter in the form of decomposing roots from the cut vegetation. The increased water-holding capacity continued through 1972 to 1973. Water-holding capacity changes were not observed on the uncut watershed.

Lawson (1967) reported that average interception on these same water-

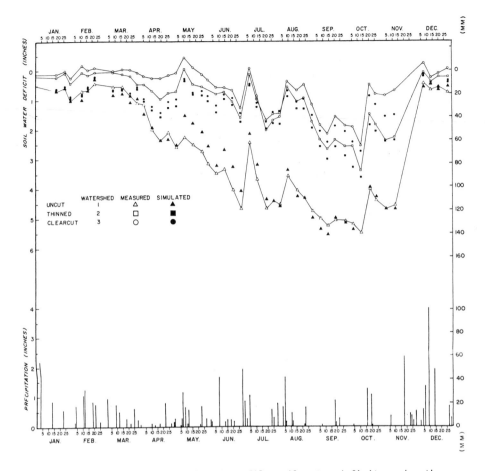

Figure 8. - Measured and simulated profile soil water deficits under three vegetation conditions with daily precipitation for 1971.

sheds was 12.7 percent of the annual precipitation. Simulated interception for a 5-year period was 16.6 percent. The difference between the measured and simulated values was considered litter interception as this was not measured in Lawson's study. The 3.9 percent value for simulated litter interception is within the 2 to 5 percent range reported by Helvey and Patric (1965).

Evapotranspiration and deep seepage were not measured directly on the watersheds, but simulated evapotranspiration values were similar to those calculated by the Thornthwaite and Mather (1957) method. Simulated deep seepage values may seem to be too high to some observers, but part of the deep seepage on these small drainages would become runoff on larger watersheds. For example, the 32-acre watershed which includes the three small drainages in this study has had about twice as much runoff as the small drainages. Rainfall for the small and large watersheds was the same, and evapotranspiration and changes in soil water were assumed to be the same since the vegetation and soils were similar. Therefore, differences in runoff between the small and large watersheds must be due to changes in the deep seepage component of the water balance.

If used cautiously, this simulation model will enable the watershed manager to better understand how altering the vegetative cover affects the hydrologic system and to predict the probable effects before initiating thinning. The computer program is available from the author.

REFERENCES

Clarke, R. T. 1973. A review of some mathematical models used in hydrology, with observations on their calibration and use. Journal of Hydrology, 19: 1-20.

Crawford, N. H., and R. K. Linsley. 1966. Digital simulation in hydrology: Stanford watershed model IV. Stanford University Department of Engineering Technical Report 39. Stanford, Calif.

Hill, R. W., A. L. Huber, E. D. Israelsen, and J. P. Riley. 1972. A self-verifying hybrid computer model of river-basin hydrology. Water Resources Bulletin 8: 909-921.

Helvey, J. D., and J. H. Patric. 1965. Canopy and litter interception of rainfall by hardwoods of Eastern United States. Water Resources Research 1: 193-206.

Holtan, H. N., and N. C. Lopez. 1971. USDAHL-70 model of watershed hydrology. United States Agriculture Technical Bulletin 1435, 84 p.

International Business Machines Corporation. 1971. General purpose simulation system V user's manual. Ed. SH20-0851-1. IBM, White Plains, N. Y.

International Business Machines Corporation. 1973. General purpose simulation system V-OS operations manual. Ed. SH20-0867-3. IBM, White Plains, N. Y.

James, L. D. 1972. Hydrologic modeling, parameter estimation, and watershed characteristics. Journal of Hydrology, 17: 283-307.

Lawson, E. R. 1967. Throughfall and stemflow in a pine-hardwood stand in the Ouachita Mountains of Arkansas. Water Resources Research 3: 731-735.

Lawson, E. R. 1975. Effects of partial and complete vegetation removal on soil water, runoff and tree growth in the Ouachita Mountains of Arkansas. PhD. thesis, Colo. State Univ., Fort Collins. 171 p.

Rogerson, T. L. 1967. Throughfall in pole-sized loblolly pine as affected by stand density. In Proceedings, International Symposium on Forest Hydrology, p. 187-190. Pergamon Press, New York.

Rogerson, T. L. 1976. Simulating hydrologic behavior on Ouachita Mountain drainages. United States Department of Agriculture, Forest Service Research Paper SO-119. 9 p.

Rogerson, T. L. 1980. Rainfall data simulation. United States Department of Agriculture, Forest Service Research Note SO-260.

Thornthwaite, C. W., and J. R. Mather. 1957. Instructions and tables for computing potential evapotranspiration and the water balance. Publications in Climatology. 10(3), 311 p. Drexel Institute Technology, Centerton, N. J.

FREQUENCY AND INTENSITY OF DROUGHT
IN NEW HAMPSHIRE FORESTS:
EVALUATION BY THE BROOK MODEL

C. Anthony Federer
Principal Soil Scientist
Northeastern Forest Experiment Station
USDA Forest Service
Durham, New Hampshire 03824

ABSTRACT

The frequency and intensity of agricultural droughts as they might affect tree growth in New Hampshire were analyzed by using the BROOK hydrologic model.

BROOK was designed to evaluate the effects of changing vegetative cover on small forested watersheds in the eastern United States. It is a lumped-parameter, water-yield model that uses a daily time interval and is driven by daily precipitation and daily mean temperature. Special features include separation of evapotranspiration into five components, provision for a variable source area, and use of leaf-area and stem-area indexes for seasonal and plant cover changes.

BROOK estimated daily soil-water deficits in hardwood forest on till soils for 50 years (1926 to 1975) at three New Hampshire locations: Berlin, Durham, and Keene. In the simulated soil, which had about 120 mm of available water, a deficit of 60 mm probably indicates moderate water stress in trees, and thus agricultural drought. This deficit was exceeded about half the time between mid-July and mid-September at all locations. Both the number of days with deficits larger than 60 mm and the mean deficit from June through August are satisfactory measures of dryness of different summers. In most years, dryness at one location was not related to dryness at the other locations, which indicates that agricultural drought in New Hampshire is a local rather than a statewide phenomenon. The string of soil-water deficits was analyzed to show the frequency of a drought of a given length, time, and intensity. Another analysis provides probability estimates of how long an existing drought will continue.

Estimates of soil-water deficits from hydrologic models are much more satisfactory than precipitation alone for studying the effects of drought on plant growth. Tree ring widths are highly correlated with soil-water deficits. In spite of New England's "well watered" reputation, diameter growth of trees in New England is probably limited by soil-water deficits in most years.

INTRODUCTION

The word "drought" has different meanings to different people depending on how the lack of water influences them. The meanings can be classified into three types: meteorologic, agricultural, and hydrologic (Wigley and Atkinson, 1977). Meteorologic drought is lower precipitation than normal for some period of time. This may be the most commonly used definition of drought but, because it fails to consider the influences of evapotranspiration, it is an inadequate measure of how drought affects human water supplies and plant growth. Hydrologic drought is a deficient water supply for human use. The water supply may be streamflow, reservoir storage, groundwater, or some combination. Agricultural drought is insufficient water for normal rapid growth or maturation of plants. It is a function of the amount of water stored in the soil within the root zone of plants.

Meteorologic and hydrologic drought are easy to study because data are readily available for several locations in New Hampshire. The National Weather Service collects and publishes precipitation data, and the U.S. Geological Survey collects and publishes streamflow data. However, data on soil water are not collected and published for any location in New Hampshire, so agricultural drought cannot be studied easily.

Federer (1980a) showed that the mean soil-water deficit in June and July was correlated with the annual diameter growth or ring width of paper birch and white oak in southern Maine (Fig. 1). There were significant differences in deficit between the weather stations at Sanford, 10 km southwest, and Saco, 20 km east of the study site. This local variability in summer drought verifies Lyon's (1943) conclusion that precipitation must be measured within 20 km of the trees for ring width studies in New England.

A hydrologic simulation model, BROOK, was used to estimate daily soil-water deficits for 50 years at each of three stations in New Hampshire (Federer, 1980b). Daily precipitation and mean temperature are required inputs for BROOK. Parameters were chosen to represent a deciduous forest growing on an upland till soil. The soil-water deficits were analyzed in several different ways to quantify the occurrence, intensity, and frequency of drought. This paper shows how a rainfall-runoff model is useful for studying agricultural drought.

THE BROOK MODEL

A hydrologic model called BROOK simulates water budgets for forest land in the eastern United States. BROOK is a deterministic, lumped-parameter water-yield model for small areas; it was not designed to simulate flood peaks or watersheds with multiple aspects. It operates with a daily time interval, and to make it maximally useful it was designed to use only daily precipitation and daily mean temperature as input variables. BROOK can simulate various forest cover types, but these types must be uniform over the watershed. Partial cuts cannot be simulated. BROOK was designed to estimate the response of streamflow on different slopes and aspects and to cover changes caused by harvesting

Figure 1. Upper part: mean radial increment for four white oak trees and five paper birch at the Massabesic Experimental Forest, Alfred, Maine. Lower part: mean daily soil-water deficit for June and July from a simple water budget calculated from daily precipitation for Sanford and Saco, Maine.

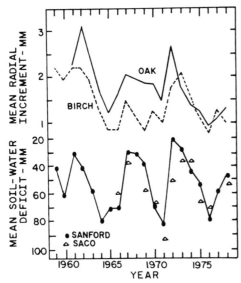

and regrowth or by conversion from hardwoods to conifers. It has been used to examine streamflow response to different hardwood transpiration characteristics (Federer and Lash, 1978b) and to estimate soil-water deficits before floods (Hornbeck and Federer, 1974). Also, it has been used as a submodel for nutrient concentration modeling and as a teaching tool at several universities.

Federer and Lash (1978a) document the theory, testing, problems, and program for BROOK. It is a fairly standard and moderately simple (28 parameters) hydrologic simulation model (Fig. 2), and, therefore, only its important or unique characteristics are given here. The root zone is a single storage--EZONE--from which evapotranspiration can occur, and the root zone has a subcompartment--EVW--that represents water that can become soil or litter evaporation. Ground water storage is the GWZONE. Below EZONE and above GWZONE is a layer of unsaturated soil--UZONE--from which interflow may occur. For fast-response watersheds, such as New Hampshire upland tills, groundwater is neglected, and UZONE is small.

In BROOK, interception by the plant canopy and overland flow from saturated source areas are subtracted from daily precipitation; the remainder is added either to soil water or to snowpack. Interception of rain and snow is determined by the amount of precipitation, the amount of potential evapotranspiration, and the characteristics of the plant canopy. The saturated source areas increase exponentially in size as EZONE gets wetter. Rain or snowmelt on the source areas becomes overland flow. Transpiration, soil evaporation, and drainage are subtracted from EZONE each day.

Drainage from EZONE increases exponentially with its water content. Zero soil-water deficit is when the water content drains 2 mm/day. This is a definition of the "field capacity" of the soil. The soil can be wetter than this, but only for short periods of time unless precipitation or snowmelt inputs are sustained.

Transpiration depends on potential evapotranspiration, soil water, and canopy cover. Potential evapotranspiration is calculated by the Hamon (1963) method, which requires only the mean daily temperature and the day length. The Hamon method does not estimate the real day-to-day changes in potential evapotranspiration because that would require solar radiation data, which is usually unavailable. But, because soil-water deficits develop slowly, simulation of the real day-

461

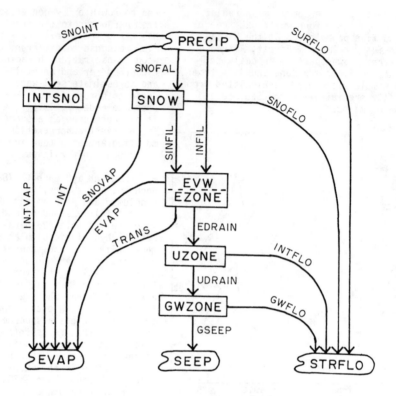

Figure 2. Block diagram of the BROOK
hydrologic model.

to-day variability of evapotranspiration is not necessary for many
studies. When the plant canopy is fully developed, transpiration is
equal to the lesser of either soil water times a constant or potential
evapotranspiration. This indicates that transpiration is limited
either by soil-water supply or by atmospheric demand.

Vegetative cover is described by two parameters, LAI and SAI. LAI
approximates the leaf area index but is only allowed to vary from 0 to 4.
The stem area index, SAI, is allowed to vary from 0 to 2. All cover
differences are handled by appropriate variation in these two parameters, through their effect on multipliers of various processes
(Fig. 3). The hypothetical shape of these relationships illustrates a
major gap in hydrologic knowledge.

The model is calibrated and verified with records over a number of
years from several experimental watersheds at the Hubbard Brook Experimental Forest in New Hampshire and the Coweeta Hydrologic Laboratory
in North Carolina (Federer and Lash, 1978a). One year's hydrograph
(Fig. 4) does not present the best or the worst data. Erroneous simulation of snow accumulation and melt is the biggest problem, but does

Figure 3. Assumed effects in BROOK of leaf-area index (LAI) and stem-area index (SAI) on rain interception, snow evaporation, transpiration, and soil evaporation.

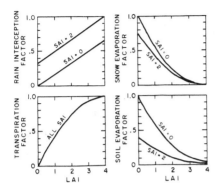

not affect the study of drought. Simulated soil-water content at Hubbard Brook agreed reasonably well with crude measurements of soil water made with Colman electrical resistance sensors. Reasonable agreement of simulated and measured streamflow for summer and autumn storms that fully recharge the soil, also confirms the accuracy of the simulated soil-water deficits.

DROUGHT IN NEW HAMPSHIRE

The BROOK model was used to estimate daily soil-water deficits for hardwood forest on till soil in Berlin, Durham, and Keene from January 1, 1926, through December 31, 1975 (Federer, 1980b). These locations are in the northern, southeastern, and southwestern, parts of the state, respectively. Daily precipitation and mean temperature for the period at each location were obtained from the National Climatic Center. Mean monthly precipitation in summer ranged from 70 to 100 mm for these stations. BROOK parameters were chosen to represent a mature hardwood forest on well-drained till soil, a combination which covers nearly 50 percent of the state's area. The only parameter that varied among the locations was the timing of the spring and autumn transitions of LAI from zero to 4 and back to zero.

The average annual cycle of soil-water deficits is nearly the same for all three stations. From December through April deficits are close to zero. The mean deficit begins to increase in May and reaches about 60 mm by late July and then holds steady around this value until late September. Late September to late November is, on the average, a period of declining deficits. The importance of the seasonal beginning and ending of transpiration is obvious, and the shorter growing season at Berlin causes its mean deficit to increase later and decrease earlier than that at the other stations.

The annual curve of deficit for any given year and location has a jagged appearance as opposed to the smooth mean curve. Deficits are reduced abruptly by storms and then increase gradually. Deficits can be reduced to zero at any time in summer by large storms. In the 50 years at each station, soil-water deficit was always reduced to negligible amounts by the end of December.

There are many possible ways to use a string of soil-water deficits to quantify the occurrence, intensity, and frequency of agricultural drought. Single values that quantify the whole growing season include both the mean soil-water deficit for the season (Wigley and Atkinson, 1977) and the fraction of days wetter than a given deficit. Deficit-duration curves show the amount of time that any given deficit is exceeded. The duration and frequency of drought require more complicated presentation and require defining "runs" of drought days and a threshold deficit value (Dracup, Lee, and Paulson, 1980). In the simplest case, a run is a sequence of days with deficits greater than the threshold.

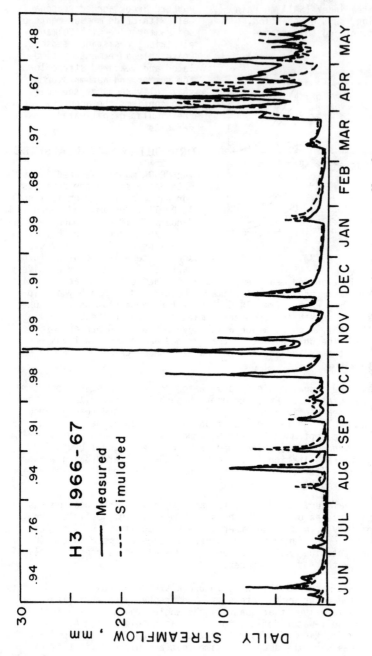

Figure 4. Measured and simulated running 3-day mean streamflow for Hubbard Brook watershed 3 in 1966-67. The line of data at the top is the correlation coefficient for running 3-day means within each month. Simulated annual flow was 770 mm while measured flow was 772 mm.

But to avoid short runs, and runs broken by only a few days less than the threshold, I have required both that the threshold must be exceeded for at least 5 consecutive days, and that the run must be ended by at least 5 consecutive days of deficits below the threshold. Using the whole record, we can calculate, for any given date, the number of years in which a drought run that exceeds any given length occurs. A similar method may be used to estimate the probability that a drought of any intensity on a date will continue for any given number of days (Federer, 1980b).

The soil parameters chosen for this study give 120 mm as the maximum amount of water available for evapotranspiration. This is also the maximum possible soil-water deficit. The amount of water-stress in a plant on a given day depends both on the soil-water deficit and on the transpirational demand. Therefore, a detailed study of plant growth and stress requires knowledge of both (Federer, 1980a). However, day-to-day variation in transpirational demand is averaged over the several weeks in which soil-water deficits develop. From Federer (1979, 1980a), for 120 mm available water, it is possible to approximate that plants will not be stressed with deficits in the 0 to 30 mm range; growth might be slightly reduced in the 30 to 60 mm range; from 60 to 90 mm stress is moderate and plant growth is reduced; and below 90 mm plant growth will be severely limited or stopped. A 60 mm criterion, therefore, is valid for defining agricultural drought in this paper.

The number of days that have deficits greater than 60 mm during a growing season of June 1 through August 31 provides a simple way to quantify the relative dryness of different years (Fig. 5). Some years such as 1929, 1949, and 1950 were dry throughout the state while others such as 1937, 1938, and 1975 were wet. But, in most years, some parts of the state were wet while others were dry. For instance, 1941 was wet in Berlin, moderate in Keene, and dry at Durham; 1967 was wet at Keene, moderate at Durham, and dry at Berlin. Because the distribution of summer convective showers varies locally throughout the state, agricultural drought in New Hampshire is usually a local rather than a statewide phenomenon.

Deficit-duration curves show the relative frequency with which any given deficit is exceeded in a given summer. For my simulated 50 years, these curves are remarkably parallel in most years, and the 60 mm criterion appears to be an optimum one for a single-valued ranking of dryness (Fig. 6). The extreme narrow range of deficits at Keene was in 1936 and the widest range of deficits in 1968.

The drought criteria discussed so far do not consider the timing of dry or wet periods. Yet, for many plants, a dry period in June may have very different effects from an equally dry period in August. A month is too long an averaging period to evaluate the effects of water deficits on plants. The half-month is a convenient compromise between daily and monthly intervals. The mean deficit for each half-month record is a convenient way to shorten the string of daily soil-water deficits (Federer, 1980b).

The probability of drought in the future can be estimated from its frequency in the past. The relative frequency with which any given deficit was exceeded on a given day can be plotted by half-months (Fig. 7). For Keene, New Hampshire, the severe stress stage of deficit greater than 90 mm was about 5 percent of the days from late July through early October.

465

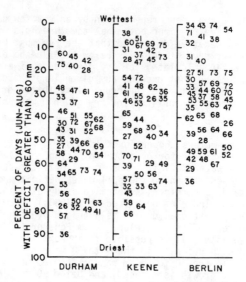

Figure 5. Years ranked by percent of days from June 1 through August 31 with simulated deficits greater than 60 mm. Initial 19 is omitted from year designation.

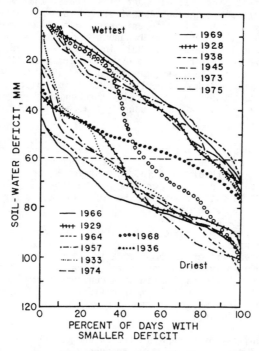

Figure 6. Simulated deficit duration curves for Keene, New Hampshire, for period June 1 through August 31. Six of the wettest and 6 of the driest years are shown along with 2 years (1936 and 1968) with extreme slopes. Dashed horizontal line is the 60mm criterion used to produce Figure 4.

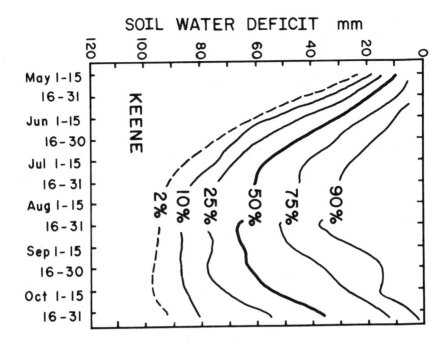

Figure 7.   Percent of days in each half-month, averaged from 1926
to 1975 at Keene, New Hampshire, that have simulated
soil-water deficits greater than a specified amount.

      More detailed analysis of drought probability can use the "run"
concept. From the record of soil-water deficits, a probability can be
estimated for any given date that a run exists that exceeds any given
deficit for any length of time (Fig. 8). Many probability estimates
can be obtained from such analysis. For example, from Figure 8, a
drought run exceeding 75 mm deficit for at least 30 days exists on
August 1 in 15 percent of the 50 years. A similar analysis can be made
for the probability that an existing run of a given deficit on a given
date will continue for any given number of days (Fig. 8). When a run
exceeding 75 mm deficit does exist on August 1 at Keene, there is a 25
percent probability that it will continue for at least 30 more days.

CONCLUSIONS

1. Soil-water deficit is a much better independent variable than precip-
    itation to use in studies of the effects of drought on plant growth.

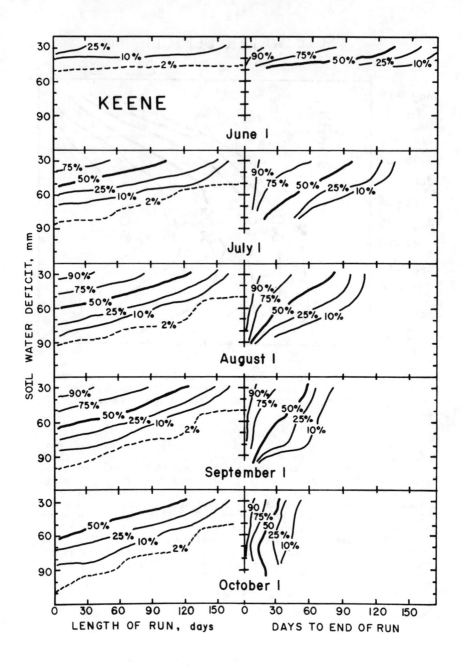

Figure 8. Left: Percent of years in which a simulated run exceeding a specified length of time exists at the beginning of each month for Keene. Right: Percent of years in which runs that exist at the beginning of each month continue for a specified number of days.

2. Any hydrologic simulation model that produces an estimate of soil-water content in the root zone is a tool for studying agricultural drought.

3. Agricultural drought in the northeastern United States is often very localized because of the heterogeneity of convective storms in summer.

4. Diameter growth of trees at lower elevations in New England is probably limited by soil-water deficits in most years.

5. The dryness of the summer can be specified by a mean soil-water deficit or by the fraction of time a certain deficit was exceeded.

6. The concept of a run, which is a string of days for which a given dryness threshold is exceeded, allows many probability statements about intensity and duration of droughts.

ACKNOWLEDGEMENTS

This study was supported in part by a grant from the USDA Forest Service, Northeastern Forest Experiment Station, to the New Hampshire Water Resource Research Center. Financial support for this publication was provided by the U.S. Department of the Interior, Office of Water Research and Technology, as authorized under the Water Research and Development Act of 1978 (Public Law 95-467).

This report is a contribution from the Hubbard Brook Ecosystem Project.

REFERENCES

Dracup, J. A., Lee, K. and Paulson, Jr., E. G. 1980. On the definition of droughts. Water Resources Research, Vol. 16, pp. 297-302.

Federer, C. A. 1979. A soil-plant-atmosphere model for transpiration and availability of soil water. Water Resources Research, Vol. 15, pp. 555-562.

Federer, C. A. 1980a. Paper birch and white oak saplings differ in responses to drought. Forest Science, Vol. 26, pp. 313-324.

Federer, C. A. 1980b. Frequency of agricultural and forest drought in New Hampshire: 1926-1975. Water Resource Research Center Research Report No. 26, 37 pp., University of New Hampshire, Durham, New Hampshire.

Federer, C. A., and Lash, D. 1978a. BROOK: A hydrologic simulation model for eastern forests. Water Resource Research Center Research Report No. 19, 84 pp., University of New Hampshire, Durham, New Hampshire.

Federer, C. A., and Lash, D. 1978b. Simulated stream-flow response to possible differences in transpiration among species of hardwood trees. Water Resources Research, Vol. 14, pp. 1089-1095.

Hamon, W. 1963. Computation of direct runoff amounts from storm rainfall. International Association of Scientific Hydrology, Publication 63, pp. 52-62.

Hornbeck, J., and Federer, C. A. 1974. Forests and floods. Forest Notes (New Hampshire). Winter 1973-1974, pp. 18-21.

Lyon, C. J. 1943. Water supply and the growth rates of conifers around Boston. Ecology, Vol. 24, pp. 329-344.

Wigley, T. M. L., and Atkinson, T. C. 1977. Dry years in south-east England since 1698. Nature, Vol. 265, pp. 431-434.

Section 6
REMOTE SENSING APPLICATION

REMOTE SENSING APPLICATION IN WATERSHED MODELING

Edwin T. Engman
Chief, Hydrology Laboratory
U. S. Department of Agriculture
Science and Education Administration
Hydrology Laboratory
Beltsville, Maryland 20705

ABSTRACT

Remote sensing is rapidly becoming an important source of data and information for hydrologic modeling and research. This paper summarizes the work that has been done to date and then discusses current research efforts and the long range potential for remote sensing in hydrologic modeling.

To date, most uses have been relatively straightforward extensions of photogrammetry. There have been a number of successful applications of Landsat for determining both urban and rural land use to estimate runoff coefficients such as the SCS runoff curve number. The extent and time of inundation by floods on major rivers has been monitored with Landsat and the meteorological satellites. Similarly, areal extent of snow cover and procedures to use this information in water supply forecasting models have been demonstrated.

Current research involves extension of these areas plus work with some of the less common wavelengths or frequencies. New work with thermal infrared and microwave data is beginning to use the unique information that specific wavelength measurements can provide. Thermal infrared measurements allow us to infer a heat budget and hence estimate crop condition, soil moisture or evapotranspiration. With microwave data we have all-weather capability plus the ability to penetrate vegetation and to directly measure soil moisture and snow water content. This research, although still very much experimental, is beginning to treat remote sensing as a unique source of data.

The potential for remote sensing and its application to hydrology is considerably greater than research has addressed thus far. Measuring the characteristics of an area rather than a point, integrating several characteristics with one composite measurement, and improving prediction models with continuous or frequent feedback from satellite measurements are just a few of the aspects that must be explored further. Considerable research is needed if we are to fully realize the potential benefits of remote sensing in hydrology. Treating remotely sensed data as a unique measurement of hydrologic characteristics offers the best chance for major advances in the field of hydrologic modeling.

INTRODUCTION

Remote sensing as it is generally known today is an outgrowth of photogrammetry. Strictly defined, remote sensing involves the collection of data by systems which are not in direct contact with the item being measured. Early remote sensing emphasized interpretation of photos and descriptive analysis of the subject. Remote sensing using images received a major boost in interest during World War II by the military. Interpretation was still emphasized although equipment was perfected and wavelengths outside the visible range began to be used. No significant progress was made in the post-war years although there was more emphasis on applications. The satellite launching of ERTS A (later renamed Landsat 1) in July of 1972 started the modern day era of remote sensing. The advent of so many data on a repetitive basis covering four spectral bands, all areas of the earth, and a relatively fine resolution, stirred the imagination of earth scientists worldwide. It is with this background that this paper addresses remote sensing applications to watershed modeling. Two major subjects are addressed—(1) current applications of remote sensing to hydrology and modeling, and (2) current research and potential uses of remote sensing in hydrology.

CURRENT APPLICATIONS

Landsat data have become a common source of information in hydrology. These data, like other remote sensing applications, are generally a fairly simple extension of photogrammetry. However, the unique characteristics of specific spectral bands and the temporal sequence of the data are beginning to be used in remote sensing applications.

Land Use and Runoff Coefficients

Land use or cover is an important aspect of hydrologic processes, particularly infiltration, erosion and evapotranspiration. Because of this, any process-oriented model (as opposed to a "black-box model") incorporates some land use data or parameters. Distributed models, in particular, need specific data on land uses identified by location within the watershed. Perhaps most of the work on adapting remote sensing to hydrologic modeling has been with the Soil Conservation Service (SCS) runoff curve number (Soil Conservation Service, 1972). The runoff curve number (RCN) is a coefficient developed from one of four soil hydrologic groups and the land cover. The RCN may be further adjusted by antecedent precipitation to account for very wet or dry conditions. Table 9.1 in the SCS National Engineering Handbook, Hydrology section, (SCS, 1972) shows how land use affects the RCN for a given soil group. The importance of land cover can be demonstrated by comparing predicted runoff for a condition where only the land use changes. For example, consider a B soil and a 10 cm rain; the calculated runoff for good pasture condition would be approximately 0.25 cm, whereas, if that same field were planted in a small grain with straight rows, the runoff calculated by the SCS procedure would be approximately 2.8 cm.

Several papers have recently been published that demonstrate the feasibility of developing the land use categories from Landsat. At first suburban and urban areas were studied because the greatest contrast would be available between the impervious and other more pervious areas. In a study on the Upper Anacostia River Basin in

Maryland, Ragan and Jackson (1980) demonstrated the suitability of using Landsat-derived land use data for calculating synthetic flood frequency relationships. The Landsat-derived results were compared to relationships developed from a conventional approach using air photos. Because the resolution of Landsat data is no finer than about one acre, this does not provide the detail necessary to use Table 9.1 (SCS, 1972) directly in determining a RCN. Consequently, one must develop a land cover table analagous to Table 9.1, but compatible to the Landsat data. Tables 1, 2, and 3 show land use. However, RCN relationships developed

Table 1. Runoff Curve Numbers for Landsat Land Use Delineations
(from Ragan and Jackson, 1980)

Land Use Description	Hydrologic Soil Group			
	A	B	C	D
Forest Land	25	55	70	77
Grassed Open Space	36	60	73	78
Highly Impervious (Commercial-Industrial-Parking Lot)	90	93	94	94
Residential	60	74	83	87
Bare Ground	72	82	88	90

Table 2. Runoff Curve Numbers for Landsat Rural Land Cover
Delineations (from Slack and Welch, 1980)

Land Use Description	Hydrologic Soil Group			
	A	B	C	D
Agricultural, vegetated	52	68	79	84
Bare Ground	77	86	91	94
Forest Land	30	58	72	78

Table 3. Pennsylvania Landsat Runoff Curve Numbers (from Bondelid
et al., 1980)

Land Cover	Hydrologic Soil Group			
	A	B	C	D
Woods	25	55	70	77
Agriculture	64	75	83	87
Residential	60	74	83	87
Highly Impervious	90	93	94	95
Water	98	98	98	98

by different workers for using Landsat data show some inconsistencies in the RCN values selected for similar land uses. This is probably the result of individual differences in how the land use is defined, the training sites chosen, and the characteristics of the study area. Although these differences do exist, it does seem feasible that a general table could be developed that could be applied just like Table 9.1.

In early work with remote sensing data, Jackson et al. (1977) demonstrated that land cover (particularly percent imperviousness) could be used effectively in the U.S. Army Corps of Engineers (USACE) STORM model (1976). In connection with the same study, Jackson and Ragan (1977) used Bayesian decision theory to demonstrate that computer aided analysis of Landsat data was highly cost effective.

Slack and Welch (1980) demonstrated that SCS RCN's could be developed in a cost effective manner for a primarily agricultural watershed in Georgia. Ragan and Jackson (1980) modified the land cover requirements for the SCS procedure for suburban areas so that Landsat data could be used. The RCN's developed from the Landsat data closely matched those obtained from a conventional approach based on air photo analysis. Synthetic flood frequencies developed from the two procedures were essentially identical. The Hydrologic Engineering Center of the USACE (1979), also had remarkable success in developing synthetic flood frequency curves from Landsat data.

One necessarily must make some compromises when abstracting land use from Landsat data. With any given pixel, one may be in error. In fact, the Corps (USACE, 1979) estimated that at the grid cell level Landsat land use designations will be incorrect about one-third of the time. However, by aggregating land use over a significant area, the misclassification of land use can be reduced from 8 to 2%. Jackson et al. (1977) estimated that Landsat-developed runoff coefficients give satisfactory results for watershed areas greater than about 2.5 km^2 (1 sq. mi.). Results of Bondelid et al. (1980) confirmed this apparent lower limit with currently available satellite data. Essentially, this means that the RCN or other runoff coefficient may not be very sensitive to a detailed land use classification for relatively large areas. In a recent study, Rawls et al. (1980) showed that the U.S. Geological Survey's LUDA (Land Use Data Archive) gave results as good as those from conventional land use surveys. The minimum mapping unit for the LUDA maps, which is derived from high altitude photography, is about 4 hectares (10 acres) for urban areas. This study also indicated that the derived RCN's are not very sensitive to the method of integrating soil and land use data, particularly for watersheds larger than about 25 km^2.

Snow Hydrology and Water Supply Forecasting

Water supply forecast models for the western United States have typically been of the multiple regression form.

$$Y = a + b_1X_1 + b_2X_2 + b_3X_3 + \ldots b_nX_n$$

where Y is the runoff volume for the forecast period, X_1 --- X_n are the snow water contents at each of $\underline{n}$ snow courses. The coefficients $\underline{a}$ and b_1 --- b_n are developed from empirical data. Other variables such as fall precipitation, base flow, etc., have been included in specific models. Areal extent of snow cover has not generally been used because data to define it have not been available except in certain case studies.

Leaf (1969) used aerial photographs to develop relationships between snow cover and accumulated runoff for some Colorado watersheds. He also showed that sequential photos showing snow cover depletion relationships could be used to help estimate the timing and magnitude of snowmelt peaks.

Since about 1973, the National Oceanic and Atmospheric Agency (NOAA) and Landsat satellites have provided a visible and infrared data base of snow cover. With these data available, procedures for analyzing the data have been developed (Meier and Evans, 1975, and Rango and Itten, 1976). NOAA has been using satellite data to map mean monthly snow cover over the Northern Hemisphere (Wiesnet and Matson, 1975).

Some of the first applications of satellite data were done by Rango et al., 1977. They developed a regression model for snow melt in the Indus River basin. This study demonstrated the utility of satellite snow cover data for large areas with little or no data base.

Aircraft and Landsat snow cover data were combined to develop a long term data base in California. The addition of snow cover area considerably reduced the seasonal runoff forecast error for the King's and Kern River Basins (see Figure 1) (Rango et al., 1977).

In the Pacific Northwest, satellite snow cover data are presently being used operationally in the Streamflow Synthesis and Reservoir Regulation (SSARR) model. In test cases for five basins over a 6-year period, the addition of satellite snow cover data to the model resulted in a definite, but statistically insignificant improvement (Dillard and Orwig, 1979).

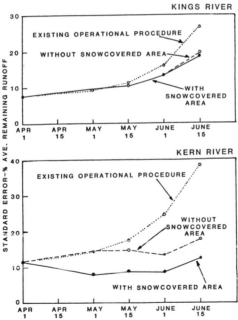

Figure 1. Standard Error of Various Forecast Procedures vs. Date During Snowmelt on Two California Watersheds (Rango et al., 1977).

Landsat imagery was used to calculate snow cover areas for six basins in Colorado over the period of 1973-1978. Shafer and Leaf (1979) concluded that the satellite imagery was of sufficient quality to accurately monitor the snow cover area. They also concluded that forecast error can be reduced on the order of 10% by using snow cover data derived from the satellite. Verification of standard forecast procedures with snow cover data late in the spring would be an additional use of the satellite data.

In California, two areas were studied by comparing satellite-derived snow cover areas with conventional snow data and by incorporating snow cover areas into the State's forecasts (Brown et al. 1979). Results indicated potential improvement in the forecast accuracy by using snow cover area, particularly in areas where conventional snow data were limited. However, certain limitations did come to the fore: These included the problems involving cloud cover and its blocking of the land or confusion between snow and clouds, the time involved in obtaining imagery because of either the 9-day pass frequency with Landsat or operational delays, and the problems associated with estimating total snow cover where no receding melt line was available for analysis.

Martinec (1970) developed a snowmelt runoff model that uses snow cover area and temperature as input data. This model was tested successfully on small watersheds in the European Alps using air photos to determine the snow cover area. Rango and Martinec (1979) have demonstrated that this model can be successfully used on basins as large as 500 km^2 by using Landsat data to determine the snow cover area. Using this approach they were able to simulate seasonal volumes within 5% of actual values and were able to explain approximately 85% of the variation in daily runoff for basins in the Wind River Mountains in Wyoming. These results are illustrated in Figure 2.

Bissell and Peck (1973) proposed using attenuation of natural gamma radiation from the soil to measure the snow water equivalent. The National Weather Service is currently using this method with low-level

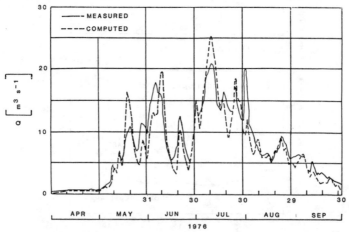

Figure 2. Daily Streamflow Simulation Produced by the Martinec Snowmelt Runoff Model Using Landsat Snowcovered Area Data on Dinwoody Creek, Wyoming (Rango and Martinec, 1979).

478

aircraft flights. The usefulness appears limited to fairly narrow bands because the aircraft must fly at fairly low elevations due to adsorption of the radiation by the atmosphere.

These studies (plus others---the reader is referred to two literature summaries for additional information and references, Rango, 1979a and 1979b) demonstrated enough promise for National Aeronautics and Space Administration (NASA) to sponsor an Applications Systems Verification and Transfer (ASVT) project on the Operational Applications of Satellite Snowcover Observations. Nine operational water management agencies (three State agencies - Arizona, California, and Colorado) participated in this project. In addition, reports from New Zealand, Switzerland, and Norway contributed to this summary (Rango and Peterson, 1979). In general, the individual reports demonstrated improved forecasting ability using satellite data.

Flood and Floodplain Mapping

The area inundated by floods and floodplains can be effectively mapped with remotely sensed data. Satellite data such as that from Landsat can be used to define coverage of an entire river basin but may have some limitations on small basins because of resolution. Infrared photography, thermal infrared data, and multispectral scanner data have all been successfully used to map the areal extent of flooding. These applications depend upon measuring changes in reflectivity caused by standing water, high soil moisture, moisture stressed vegetation, and temperature changes from ambient. These effects last for some time after inundation and can be detected up to 2 weeks after the passage of a flood. A number of studies using Landsat data and infrared photography have been reported by Rango and Anderson (1974), Williamson (1974), Morrison and Cooley (1973) and Hoyer et al. (1973). A series of papers related to this subject can be found in the 1974 Volume 10, No. 5 issue of the Water Resources Bulletin. In spite of the coarser resolution (900 m vs. 80 m for Landsat), the NOAA satellite thermal infrared sensor has proved effective in measuring areas of flood innundation (Berg et al., 1979 and Wiesnet et al., 1974). In addition, the NOAA satellites have the advantage of more frequent coverage (twice daily average vs. 18-day coverage for Landsat).

Floodplains have been delineated using remotely sensed data and inferring the extent of the floodplain from vegetation changes or some other features commonly associated with floodplains. Rango and Anderson (1974) have developed a list of indicators that can be used to infer floodplains from Landsat data. In a more recent study, Sollers et al. (1978) examined multispectral aircraft and satellite classifications of land cover features indicative of flood plain areas. They concluded that satellite data can be used to delineate flood-prone areas in agricultural and limited development areas but may not give good results in areas with a heavy forest canopy. The remotely sensed data may best be used for preliminary planning and for monitoring flood plain activities with time.

Wetlands

The environmental importance of wetlands has resulted in an increase awareness of this resource. Primary to management and protection of these areas is an accurate map and inventory. Remote sensing has provided the ideal tool to do this because of the general inaccessibility to these areas. Much of this work has been done by using vegetation characteristics to delineate the wetland boundaries from air photos. A very complete summary of this type of analysis has been prepared by Carter (1978). Landsat data have also been successfully used in wetland studies. Samples of this type of application have been reported by Carter et al. (1977), Morrow and Carter (1978), and Carter et al. (1979).

CURRENT RESEARCH AND POTENTIAL USE IN HYDROLOGY

To date, most remote sensing applications have consisted of fairly direct extensions of photogrammetry. However, using information from specific spectral bands to infer land use properties is an example of using remote sensing information as a unique data source or measurement. The spectral classification used by Bondelid et al. (1980) is an example of this. Fortunately, we appear to be on the threshold of major new applications and uses of data. These fall into four areas:

(1) Use of electromagnetic radiation outside of the visible range such as thermal infrared and microwave for their unique responses to properties important to hydrology.

(2) The potential for frequent measurement to develop time series of changes in given parameters and to monitor the dynamic properties in hydrology.

(3) The use of data representing an area in which the spatial variability of specific parameters of the area have been integrated.

(4) The merging of several data sets of different wavelengths, polarizations, look angles, etc., to provide specific measurements of hydrologic parameters developed from the unique characteristics of remote sensing.

Each of these areas presents a unique opportunity for hydrologists to apply remote sensing in ways other than simple extensions of photogrammetry. It is giving us a complex measurement that is simultaneously observing several factors. It is also giving us a view that is uncommon to our way of thinking in that it looks at a relatively large area and somehow integrates information from the entire scene. The challenge is to learn how to use this information and to understand it. To do this we must develop new concepts and challenge our usual way of conceptualizing hydrologic processes. Some areas of current research and areas of opportunity are discussed below.

Soil Moisture

One of the more exciting aspects of remote sensing for hydrologists is the potential for direct measurement of soil moisture. Successful measuring of soil moisture by remote methods depends upon the type of reflected or emitted electromagnetic energy, noise or interference problems, or other limitations. Table 4 summarizes the various remote sensing techniques based on wave length and property measured and lists both the advantages and disadvantages. Schmugge et al. (1979) presented a good comprehensive summary of remote sensing approaches for measuring soil moisture and their theoretical basis. That publication also presents a comprehensive bibliography for those who may want to pursue the subject further.

Reflected solar radiation is not a potential method for measuring soil moisture because there are so many noise elements that confuse the

Table 4. Summary of Remote Sensing Techniques for Measuring Soil Moisture

Wavelength Region	Property Observed	Advantages	Disadvantages or Noise Sources
Reflected Solar	Albedo; index of refraction.	Data available	No unique relationship between spectral reflectance and soil moisture; thin surface layer; only cloud interference.
Thermal infrared	Surface temperature (measure diurnal range of surface temperature or crop canopy temperature).	High spatial resolution, large swath. Relationship between temperature and soil water pressure is independent of soil type.	Bare soil only; cloud interference; surface topography and local meteorologic conditions can cause noise; surface layer only (2-4 cm).
Active microwave (1 - 100 cm)	Backscatter coefficient; dielectric constant.	All weather high resolution; limited swath width	Surface roughness; vegetation; topography.
Passive microwave (1 - 100 cm)	Brightness temperature (microwave emission); dielectric constant; soil temperature.	All weather; penetrates some vegetation; large areal coverage.	Limited spatial resolution; soil temperature; surface roughness; vegetation; interference from communications.

interpretation of the data. These include organic matter, roughness, texture and angle of illumination (Jackson et al., 1978).

Thermal infrared measurements have been successfully used to measure surface (0-4 cm) soil moisture. After meteorological inputs to the soil surface have been accounted for, surface temperature is primarily dependent upon the thermal inertia of the soil. The thermal inertia in turn is a function of both the thermal conductivity and the heat capacity which increase with soil moisture. Thus, by measuring the amplitude of the diurnal temperature change, one can develop a relationship between the temperature change and soil moisture. Figure 3 illustrates the results of research at the USDA Water Conservation Laboratory in Phoenix (Idso et al., 1975).

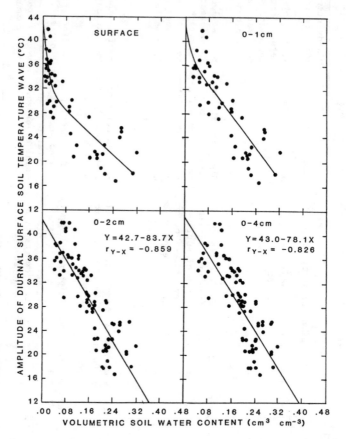

Figure 3. Summary of Results for the Diurnal Temperature Variation vs. Soil Moisture (Idso et al., 1975).

The dielectric constant of a medium describes the propagation characteristics for electromagnetic radiation. These propagation characteristics can be described by either the emissivity of the medium or its reflectance. Because the dielectric constant of water is roughly an order of magnitude larger than that of dry soil, change in

the dielectric constant of a soil is directly related to its water
content.

There are two basic approaches to the microwave measurement of soil
moisture. Passive microwave is a radiometric approach that measures
the emissivity of the soil. Active microwave is a radar approach that
measures the backscatter or reflectance of the soil. In passive
measurements, a brightness temperature is measured with a microwave
radiometer. This brightness temperature is proportional to the surface
temperature and the emissivity (dielectric property) of the soil.
Figure 4 illustrates the relationship between brightness temperature
and soil moisture.

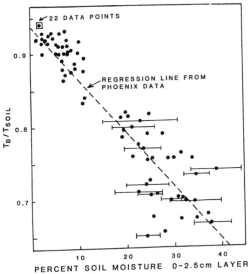

Figure 4. Aircraft Observations of Microwave Brightness Temperature
(T_B) during 1976 and 1977 Flights over Agricultural Fields
in Hand County, South Dakota. (Schmugge, et al., 1978).

Active microwave techniques use a radar approach in which the
reflected pulse (referred to as the backscattering coefficient) is a
function of the surface roughness incidence angle and water content.
Ground-based experiments using truck-mounted equipment at the
University of Kansas have examined the effects of these various target
characteristics of the backscatter coefficient. Figure 5 illustrates
the active microwave dependence on soil moisture. Only recently has
active microwave equipment been demonstrated with aircraft flights
(Jackson et al., 1980b).

Area vs. Point Data

Remote sensing allows viewing an area, rather than a point. To
some this is a deficiency because they would like to reproduce the
point data they are comfortable with. I would suggest this is because
our concepts and models have been developed from a point concept; i.e.,
raingage, soil column, and soil moisture access tube. Apparently there
is much more information in a remote sensing scene, and it may be much
more valuable than a point measurement. We simply have to learn what
information is there and how to use it. This may require developing

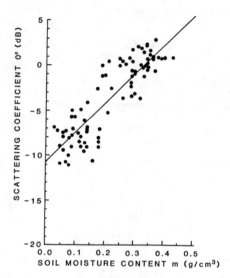

Figure 5. Scattering Coefficient as a Function of Moisture Content in Top cm of soil for 84 Data Sets (data from all five fields were included). (Ulaby and Batlivala, 1976).

new concepts and models to accommodate this type of information. As a mental exercise, consider how you would develop a hydrologic model if you had only remotely sensed data, had no schooling in traditional hydrology, and had no awareness of raingages, soil column models, and similar point concepts or measurements.

I suggest that by concentrating only on detail in the vertical direction you have been looking at the wrong question. It seems to me that variability in the horizontal plane may be hydrologically made more significant than anything we have been studying in the last few decades. Remote sensing, and its ability to measure the response from an area, is potentially one way to approach this problem.

Runoff Coefficients

The use of remote sensing to determine land use for the SCS RCN and other runoff coefficients was reviewed in an earlier section. Although used as an extension of photogrammetry, these examples did illustrate the use of characteristics of specific spectral bands to infer the land use properties. We may be able to determine runoff coefficients directly because a remote sensing measurement potentially can integrate several features into one response.

The response of different wavelengths is determined by the surface and near surface of the target. This measured response is a composite response of several individual features. For example, microwave brightness temperature is affected by surface roughness, grain size of the soil, vegetal cover, and soil moisture. Each of these has a different effect, and this effect varies with wavelength, angle of incidence, and polarization. Therefore, any one microwave measurement is an integrated measure of these effects as well as a spatial sample. These features are the same watershed characteristics that are used to determine an SCS runoff curve number (SCS, 1972). The analogy is approximately as follows:

grain size of the soil = hydrologic soil group A, B, C, or D

vegetal cover and
surface roughness = hydrologic soil cover complex

soil moisture = antecedent moisture condition, I, II or
III

It is possible that remote sensing in the microwave area can give us a direct measure of runoff potential. Blanchard et al. (1975) have had some success in determining a RCN for some watersheds in Texas. In their study using an airborne passive microwave imaging scanner, they investigated the relationship between runoff curve number and the antenna temperature differences for two flights over the same watersheds. Of the various combinations tried, the horizontal polarization (see Figure 6) seems the most promising as a direct measure of runoff curve number. Perhaps we should consider the remote sensing measurement as a direct measure of runoff potential in the same sense as an infiltrometer directly measures infiltration. Use of several different wavelengths, polarizations, etc., may provide all the information we need to predict runoff for a fairly large area. To do so may require that we hydrologists develop some new concepts or models to use this type of information; but that is the challenge we should accept.

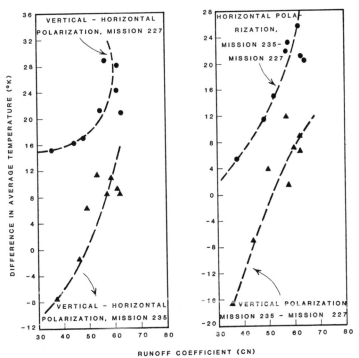

Figure 6. The Relation between Cross- and Like-Polarized Temperature Differences and Storm Runoff Coefficients Determined from Rainfall-Runoff Data. Numbers Refer to Watersheds Chosen for Study (Blanchard et al., 1975).

Frozen Soils

With microwave remote sensing we can differentiate between frozen and nonfrozen soils. For a given soil moisture, the dielectric constant changes dramatically when the soil water changes from a frozen to a liquid state. Thus we have the potential for remotely determining whether or not a soil is frozen. This should greatly benefit the people responsible for flood forecasting, particularly in the upper midwestern United States. However, I'm not sure we know how to use this information. How many of our prediction models can use information that tells us the soils in certain areals of a watershed are frozen? How do we change our runoff coefficients or infiltration models to account for frozen soils? How can we treat areas that are partially frozen and partially frost free (like south-facing slopes)? Can we determine what type of frost is present (concrete vs. columnar) and assign infiltration rates to each?

Drought

In 1977, much of the midwest and western U. S. experienced severe droughts. In many areas the droughts were very obvious—dry streams and empty reservoirs.

There were other drought areas which were more difficult to assess. Below normal rainfalls and drought indices all indicated serious moisture deficiencies, but at harvest time the grain had to be piled up in the streets. What happened? Presumably, barely enough rain fell at precisely the right time to prevent a production disaster. Out of this case came a need to better understand drought (agricultural vs. meteorological) and the role of soil moisture in crop production. The SCS is establishing ten soil moisture index stations in important agricultural areas of the United States. Data from these stations will allow them to monitor the available water in the soil at these points. How far they can extrapolate these data, we do not know. However, I do think we could devise ways, through remote sensing, to extend point data to vast areas of agricultural importance. This will require some research—but the potential is excellent for developing a soil-moisture-based drought measurement.

Energy Budget and Evapotranspiration

Satellite measurements of surface temperature fields offer the potential for energy budget studies over large, complex areas. Dodd (1979) used HCMM (Heat Capacity Mapping Mission) data in combination with a numerical model of the boundary layer proposed by Carlson and Boland (1978) to estimate the spatial distribution of thermal inertial, moisture availability, and the sensible and evaporative heat fluxes. The approach was tested over two urban areas, Los Angeles and St. Louis. The spatial variability of thermal inertia and moisture availability and the surface heat and moisture flux were determined from the model and temperatures derived from the HCMM data. Recent research by Price (1980) suggests the potential for using remotely sensed thermal data for assessing the surface moisture budget. In this study analytical expressions were derived, with a diurnal correction, that relate mean evaporation rate and a soil moisture parameter to surface temperature of bare soils.

These studies have demonstrated a potential for use of satellite data in energy budget work. The possibility of determining the spatial distribution of evaporative flux or moisture availability for complex areas has many potential uses in hydrology, agriculture, forestry and climatology. Much work needs to be done, particularly in improving the boundary layer models and understanding the edge effects caused by land use changes and the spatial variation of roughness length.

Feedback for Simulation Models

Hydrologists have built a large number and variety of continuous simulation models. Most are mass balance-type models, taking rainfall (or snowmelt) as input and routing it to stream flow; often a portion is temporarily stored. The stored water defines the state of the system and, as such, controls the rate of sequential processes and events. Since each successive computation is based on the previous state of the system, errors in the predicted output often get larger with time. How well could we improve our prediction if we could check our system periodically and update our predictions? Repetitive measures of soil moisture use as feedback to the model could do this. Improved prediction accuracy may have large tangible benefits.

A recent study by Jackson et al. (1980a) demonstrated how possible applications of repetitive, remote measurements of soil moisture might be used. They discussed how these areal data may be used to calibrate soil and vegetation parameters and to correct errors resulting from point measurements of precipitation. In the study they demonstrated how soil moisture observations are useful in calibration and updating the state of the system. However, it was also pointed out that the model structure itself may preclude a valid analysis of the value of soil moisture measurements or the frequency needed to improve the simulations. One must carefully choose the model to be used in this type of study; it may be necessary to develop a new model or make significant modifications to existing models.

Soil Moisture Profiles

We can measure soil moisture by remote sensing. We are not sure how deep we measure it and what to do with the numbers yet, but we should be able to develop profile models that use information from several shallow layers near the surface. The changes in the surface soil moisture should reflect the total profile moisture to some extent. In a study simulating a soil moisture profile, Jackson (1980) demonstrated that under certain conditions one could predict the total profile from the type of measurement one may obtain from remote sensing. Simulated measurements of the top 0.1 m of soil were used to simulate the total soil water profile to about 1.0 m within an acceptable error. Perhaps, this is accurate enough for practical purposes—especially when one considers the problems of spatial variations. Good, simple, soil moisture profile models that use repetitive surface and shallow layer measurements could provide a major improvement to both hydrologic and crop yield predictions.

Snow

Successful applications of satellite data for snowmelt and water supply forecasting were discussed in an earlier section. Although promising, current efforts also pointed out limitations in the snow cover area approach. These include lack of direct information on depth, water equivalent, ripeness, and other fundamental properties of snow packs. However, a considerable amount of work remains; research is in progress to provide more useful information on the snow pack with remote sensing.

Various researchers have been using the existing satellite data to try to infer depth, water equivalent, and density. McGinnis et al. (1975) used the brightness of reflected radiation, as measured by very high resolution radiometer (VHRR) on board the NOAA Z satellite, as an indicator of snow depth. Using a relationship that relates increasing brightness to increasing snow depth, they achieved a correlation of 0.86 by pairing the brightest value in a 32 x 32 km square with the ground measured depth. However, this approach seems limited to fairly shallow snow packs over large areas because the rate of increase becomes quite small at depths greater than about 30 cm.

Like soil moisture and other hydrologic applications, the microwave region of the spectrum offers the snow hydrologist a potentially powerful tool. Not only can a microwave sensor be an all-weather instrument because it penetrates cloud cover; it can also penetrate the snow pack, which presents one with the opportunity of inferring many of the properties of the snow pack. These include depth and water content as well as the degree of ripeness, crystal size, and the presence of liquid water in a melting snow pack. As with soil moisture, the microwave measurement reflects several characteristics at once. Moreover, a considerable amount of experimental evidence indicates that the technique holds great promise.

Rango et al., (1979) developed a good snow depth-microwave brightness temperature relation for dry snow in the Canadian high plains. There, data were taken from the Electrically Scanning Microwave Radiometer on the Nimbus 6 satellite. These results are summarized in Figure 7 and represent the snow cover over an area covering approximately the southern one-third of Alberta and

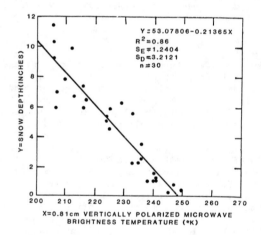

Figure 7. Nimbus 6 Vertically Polarized Microwave Brightness Temperature vs. Snow Depth on the Canadian High Plains (Rango et al., 1979).

Saskatchewan. More detailed studies from truck mounted equipment show a similar strong relationship between the scattering coefficient or brightness temperature and the snow water equivalent (Figure 8). An example of how the microwave responses change as the amount of liquid water in the snow increases is shown in Figure 9. These types of experimental evidence illustrates the great potential for improving

488

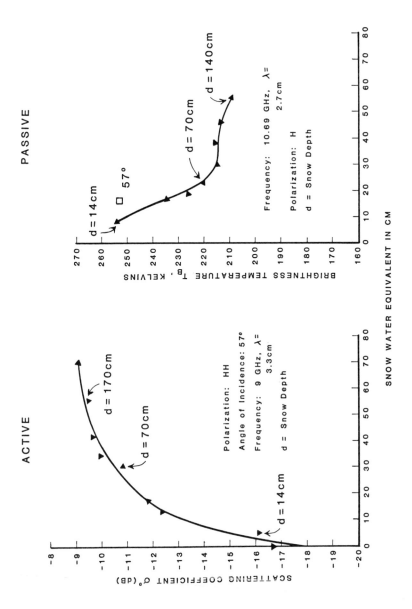

Figure 8. Variation of Microwave Response with Snow Water Equivalent
for Artificially Piled Snowpacks: (a) Active Case,
(b) Passive Case (Ulaby and Stiles, 1979).

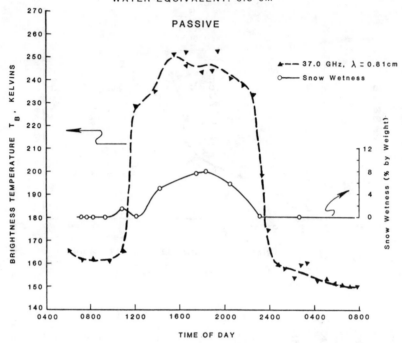

DATE: 2/17 – 2/18/77
ANGLE OF INCIDENCE: 55
WATER EQUIVALENT: 6.3 cm

Figure 9. Diurnal Variation of Microwave Response and Snow Wetness:
(a) Active Case (note that snow wetness scale has been
reversed for ease of comparison with 0°); (b) Passive Case
(Ulaby et al., 1978).

predictions of snow water content and snow melt with remote sensing.

SUMMARY

Remote sensing is currently being used as an important source of
data and information for hydrologic modeling and research. This paper
has summarized the more important applications which include developing
runoff coefficients, determining the area and extent of flooding, and
determining the areal extent of snow cover for predicting water yield.

Current research involves extension of these areas plus work with
some of the less common wavelengths or frequencies. Thermal infrared
measurements allow us to infer a heat budget and hence estimate crop
condition, soil moisture, or evapotranspiration. With microwave data
we have all-weather capability plus the ability to penetrate vegetation
and to measure soil moisture and snow water content directly. This
research, although still very much experimental, is beginning to treat
remote sensing as a unique source of data.

The potential for remote sensing and its application to hydrology
is considerably greater than research has addressed thus far.

490

Measuring the characteristics of an area rather than a point, integrating several characteristics with one composite measurement, and improving prediction models with continuous or frequent feedback from satellite measurements are just a few of the aspects that must be explored further. Considerable research is needed if we are to fully realize the potential benefits of remote sensing in hydrology. Treating remotely sensed data as a unique measurement of hydrologic characteristics offers the best opportunity for major advances in the field of hydrologic modeling.

REFERENCES

Berg, C. P., McGinnis, D. F., and Forsyth, D. G. 1980. Mapping the 1978 Kentucky River Flood from NOAA-5 satellite thermal infrared data. Technical Paper, ACSM-ASP Convention, pp. 106-111, American Society of Photogrammetry, St. Louis, Missouri.

Berg, C. P., Matson, M., and Wiesnet, D. R. 1979. Assessing the Red River of the North 1978 flooding from NOAA satellite data. Proceedings, Pecora 5 Symposium, Sioux Falls, South Dakota.

Bissell, V. C. and Peck, E. L. 1973. Monitoring snow water equivalent by using natural soil radioactivity. Water Resources Research, 9(4):885-890.

Blanchard, B. J., Rouse, J. W., Jr., and Schmugge, T. J. 1975. Classifying storm runoff potential with passive microwave measurements. Water Resources Bulletin, 11(5):892-907.

Bondelid, T. R., Jackson T. J., and McCuen, R. H. March 1980. Comparison of conventional and remotely sensed estimates of runoff curve numbers in Southeastern Pennsylvania. Technical Paper, ASCM-ASP, American Society of Photogrammetry, pp. 81-96, St. Louis, Missouri.

Brown, A. J., Hannaford, J. F., and Hall, R. L. 1979. Application of snow covered area to runoff forecasting in selected basins of the Sierras, Nevada, California. Proceedings, Workshop on Operational Applications of Satellite Snow Cover Observations, NASA Conference Publication 2116, pp. 185-200, Sparks, Nevada.

Carlson, T. N. and Boland, F. E. 1978. Analysis of urban-rural canopy using a surface heat flux/temperature model. Journal of Applied Meteorology, 17(7):998-1013.

Carter, V. 1978. Coastal wetlands: role of remote sensing. Proceedings, Symposium on Technical, Environmental, Socioeconomic and Regulatory Aspects of Coastal Zone Management, San Francisco, California.

Carter, V., Garrett, M. K., Shima, L., and Gammon, P. 1977. The great dismal swamp: management of a hydrological resource with the aid of remote sensing. Water Resources Bulletin, 13(1):1-12.

Carter, V., Malone, D., and Burbank, T. H. 1979. Wetland classification and mapping in western Tennessee. Photogrammetric Engineering, Remote Sensing, 45(3):273-284.

Dillard, J. P. and Orwig, C. E. 1979. Use of satellite data in runoff forecasting in the heavily forested, cloud covered Pacific

Northwest. Proceedings, Workshop on Operational Applications of Satellite Snowcover Observations, NASA Conference Publication 2116, pp. 127-150, Sparks, Nevada.

Dodd, J. K. 1979. Determination of surface characteristics and energy budget over an urban-rural area using satellite data and a boundary layer model. Unpublished Master's Thesis, 87 pp., The Pennsylvania State University.

Hoyer, B. E., Hallberg, G. R., and Taranik, J. V. 1973. Seasonal multispectral flood innundation mapping in Iowa. In Management and Utilization of Remote Sensing Data; Proceedings, Symposium of the American Society of Photogrammetry, pp. 130-141, Sioux Falls, South Dakota.

Idso, S. B., Schmugge, T. J., Jackson, R. D., and Reginato, R. J. 1975. The Utility of Surface Temperature measurements for remote sensing of soil water status. Journal Geophysical Research, 80:3044-3049.

Jackson, R. D., Ahlar, J., Estes, J. E., Heilman, J. L., Kakle, A., Kannemasu, E. T., Millard, J., Price, J. C., and Wiegand, C. 1978. Soil moisture estimation using reflected solar and emitted thermal radiation. Chapter 4: Soil Moisture Workshop, NASA Conference Publcation 2073, 219 pp.

Jackson, T. J. 1980. Profile soil moisture from surface measurements. ASCE Journal of Irrigation and Drainage Division, ASCE, Proceedings of Paper 15474, 105(IRZ):81-92.

Jackson, T. J. and Ragan, R. M. 1977. Value of Landsat in urban water resources planning. Journal of Water Resources Planning and Management, Division 103 (WR1):33-46.

Jackson, T. J., Ragan, R. M., and Fitch, W. N. 1977. Test of Landsat-based urban hydrologic modeling. ASCE Journal of Water Resources, Planning and Management Division 103 (WR1):141-158.

Jackson, T. J., Schmugge, T. J., Nicks, A. D., Coleman, G. A., and Engman, E. T. 1980a. Soil moisture updating and microwave remote sensing for hydrologic simulation. Bulletin International Association of Scientific Hydrology.

Jackson, T. J., Chang, A. and Schmugge, T. J. 1980b. Aircraft Active Microwave Measurements for Estimating Soil Moisture. In: Technical Papers, American Society Photogrammetry, Fall Technical Meeting, RS-3-C-1 to RS-3-C-10 pp., Niagara Falls, New York.

Leaf, C. F. 1969. Aerial photographs for operational streamflow forecasting in the Colorado Rockies. Proceedings, 37th Western Snow Conference, Salt Lake City, Utah.

Martinec, J. 1970. Study of snowmelt-runoff process in two representative watersheds with different elevation range. In: Results of Research and Experimental Basins. Proceedings, Wellington Symposium, Pub. 96, pp. 29-39.

McGinnis, D. F., Jr., Pritchard, J. R., and Wiesnet, D. R. 1975. Determination of snow depth and snow extent from NOAA Z satellite very high resolution radiometer data. Water Resources Research, 11(6):897-902.

Meier, M. J. and Evans, W. E. 1975. Comparison of different methods for estimating snow cover in forested, mountainous basins using LANDSAT (ERTS) images in Operational Applications of Satellite Observations, NASA SP-391, pp. 215-234, Washington, D.C.

Morrison, R. B., and Cooley, M. E. 1973. Assessment of flood damage in Arizona by means of ERTS-1 imagery. Proceedings, Symposium on Significant Results Obtained from ERTS-1, Vol. 1, pp. 775-760, New Carrollton, Maryland.

Morrow, J. W. and Carter, V. 1978. Wetland classification on the Alaskan North Slope. 5th Canadian Symposium on Remote Sensing. Victoria, British Columbia.

Price, J. C. 1980. The potential of remotely sensed thermal infrared data to infer surface soil moisture and evaporation. Water Resources Research, 16(4):787-795.

Ragan, R. M. and Jackson, T. J. 1980. Runoff synthesis using Landsat and the SCS model. Proceedings, ASCE, Paper 15387, 106 (HY5):667-678.

Rango, A. April 1979a. Review of remote sensing capabilities in snow hydrology. ASCE Preprint 3511.

Rango, A. February 1979b. Remote sensing of snow and ice: A review of research in the United States 1975-1978. NASA Technical Memorandum 79713.

Rango, A. and Anderson, A. T. 1974. Flood hazard studies in the Mississippi River Basin using remote sensing. Water Resources Bulletin, 10(5):1060-1081.

Rango, A., Chang, A. T. C., and Foster, J. L. 1979. The utilization of spaceborne microwave radiometers for monitoring snowpack properties. Nordic Hydrology, 10(1):24-40.

Rango, A., Hannaford, J. F., Hall, R. L., Rosenzweig, M., and Brown, A. J. 1977. The use of snowcovered area in runoff forecasts. NASA, Goddard Space Flight Center, Greenbelt, Maryland Document X-913-77-48.

Rango, A. and Itten, K. I. 1976. "Satellite potentials in snowcover monitoring and runoff prediction", Nordic Hydrology, 7:209-230

Rango, A. and Martinec. J. 1979. "Application of a Snowmelt-Runoff Model using Landsat Data", Nordic Hydrology, 10:225-238.

Rango, A. and Peterson, R., eds. 1979. Proceedings, Workshop on Operational Applications of Satellite Snow Cover Observations. NASA Conference, Publication 2116, Sparks, Nevada.

Rawls, W. J., Shalaby, A., and McCuen, R. H. 1980. Comparison of methods for determining urban runoff curve numbers. American Society of Agricultural Engineering, Paper 80-2565, 14 pp., Winter Meeting, Chicago, Illinois.

Schmugge, T. J., E. G. Njoku, E. Peck, and F. T. Ulaby. 1978. "Microwave and Gamma Radiation Observations of Soil Moisture," Chapter 5: Soil Moisture Workshop, NASA Conf. Pub. 2073, 219 pp.

Schmugge, T. J., Jackson, T. J., and McKim, H. L.   1980.   Survey of
    methods for soil moisture determination.   Water Resources Research,
    Vol. 16, No.6, pp. 961-979.

Shafer, B. A. and Leaf, C. F.   1979.   Landsat derived snowcover as an
    input variable for snowmelt runoff forecasting in central
    Colorado.   Proceedings, Workshop on Operational Applications of
    Satellite Snowcover Observations.   NASA Conference Publication
    2116, pp. 151-169, Sparks, Nevada.

Slack, R. B. and Welsh, R.   1980.   Soil Conservation Service Runoff
    Curve Number Estimates From Landsat Data.   Water Resources
    Bulletin, 16(5):887-893.

Soil Conservation Service.   1972.   SCS National Engineering Handbook,
    Section 4: Hydrology, U.S. Department of Agriculture, Washington,
    DC.

Sollers, S. C., Rango, A., and Henninger, D. L.   1978.   Selecting
    Reconnaissance Strategies for Flood Plain Surveys.   Water Resources
    Bulletin, 14(2):359-373.

Ulaby, F. T., and P. P. Batlivala.   1976.   "Optimum Radar Parameters for
    Mapping Soil Moisture," IEEE Trans. on Geoscience Electronics,
    V. GE-14, n.2, pp 81-93.

Ulaby, F. T., Fung, A. K., and Stiles, W. H.   1978.   Backscatter and
    Emission of Snow: Literature Review and Recommendations for Future
    Investigations, R. S. L. Technical Report 369-1, University of
    Kansas Center for Research, Lawrence, Kansas.

Ulaby, F. T. and Stiles, W. H.   1979.   The Active and Passive Microwave
    Response to Snow Parameters, Part II: Water Equivalent of Dry Snow.
    Journal Geophysical Research.

U.S. Army Corps of Engineers.   November 1979.   Determination of land
    use from LANDSAT imagery: applications to hydrologic modeling.
    Research note 7, The Hydrologic Engineeering Center, U.S. Army
    Corps of Engineers.

U.S. Army Corps of Engineers.   July 1976.   Urban storm water runoff
    "STORM." Computer Program 723-58-L2520, Hydrologic Engineering
    Center, Davis, California.

Wiesnet, D. R. and Matson, M.   1975.   Monthly winter snowline variation
    in The Northern Hemisphere from Satellite Records, 1966-1975, NOAA
    Technical Memo NESS 74, 21 pp., National Environmental Satellite
    Service, Washington, DC.

Wiesnet, D. R., McGinnis, D. F., and Pritchard, J. A.   1974.   Mapping of
    the 1973 Mississippi River floods by the NOAA-2 satellite.   Water
    Resources Bulletin, 10(5):1040-1049.

Williamson, A. N.   1974.   Mississippi River flood maps from ERTS-1
    digital data.   Water Resources Bulletin, 10(5):1050-1059.

RESEARCH FOR RELIABLE QUANTIFICATION OF WATER SEDIMENT CONCENTRATIONS FROM MULTISPECTRAL SCANNER REMOTE SENSING DATA

Charles H. Whitlock
Senior Research Engineer
NASA Langley Research Center

William G. Witte
Research Engineer
NASA Langley Research Center

Theodore A. Talay
Research Engineer
NASA Langley Research Center

W. Douglas Morris
Research Engineer
NASA Langley Research Center

Jimmy W. Usry
Research Engineer
NASA Langley Research Center

Lamont R. Poole
Research Engineer
NASA Langley Research Center

ABSTRACT

During F.Y. 1980, a 6-year joint program known as AgRISTARS (Agriculture and Resources Inventory Surveys Through Aerospace Remote Sensing) was initiated between five United States federal agencies. AgRISTARS is directed toward developing the technology and testing the capability to employ remote sensing in an economical manner in early warning and crop condition assessment; foreign commodity forecasting; yield model development; soil moisture mapping; domestic crop and land cover assessment; renewable resources inventory and conservation and pollution assessment. It is the latter area, conservation and pollution, which is of prime importance to agricultural rainfall runoff modeling. This paper describes research being conducted in the conservation and pollution element of AgRISTARS for assessing conservation practices through remote monitoring of sediment concentration in drainage basin waters. More specifically, the problem of large sediment concentrations is discussed, laboratory spectral signature results are presented, remote sensing penetration depth knowledge is assessed, and research aimed at quantifying atmospheric transmission, specular reflection, and path radiance is described. The results of this research are expected to be improved experimental techniques and more reliable data analysis procedures for quantification of water sediment concentrations in future years. Such data should be useful in validating rainfall-runoff models of large drainage basins.

INTRODUCTION

It is well known that satellite and aircraft multispectral scanner instruments can observe turbidity patterns in water bodies. Statistical methods have been used (Johnson, 1975) to separate chlorophyll and dissolved carbon effects in attempts to quantify surface sediment concentrations. Unfortunately, these techniques require numerous ground truth samples which must meet several mathematical criteria with specific limitations on water and atmospheric conditions for best accuracy (Whitlock and Kuo, 1979). Numerous other authors have developed simplified methods for sediment quantification at specific sites (Yarger and McCauley, 1975, for example); however, most approaches are site specific and cannot be applied on a national basis. To date, no technique has been demonstrated to have national applicability over the wide range of concentrations, mineral types, and water mixtures which exist within the United States. Little is known concerning the remote sensing of sediments at very high concentrations (50 ppm to 1000 ppm), nor how deep remote sensing sees into turbid waters, particularly at near-infrared wavelengths.

It is the purpose of the AgRISTARS Program to assess the effectiveness of conservation practices through remote monitoring of sediment concentration in drainage basin waters. This task is expected to be achieved in the following manner:

1. Development of reliable "sediment-only" algorithms for remote sensing of turbid reservoir waters.

2. Development of basin-scale pollution models which utilize remote-sensing data for both input and verification of output.

3. Demonstration of remote-sensing algorithms by quantifying surface sediment concentrations at five national test sites.

4. Utilization of remotely sensed sediment concentrations in models to assess water pollution and drainage basin conservation practices.

The task is a cooperative effort between the National Aeronautics and Space Administration (NASA) and the United States Department of Agriculture (USDA) with both agencies participating in and transferring technology in each of the above activities. It is the purpose of this paper to describe the research effort being conducted to develop reliable "sediment-only" algorithms. Fundamental laboratory investigations, remote sensing penetration depth quantification, atmospheric correction research, and field experiments are discussed. It is expected that results from this research will be useful in future years for validating rainfall-runoff models of large drainage basins, whether they have agricultural, forestry, or urban characteristics.

FUNDAMENTAL LABORATORY INVESTIGATIONS

Laboratory studies are being conducted in the Marine Upwelled Spectral Signature Laboratory (MUSSL) at the NASA Langley Research Center. Figure 1 is a photograph of the facility which consists of a 11,600 ℓ water tank illuminated by 32,280 ℓm/m² of artificial sunlight. (Clear weather total illumination on the Earth surface typically ranges from 30,000 ℓm/m² to 130,000 ℓm/m² depending on solar elevation angle and atmospheric haze.) Upwelled radiance is measured above the water surface by a rapid scanning spectrometer. (The same instrument is used

Figure 1.- Marine Upwelled Spectral Signature Laboratory.

off the side of a boat for field experiments.) Input light is monitored by measuring upwelled radiance from a horizontal, near-Lambertian
99 percent reflector (EASTMAN 6080 paint) just above the surface of the
water. The ratio of diffuse radiance upwelled from the water column to
that from the near-Lambertian reflector is the inherent upwelled
reflectance. Spectral variation of upwelled reflectance causes changes
in water color and near-infrared signals which are observed by satellite and aircraft multispectral scanners. In order to interpret
remote-sensing signals, it is necessary to understand how upwelled
reflectance varies with water column chemistry. It is desirable to
relate chemical variations to underwater optical coefficients which, in
turn, create changes in upwelled reflectance spectra. The trailer
shown in figure 1 contains prototype instruments which are used to
measure beam attenuation coefficient, absorption coefficient, and
volume scattering function spectra to complement reflectance measurements from the large tank. For most laboratory tests, water samples
are analyzed for Total Suspended Solids (TSS), Dissolved Organic Carbon
(DOC), total organic carbon, particulate organic carbon, chlorophyll a,
iron, copper, sediment mineralogy, and particle size distribution.
More detailed descriptions of the laboratory apparatus and optical
instruments are given in Whitlock (1977), Witte, et al. (1979),
Friedman, et al. (1980), and Whitlock, et al. (1981).

Two national test sites have been selected for initial work.
These sites are Lake Chicot, Arkansas, and John H. Kerr Reservoir,
Virginia (fig. 2). Three additional sites will be selected at a later
date. Bottom sediments from both Lake Chicot and Kerr Reservoir were
mixed with filtered-deionized tap water (TSS < 0.5 ppm, DOC < 0.2 ppm)
to achieve TSS concentrations between 5 ppm and 783 ppm in the MUSSL
tank. Figure 3 shows preliminary data for upwelled reflectance for the
two sediments. In both cases, signals tend to saturate having little

497

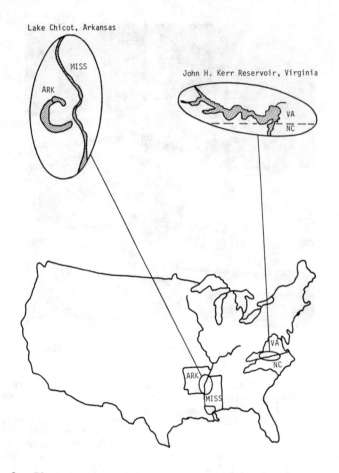

Figure 2.- First two test sites for AfRISTARS Conservation Assessment Task.

change with concentration for TSS > 100 ppm in the wavelength (λ) range between 400 nm and 600 nm. For 750 nm $\leq \lambda \geq$ 950 nm, there appears to be good signal separation for all concentrations. It must be noted, however, that the sediment signal beyond 900 nm may be of limited value because of the atmospheric water vapor band near 940 nm which reduces both surface solar illumination and transmission of upwelled radiance at low solar elevation angles or under conditions of high atmospheric haze. The effect of sediment concentration on upwelled reflectance is more clearly shown for several different λ values in figure 4. From these results, it is clear that upwelled radiance is not a linear function of TSS except over limited concentration ranges. The degree of nonlinearity is a function of λ. While these results are for only two particular sediments, it must be noted that similar non-linear characteristics have been previously observed for other sediments by Yarger, et al. (1973), Whitlock, et al. (1977), and Holyer (1978). Previous laboratory results (Whitlock, et al., 1977) have shown that various inorganic sediments have different upwelled reflectance spectra. At this time, it is not known which factors are most significant in causing the upwelled spectral signature of a particular sediment. The

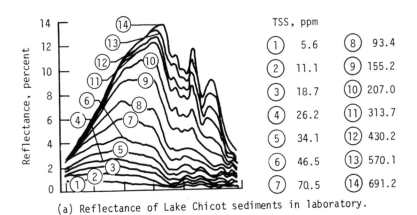

(a) Reflectance of Lake Chicot sediments in laboratory.

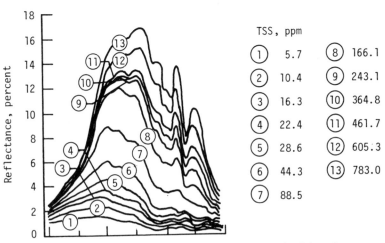

(b) Reflectance of Kerr Reservoir sediments in laboratory.

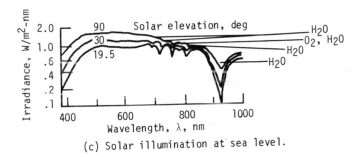

(c) Solar illumination at sea level.

Figure 3.- Spectral characteristics of remotely-sensed upwelled reflectance.

AgRISTARS Conservation Assessment Task plans to examine the effects of clay mineralogy, quartz content, iron and metal content, and particle size distribution relative to reflectance in order to determine those factors which most influence spectral characteristics for particular soil mixtures. It must be known whether or not usual changes in particle mixture due to reservoir settling have a significant influence on spectral characteristics if reliable data analysis procedures are to be developed.

A major concern in the development of "sediment-only" algorithms is the effect of natural DOC in the form of yellow-colored humic materials from soil leaching and decomposition of biological materials. To date, the effects of humic material have been examined in the MUSSL facility under one set of controlled conditions. Results of these tests are shown in figure 5. The upper chart shows reflectance increasing as sediment is added to filtered-deionized tap water (mined Calvert clay from Whitlock, et al., 1977). The lower chart shows reflectance decreasing as humic material is successively added to a Calvert clay-deionized water mixture. (It should be noted that the 1977 Calvert clay spectra do not precisely agree with 1980 data because of changes in spectral resolution, data presentation technique at near-infrared wavelengths, laboratory illumination, and soil batch.) A commercial humic material (Aldrich HI, 675-2) was used in the laboratory simulation. Of concern is how well the commercial material simulates optical

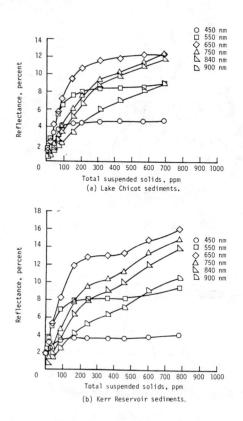

(a) Lake Chicot sediments.

(b) Kerr Reservoir sediments.

Figure 4.- Signal linearity characteristics.

characteristics of natural DOC. A comparison of differential absorption coefficient was made at NASA which indicated that the commercial material has similar optical spectra to that of natural DOC materials from five Georgia rivers for 400 nm $\leq \lambda \leq$ 750 nm. A difference in absorption was observed at 800 nm which suggests that the laboratory reflectance changes observed at near-infrared wavelengths above 750 nm may not be similar to those in nature. Natural DOC from the Georgia rivers had lower absorption than the commercial material at 800 nm. This would imply that reflectance for $\lambda >$ 750 nm is influenced little by natural DOC, not as the 50 percent reduction shown in figure 5(b). Additional tests to compare optical properties of the commercial material with that of natural DOC in reservoirs are required to assess laboratory test validity at near-infrared wavelengths. Laboratory results for 400 nm $\leq \lambda \leq$ 750 nm are considered valid. Upwelled reflectance decreased by as much as a factor of 4 for 500 nm $\leq \lambda \leq$ 700 nm. Unfortunately, this is the same wavelength range as maximum sediment signals and near-linear response when TSS < 100 ppm. The fact that increases in natural DOC seem to cancel the effect of sediment increases for 400 nm $\leq \lambda \leq$ 720 nm causes a problem in the analysis of remote sensing data in water bodies where significant natural DOC variations occur. Present indications from both Georgia river data and traditional oceanographic results (Jerlov, 1968) suggest that natural DOC, unlike

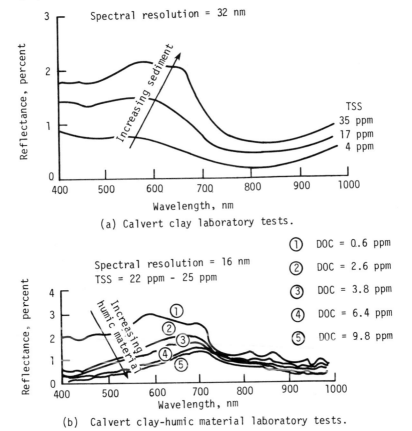

(a) Calvert clay laboratory tests.

(b) Calvert clay-humic material laboratory tests.

Figure 5.- Effects of humic material on sediment reflectance.

501

the commercial material, may not have a significant effect on reflectance at near-infrared wavelengths above 720 nm. Laboratory results shown in figure 3 suggest that Band 4 (760 nm - 900 nm) of the future Thematic Mapper instrument offers potential for observing sediment without the counter-effect of natural DOC if chlorophyll reflectance problems associated with algae content can be resolved.

The AgRISTARS effort has not yet begun to actively address the algae problem. Review of oceanographic and coastal processes literature shows that little optical coefficient data for algae exist for λ > 700 nm (Duntley, et al., 1974, Morel, 1980, for example). A further complication is that most literature is for non-bloom growth conditions at which reflectance values are known to be low. Holyer (1978) states that chlorophyll absorption is so weak relative to scattering by sediment particles that it can be ignored when TSS > 25 ppm, except in cases of extreme algae bloom. Bloom reflectance in waters with significant sediment concentration requires future investigation. At this time, only a simplistic analogy in combination with one set of previous MUSSL tests can be used to speculate on near-infrared bloom optical characteristics. Whitlock, et al. (1979) tested a golden-brown algae (Phaeodactylon tricornutum) in artificial seawater in which TSS values were reasonably constant (7 ppm to 11 ppm). Figure 6(a) shows results of those tests. As chlorophyll a concentration increased from 20 μg/ℓ to 31 μg/ℓ by overnight growth, reflectance increased by a factor of 2 for all λ < 850 nm. Little change occurred for λ > 850 nm. If the observed changes were caused by algae bloom, they might be explainable by the following analogy. Green leaf chlorophyll is known to have high reflectance in air at near-infrared wavelengths (fig. 6(b) from Malila, et al., 1977). Reflectance characteristics of suspended chlorophyll-bearing particles are reduced by the absorption properties of water (fig. 6(c) from Curcio and Petty, 1951, and Smith and Baker, 1981). Comparing figures 6(b) with 6(c), it is evident that constant-magnitude near-infrared diffuse scattering would change to wavelength dependent reflectance if photons had to pass through a water layer before exiting to the atmosphere. The increasing magnitude of absorption with increasing near-infrared wavelength is at least a partial explanation for the lack of chlorophyll reflectance change observed for λ > 850 nm in figure 6(a). (The same type of analogy cannot be used to explain reflectance characteristics at visible wavelengths because of fluorescence processes known to occur in that region.) More extensive testing is required to firmly establish chlorophyll reflectance characteristics at near-infrared wavelengths. Reservoir field data must be compiled to establish typical sediment-chlorophyll over the yearly cycle, and controlled-water tests (MUSSL or outside tank) are required to establish at what sediment concentration levels chlorophyll (both bloom and non-bloom conditions) is no longer a significant influence on the sediment signal. Without such research, it is difficult to differentiate between moderate-sediment and algae-bloom conditions at near-infrared wavelengths.

REMOTE SENSING PENETRATION DEPTH

It is necessary to know the depth from which the remote-sensing signal originates at both visible and near-infrared wavelengths, and how these values are related to the traditional Secchi depth. It is also a requirement that simplified techniques be developed which enable instant assessment of field Remote Sensing Penetration Depth (RSPD) values so that ground-truth water samples can be taken at correct depths. In water bodies with significant vertical concentration gradients, knowledge of RSPD is vital to the correct application of

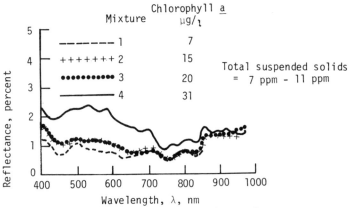

(a) Laboratory tests of artificial seawater
plus Phaeodactylon tricornutum.

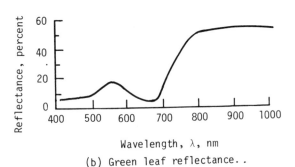

(b) Green leaf reflectance..

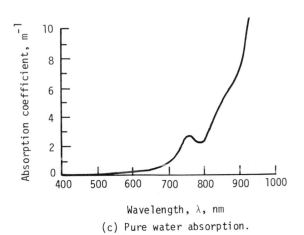

(c) Pure water absorption.

Figure 6.- Effects of chlorophyll on upwelled reflectance.

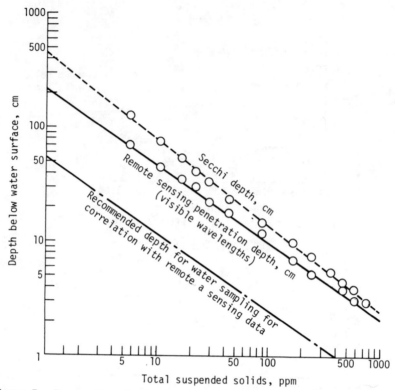

Figure 7.- Penetration depths at visible wavelengths from Kerr sediment laboratory tests.

remote sensing results for validation of computerized water quality models.

The AgRISTARS Conservation Assessment Task will address the penetration depth problem with particular emphasis on near-infrared characteristics. Recent coastal zone data (Morris, et al., 1980) show that under certain conditions, RSPD may be greater at near-infrared λ than in the visible range. To date, one MUSSL test and two field experiments (Kerr Reservoir) have been conducted, but data have not yet been completely analyzed. Figure 7 shows some preliminary data that are limited to visible wavelengths between 400 nm and 700 nm. From this chart, it is seen that RSPD is less than Secchi depth. As TSS increases from 10 ppm to 1000 ppm, visible-range RSPD increases from 56 percent to 88 percent of Secchi depth. The recommended depth for water sampling is that station which divides the RSPD layer in such a manner that 50 percent of upwelled reflectance originates above that point. Careful examination of optical model calculations by McCluney (1974) indicates that the 50 percent signal origination depth is approximately 25 percent of RSPD. These results suggest that it is very easy to take water samples from too deep a depth in highly turbid waters. Such an error may have a serious effect on correlations of remote-sensing results with ground truth data. It must be noted that results shown in figure 7 are human-eye averages strictly limited to visible wavelengths. Future spectral penetration depth data (400 nm $\leq \lambda \leq$ 1100 nm) are expected to give much insight into near-infrared characteristics relative to those in the visible region.

ATMOSPHERIC CORRECTION RESEARCH

It is well known that remote-sensing signals received by aircraft or satellite instruments contain atmospheric effects that modify upwelled radiance emitted at the Earth's surface. For water bodies, diffuse skylight is specularly reflected upward from the air-water surface which adds to upwelled radiance transmitted through the surface from the water column. Both water column and reflected skylight radiances are then attenuated by atmospheric haze before they reach the remote-sensing platform. Finally, there is a large component (known as path radiance) added to the attenuated surface signal because of light backscattered upward into space from atmospheric particles and water vapor. Path radiance is mostly a function of solar elevation angle as well as atmospheric haze conditions. Usually, less than 30 percent of the signal received at the satellite or aircraft contains information related to the water column, the remainder being caused by reflected skylight, attenuation, or path radiance effects.

Numerous techniques and modeling efforts have been proposed to account for atmospheric effects and varying solar illuminations. Ratios of Landsat band radiances appear to be nearly independent of solar and atmospheric scattering effects on concrete-dam targets with no skylight reflection component (Yarger and McCauley, 1975). Dark targets in the remote-sensing scene can serve as backgrounds against which the additive atmospheric path radiance can be estimated if one assumes constant atmospheric conditions over the entire scene. The use of clear lakes (Ahern, et al., 1977), scene shadows (Piech and Schott, 1974) and cloud shadows (Langham, 1977) represent variations of this technique. Dark target subtraction followed by band ratioing has been used by Scarpace, et al. (1979) as an atmospheric correction to Landsat scenes of Wisconsin lakes. Gordon (1980) has corrected for atmospheric effects in Coastal Zone Color Scanner (CZCS) data using a technique which assumes nonreflectance of ocean water at near-infrared wavelengths. Holyer, et al. (1980), Bukata, et al. (1980) and figure 3 MUSSL results suggest that such a technique is not generally applicable to coastal and inland waters. A technique is desired which accounts for atmospheric effects over the specific reservoir or water body under analysis without assumption of constant conditions over a large geographical area. Recent AgRISTARS field measurements at Kerr Reservoir and coastal data from Holyer, et al. (1980) indicate that significant variations in atmospheric haze (which are invisible to the human eye) can occur in a matter of minutes on days which are observed to be uniformly clear. Incorrect analysis of such haze patches may cause generation of artificial sediment patterns during analysis of remote-sensing data.

One method for making local atmospheric corrections is to make ground measurements of atmospheric transmittance and path radiance parameters at the water body of interest. These parameters are input to various radiative transfer models (such as those developed by Turner (1977) and Kneizys, et al. (1980)) for computation of skylight, path radiance, or transmittance in the local area. The major disadvantage of such techniques is that local ground truth observations are required at each water body under analysis, and the problem of high-frequency, localized haze patches may not be completely eliminated.

AgRISTARS atmospheric correction research is scheduled to begin during the latter half of F.Y. 1981. The general approach is one in which precise measurements of the spectral variation of atmospheric optical depth, sky radiance, and water vapor parameters will be made under repeated Landsat and simulated aircraft Thematic Mapper overpasses of Kerr Reservoir. Upwelled radiance at the water surface and underwater optical coefficients will also be monitored. The data will be input to the various ratio techniques as well as radiative transfer models to establish which of the presently developed concepts is most accurate for monitoring of highly turbid waters. A highly accurate atmospheric spectrophotometer is presently in final stages of construction and checkout. This instrument is a duplicate of one constructed several years ago by the Scripps Institution of Oceanography for the U.S. Navy and is expected to yield stable, accurate absolute measurements superior to those possible with most low-cost instruments presently available. The device is most appropriate for examining atmospheric stability under different weather conditions because it does not rely on day-long Langley method readings for application. It is expected that such experiments will either confirm one or more existing techniques or provide insight into the development of new concepts which allow correction of near-infrared signals to water-column upwelled radiance values.

FIELD EXPERIMENTS

Throughout previous sections, various field experiments have been mentioned. Such activities are required to guide the MUSSL test program and to perform atmospheric correction research. A background study of previous Landsat images of both Kerr Reservoir and Lake Chicot has proved quite useful. While limited to qualitative analysis because of the lack of ground truth data, historical images provide knowledge of circulation patterns, which aid future field experiments, and are useful in validating trends of laboratory data. Figure 8 shows one early scene (October 11, 1972) of Kerr Reservoir. Band 5

(a) Band 5 (600 nm to 700 nm).

(b) Band 7 (800 nm to 1100 nm).

Figure 8.- Landsat images of Kerr Reservoir on October 11, 1972.

(600 nm $\leq \lambda \leq$ 700 nm) shows near saturation signals at the western end of the reservoir which suggests high TSS values. The highly turbid waters are also visible (but not near saturation) at near-infrared wavelengths in band 7 (800 nm $\leq \lambda \leq$ 1100 nm). This tends to confirm laboratory-measured spectral trends. Close examination of original images shows that plume and eddy circulation patterns are different for each of the bands. (The differences are not observable in image reproductions.) These differences probably represent sediment concentration pattern differences as a result of gray-scale count grouping and threshold differences between the Landsat bands. Without improved knowledge of algae bloom spectra and near-infrared DOC effects, one cannot be entirely confident of the sediment analysis, however. Future field experiments will focus on atmospheric correction research and attempt to confirm laboratory spectral results in a quantitative manner. A combined laboratory and field experiment approach is believed to be the fastest method for development of "sediment-only" algorithms.

CONCLUDING REMARKS

During F.Y. 1980, a 6-year program known as AgRISTARS was initiated. One small element of that program is the assessment of conservation practices through remote sensing of sediment concentration in drainage basin waters. This task requires the development of "sediment-only" algorithms for TSS values up to 1000 ppm. Two of five national test sites have been selected to date. Laboratory tests have been conducted with sediments from John H. Kerr Reservoir, Virginia, and Lake Chicot, Arkansas. Preliminary results show nonlinear upwelled reflectance changes with sediment concentration. At visible wavelengths, signals tended toward saturation at TSS values near 100 ppm. Signal discrimination was good for TSS up to 800 ppm at near-infrared wavelengths greater than 750 nm. Improved knowledge of natural DOC and algae reflectance characteristics is required to fully assess the potential of near-infrared "sediment-only" measurements. Improved knowledge of near-infrared remote sensing penetration depth values is also required. Much work has been done concerning atmospheric correction procedures, but most have pitfalls which preclude application to the turbid water situation in which the atmosphere is allowed to have horizontal variations over the scene. Detailed measurements are required to compare various concepts and give insight into advanced strategies for dealing with near-infrared wavelengths. A parallel laboratory and field experiment program is believed to be the fastest approach for development of "sediment-only" remote sensing data analysis algorithms.

ACKNOWLEDGEMENTS

We wish to thank Drs. Jerry C. Ritchie, John R. McHenry, and Frank R. Schiebe from USDA for their guidance and aid in development of the AgRISTARS Conservation Assessment effort. We are further indebted to Dr. Charles M. Cooper, USDA, who furnished Lake Chicot sediments for the tests described herein. We look forward to their council and the performing of joint experiments during algorithm development activities such that remote sensing results can be more

easily integrated into USDA-development models. Finally, we thank
Dr. Vijay P. Singh and Mr. David T. Williams for their invitation
to this symposium.

REFERENCES

Ahern, F.J., Goodenough, D.G., Jain, S.C., Rao, V.R. and Rochon, G.
1977. Use of Clear Lakes as Standard Reflectors for Atmospheric
Measurements. Proceedings of the Eleventh International Symposium
on Remote Sensing of the Environment, Ann Arbor, Michigan.

Bukata, R.P., Jerome, J.H., Bruton, J.E. and Jain, S.C. 1980. Nonzero
Subsurface Irradiance Reflectance at 670 nm from Lake Ontario
Water Masses. Applied Optics, Vol. 19, pp. 2487-2488.

Curcio, J.A. and Petty, C.C. 1951. The Near Infrared Absorption
Spectrum of Liquid Water. Journal of the Optical Society of
America, Vol. 41, pp. 302-304.

Duntley, S.O., Austin, R.W., Wilson, W.H., Edgerton, C.F. and Moran,
S.E. 1974. Ocean Color Analysis. SIO Ref. 74-10, 70pp., Scripps
Institution of Oceanography, San Diego, California.

Friedman, E., Poole, L., Cherdak, A. and Houghton, W. 1980. Absorption
Coefficient Instrument for Turbid Natural Waters. Applied Optics,
Vol. 19, pp. 1688-1693.

Gordon, H.R. and Clark, D.K. 1980. Initial Coastal Zone Color Scanner
Imagery. Proceedings of the Fourteenth International Symposium on
Remote Sensing of the Environment, San Jose, Costa Rica.

Holyer, R.J. 1978. Toward Universal Multispectral Suspended Sediment
Algorithms. Remote Sensing of the Environment, Vol. 7,
pp. 323-338.

Holyer, R.J., LaViolette, P.E. and Clark, J.R. 1980. Satellite
Oceanography Research, Development, and Technology Transfer.
Proceedings of the Fourteenth International Symposium on Remote
Sensing of the Environment, San Jose, Costa Rica.

Jerlov, N.G. 1968. Optical Oceanography. Elsevier Publishing Co.,
New York, pp. 56.

Johnson, R.W. 1975. Quantitative Suspended Sediment Mapping Using
Aircraft Remotely Sensed Multispectral Data. Proceedings of the
NASA Earth Resources Survey Symposium, Houston, Texas. (NASA
TMX-58168, National Technical Information Service, Springfield,
Virginia.)

Kneizys, F.X., Shettle, E.P., Gallery, W.O., Chetwynd, J.H., Abreu, L.W.,
Selby, J.E.A., Fenn, R.W. and McClatchey, R.A. 1980. Atmospheric
Transmittance/Radiance: Computer Code LOWTRAN 5. AFGL-TR-80-0067,
233pp., National Technical Information Service, Springfield,
Virginia.

Langham, E.J. 1977. Diffuse Backscatter of Solar Radiation.
Presented at the Fourth Canadian Symposium on Remote Sensing,
Quebec City, Canada.

Malila, W.A., Gleason, J.M. and Cicone, R.C. 1977. Multispectral
System Analysis Through Modeling and Simulation. Proceedings of
the Eleventh International Symposium on Remote Sensing of the
Environment, Ann Arbor, Michigan.

McCluney, W.R. 1974. Estimation of Sunlight Penetration in the Sea
for Remote Sensing. NASA TMX-70643, 29pp., National Technical
Information Service, Springfield, Virginia.

Morel, A. 1980. In-Water and Remote Measurements of Ocean Color.
Boundary-Layer Meteorology, Vol. 18, pp. 177-201.

Morris, W.D., Witte, W.G., and Whitlock, C.H. 1980. Turbid Water
Measurements of Remote Sensing Penetration Depth at Visible and
Near-Infrared Wavelengths. Proceedings of the Symposium on
Surface-Water Impoundments, Minneapolis, Minnesota. (NASA TM
81843, National Technical Information Service, Springfield,
Virginia.)

Piech, K.R. and Schott, J.R. 1974. Atmospheric Corrections for
Satellite Water Quality Studies. Proceedings of the Society of
Photo-Optical Instrumentation Engineers, Vol. 51, San Diego,
California.

Scarpace, F.L., Holmquist, K.W. and Fisher, L.T. 1979. LANDSAT
Analysis of Lake Quality. Photogrammetric Engineering and
Remote Sensing, Vol. 45, pp. 623-633.

Smith, R.C. and Baker, K.S. 1981. Optical Properties of the Clearest
Natural Waters (200-800 nm). Applied Optics, Vol. 20, In press
for January 15 issue.

Turner, R.E. 1977. Atmospheric Transformation of Multispectral Remote
Sensor Data. NASA CR-135338, 92pp., National Technical Information Service, Springfield, Virginia.

Whitlock, C.H. 1977. Fundamental Analysis of the Linear Multiple
Regression Technique for Quantification of Water Quality
Parameters from Remote Sensing Data. NASA TM X-74600, 176pp.,
National Technical Information Service, Springfield, Virginia.

Whitlock, C.H., Usry, J.W., Witte, W.G. and Gurganus, E.A. 1977.
Laboratory Measurements of Upwelled Radiance and Reflectance
Spectra of Calvert, Ball, Jordan, and Feldspar Soil Sediments.
NASA TP 1039, 32pp., National Technical Information Service,
Springfield, Virginia.

Whitlock, C.H. and Kuo, C.Y. 1979. A Regression Technique for
Evaluation and Quantification for Water Quality Parameters from
Remote Sensing Data. Proceedings of the Thirteenth International
Symposium on Remote Sensing of the Environment, Ann Arbor,
Michigan. (NASA TM 80101, National Technical Information
Service, Springfield, Virginia).

Whitlock, C.H., Usry, J.W., Witte, W.G., Farmer, F.H. and Gurganus, E.A.

1979. Investigation of the Effects of Background Water on Upwelled Reflectance Spectra and Techniques for Analysis of Dilute Primary-Treated Sewage Sludge. NASA TP 1446, 33pp., National Technical Information Service, Springfield, Virginia.

Whitlock, C.H., Poole, L.R., Usry, J.W., Houghton, W.M., Witte, W.G., Morris, W.D. and Gurganus, E.A. 1981. Comparison of Reflectance with Backscatter and Absorption Parameters for Turbid Waters. Applied Optics, Vol. 20, In press for February 1 issue.

Witte, W.G., Usry, J.W., Whitlock, C.H. and Gurganus, E.A. 1979. Spectral Measurements of Ocean-Dumped Wastes Tested in the Marine Upwelled Spectral Signature Laboratory. NASA TP 1480, 31pp., National Technical Information Service, Springfield, Virginia.

Yarger, H.L., McCauley, J.R., James, G.W. and Magnuson, L.M. 1973. Water Turbidity Detection Using ERTS-1 Imagery. Symposium on Significant Results Obtained from the Earth Resources Technology Satellite-1, New Carrollton, Maryland. (NASA SP-327, National Technical Information Service, Springfield, Virginia.)

Yarger, H.L. and McCauley, J.R. 1975. Quantitative Water Quality with LANDSAT and SKYLAB. Proceedings of the NASA Earth Resources Survey Symposium, Houston, Texas. (NASA TM X-58168, National Technical Information Service, Springfield, Virginia.)

USE OF LANDSET DATA TO IMPROVE WATER STORAGE INFORMATION IN CONSERVATION AREA, FLORIDA

S. F. Shih

Associate Professor of Agricultural Engineering
University of Florda, IFAS
Agricultural Research and Education Center
Belle Glade, Florida 33430, U.S.A.

ABSTRACT

The Conservation Areas in south Florida have been considered as a major source of water supply for the Everglades National Park and the Lower East Coast. The South Florida Water Management District has only limited knowledge of the total amount of water stored in these areas as they are large, remote, and almost flat. In order to improve the water storage information in the Conservation Areas, LANDSAT radiances in Conservation Area 3A were correlated with the water depth. The results showed that the multiple correlation coefficient was as high as 0.75 in multiple regression analysis between LANDSAT radiances and the water depth. An alphanumeric map of water depth was generated. The water surface area corresponding to each water depth was computed. From this, the volume of storage was developed. This newly developed stage-storage relation was compared with the conventional data. The results showed that the storage capacity computed from the remotely sensed data is always less than that from the conventional data and that about 50% of the total Conservation Area 3A was found in the elevation range between 220 to 250 cm MSL. The storage capacity has been reduced about 30% at some levels. Part of this storage capacity modification could be caused by the peat development and losses over past 50 years. The estimation of peat development was about 4 to 6 cm in the past 50 years. If the results of this study are found acceptable, new stage-storage relationships should be established as soon as possible, not only for Conservation Area 3A but also for other areas such as Conservation Areas 1, 2A, 2B and 3B.

INTRODUCTION

In south Florida large tracts of grassy marshland have been set aside to act as natural water retention areas. These Everglades areas consist of wet prairies broken by elevated ridges of trees called tree islands. These areas are diked, and most of the flow into and out of them is controlled by pumps and structures.

511

Vast amounts of water stored in the three major Conservation
Areas (as shown in Fig. 1) flow slowly from the agricultural region
south of Lake Okeechobee to the Everglades and finally to the Gulf
of Mexico. The Conservation Areas in south Florida have been considered as one of the major water storage areas to provide a water
supply for Everglades National Park and the Lower East Coast (LEC).
Due to the increasing water demand of the area, additional backpumping
of the surplus runoff from the LEC area into the Conservation Areas has
been considered as one of several alternative plans by the South
Florida Water Management District (SFWMD, 1977). The Receiving Water
Quantity Model developed by the Environmental Protection Agency (EPA,
1971) was adopted and modified by Lin and Shih (1979) to be applicable
in the Conservation Areas to investigate the possible impact of additional inflow under various backpumping cases. However, the SFWMD has
only limited knowledge of the total amount of water stored in those
areas because they are large, remote, and almost flat. Consequently,
in 1978 the SFWMD entered into a joint project with the University of
Florida and National Aeronautics and Space Administration (NASA) to
examine the possibility of using LANDSAT data to correlate the water
depth in Conservation Area 3A (Fig. 1).

Gervin and Shih (1980) used the LANDSAT data to improve the relationship between lake stage/lake surface and lake storage capacity.
In previous studies reported by Higer (1975), tried to relate LANDSAT

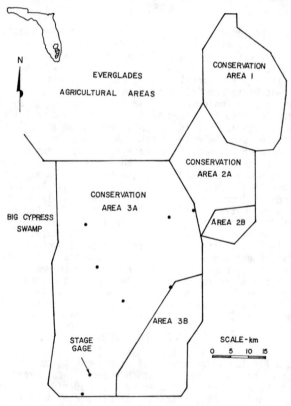

Fig. 1. Map of Conservation Areas showing locations of selected stage gages.

radiance to water depths in Conseravtion Area 1 with some success. In more recent unpublished work, he had investigated the correlation between LANDSAT radiance and water depths in Conservation Area 3A on two dates using 100 field measurements. He had achieved good correlations but, because of manpower required to gather field measurements, this technique is expansive to become an operational tool. Therefore, the purposes of this study were: 1) to describe the general concept of using an automated computer technique to correlate between LANDSAT radiance and water depth; and 2) to compare the difference in Conservation Area's storage prediction between conventional data and that of LANDSAT data.

METHODOLOGY

Remote Sensing Data Analysis

Twelve relatively cloud-free LANDSAT tapes, representing the full range of water levels from minimum to maximum, were identified (Fig. 2). From these, two sets of three dates (each set covering a minimum to maximum cycle) were selected for more intensive study. The first set covers the dates of 10/19/74, 12/12/74, and 3/3/75. The second set covers the dates of 4/12/76, 10/17/76, and 2/20/77. Sets of consecutive dates were chosen to minimize the likelihood that long term changes in vegetation distribution would affect the data analysis. SFWMD then provided water elevation data, ground elevations, and latitudes and longitudes for the eight water depth stations as shown in Fig. 1 in Conservation Area 3A.

Meanwhile, a preliminary test of the technique was performed with the existing historical data. After examining each date carefully, it was found that there was very little water in the Conservation Area 3A on the two days of minimum water level, i.e. the dates of March 3, 1975 and April 12, 1976. Therefore, these two dates were removed from the test, leaving four dates representing medium and high water levels. In order to analyze the water depth

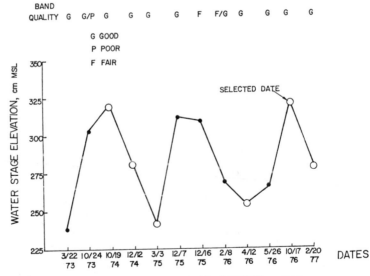

Fig. 2. Selection for available LANDSAT scenes.

of the area, digital data from the LANDSAT earth-orbiting satellite was analyzed on General Electric's multispectral image analyzer, the Image 100. The performance of Image 100 was described in detail by Gervin and Shih (1980).

Due to differences in the rate of scan, altitude and other factors, LANDSAT digital data do not precisely represent the relative geographic location of points on the ground. Therefore, the Image 100 was used to develop a geometric correction matrix for the four remaining dates, using the method described by Gervin and Shih (1980). This matrix was used to locate the water depth stations on the LANDSAT tape from their latitudes and longitudes since very few landmarks or other distinct features are available in a marsh. The Image 100 was then used to measure the LANDSAT radiance of the four pixels closest to each gauging station on each date. The size of pixel is about 79x79 m. The LANDSAT radiance includes four bands in Band 4 (green), Band 5 (red), Band 6 (near infrared), and Band 7 (middle infrared). If a cloud obscured the gauge location, the nearest group of clear pixels was used.

A multiple regression analysis was used to establish the relationship between LANDSAT radiance as measured on the Image 100 and the water depth as measured by the gauging stations. All four bands, and each possible combination of two bands, were used. A specific date choice was based on not only the result of regression analysis which had a high multiple correlation coefficient but also the range of water depths available and freedom from clouds. After the map of water depth was generated, the areas occupied by each water depth class were measured automatically on the Image 100.

Storage Capacity Prediction

The water surface area occupied by each depth was used to compute the storage capacity in Conservation Area 3A. The equation used to calculate the increase in storage volume associated with each increase in stage was

$$V_{s + \Delta s} = V_s + \Delta s \left[1/2 (A_s + A_{s + \Delta s}) \right] \tag{1}$$

where

$V_{s + \Delta s}$ = storage volume at stage $s + \Delta s$ m, in m^3;

V_s = storage volume at stage s m, in m^3;

$A_{s + \Delta s}$ = water surface area at stage $s + \Delta s$ m, in hectares; and

A_s = water surface area at stage s m, in hectares.

RESULTS AND DISCUSSION

LANDSAT Radiance Analysis

The four bands individually and each possible combination of two bands were used to establish relationship between LANDSAT radiance and water depth. The resulting of multiple correlation coefficients for the multiple regression analysis are shown in Table 1. The four band combinations and some of the two band combinations produced acceptable correlations but the small number of samples (eight measure stations in 1,580 km^2) may reduce the reliability of the predicted relationships. The addition of some gauging stations and/or supplemental field measurements are recommended to truly test the technique. However, the applicability of this technique can be demonstrated by using a map of water depth which was generated using the Band 5 and Band 7 relationship for October 17, 1976. This date was chosen because its water stage was 317.6 cm and it is also cloud-free. This stage of 317.6 cm is the one of the highest stages available from the four dates tested. There is also a good correlation between the two-band combination and water depth.

Table 1. Multiple correlation coefficients for the multiple regression analysis between LANDSAT radiance and water depth in Conservation Area 3A.

LANDSAT Bands	Dates			
	Oct. 19, 1974	Dec. 12, 1974	Oct. 17, 1976	Feb. 20, 1977
Bands 4 - 7	.97	.98	.83	.91
Bands 4 & 5	.40	.60	.74	.73
Bands 4 & 6	.03	.60	.64	.69
Bands 4 & 7	.36	.49	.70	.78
Bands 5 & 6	.19	.84	.71	.54
Bands 5 & 7	.28	.73	.75	.57
Bands 6 & 7	.70	.51	.56	.60

After the water depth made for the entire Conservation Area 3A was constructed, the areas occupied by each water depth were measured automatically on the Image 100. The resulting of water surface area and percentage of surface area associated with the different water depths are listed in Table 2. About 53% of the area falls in the depth interval of 45 to 75 cm. The area with a depth of 75 cm and deeper is 21% only. Meanwhile, the area associated with a depth of 45 cm and shallower is about 26%. This means that a large area of uniform depth platform (about 840 km^2) varied within 30 cm is existed in the Conservation Area 3A. This large platform needs further investigation because it will not only influence the scheme of water resources management but also affect the ecosystem of the area.

Storage Capacity Elevation

The water surface area occupied by each water depth (Table 2) was used to compute the storage capacity in Conservation Area 3A. The average stage from eight recorded stations on October 17, 1976 was 317.6 cm MSL. This average elevation was used as a bench mark to compute the relationship between stage elevation and water surface area. Before computing the storage capacity, the water surface area associated with each increase in elevation needs to be computed based on the water surface area related to water depth as given in Table 2. As Table 2 shows, the surface areas are undefined when the water depth is either shallower than 15.2 cm or deeper than 121.9 cm. Thus, three assumptions should be made. First, the water surface area between 0 and 15.2 cm of water depth is near zero. Second, the water surface area at the water depth of 121.9 cm is assumed to be

515

Table 2. Water surface at different depth in Conservation Area 3A.

Water Depth (cm)	Water surface (hectares)	%
15.2 - 30.5	9,067	5.7
30.5 - 45.7	31,297	19.8
45.7 - 61.0	41,671	26.4
61.0 - 76.2	42,353	26.9
76.2 - 91.4	21,471	13.6
91.4 - 106.7	7,094	4.5
106.7 - 121.9	4,910	3.1
Total	157,863	100.0

the same as the water surface area in between 106.7 and 121.9 cm of water depth because the conventional data shows that there is some water stored at greater than 121.9 cm water depth. Third, the average stage on October 17, 1976 represented a well average condition of the stage elevation. Based on those three assumptions and 317.6 cm of the average stage, the water surface areas related to stage elevation were computed and the results are listed in Table 3.

The storage capacity as a function of stage can be computed

Table 3. Total surface area at different stages in Conservation Area 3A.

Elevation (cm, MSL)	Total water surface area (hectares)
302.4 - 317.6	157,863
287.1 - 302.4	157,863
271.9 - 287.1	148,796
256.6 - 271.9	117,499
241.4 - 256.6	75,828
226.2 - 241.4	33,475
210.9 - 226.2	12,004
195.7 - 210.9	4,910
< 195.7	4,910

based on the remotely sensed data as listed in Table 3. The method as
defined in Equation 1 is used to calculate the increase in volume associated with a small increase in stage. The initial volume at stage
195.7 cm is 22.4 million m³ which was obtained from the conventional
data. The resulting differences between the conventional and remotely
sensed storage capacities were 99.9, 49.3, 24.7, 35,8, 45.6, 32.1, and
8.6 million m³ for the stage at 310.0, 294.7, 279.5, 229.1, 249.0,
233.8, and 218.5 cm MSL, respectively, as shown in Figure 3. These
differences are approximately equivalent to the errors of 10, 6, 5, 11,
23, 30 and 19% of the conventional data in the same order of stages.
There seems to be a large error existing between 220 and 250 cm MSL of
the stages.

As discussed in previous section, the number of ground recorded
stage sites is not large enough to draw a final conclusion. However,
using the Band 5 & Band 7 relationship for October 17, 1976, the multiple correlation coefficient between LANDSAT radiance and water depth
has been found to reach 0.75. In other words, this technique is quite
encouraging to detect the water depth. If this is the case, three conclusions may be made. First, there probably is an error in the
conventional data in the relationship between stage and storage.
Second, in between the stages around 220 and 250 cm MSL, the storage
volume is extensively reduced as compared with the conventional data.
The reason for this reduction should be investigated further. Third,
the storage capacity computed from the remotely sensed data is always
less than from the conventional data. This implies that the volume of
water stored in the Conservation Area 3A has been overestimated in the
conventional data. Therefore, a correct relationship between stage and
storage should be established as soon as possible.

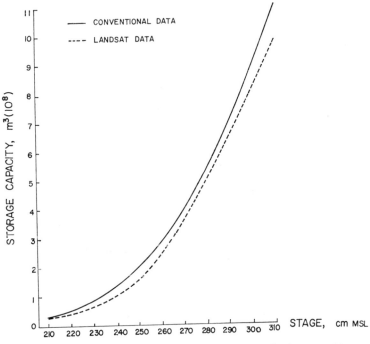

Fig. 3. Relationship between stage and storage capacity in Conservation
Area 3A.

Part of the reason for this overestimation of the volume of water stored in the Conservation Area 3A could be explained as one of the results of peat development. According to the study reported by Stephens (1969), the average peat development in the Everglades Agricultural Area (Fig. 1), before drainage for agricultural production, is about 8.54 cm per century. Forbes (1980) also reported that the peat development in the Apopka, Florida area is about 11.28 cm per century. As Fig. 1 shows, the Conservation Areas are adjacent to the Everglades Agricultural Area (EAA). The vegetation grown in Conservation Areas should be similar to the EAA condition (before drainage for agricultural production). In other words, the peat developed in Conservation Areas is also somewhere between 8.54 cm and 11.28 cm per century. The conventional data used by the SFWMD (1977) and also used in this study was established in the early 20th century. Therefore, the peat in Conservation Area 3A could be developed as much as 4 to 6 cm in the past 50 years. This value may be even higher because after the levees were constructed around the Conservation Areas, the loss of organic material should be significantly reduced in areas not subjected to fire. On other hand, the rate of peat development could be increased. The consequences of the peat developed in the area is that the storage capacity is reduced as the time is progressed. Particularly, the development of the area with a large flat platform (stage between 220 to 250 cm) could be attributed to fire losses in the north combined with peat development in the south. A 20% reduction in storage capacity could be attributed to a net 6 cm increase in peat. According to the result of this study, new stage-storage relations should be established as soon as possible not only for the Conservation Area 3A but also for other areas such as Conservation Areas 1, 2A, 2B and 3B as shown in Fig. 1.

ACKNOWLEDGEMENTS

This research was supported by NASA Grant Contract NASA 10-9348. The author wishes to thank Ms. J. C. Gervin for analyzing the LANDSAT data, and to the South Florida Water Management District for its support, advice, and assistance in this study.

REFERENCES

1. Environmental Protection Agency. 1971. Storm Water Management Model. EPA Report No. 11024, Doc:07-10.

2. Forbes, R. B. 1980. Management of Florida's Histosols for Crop Production. Soil and Crop Science Society of Florida Proc. Vol. 40. (in press).

3. Gervin, J. C. and Shih, S. F. 1980. Improvements in Lake Volume Predictions Using LANDSAT Data. In Satellite Hydrology, Amer. Water Resources Assoc. (in press).

4. Higer, A. L. 1975. Water Management Models in Florida from ERTS-1 Data. Final Report, United States Geological Survey Water Resources Division, Miami, Florida.

5. Lin, S. T. and Shih, S. F. 1979. Modified Water Quantity Receiving Model for Florida Conservation Areas. Water Resources Bulletin, Vol. 15(1), pp. 155-166.

6. South Florida Water Management District. 1977. Water Use and Supply Development Plan. Vol. 111A, Lower East Coast Planning Area, Technical Exhibits, "I-K", West Palm Beach, Florida.

7. Stephens, J. C. 1969. Peat and Muck Drainage Problems. Journal of Irrigation and Drainage Division, Amer. Soc. of Civil Engr. Vol. 95(IR2), pp. 285-305.

ESTIMATING RUNOFF CURVE NUMBERS USING REMOTE SENSING DATA

T.R. Bondelid
Department of Civil Engineering
University of Maryland
College Park, Maryland 20742

T.J. Jackson
Hydrology Laboratory
USDA-SEA-AR
Beltsville, Maryland

R.H. McCuen
College of Engineering
University of Maryland
College Park, Maryland 20742

ABSTRACT

The curve number is an index that is based on land use, soil type, hydrologic condition, and antecedent soil moisture. Models that utilize the curve number are widely used and are sensitive to errors in the estimated curve number. The estimation of curve numbers can be a laborious and time consuming procedure because detailed land cover and soils data are required. The objective of this study was to develop and test procedures for using more cost-effective remote sensing data bases for estimating curve numbers. Data were tested on three watersheds in Pennsylvania using three different land cover data sources: conventional surveys, land cover maps developed by the U.S. Geological Survey and based on remotely sensed data, and Landsat data. The results indicate that curve number estimation is not highly sensitive to the land cover data source. A computer program was developed for streamlining the estimation of curve numbers when a digital land cover data base, such as Landsat, is used. The program integrates watershed boundary, soils, and land cover data for computing curve numbers. The results of this study indicate that remotely sensed curve number estimation is an acceptable and practical alternative to conventional curve number estimation methods.

INTRODUCTION

The series of hydrologic models developed by the Soil Conservation Service (SCS) are widely used in water resource planning and design. The SCS models are appealing because the major input parameters are defined in terms of land cover and soil type, both of which are easily

519

obtained at ungaged locations. The "heart" of the SCS models is the runoff curve number (CN), which is a function of land cover, soil type, and antecedent moisture conditions. The determination of CN's is often a time consuming and labor intensive procedure because detailed land cover and soils data are required. The objective of this study was to develop and test less costly remote sensing procedures for estimating CN's.

Ragan and Jackson (1980) and Slack and Welch (1980) have investigated the use of Landsat for estimation of CN's. The Landsat analysis in this study is an extension of this previous work.

Ragan and Jackson (1980) found that Landsat data could not be used to identify land cover at the level of detail that is used in the SCS procedures. However, they found that by using the less detailed set of land covers, which can be identified using Landsat data, the CN could be estimated satisfactorily. The Landsat compatible methodology was tested on a 54 sq km (21 sq mi) subwatershed of the Anacostia River. The land cover was defined for 109 subwatersheds ranging in size from 0.01 to 6.1 sq km (0.005 to 2.4 sq mi). For the SCS land use categories this operation required about 2 man-months to complete. They determined an overall CN of 63.5. A team operating at the University of Maryland used the remote sensing land cover classification system, and digital geographic data was developed using Landsat data tapes. This digital data base was interfaced with the land cover classifications and an overall CN of 64 was computed. This effort required less than 0.5 man-day and showed excellent agreement with the CN computed by the conventional approach.

Slack and Welch (1980) applied a methodology similar to that developed by Ragan and Jackson (1980) to the Little River watershed in south Georgia. The Landsat data were analyzed using a computer package developed by the Office of Remote Sensing of Earth Resources (ORSER) of Pennsylvania State University. To test the system, subwatershed CN's were analysed. These agreed within two CN's of the SCS field survey results. The computer analysis of the Landsat data was very cost effective; the land cover map used in the study cost $0.90/sq km ($2.25/sq mi) excluding labor.

The studies cited above indicate that Landsat data may be an acceptable and cost-effective substitute for conventional data sources for CN estimation. However, several potential problems need to be addressed before Landsat can be fully accepted as a viable CN estimation method. To meet the objectives of this study, these problems will be addressed.

One potential problem with using Landsat is that even if there may be good agreement between Landsat and conventionally derived CN's for a watershed as a whole, there may be large differences in CN's for individual subwatersheds within the watershed. The differences in CN estimates for individual subwatersheds needs to be examined.

A second potential problem is that Landsat may not be acceptable in areas with certain land cover types. This problem can be addressed by analyzing areas with diverse land cover types.

In order to address these two potential problems, CN's for the following three watersheds in southeastern Pennsylvania are analyzed in this study: the Quittapahilla, Chickies Creek, and Little Mahanoy River basins. The sub-areas of the watersheds ranged in size from 0.425 to 16.819 sq km (0.164 to 6.494 sq mi). The watersheds contain diverse land covers including forest, agriculture, urban, strip mines, and wetlands. The diversity of land cover allows the CN procedures to be tested under a variety of conditions.

A third potential problem may exist because many different Landsat
land cover delineations are possible. This problem will be addressed by
evaluating two different Landsat-based land cover data sources.

CURVE NUMBER ESTIMATION PROCEDURES

Three methods for estimating CN's are evaluated in this study:
(1) the conventional method, which uses aerial photos and site investigations; (2) Land Use Data Archive Method (LUDA), which uses land use
maps developed by the United States Geological Survey (USGS); and (3)
the Landsat method, which utilizes data from satellites.

The CN for a particular watershed is the weighted average of the CN's
for each land cover-soil type complex in the watershed. The soils data
used for the three techniques in this study were based on county soil surveys. All three techniques use the same soils data base. The primary
difference between them was how the land cover was determined.

Conventional Method

The conventional method CN definition was performed by Sprague (1979).
He used low altitude aerial photography and detailed soil surveys. The
aerial photographs were enlarged to the scale of the base maps for the
area under study. Watershed boundaries were then transferred to the enlarged aerial photographs. Various types of land cover, i.e., urban,
cropland, woodland, etc., were delineated on the aerial photographs, with
support from onsite investigations.

In the conventional method, two methods can be used to integrate land
cover and soils data. In one approach, hydrologic soil maps are overlaid
on the aerial photographs. The area for each land cover-soil type complex
for each subwatershed is planimetered, and a CN is calculated. The overall subwatershed CN is computed as a weighted average of the subwatershed
values. An alternate technique is to planimeter the land cover and soil
types separately for each subwatershed. The CN for each land cover in the
subwatershed is then computed based on a weighted average of the soil types.
Comparisons between these two techniques on the watersheds under study
have shown no significant difference between the techniques (Sprague, 1979).

LUDA Method

The USGS is currently producing land use and land cover maps for the
United States with a base of 1:250,000, which are called LUDA maps. The
data are being digitized and will be made available to the public in both
graphic and digital form. This Geographic Information and Retrieval System (GIRAS) is designed to input, manipulate, analyze, and output the land
use and land cover data. The system is designed to serve a wide variety
of data needs, including water resource inventories.

The USGS specified four levels of land use and land cover data.
Level I, the most general, classifies land use into nine types. These
level I classifications are: (1) urban or built-up land (2) agricultural
land; (3) range land; (4) forest land; (5) water; (6) wetland;
(7) barren land; (8) tundras; and (9) perennial snow or ice.

Each of the nine classifications is further divided into Level II
classifications, the Level II classifications are divided into Level III

classifications, and Level III into Level IV. For instance, the Level I classification of urban or built-up land has Level II classifications such as residential, commercial, and industrial.

A minimum mapping unit of about 4 ha (10 acres) is used for all urban and water covers and in a few other categories. For all other cover types a minimum unit of 16 hectares (40 acres) is used. Each spatial mapping unit must be assigned to a single category.

The CN estimation portion of the LUDA study was conducted by Murrell (1979) with the LUDA data enlarged to 1:24,000. A set of CN's was developed for the USGS Level II categories and is shown in Table 1. The CN's were not developed for hydrologic soil group A because no A soil existed in the watersheds.

TABLE 1. LUDA LAND COVER CURVE NUMBERS

Land Cover Description	Hydrologic Soil Group		
	B	C	D
Residential	85	90	92
Commercial & Services	85	90	92
Industrial	85	90	92
Transportation, Communications, Utilities	85	90	92
Mixed Urban or Built-up Land	85	90	92
Other Urban or Built-up Land	85	90	92
Crop or Pasture	77	84	88
Orchards, Groves, Nurseries, and Ornamental Horticultural areas	77	84	88
Confined Feeding Operations	85	90	92
Deciduous Forest Land	55	75	80
Evergreen Forest Land	55	70	77
Mixed Forest Land	55	75	80
Lakes	99	99	99
Reservoirs	99	99	99
Forested Wetland	85	90	92
Strip mines, Quarries and Gravel Pits	85	90	92

Polygons of land cover-soil type were delineated and assigned a CN for each watershed. These were then aggregated to establish the sub-watershed curve numbers.

Landsat Method

Preliminary spectral responses were analyzed and final 1:24,000 land cover maps were performed using the Office of Remote Sensing for Earth Resources (ORSER) of Pennsylvania State University computer software package described by Borden, et al., (1975). The ORSER package is used to extract the data for the study areas from the Landsat scene, geometrically correct and rescale it to 1:24,000, perform statistical analyses on selected areas of known land cover, and produce final classification maps.

After preliminary statistical analyses were performed with the ORSER package, the classifications were tested and refined using a special pur-pose interactive processor, called the Atmospheric and Oceanographic Image Processing System, at Goddard Space Flight Center. This system includes a color cathode ray tube monitor and supporting software that permits a sunoptic, visual interpretation of the land cover classifications. These classifications were visually calibrated to high altitude false-color infrared photographs.

Landsat scenes for two dates, July 1973 and October 1973, were analyzed. The July 1973 scene was found to be unacceptable because large areas of the urban and agricultural land were not spectrally separable at this time of year. The October 1973 scene was acceptable and a parallelepiped classification based on bands 5 and 7 was developed. The land cover categories used in this classification were forest, agriculture, water, residential, and highly impervious.

The final 1:24,000 land cover maps based on the interactive analysis were produced using the ORSER package. The watershed and subarea boundaries were superimposed on those land cover maps and the maps were then used for determining land cover percentages for each subarea. The CN's were calculated on the basis of a modified CN shown in Table 2.

TABLE 2. PENNSYLVANIA LANDSAT CURVE NUMBERS

Land Cover	Soil Group			
	A	B	C	D
Woods	25.0	55.0	70.0	77.0
Agriculture	64.0	75.0	83.0	87.0
Residential	60.0	74.0	83.0	87.0
Highly Impervious	90.0	93.0	94.0	95.0
Water	98.0	98.0	98.0	95.0

COMPARISON OF CURVE NUMBER ESTIMATES

If the LUDA and Landsat CN estimation methods are to be accepted, then these procedures should produce CN estimates that are comparable to CN estimates derived by the conventional method. The accuracy of the two alternatives can be evaluated by comparing the estimates to the conventionally derived CN's.

Accuracy consists of two sources of variation, both of which reflect different causes of error. Systematic error, which is referred to as a bias, is the average difference of the estimate from the true value, which for this study will be the CN estimated by conventional means. A significant bias will suggest an error in the model that can often be related to another factor and corrected for. Bias can be measured using the mean of the residuals. Nonsystematic error variation, which is referred to as precision, is the variation of the estimated values about the true values; thus, it is assessed as the standard deviation of the residuals.

Bias and precision are two criteria that can be used to assess the validity of a proposed estimation technique. In this analysis, the residuals are evaluated for both the LUDA and Landsat methods for each subwatershed. The principle objectives of this analysis are to first quantify the bias and lack of precision and, second, to examine the probable sources of the inaccuracies. If the sources of the error can be determined, then this information can be used in the future to possibly increase the accuracy of remotely sensed CN estimates.

Summary of CN's

Tables 3, 4, and 5 present the CN's and residuals for the three techniques for each subarea. All CN's are rounded to the nearest whole value. The residuals for the CN's derived from LUDA and Landsat were computed relative to the conventionally derived CN's. A positive residual indicates overestimation as compared with the conventional CN, and a negative residual underestimation. The CN's ranged from 58 to 87, and the residuals ranged from -8 to +8. Only 4 residuals were greater than 4 in absolute value.

Table 6 presents the means and standard deviations of the residuals for each watershed and for the three watersheds combined. The residual means for LUDA are all positive, ranging from 1.19 to 2.00, which indicates a slight positive bias (i.e., overestimation) for the LUDA method. The

TABLE 3. QUITTAPAHILLA CREEK CURVE NUMBERS

Subarea	Area (acres)	Conventional CN	LUDA CN	Landsat CN	LUDA Residual	Landsat Residual
1	1,780	84	83	81	-1	-3
2	1,458	84	85	81	1	-3
3	3,256	83	83	80	0	-3
4	978	70	72	71	2	2
5	540	77	76	75	-1	-2
6	316	72	73	72	1	0
7	321	81	83	79	2	-2
8	117	85	86	82	1	-3
9	3,762	82	83	80	1	-2
10	1,442	72	72	70	0	-2
11	1,134	85	86	83	1	-2
12	3,222	82	83	80	1	-2
13	837	72	76	73	4	1
14	1,076	85	88	84	3	-1
15	601	81	85	78	4	-3
16	791	84	86	76	2	-8
17	803	78	75	78	-3	0
18	2,876	81	81	79	0	-2
19	3,923	70	69	69	-1	-1
20	737	75	77	74	2	-1
21	382	81	84	75	3	-6
22	185	77	79	76	2	-1
23	728	81	81	77	0	-4
24	103	73	81	74	8	1
25	1,088	84	83	82	-1	-2
26	4,156	87	87	86	0	-1

TABLE 4. LITTLE MAHANOY CREEK CURVE NUMBER

Subarea	Area (acres)	Conventional CN	LUDA CN	Landsat CN	LUDA Residual	Landsat Residual
1	308	62	59	60	-3	-2
2	795	63	65	64	2	1
3	623	67	66	67	-1	0
4	626	74	70	72	-4	-2
5	461	74	73	71	-1	-3
6	336	71	72	70	1	-1
7	1,388	73	70	70	-3	-2
8	361	73	68	73	-5	0
9	527	75	72	73	-3	-2
10	896	84	81	82	-3	-2
11	633	73	75	72	2	-1

Landsat means are mostly negative, ranging from -1.92 to +0.12, which indicates a slight negative bias (i.e., underestimation).

The precision of the LUDA and Landsat methods is reflected by the standard deviations of the residuals. The LUDA standard deviations range from 1.59 to 2.42 and the Landsat standard deviations range from 1.19 to 2.07. The Landsat standard deviations are consistently less than the LUDA standard deviations indicating that the Landsat method is consistently more precise than the LUDA method.

TABLE 5. CHICKIES CREEK CURVE NUMBERS

Subareas	Area (acres)	Conventional CN	LUDA CN	Landsat CN	LUDA Residual	Landsat Residual
1	2,212	68	68	69	0	1
2	1,304	59	59	58	0	-1
3	1,916	59	59	59	0	0
4	939	74	76	74	2	0
5	2,765	70	71	72	1	2
6	1,352	78	78	74	0	-4
7	990	79	83	81	4	2
8	860	79	82	79	3	0
9	1,971	79	82	79	3	0
10	537	75	79	74	4	-1
11	1,433	76	79	77	3	1
12	1,320	78	80	77	2	-1
13	548	79	83	79	4	0
14	1,667	79	83	80	4	1
15	2,077	77	78	76	1	-1
16	1,463	77	78	76	1	-1

TABLE 6. SUMMARY OF RESIDUALS

Watershed	Number of Subareas	LUDA mean	Landsat Mean	LUDA St. Dev.	Landsat St. Dev.
Quittapahalla	26	1.19	-1.92	2.15	2.07
Little Mahanoy	11	1.64	-1.27	2.42	1.19
Chickies Creek	16	2.00	0.12	1.59	1.45
Combined	53	1.30	-1.66	2.06	1.74

Sources of Variation in Curve Numbers

Although there was generally good agreement in the CN's developed by the three methods, identification of possible causes of the variations is important. To guide the search for sources of variance, two factors were examined: the spatial distribution of the residuals and the land cover classifications.

Spatial Distribution of Residuals

An examination of the spatial distribution of the residuals showed one significant trend. Three of the residuals that were greater than four in absolute value were in the northern part of the Quittapahilla watershed. This suggested that some characteristic of this area produced less consistency in CN estimates. The large negative Landsat residual in one of the subareas can, at least in part, be attributed to a misclassification by

525

Landsat of a large area of forested wetland as simple forest. The CN's for forested wetlands are significantly higher than those for forest. Also because this subarea is relatively small the effects of the misclassification are pronounced. Murrell (1979) noted difficulties in analyzing small subareas by the LUDA method because of pronounced edge effects that occur from enlarging the 1:250,000 LUDA maps to a 1:24,000 scale. The heterogeneous nature of the northern part of the Quittapahilla watershed, combined with the small size of the subarea, probably account for the large LUDA residual. The lack of precision in the northern portion of the Quittapahilla watershed can, therefore, be attributed to the combination of small subareas and heterogeneous land cover.

One would expect that as the subarea size increases, the accuracy would improve. Thus, the residuals were plotted against the subarea size. Figures 1, 2, and 3 show the residuals vs. subarea size for the Quittapahilla, Little Mahanoy, and Chickies Creek watersheds, respectively. An examination of these figures indicates that there is at best a slight improvement in CN estimation as subarea size increases. Therefore, subarea size by itself cannot be considered a significant factor in CN estimation accuracy. However, as indicated above, small subarea size combined with heterogeneous land covers is a significant factor in CN estimation accuracy.

Comparison of Land Cover Classifications

Differences in land cover classifications are a potentially significant source of both bias and lack of precision. A consistent misclassification of land cover types can produce a bias, and random error in land cover classification can produce a lack of precision.

Table 7 summarizes the conventional and Landsat land cover classification for the Little Mahanoy and Chickies Creek watersheds. In Little Mahanoy there are significant differences between the two methods in the

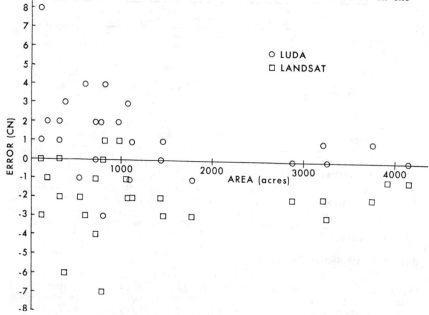

FIGURE I. Curve Number Prediction Error Versus Subwatershed Size for Quittapahilla Creek

526

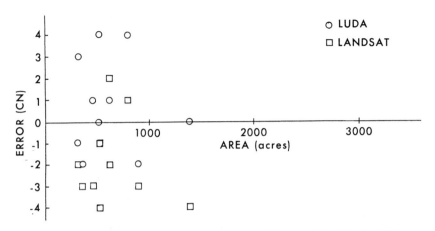

FIGURE 2. Curve Number Prediction Error Versus Subwatershed Size
for Little Mahanoy Creek

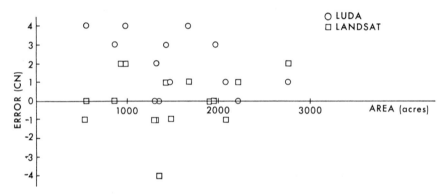

FIGURE 3. Curve Number Prediction Error Versus Subwatershed Size
for Chickies Creek

forest and agriculture classification percentages. In Chickies Creek there
are significant differences in the agriculture and urban percentages.

There is an overall negative bias in the Landsat CN estimates in the
Little Mahanoy watershed. This bias coincides with the differences in land
cover percentages because agriculture has higher CN's than forest. Therefore,
the differences in land cover classification in Little Mahanoy are a probable source of bias.

In Chickies Creek the overall bias between the conventional and Landsat
methods is near zero, although there are significant differences in the
agriculture and urban land cover classification percentates. A large proportion of the conventional urban percentage is residential, which is almost
equivalent to agriculture in terms of CN value. The higher Landsat agriculture percent is thus offset by the higher conventional residential percent, effectively removing the bias.

The analysis of CN estimates indicates that LUDA and Landsat provide
generally good agreement with conventionally derived CN's. However, a

TABLE 7.   SUMMARY OF CONVENTIONAL VS LANDSAT LAND COVER CLASSIFICATIONS

Watershed	Forest (%)		Agriculture (%)		Urban (%)*	
	Conven-tional	Landsat	Conven-tional	Landsat	Conven-tional	Landsat
Little Mahanoy	76.0	80.8	6.9	0.1	16.0	16.1
Chickies Creek	29.0	29.8	57.0	63.4	14.0	6.8

*Includes residential

significant loss in precision occurs in heterogeneous regions with small subareas. Also, a consistent land cover misclassification may or may not produce a bias, depending on the land cover types misclassified.

CONCLUSIONS AND RECOMMENDATIONS

Three methods for estimating CN's were compared by analyzing three watersheds in southeastern Pennsylvania. The results indicate that CN estimates are not highly sensitive to the source of the land cover data base. Remotely sensed data bases, specifically LUDA and Landsat, are, therefore, acceptable for use in CN estimation.

The causes of the largest CN errors was investigated. The results indicate that special care should be exercised in analyzing small watersheds that contain heterogeneous land cover.

ACKNOWLEDGEMENT

The authors are indebted to Bill Sprague and George Murrell of the SCS. Bill supplied us with the conventional method information and George provided the LUDA data.

We also wish to thank Al Rango and Jim Ormsby of NASA Goddard Space Flight Center for providing the time on the interactive analysis computer.

REFERENCES

1. Borden, F.Y., Applegate, B.J., Turner, B.J., Lachowski, H.M., Merenbeck, B.F., and Hoosty, J.R., 1975, Satellite and Aircraft Multispectral Scanner Digital Data Users Manual; Technical Report 1-75, Office for Remote Sensing of Earth Resources, Pennsylvania State University, University Park, Pennsylvania.

2. Murrell, G., 1979, Personal Communication, SCS Special Mapping Unit, Reston, Virginia.

3. Ragan, R.M. and Jackson, T.J., 1980, Runoff Synthesis Using Landsat and the SCS Model, accepted for publication in the Journal of the Hydraulics Division of the ASCE.

4. Slack, R.B. and Welch, R., 1979, Soil Conservation Service Runoff Curve Number Estimates From Landsat Data; ASP Fall Convention, Souix Falls, South Dakota.

5. Sprague, W., 1979, Personal Communication, Soil Conservation Service, Harrisburg, Pennsylvania.

LUMPED WATER BALANCE MODEL USING COVER AREAS DETERMINED BY LANDSAT REMOTE SENSING

Donald L. Chery, Jr., Ph.D.
Senior Hydrologist
Dames & Moore
7101 Wisconsin Ave.
Washington, D.C. 20014

John R. Jensen, Ph.D.
Professor of Geography
University of Georgia
Athens, Georgia 30702

ABSTRACT

A lumped watershed water balance model was developed for use with remotely sensed (Landsat) land cover determinations. The structure of the model and functioning of the evapotranspiration, bare field, cropped field, pond, and forested-alluvial subroutines are described. The daily outflow predictions of the model are compared with measured responses of a 6-mi^2 (15.5-km^2) Southeastern United States watershed for a period about June 11, 1977. Landsat data from this data were used to identify the land cover and measure the areal extent of each. The model gave reasonable predictions and has potential for assessing demand on the groundwater supplies of the Southeast or crop yields due to water stress. The model performance is discussed with respect to available real watershed input and internal system measurements.

INTRODUCTION

In many agricultural areas an expedient and simple method is needed for assessing the effects of agricultural practices on the water balance of an extended area. A specific concern exists about the groundwater draft in southern Georgia due to irrigation supplementing the natural rainfall. This requires an estimate of type and areal extent of different crops in a given area, along with some prediction of the state of available soil water. One method is to inventory the crops with Landsat remote-sensed data and then use this information in a water balance model to predict the status of the soil water. The inventory of crops entails selecting propitious dates and making crop cover determinations for relatively small irregularly shaped fields. These problems and procedures were discussed in a previous paper (Jensen and Chery, 1980). This paper describes the water balance model used in conjunction with the land cover identification to assess the state of the soil water in a watershed.

SYSTEM CONCEPT

The watershed system was idealized as shown in Figure 1. The watershed soil mantle is divided into two components--the well-drained

RAINFALL

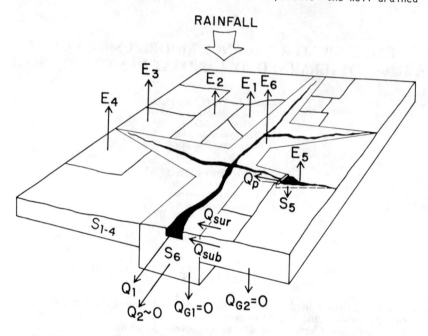

Figure 1.   Conceptual watershed system for water balance model.

field soils, and the deeper, poorly drained valley soils that intrude into the field system. Earthern dams have been built in some of the drainages, forming small ponds or lakes. Rainfall (R) input to the entire watershed leaves the system by evaporation from various agricultural field covers (E_1 to E_4), ponds (E_5), forests (E_6), and by surface discharge (Q_1). The vertical discharges (Q_{G1} and Q_{G2}) and subsurface discharge (Q_2) were assumed to be negligible. Intrawatershed transfers of surface (Q_{sur}) and subsurface (Q_{sub}) flow were from the field to the forest block. Also, pond overflow (Q_p) flows to the forest block. Water is stored in each of the field blocks (S_{1-4}), the ponds (S_5), and the forest soil (S_6). Fields with the same crop or cover were lumped together as one unit.

This conceptualization of the natural system was developed for watersheds in the Coastal Plain of southern Georgia. Information from the Little River Watershed, a 126-mi^2 (326.3-km^2) area maintained by the Science and Education Administration of the U.S. Department of Agriculture (Figure 2) was used for determination of dimensions and parameters in the model.

This watershed is "fairly representative of a large area" (Yates

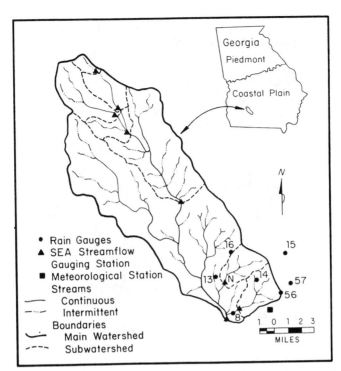

Figure 2.   Location and watershed boundary map of Little River Watershed,
            Location 74, USDA, Science and Education Administration.
            Subwatershed N is the area used in this study.

and Stephens, 1968, p. 61), that is, a mosaic of small agriculture
fields with mixed deciduous and evergreen forest areas along the drainage
system (Figure 3). The rolling topography has slopes from 0 to 8 percent
and has generally deep, well drained, moderately to very permeable
soils. Among the undulations are nearly level flat depressions and
drainage ways that have deep, poorly drained, moderately permeable
soils (USDA, 1959; National Cooperative Soil Survey, 1976, 1978, 1979).
Beneath the soil mantle is an indurated layer that restricts vertical
movement of water; thus, the loss of surface water to deep seepage
is relatively insignificant (Yates and Stephens, 1968, p. 61). Additional
information about the watershed and its instrumentation may be found
in Yates and Stephens (1968), USDA (1971), and USDA-University of Georgia
(1969).

     The introduction of inputs, routing of flows through the system,
and disposition of outputs is diagrammed in Figure 4. The structure
of the computer program follows this schematic. The model will be
discussed in the order of inputs and then evapotranspiration, field,
urban-bare, pond, and forest subroutines. The model is an accounting
of amounts of water held in various compartments of the watershed and
passed from one to another or out of the watershed (the system). This
accounting is done in increments of one day.

Inputs

     The beginning and ending days of any period within a year and

531

Figure 3.   Aerial photograph of Little River Subwatershed N.   Note
the forests and irregular heterogenous fields.

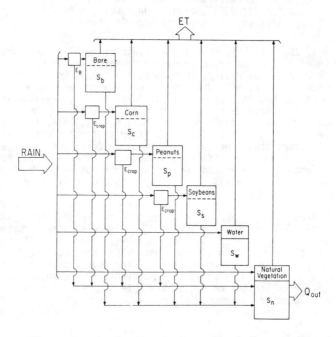

Figure 4.   Logic diagram for water balance model.

the number of covers (i.e., corn, pond, forest, bare, etc.) that have been designated in the watershed are specified first to control further operations of the program. The simulation is then made over the specified period and a line of input read for each cover. The information required for each cover is area (acres), soil depth (inches), soil water capacity (inch/inch), initial soil water amount (inches), and the Julian dates that determine periods of four growth stages for the cultivated crops.

Once the evapotranspiration subroutine is entered, then meteorological data, daily rainfall depth (inches), maximum and minimum daily air temperature ($^{\circ}$F), total daily wind run (miles), maximum and minimum relative humidity (percent), a cloudiness code, and radiation (Langleys) are read. Using this information, the program proceeds with the accounting calculations.

When the model is utilized for watersheds of any significant proportions, then the area of the different covers may be determined by remote sensing, using such procedures as those incorporating Landsat digital data reported by Jensen and Chery (1980).

Evapotranspiration Subroutine

Evapotranspiration is calculated by procedurers of Doorenbos and Pruitt (1975), with evapotranspiration determined for a reference surface (grass) using a modified Penman equation. Its form is:

$$ETO^* = W \times R_n + (1-W)*f(u)*(ea - ed) \tag{1}$$

where ETO^* is the unadjusted reference crop evapotranspiration in mm/day, W is a temperature-related weighting factor dependent upon station elevation, R_n is net radiation expressed in mm/day of evaporation, f(u) is a function of the average daily wind speed, ea is the saturation vapor pressure of the mean air temperature, and ed is the saturation vapor pressure of the mean dew point temperature.

Elevations in the southern Georgia region where the model was evaluated varied between 300 and 400 feet (91. 4 to 121.9 meters). Doorenbos and Pruitt (1975, p. 43) have a table of weighting coefficient (W) values for elevations of 0, 500, 1000, and 4000 meters for a series of temperatures ranging from 2 to 40 $^{\circ}$C. For the 0-meter elevation, a linear relation was fit to the table values of temperature, giving:

$$W_0 = 0.4389 + 0.01124 \; (\overline{T}) \tag{2}$$

where $\overline{T} = (t_{max} + t_{min})/2$ in $^{\circ}$C.

Net radiation, R_n, is a combination of net solar radiation, R_s, less net long wave radiation, R_{n1}, as expressed in the relation:

$$R_n = (1 - \alpha) R_s - R_{n1} \tag{3}$$

where α = cover albedo and is taken as 0.25 for the reference grass.

R_s is the measured solar radiation in equivalent millimeters per day of evaporated water. Using the latent heat of vaporization of water at 40°C, a conversion factor of 0.0174216 is obtained that when multiplied times the measured radiation in Langleys gives a measure of radiation in equivalent millimeters of water.

The net long wave radiation is determined from the product of a function of temperature, $f(t)$; a function of vapor pressure, $f(ed)$; and a function of sunshine hours, $f(n/N)$:

$$R_{nl} = f(t)*f(ed)*f(n/N) \qquad (4)$$

The function, $f(t)$, is given in tabular form by Doorenbos and Pruitt (1975, Table 15, p. 46). The following linear relation was fit to those tabular values:

$$f(t) = 10.8 + 0.1965 \ (\overline{T}) \qquad (5)$$

The function $f(ed)$ is given as:

$$f(ed) = 0.56 - 0.079(ed)^{\frac{1}{2}} \qquad (6)$$

where:

$$ed = \frac{ea \ (\overline{RH})}{100}$$

$$ea = 0.81358 + 0.027 \log \ (\overline{T})$$

$$\overline{RH} = (RH_{max} + RH_{min})/2$$

(Doorenbos and Pruitt, 1975, p. 46). The function $f(n/N)$ is expressed as:

$$f(n/N) = 0.1 + 0.947 \ (n/N)$$

which is obtained from Table 17 in Doorenbos and Pruitt (1975, p. 46) and where n/N could only be crudely estimated as 1, 0.5, and 0.05 for clear, partly cloudy, and completely cloudy as supplied in the available meteorological data.

The weighting function $(1-W)$ was evaluated by fitting a linear relation to tabular values for the 0-meter elevation in Table 10 of Doorenbos and Pruitt (1975, p. 43) and is:

$$(1 - W)_0 = 0.561 - 0.01124 \ (\overline{T}) \qquad (7)$$

The influence of wind on the evapotranspiration is given by:

$$f(u) = 0.27 \ (1 + U2/100) \qquad (8)$$

where $U2$ = wind run at 2 meters elevation in kilometers per day.

The measured wind run at 24 inches (61 cm) above the ground surface was adjusted and units converted by the following relation:

$$U2 = (wind) * (0.6214) * (1.31) \qquad (9)$$

The difference between the saturated vapor pressure at mean air temperature, ea, and the saturated vapor pressure at the mean dew point temperature, ed , is obtained from these two relations:

$$ea = 10^{(0.81358 + 0.027 \ \overline{T})} \qquad 10)$$

$$ed = ea \ (\overline{RH}/100) \qquad (11)$$

Doorenbos and Pruitt (1975) made an adjustment to the evapotranspiration (ET) during the first stage of growth by reducing the ET for increasing frequency of rainfall or irrigation. Since the model is a deterministic daily accounting, the relations given by Doorenbos and Pruitt were interpreted as time since last rainfall. The graph of empirical reduction coefficients (Doorenbos and Pruitt, 1975, p. 64) could not be fit by a single continuous function so they were fit by the sliding polynomial method of Snyder (1976). These interpolations were used to calculate an adjusted evapotranspiration amount which decreased as time since the last rainfall increased.

The calculation of a specific day's reference grass evapotranspiration was then available to the other subroutines for their accounting purposes.

Field Subroutines

A field is any subarea of the watershed above the forested-alluvial drainage system (see Figures 1 and 3). As shown in Figure 4, all fields drain both surface and subsurface flow to the alluvial portion of the watershed. A field may be either bare or cropped. If a field has a crop and if the specified simulation period extends either before or after the growing period for that crop, then the field is assumed to be bare for the time before or after the growing period. The bare subroutine is called by the particular crop subroutine during these pre- or post-growing season periods for the accounting calculations.

Bare Field Subroutine

The bare subroutine is typical of all the field subroutines in that the first step is a check for rain on the day of the calculation. If there is rain, then the daily amount is compared to a threshold amount of 0.2046 inch (0.52 cm) and if it is greater than that amount, an excess (surface flow passed directly to the forest-alluvial block) is calculated by the linear relation:

$$E = A + B*R \tag{12}$$

where E = daily excess (inches)

R = daily rainfall depth (inches).

Some data were available on an 0.85-acre (0.344-hectare) area in the Little River watershed and, as best as could be determined by a regression analysis, A was evaluated as -0.01776 and B as 0.0868.

An amount equal to the rainfall less the excess is then added to the soil water storage unless the soil water capacity is completely filled in which case all the rain in excess of the filled capacity is moved from the field as surface discharge.

Next, the reference evapotranspiration (ET) is adjusted by a coefficient (0.7) for bare ground for two days following a rain, as recommended by Doorenbos and Pruitt (1975, p. 82). After two days, ET is set equal to zero for bare ground. These amounts of ET are then deducted from the soil water storage.

Finally, the subsurface drainage is modeled by an exponential storage decay, expressed as:

$$q = q_0 e^{-bt} \tag{13}$$

where   q = unit discharge rate, inches^2/day

         q_o = initial unit discharge rate, inches^2/day

         t = time, days

         b = coefficient unit, S,

and where the total storage, S, is expressed as:

$$S = \int_0^\infty q_o\, e^{-bt} dt \tag{14}$$

Using a Darcy relation, the unit discharge, q, of Figure 5 is expressed as:

$$q = Kh \frac{dh}{dx} \tag{15}$$

where h = $\frac{S}{p}$.

The slope dh/dx will be taken as equal to the surface slope. From a sampling of field slopes as measured from a quadrangle map (USGS, 1973), dh/dx averaged 0.03 ft/ft (m/m).

From soils information for the same area, K equaled 10 inches/hour (25.4 cm/hr) and porosity, p, generally equaled 0.10 inch/1.0 inch (0.254 cm/2.54 cm). However, from an evaluation of the March 17, 1969, and September 21, 1969, subsurface flow events reported in Snyder and Asmussen (1972), it was determined that using the given permeability values for K gave peak subsurface discharges an order of magnitude less than those reportedly measured by Snyder and Asmussen (1972); thus, for the model the K values were increased by 10.

In order to obtain an initial unit discharge, q_o, the length, L, (Figure 5), had to be estimated. This was done by taking the aggregated area, A, of each identified field type and dividing by an average width, $\overline{W}$, or depth of fields (Figure 4):

$$L = A/\overline{W} \tag{16}$$

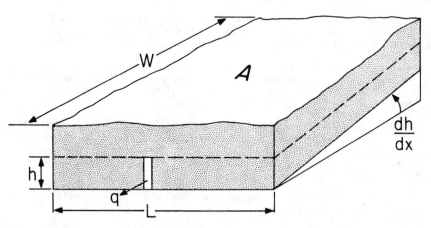

Figure 5.   Definition sketch for a field block.

From a sampling of W on the topographic map (USGS, 1973), the average field width, $\overline{W}$, was measured as 875 feet (266.7 m).

The initial discharge, $\hat{q}_0$, from the Darcy relation with K increased by 10, is:

$$\hat{q}_0 = 240(10) \; \frac{S}{p} \frac{dh}{dx} \quad \frac{in^2}{day} \tag{17}$$

$$\frac{dh}{dx} = 0.03$$

$$\hat{q}_0 = \frac{2400 \; (0.03)}{144} \quad \frac{S}{p} \quad \frac{ft^2}{day}$$

$$\hat{q}_0 = \frac{1}{2} \; \frac{S}{p}$$

In terms of total discharge, $\hat{Q}_0$:

$$\hat{Q}_0 = L\hat{q}_0 \qquad \frac{ft^3}{day} \quad (0.02832 \; m^3/day) \tag{18}$$

$$\hat{Q}_0 = \frac{A}{W} \; \frac{1}{2} \; \frac{S}{p} \qquad \frac{ft^3}{day} \quad (0.02832 \; m^3/day)$$

Converting to discharge per unit surface area, q_0:

$$q_0 = \frac{Q_0}{A} \qquad\qquad ft/day \; (0.3048 \; m/day) \tag{19}$$

$$= \frac{A}{W} \frac{1}{2} \frac{S}{p} \frac{1}{A} \quad ft/day \; (0.3048 \; m/day)$$

$$= \frac{1}{2} \; \frac{S}{Wp} \quad (12) \quad in/day \; (2.54 \; cm/day)$$

$$= 6 \; \frac{S}{Wp} \qquad\qquad in/day \; (2.54 \; cm/day)$$

where S = storage (inches, 2.54 cm)

W = field width (feet, 0.3048 m)

p = porosity (inches/inch, cm/cm)

The total discharge for any given day following a rain, Q_n, is obtained by integrating the instantenous discharge relation over a one-day period:

$$Q_n = \int_n^{n+1} q_0 \; e^{-bt} dt \tag{20}$$

$$Q_n = q_0 \left[-\frac{1}{b} \; e^{-bt} \right]_n^{n+1}$$

where n = number of days since the last day with rain.

$$Q_n = q_0 \left[-\frac{1}{b} \; e^{-b \, (n+1)} + \frac{1}{b} \; e^{-bn} \right] \tag{21}$$

The exponent, b, was estimated using data from the March 17, 1969, event reported in Snyder and Asmussen (1972). The initial discharge, q_0, was determined as follows:

537

$$q_o = \frac{6 \, (S)}{W} p \tag{22}$$

where S = 2.07 inches (5.26 cm)

W = 193.1 feet (58.86 m)

p = 0.1 inch/inch (cm/cm)

therefore, q_o = 0.643 inch/day (1.63 cm/day)

and then b determined from the relation:

$$S = \int_0^\infty q_o \, e^{-bt} dt \tag{23}$$
$$S = q_o \left[\frac{1}{b}\right]$$

and thus for S = 2.07 inches (5.26 cm) and q_o = 0.643 inch/day (1.63 cm/day), b = 0.3106, the value used in the model of drainage given in Equation 21. The daily amount of drainage is subtracted from the soil water and passed to the forest-alluvial soil water storage (see Figure 4), unless the amount exceeds what remains in storage; then only the remaining amount is drained out.

Cropped Field Subroutines

Three crops were distinguished using the four-band digital data for a June 11, 1977, Landsat record of Watershed N in the Little River watershed. The three crops were corn, soybeans, and peanuts. The details of the identification were reported by Jensen and Chery (1980).

A separate subroutine was written for each crop. Each crop subroutine function was similar to the bare subroutine with respect to determination of excess, drainage, and how the storage accounting is done. They differ in the calculation of evapotranspiration for the particular crop.

The reference evapotranspiration (ET) is adjusted by actual ET for the crop by a set of coefficients provided by Doorenbos and Pruitt (1975). They differ for each portion of the four-part growing season and also are different for each part of a two-part division of both relative humidity and wind for each day.

Pond Subroutine

A pond is assumed to be a tightly contained body of water that loses water only by evaporation until it fills above a spillway elevation. In this version, ponds intercept only their area of the rainfall over the watershed. Evaporation is determined by multiplying coefficients that also depend on minimum relative humidity and wind times the reference grass evapotranspiration (Doorenbos and Pruitt, 1975). If the water in the net lumped pond rises above an average spillway elevation, then water is passed to the forest-alluvial unit by the

$$q_p = S_d \left[\frac{1}{e^n} - \frac{1}{e^{n+1}} \right] \tag{24}$$

decay function where S_d is the difference between the maximum storage and the net pond-full storage on the day of the rain. The exponent, n, is an integer count of days since the day with the rain event.

Forest-Alluvial Subroutine

The forest-alluvial subroutine collects all the surface and subsur faces flows from fields and ponds and intercepts its proportion of the rainfall. It is assumed that no water flows vertically downward to the regional groundwater reservoir and that an unmeasured negligible amount passes the outlet of the watershed underground. The model then calculates outflows of ET and surface discharge.

The recessions of the 1977 flow events were plotted on semilog graph paper. From these plots it is obvious that there were two distinct stages of recession flow--a very rapid recession in the first two days following a rainfall event and then a slower recession for the following days. The rapid recession immediately following the rainstorm was divided into two-day responses depending on the initial discharge rate, q_o. Three relations are thus developed with $t \leq 2$ days after rainfall:

for $q_o \geq 0.2$ inch/day (0.51 cm/day):

$$q_1 = q_o \, e^{-0.992t} \tag{25}$$

for $q_o < 0.2$ inch/day (0.51 cm/day):

$$q_2 = q_o \, e^{-0.5493t} \tag{26}$$

with $t > 2$ days after rainfall:

$$q_3 = q_o' \, e^{-0.113t} \tag{27}$$

where q_o' is the value of q_1 or q_2 at $t = 2$, whichever applies.

The initial discharge, q_o, was assumed to be exponentially related to the storage depth above the depth of complete soil saturation, S_d, which is expressed as:

$$q_o = Ae^{BS_d}$$

In a series of runs of the model, the predicted model free water storage depth, S_d, was plotted versus the measured plot discharge and after several iterations the relation:

$$q_o = 0.01 \, e^{0.5697 \, S_d}$$

was selected as the most satisfactory with respect to the rainfall record being used and the uncertainties about the evapotranspiration amounts.

The evapotranspiration (ET) for the forest-alluvial block was initially determined by adjusting the reference grass ET with coeffi- cients given by Doorenbos and Pruitt (1975, p. 77) for full-grown deciduous trees in the months of June and July. As modeling of the discharge from the forest-alluvial block was refined, it became apparent that not enough water was being removed from this subsystem. If the assump- tion of negligible losses to deep groundwater storage and subsurface flow past the watershed outlet are valid, then the amount of water being removed by ET needed to be increased. It is reasonable that the coefficients adopted from Doorenbos and Pruitt underestimated the ET for the very dense, wet forest-alluvial areas that were being modeled.

Thus, the coefficients from Doorenbos and Pruitt were increased in an empirical fashion by adding 0.5 and then 0.8. The effect of these two increases in the ET from the forest is shown in Figure 6. The

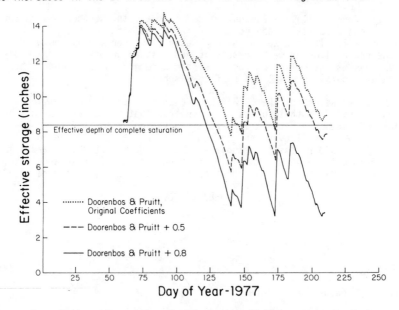

Figure 6. Comparative plots of effective soil/free water depth in the forest block for original plus two modifications of the Doorenbos and Pruitt evaporation coefficients.

initial Doorenbos and Pruitt coefficients, in conjunction with the modeling routines, predicted a sequence of small flow events between days 125 and 210. Between these days was a period of no flow in the measured natural system. As is seen in Figure 6, the effective storage depth was above the threshold level of 8.4 inches, and then dropped completely below the threshold when the coefficients were increased by the addition of 0.8. The model prediction with the initial ET values is shown in Figure 7, and the prediction with the ET increased by 0.8 is shown in Figure 8.

DISCUSSION AND CONCLUSIONS

To evaluate the water balance model, we initially selected a 90-day period (days 121-211) around June 11, 1977, the date of the satellite imagery from which the covers were determined. However, there was no flow from the natural system during this 90-day period, so the simulation period was started earlier at day 60. This expansion encompassed measured flow between days 60 and 120. To make this expansion, the cover prior to the planting of peanuts and soybeans was assumed to be bare. The growing seasons of the crops, along with the measured meteorological data needed for the model, are shown in Figure 9. The model also needed average block values for soil depth and water capacity and the initial saturated soil water storages. These estimated values are given in Table 1.

The performance of the model is shown by Figures 7 and 8 where

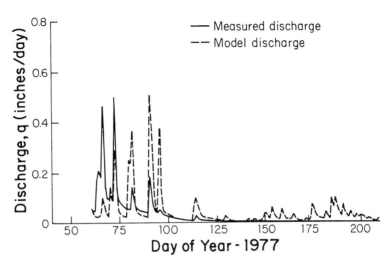

Figure 7.   Model versus measured discharge for Little River Subwatershed
N, with initial Doorenbos and Pruitt ET coefficients for
the forest block.

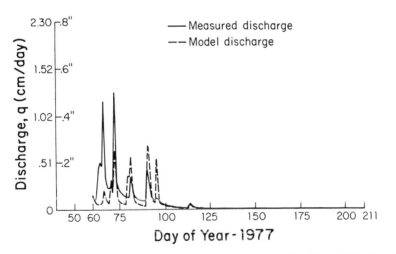

Figure 8.   Model versus measured discharge for Little River Subwatershed
N, with Doorenbos and Pruitt + 0.8 ET coefficients for
the forest block.

the predicted daily discharge, q, is compared with the measured discharge.
In Figure 8, it is seen that the model begins at the proper initial
condition, then underpredicts the first two peak flow amounts, then
overpredicts the following three peak flow amounts, and ends by coinciding
with the last small flow event and having no flow during the dry period
of days 120 to 211. Note that the recessions between events are shaped
like the measured recessions.

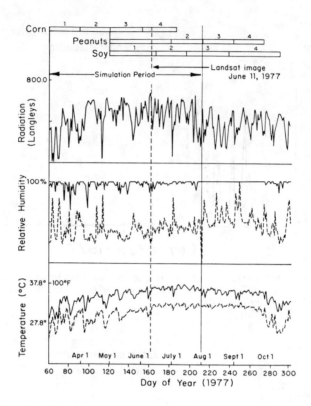

Figure 9.   Growing seasons and meteorological data used as input for testing the water-balance model.

Table 1.   Initial Values of Parameters of Water Balance for Little River Subwatershed N, March 1 to July 30, 1977

Cover	Area (hectares)	Soil depth (cm)	Water capacity (cm/cm)	Initial saturated water depth (cm)	Julian day of growth periods				
Bare	117	152.4	.254	7.62					
Corn	594	152.4	.254	7.62	60	91	122	154	188
Peanut	274	152.4	.254	7.62	121	182	213	244	274
Soybeans	42	152.4	.254	7.62	121	167	197	240	290
Water	21	152.4	2.54	127.00					
Forest	519	213.4	.254	21.59					

The differences between prediction and measured responses may depend on magnitudes of rainfall input to the watershed. The rainfall input was obtained from a daily standard gage located 4.54 miles (24, 000 ft; 7.3 km) southeast from the center of the watershed. A better measurement of rainfall to the watershed could be obtained from an average of raingages 13, 14, and 16, maintained by the U.S. Department of Agriculture, Science and Education Administration, but these data were not processed and available at the time of the study. As seen in Figure 8, the watershed begins responding to. rainfall before the model, and then added to this early response is apparently a larger rain than was used as input to the model. After the first two major peaks, this situation reverses and apparently larger rains were used as input to the model than the watershed.

The model cannot be refined further until questions about the input amounts, internal flow transfers, soil depths, and forest vegetation ET are resolved. Even with the less than ideal measurements and information about the real system, the model gave a reasonable prediction and demonstrated the potential of the overall procedure.

ACKNOWLEDGEMENT

The model was initially developed while the senior author was employed as Research Hydraulic Engineer with the Southeast Watershed Research Program, U.S. Department of Agriculture, Science and Education Administration, Agricultural Research, Athens, Georgia.

REFERENCES

Doorenbos, J., and Pruitt, W.O., 1975. Guidelines for predicting crop water requirements. Irrigation and Drainage Paper, No. 24, United Nations, Food and Agriculture Organization, Rome.

Jensen, J.R., and Chery, D.L., Jr. 1980. Landsat crop identification for watershed water balance determination. Proceedings of the American Congress of Survey and Mapping and American Society of Photogrammetry Convention, St. Louis, Missouri, pp. 65-80, American Society of Civil Enginers.

National Cooperative Soil Survey. 1976. Pelham series, Rev. GLB:RR. U.S. Department of Agriculture, Soil Conservation Service.

National Cooperative Soil Survey. 1978. Tifton Series, Rev. MES, U.S. Department of Agriculture, Soil Conservation Service.

National Cooperative Soil Survey. 1979. Kershaw Series, Rev. RHA:MES, U.S. Department of Agriculture, Soil Conservation Services.

Snyder, W.M. 1976. Interpolation and smoothing of experimental data with sliding polynomials. 34 pp. ARS-S-83, United States Department of Agriculture, Agricultural Research Service.

Snyder, W.M., and Asmussen, L.E. 1972. Subsurface hydrograph analysis by convolution. Journal of the Irrigation and Drainage Division, Vol. 98, No. IR3, pp. 405-418, American Society of Civil Engineers.

U.S. Department of Agriculture, Soil Conservation Service. 1959.
Soil Survey of Tift County, Georgia. Series 1946, No. 3, 28 pp.

Mills, W.C. 1971. Tifton Georgia. Chapter 3 in Agricultural Research
Service Precipitation Facilities and Related Studies, ARS 41-176, 117
pp., U.S. Department of Agriculture, Agricultural Research Service.

U.S. Department of Agriculture and University of Georgia. 1969. Final
Report of Hydrology Study of Upper Little River Watershed for January
1, 1966 - March 10, 1969. 41 pp., Research and Service Contract No.
12-14-100-8894, Athens, Georgia.

U.S. Geological Survey. 1973. Chula, Georgia, 7½-minute series topographic
map. National Mapping Program.

Yates, P., and Stephens, J.C., 1968. Watershed research in the Southeast.
Proceedings of the Conference on Hydrology in Water Resources Management,
pp. 54-76, Clemson University, Council on Hydrology, Clemson, South
Carolina.

SOURCE DETERMINATION OF RICE IRRIGATION WATER IN SOUTHWESTERN LOUISIANA FROM LANDSAT IMAGES

J.M. Hill
Assistant Professor
Remote Sensing and Image Processing Laboratory
Civil Engineering Department
Louisiana State University
Baton Rouge, LA 70808

Dale J. Nyman
Hydrologist
U.S. Geological Survey

and

Michael E. Neal
Environmental Engineer
Walk, Haydel and Associates, Inc.

ABSTRACT

The irrigation of rice in southwestern Louisiana constitutes the largest water use in this part of the State. Data for water resources withdrawn for irrigation are now obtained every 5 years; but because of economic and climatic conditions, it is desirable to acquire these data on a yearly basis. As part of an extended research project by Louisiana State University in cooperation with the U.S. Geological Survey, a method for determining sources of irrigation water (ground water or surface water) in rice fields was developed, using Landsat data obtained April 14, 1979. Irrigated fields were selected for determining the source of irrigation water in each of three parishes (counties); irrigation data for each farm (compiled by the U.S. Geological Survey) served as primary ground truth. Turbidity samples from and visual observations of many fields provided additional ground-truth data. Radiance values derived from Landsat data for these irrigation waters were analyzed using discriminant-analysis techniques. The results indicate that the source of irrigation water may be determined with an accuracy of 70-80 percent.

INTRODUCTION

Rice irrigation is the largest single water use in southwestern Louisiana; essential to determining this use is an estimation of the irrigated acreage. Monitoring annual changes in rice acreage caused by fluctuating farm markets and climatic conditions requires a type of surveillance that frequently views the entire area and provides a variety of quantitative data. Landsat data can provide this type of monitoring capability.

A concern at present is that during the next decade water resources will be taxed to their limit by water demands for industrial, agricultural, and domestic use and even interstate export of water. Consideration needs to be given to the determination or prediction of future minimum usage requirements for water resources from a regional perspective.

About 500,000 acres (200,000 hectares) of rice is grown each year in southwestern Louisiana. In this area, rice is the most important crop and was valued at 131 million dollars for 1976. Also, because rice is irrigated, more water is pumped to grow this crop than is pumped in the area for industrial, public supply, and other water uses combined. Irrigation pumpage in southwestern Louisiana averaged about 4 billion gallons per day during the 1974 irrigation season.

The goals of this research were to: (1) Determine the feasibility of calculating total irrigated rice acreage, and (2) differentiate the acreage irrigated by ground water (wells) from the acreage irrigated from surface-water sources (rivers, bayous, and lakes), using Landsat data. The eventual goal is to estimate total water use and the distribution of water from ground- and surface-water sources so that these data may be incorporated into a water-management oriented hydrologic model.

Southwestern Louisiana is a water-rich area with ample sources of fresh ground water and surface water. Because of the plentiful supply, water use is unregulated; the only limitation is the amount an individual or company needs or can afford to use for irrigation. An acre of rice receives either an average of nearly 2 acre-feet (0.6 meters per hectare) of ground water, or an average of more than 4 acre-feet (1.2 meters per hectare) of surface water each year. There has been a steady increase in ground-water use because irrigation systems using wells can be designed to be more water efficient than surface-water systems. Surface water is obtained from three major river systems in southwestern Louisiana and is distributed to the farmers primarily through an extensive system of elevated canals. Each farmer receives an unlimited amount of water, for which he pays the canal company one-fifth of the rice grown. Elevated canals are becoming very expensive to operate and maintain; therefore, the acreage irrigated by ground water is steadily increasing.

Environmental problems may occur with changes in pumping patterns. During low-flow periods, many canal companies have a pumping capacity of several times the flow of the river. This means that river flow is reversed, particularly during the summer and fall months. Aside from saltwater intrusion, the disposal of municipal and industrial effluents also becomes a real problem.

BACKGROUND

Acquisition of water-resources information concerning water type, irrigated acreage, and various other data is accomplished through the cooperation of the U.S. Agricultural Stabilization and Conservation Service (ASCS). Every 5 years farmers are interviewed when they certify their rice acreage with the ASCS. This method and the associated time frame have proven to be incomplete and unsatisfactory with respect to the changes (for example, rotation of cash crops) occurring during years between data collection.

Landsat images have revealed the potential to not only detect

water in irrigated fields, but preliminary research efforts also have
been able to distinguish surface water from ground water used for rice
irrigation in several parishes during one season. This is because
river water generally is more turbid and, therefore, more reflective
than ground water. Landsat imagery shows which fields are flooded;
and because the canal system is stationary, the farms using river
water are predictable due to water access. Thus, the pumping demand
on the rivers can be estimated, assuming the other hydrologic
parameters needed are available.

Landsat satellites provide a means for the collection of synoptic
and accurate data over large areas at one time and at relatively
consistent intervals; the imagery also provides a historical record.
The use of Landsat data for discrimination of water types in shallow
water environments (2-8 inches or 5-20 centimeters) is unprecedented
in current literature. However, previous research has established
techniques for irrigated-acreage determination for other crops in
various areas of the United States (Williams and Poracsky, 1979). An
image from Landsat, as seen in figure 1, covers the entire
rice-growing area of southwestern Louisiana in one scene. The
irrigated rice fields appear on the image as small rectangles in
shades of light to dark blue (black to gray in black and white images)
scattered throughout this scene.

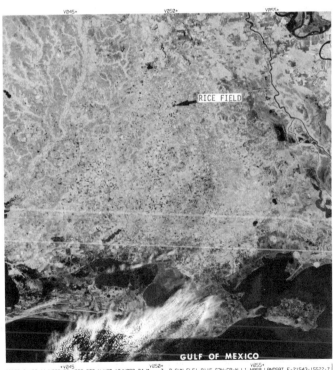

Figure 1. Landsat image of southwestern Louisiana. (Scale,
1:1,000,000). The small, dark rectangles in the middle of the
image denote irrigated rice fields.

Landsat satellites record data as they orbit at an altitude of approximately 570 miles (920 kilometers) above the earth in a sun-synchronous, nearly polar orbit. In this research the multispectral scanner (MSS) system was used for data collection (U.S. Geological Survey, 1979). The MSS picture element is the smallest ground-resolution element (1.1 acres or 0.45 hectare) and is termed a "pixel". Thus, each pixel is a spectral blend of all targets sensed in a 1.1-acre (0.45-hectare) area. The MSS contains six detectors for each of four bands [Band 4 - green (500-600 nanometers), Band 5 - red (600-700 nanometers), Band 6 - infrared (700-800 nanometers), and Band 7 - infrared (800-1,100 nanometers)]. Every object and material on earth (and in the atmosphere) emits, reflects, absorbs, or transmits electromagnetic radiation; these characteristics enable investigators to use remote-sensing devices to detect and identify materials and objects. For example, water intensively absorbs energy in the near-infrared region of the spectrum; whereas, soil and vegetation are highly reflective in the same spectral interval. That particular distinction is important in this research, because the irrigated fields need to be readily distinguished and identified before they can be analyzed spectrally to determine the source of irrigation water.

The most important energy interaction with the target (irrigated rice fields) in this research is the volume reflectance, because it is the part of the remotely sensed signal that is generated by reflectance in the water column or volume. Thus, this part of the signal reveals the water quality or the characteristics of the water being considered. The interactions that occur in the atmosphere, or on the surface of the water, and reflections from bottom materials can be dominant factors in some cases. These factors obscure the signal resulting from water interaction; however, the only factors that attenuate the signal levels generated from the water are color and turbidity (Yarger et al., 1973; Egan 1974 Scherz et al.; 1975; Moore, 1978). When bottom reflectance is present, it may nearly dominate the signal. Bottom reflectivity has been shown to increase from about 20 percent at 500 nanometers (Band 4) to 50 percent at 1,100 nanometers (Band 7) over a uniform bottom (Moore, 1978). The maximum depth of bottom detection using Landsat multispectral scanner systems are 125 feet (38 meters) for Band 4, 20 feet (6 meters) for Band 5, 6 feet (2 meters) for Band 6, and 0.3 foot (0.1 meter) for Band 7. Thus, detection of the bottom should be inherent and strong in this study due to the use of Bands 4 and 5 and the shallow depth of the irrigation water. Turbidity is caused by the presence of clay, silt, organic and inorganic "fines", and planktonic organisms in the water. In general, it has been shown that an increase in turbidity increases reflectance levels as measured by Landsat and other remote-sensing techniques (Klemas et al., 1973; Harris et al., 1976).

METHODOLOGY

One of the major phases of this project was establishing the necessary ground truth, which entailed the selection of parishes and farms and collection of various data in selected test areas based on a 1979 survey of farmers in the southwestern Louisiana region. Three parishes (Allen, Jefferson Davis, and St. Landry) were chosen for further study on the basis of the quantity of irrigation and variation of major soil types. In each of the selected parishes, about 50 to 100 farms were evaluated to determine the source of irrigation water. The selection of farms using either ground water or surface water was based on location in the parish (even distribution) and size of the farm (as seen on Landsat data). Parish ASCS offices provided: (1) Farm identification number; (2) verification of the reported acreage from rice certification reports; (3) enlarged photographs of each test farm; (4) time of irrigation; (5) stage of irrigation by April 14, 1979; and (6) identification and outline of the farm boundaries on ASCS photomosaics of the parish through use of the enlarged photographs.

Various weather conditions postponed the usual March to early April irrigation practices; and by April 14, 1979 (time of satellite overpass), the rice fields were only approximately 65 percent irrigated in relation to previous years. Selection of at least 10 farms in each parish for each water type (ground or surface) was the next task. ASCS personnel were used to contact the farmers to obtain the required information concerning: (1) The time of irrigation; (2) whether irrigation of the study area was complete or incomplete by April 14, 1979; (3) the number of acres irrigated, if not completely flooded; and (4) the source of irrigation water. Water samples for turbidity measurements were acquired from pumping plants, irrigation canals, ground-water wells, rivers, and various fields irrigated by surface and ground water.

A major part of the digital-data processing was conducted using the facilities of the College of Engineering's Remote Sensing and Image Processing Laboratory (RSIP). The equipment included a computer for image processing and analysis and a display unit that allows color presentation of data and some processing capability. Image analysis usually follows the general procedure of: (1) Selecting homogeneous areas representative of desired spectral classes (ground or surface water), and (2) classifying the remaining data by statistical programs that use the areas selected in part (1) as classification criteria.

The areas selected from Landsat imagery as being representative of a particular spectral class are called training fields or polygons. Two different techniques of selecting training fields were implemented, without edge effects (homogeneous) and with edge effects. A homogeneous training field is outlined within the irrigated field boundaries so that no bordering material is included in these data. The training fields with edge effects include the bordering pixels of the irrigated field that produce direct and composite effects from the surrounding material. A standard unsupervised classification program using a 3x3 pixel classification "window" to denote homogeneity was unable to select homogeneous training samples because the rice fields were randomly distributed and too small. A manual, supervised classification technique was then used to outline fields and acquire spectral Landsat data.

The goal was to classify the source of irrigation water, by training field, because each field was irrigated by only one kind of water. A question was raised concerning the possibility of a relationship existing for an increase in average count (reflectance) value with an increase in the size of a training field. A relationship would indicate the necessity to weight these data. Regressions were applied to these data for ground and surface water in two Landsat bands, and the results indicated that only a very small variation in count value is due to field size.

The technique of discriminant analysis was selected for classification of the Landsat data. This method analyses the input

data and builds a model which discriminates (or classifies) these data
into the categories described by the investigator (in this instance,
surface or ground water). Discriminant analysis is the appropriate
statistical technique when the dependent variable is categorical and
the independent variables are numeric (Hair et al., 1979). In this
study, the dependent variables are ground water and surface water, and
the independent variable(s) are the average count value(s) derived
from a training field in a selected band(s). This technique also is
appropriate for testing the hypothesis that the group means of two or
more groups (water types) are equal (Harris, 1976). Discriminant
analysis begins with the derivation of a linear combination of the
independent variable(s), or average count values, that will
discriminate best between the defined groups (water types). The
program input consists of the average count (reflectance) value(s) for
each training field and a code to inform the program which category
these value(s) belong to. The combining of the independent variables
(that is, two bands of input per type) into one function on a single
axis is illustrated in figure 2. Output consists of a listing for
each training field indicating the actual category (water type)
represented, the category into which it was classified, and the
probability of classification in either category. Also produced is a
summary of the accuracy of classification overall and by category.
The assumptions for discriminant analysis are multivariate normality
and equal dispersion. Landsat data have been demonstrated to approach
the requirements of normality by many investigators (Swain and Davis,
1978; National Aeronautics and Space Administration, 1979). The
conventional criteria of 0.05 and 0.01 significance levels are
maintained in evaluating results.

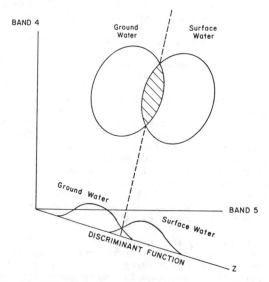

Figure 2. Graphic illustration of the discriminant function (adapted
    from Hair et al., 1979). The discriminant function reduces the two
    axes to one single axis (Z), and one distribution for each category
    (ground or surface water).

Water types were examined using three approaches--all parishes
combined, paired parishes, and individual parishes. Most of the
analyses were accomplished with a probability (cost of

misclassification into either group) set to the proportion of each category (ground or surface water) in the data set. For a more detailed explanation of data analysis see Neal (1980).

Considering a combined approach using all training fields, an overall classification accuracy (training fields with or without edge effects) of 70-80 percent may be achieved (table 1). The classification accuracy for surface water in most cases is considerably less than that derived for ground water. This is due to the probabilities (cost of misclassification) being set to proportional group size. The model indicates a greater accuracy for the larger group (ground water) in such cases.

Table 1. Summary of results of discriminant analyses of water types for Allen, Jefferson Davis, and St. Landry Parishes

Data set	Ground Water		Surface Water		Overall % correctly classified	n	Signifi- cance level[1]
	% correctly classified	n	% correctly classified	n			
(a) Training fields without edge effects (Band 4)	95.8	71	30.4	23	79.8	94	0.0001
[2] Holdout group of samples	92.3	26	55.6	9	82.9	35	
(b) [3] Training fields without edge effects and equal probability (Band 4)	71.4	49	63.2	19	69.1	68	.0002
(c) Training fields with edge effects (Band 4)	89.7	39	55.6	18	75.4	57	.0018
(d) [3] Training fields without edge effects and equal probability (Band 5)	70.0	49	63.6	14	68.5	92	.0002
(e) Training fields without edge effects (Band 5)	98.6	69	31.8	7	82.6	92	.0002
[2] Holdout group of samples	100.0	24	60.0	6	80.0	34	

[1] The F test (for testing = 0) indicated there is a highly significant difference in the group means (0.01 level).

[2] A randomly selected group of training fields that were used to test the model.

[3] Data sets analyzed with equal cost of misclassification in either group (probability = 0.50). Data sets a and c were analyzed with probability equal to the proportion of each group (ground or surface water) in the data set.

An average of six runs for training fields without edge effects (Bands 4, 5) indicates greater classification accuracies for all categories than the case with only Band 5. Using training fields having edge effects and using only Band 5 produces less accuracy in each category than the same data set using Bands 4 and 5. Training fields without edge effects produce a greater classification accuracy than the same fields with edge effects. The results also indicate that the use of Bands 4 and 5 produce greater accuracies than the same case using only Band 5.

RESULTS AND DISCUSSION

Accurate assessment is contingent upon the factors of turbidity, bottom reflectance, variation in soils, and statistical parameters. In parts of Allen Parish anomalous concentrations of iron (greater than 6 milligrams per liter) are present in ground water. These concentrations of iron produce ferric hydroxide and result in relatively turbid waters. Classification results indicate that surface water and ground water reverse their typical spectral

551

appearance in this area due to turbidity anomalies.

Sources of irrigation water in Jefferson Davis Parish were classified with an accuracy ranging from 70 to 80 percent. The significant factors for training field classification appear to be linked to the turbidity evident in surface-water sources and possibly to the effects of soil type. Scherz and Van Domelen (1973) found in laboratory tests that turbidity overrides bottom reflectance at approximately 15-20 turbidity units (Band 5). For values greater than 15-20, an increase in turbidity increases reflectance proportionately; and for values less than 15-20, the reflectance from bottom effects is asymptotically related to the soil reflectance value. Therefore, turbidity and bottom reflectance are both discernible and significant factors in the classification of water types.

St. Landry Parish is quite diverse in actual soil types and spectrally they range in appearance from light to dark. The spread of radiance values from the small sample of fields (all but two were irrigated by ground water) is so great that some of the values match fields from either Allen or Jefferson Davis parishes. The fields separate and cluster in distinct groups over a large range.

CONCLUSIONS

In conclusion, it is appropriate to state that the sources of irrigation water in rice fields can be determined with reasonable accuracy in southwestern Louisiana. Therefore, it appears feasible to determine the total acreage of irrigated fields and to estimate the acreage irrigated by ground water and by surface water. Except for the unique area of turbid ground water in Allen Parish, the two categories of waters tend to produce significantly different group parameters and classification accuracies averaging from 90 to 60 percent (ground and surface water, respectively). In other areas of southwestern Louisiana, such as those in Jefferson Davis Parish, greater classification accuracies possibly may be obtained by analyzing the data by parish. Classification of water types over the three-parish area produced an overall accuracy of approximately 70-80 percent. These results indicate that sources of irrigation water can be correctly classified even over larger areas having considerable variations in soil types.

The major limitations of the overall technique are in Landsat data collection and acceptable climatological conditions. Southwestern Louisiana typically is obscured by cloud cover during late April and May, and all irrigated fields generally are flooded by late April. These factors may present a problem in some years, because a cloud-free image of the entire rice-growing area is necessary for analysis. A Landsat related problem may be in the time interval of coverage. If only one satellite is functional, the interval of data collection is 18 days; when two satellites are functional, the interval is decreased to 9 days.

The technique of discriminant analysis indicates a great amount of versatility and application in classification of water types. This technique has the ability to discriminate several classes (four or more is not unusual) according to Hair et al., (1979). This method may be applicable for other spectral classes (for example, forest types, rural versus urban). The technique of discriminant analysis also has proven its utility in the analysis of spectral entities that occupy only a very small percentage of the total classes existing in one image.

Our future research direction is to process April 1978 Landsat data in an effort to verify this technique for 2 years. At the same time, water-management oriented hydrologic models are being surveyed and evaluated with the eventual goal of using these Landsat and related data to drive such a model.

REFERENCES

Egan, W. G., 1974. Boundaries of ERTS and Aircraft Data Within Which Useful Quality Information Can Be Obtained. Proceedings, 9th International Symposium on Remote Sensing of Environment, Environmental Research Institute of Michigan, Ann Arbor, Vol. 2, pp. 1319-1343.

Hair, J. F., Jr., R. E. Anderson, R. L. Talham, and B. J. Grablowsky, 1979. Multivariate Data Analysis. Petroleum Publishing. Tulsa, Oklahoma. pp. 82-122.

Harlan, J. C., J. M. Hill, C. Bonn, H. A. El-Reheim, 1975. A Biological and Physical Oceanographic Remote Sensing Study Aboard the Calypso. Presented, 10th International Symposium on Remote Sensing of Environment, Environmental Research Institute of Michigan, Ann Arbor. Vol. 2, pp. 661-670.

Harris, R. J., 1976, A Primer of Multivariate Statistics: Academic Press, New York, N. Y.

Harris, G. P., R. P. Bukata, and J. E. Bruton, 1976. Satellite Observations of Water Quality, Transportation Engineering Journal. Proceedings of the American Society of Civil Engineers, Vol. 102, No. TE3.

Horstmann, U., K. A. Ulbright, and D. Schmidt, 1977. Detection of Eutrophication Processes from Air and Space. Proceedings, 12th International Symposium on Remote Sensing of Environment, Environmental Research Institute of Michigan, Ann Arbor, Vol. 2, pp. 1379-1389.

Klemas, V., J. F. Borchardt, and W. M. Treasure, 1973. Suspended Sediment Observations from ERTS-1, Remote Sensing of Environment. Vol. 2, pp. 205-221.

Moore, G. K., 1978. Satellite Surveillance of Physical Water-Quality Characteristics. Proceedings, 12th International Symposium on Remote Sensing of Environment, Environmental Research Institute of Michigan, Ann Arbor, Vol. 1, pp. 445-462.

National Aeronautics and Space Administration, 1979. Earth Resources Orientation and Training Course in Remote Sensing Technology, Landsat Series, NASA Goddard Space Flight Center, Greenbelt, Md.

Neal, M. E., 1980. Discrimination of Water Types for Irrigated Rice Acreage Using Landsat Data in Southwestern Louisiana. Unpublished masters thesis. Institute of Environmental Studies. Louisiana State University.

Scherz, J. P., and J. F. Van Domelen, 1973. Water Quality Indicators from Aircraft and Landsat Images and Their Use in Classifying Lakes. Proceedings, 10th International Symposium on Remote Sensing of Environment, Environmental Research Institute of Michigan, Ann Arbor, Vol. 1, pp. 447-460, 154 p.

Scherz, J. P., D. R. Crane, and R. H. Rogers, 1975. Water Quality by

Use of Satellite Imagery. Proceedings of the American Society of Photogrammetry, Phoenix, Arizona, pp. 320-343.

Strong, A. E., 1974. Remote Sensing of Algal Blooms by Aircraft and Satellite in Lake Erie and Utah Lake, Remote Sensing of Environment, Vol. 3, pp. 99-107.

Swain, P. H., and S. M. Davis, 1978. Remote Sensing, The Quantitative Approach, Laboratory for Applications of Remote Sensing.

U. S. Geological Survey, 1979. Landsat Data Users Handbook, Washington, D.C. Government Printing Office.

Wezernak, C. T., 1974. The Use of Remote Sensing in Limnological Studies. Proceedings, 9th International Symposium on Remote Sensing of Environment, Environmental Research Institute of Michigan, Ann Arbor, Vol. 2, pp. 963-980.

Williams, T. N. L., and J. Poracsky, 1979. Mapping Irrigated Lands in Western Kansas from Landsat. Proceedings, 5th Annual William T. Pecora Mem. Symposium - "Satellite Hydrology," pp. 707-714.

Yarger, H. L., J. R. McCauley, G. W. James, and L. M. Magnuson, 1973. Quantitative Water Quality with ERTS-1. Third Earth Resources Technology Satellite-1 Symposium, Vol. 1, pp. 1637-1651.

SNOWCOVER AND SNOWMELT RUNOFF RELATION IN BEAS CATCHMENT, INDIA - USING LANDSAT IMAGES

A.J. Duggal[1]
Research Fellow
Department of Earth Sciences
University of Roorkee

R.P. Gupta
Reader
Department of Earth Sciences
University of Roorkee

G. Sankar[2]
M. Tech. Student
Department of Earth Sciences
University of Roorkee

B.B.S. Singhal
Professor and Head
Department of Earth Sciences
University of Roorkee
Roorkee, 247672, India

ABSTRACT

At higher altitudes and latitudes precipitation is mostly in the form of snow. It lies dormant for a considerable time and affects the river discharge at a much later date, when the snow melts. River Beas, in the NW part of India, has been selected for studying the relation between snowcover area and snowmelt runoff. Snowcover area was mapped from Landsat images of the months March/April of the years 1972-73, 1974-75, 1975-76 and 1976-77 and daily discharge data of Beas and its tributaries for corresponding years was collected from the hydroelectric project authorities. It has been found that in each catchment/ subcatchment, the snowcover area and subsequent snowmelt runoff are well correlated. However, the slope of the regression curve seems to depend on lithology, valley slopes and size of subcatchment area. Hence, for each type of geologic subcatchment, a different empirical curve should exist.

---

[1] Presently with the Oil and Natural Gas Commission, Assam, India.

[2] Presently at Centre for Earth Science Studies, Trivandrum, India.

INTRODUCTION

At higher latitudes and altitudes, most of the precipitation takes the form of snow, which may lie dormant for a considerable time and may affect the discharge of a river at a much later date, as and when the snow melts. Hence, estimation of snowmelt runoff is of great importance for proper management of water resources.

Beas, a tributary in the Indus River system, has been selected for the present investigations, as it is known to carry a substantial amount of snowmelt. It has a rather well-defined catchment with few glaciers, and is under intensive river valley development. It may be mentioned here that Singh and Mathur (1976) made an attempt to estimate snowmelt in the Beas catchment using meteorological parameters and applying Corps of Engineers (1973) formulae. Their work considered snowmelt only above 14,000 feet (4000 meters), i.e. from permanently snowcovered zone. However, the snowline in this area decends to about 8000 feet (2500 meters) in winters and hence their calculated snowmelt would only be a part of the total snowmelt runoff. In the present study, an attempt has been made to correlate the snowcover areas mapped on the Landsats with subsequent snowmelt runoff in different subcatchments of Beas.

Location

The river Beas originates in the Higher Himalayas near Rohtang Pass (Fig. 1). Total length of its course up to its confluence with the Sutlej is about 395 km; catchment area of river Beas is approximately 12916 sq. km out of which nearly 763 sq. km lies under permanant

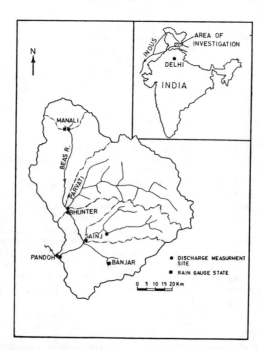

1. Location map of the Beas river, Pandoh dam site and hydro-meteorological measurement stations.

556

snowcover. A diversion dam has been constructed at a site called Pandoh. The area of Pandoh catchment, which has been investigated in the present study, is nearly 5460 sq. km. The various subcatchments of the river Beas, which have been mentioned in the following text, are named after discharge measurement sites located at Manali, Parvati, Bhunter, Sainj and Pandoh (Fig. 1).

Hydrological Year

As a first step to any hydro-meteorological analysis, it is necessary to define the hydrological year for the area. For this purpose the precipitation and discharge data of this area for a number of years have been compared. It is found that the precipitation is fairly high in the months of December-February but daily discharge values in these months keep low and touch a minimum during this period. Because it is winter, the precipitation falling in the form of snow gets stored in the catchment and causes no rise in discharge. During March-April-May, there is very little precipitation but hydrographs show a steady increase in discharge, implying that the snow that had accumulated in the preceding months is undergoing melting. During June-July-August, the area receives abundant precipitation (monsoon rain) and the corresponding very high values in daily discharge are quite understandable. From September to November, the daily discharge values show a general decreasing trend, the catchment receiving little precipitation, until the discharge stabilizes to a minimum which persists through December to February. With the above background, the hydrological year in the area has been taken as from November to October.

METHODOLOGY

Landsat images of the Beas catchment have been used for mapping the snowcover. The coverages of March/early April have been used. At this time the snowcover is at a maximum and the snowmelt assumes significance only during the following months of April, May and June. Snowcover area was mapped manually as it is considered to be more accurate than by the automatic digital method based on gray-tone slicing. The daily discharge data from a number of discharge sites (Manali, Parvati, Bhunter, Sainj and Pandoh) in the Beas catchment (Fig. 1) have been used. The analysis has been done for hydrological years 1972-73, 1974-75, 1975-76 and 1976-77.

Method of Snowcover Mapping

Mapping of snowcover area was carried out on enlarged Landsat images. Snow appears bright white on images owing to its very high albedo. The various subcatchments were delineated using survey of India regional topographical maps of 1:253,440 scale (1" = 4 miles). The catchment plan and the snowcover maps were exactly superimposed over each other using Bausch and Lomb Zoom Transferscope. The snowcover maps were prepared for the date March 3, 1973; March 2, 1975; April 1, 1976; and March 27, 1977, to serve for the respective hydrological years.

Effect of Shadows Due to Relief on Snowcover Mapping

The Landsat views the earth, in an oblique illumination condition. Thus, the inclination of the suns rays projects shadows in the rugged Himalayan terrain. The average sun azimuth on the Landsat images used is nearly 127° and the sun angel about 42° and hence the snow on the NW facing slopes of the hills are shadowed or receive less illumination and

2.   The south-facing slopes are shown bright white covered with
      snow whereas the adjacent north-facing slopes are seen dark
      due to shadow cast owing to oblique solar illumination.
      Landsat image, dt. March 2, 1975; MSS 7 (infrared).

are darker in comparison to the adjacent SE facing slopes (Fig. 2).
This effect has also been considered while mapping the snowcover area
from the Landsat images.

     Therefore, in this study, two types of snowcover maps have been
prepared, namely, (a) actual snowcover maps and (b) apparent snowcover
maps. Actual snowcover maps are those which have been prepared after
interpreting the shadows and proper extrapolation of the snowline,
whereas apparent snowcover maps are prepared by considering that all
that is white, is snow and all that is non-white is not snow. This
situation would be similar to the one where only gray-level slicing is
used for mapping the snowcover.

     The relationship between actual snowcover area and apparent
snowcover area in various subcatchments (Manali, Bhunter, Parvati, Sainj
and Pandoh) was investigated for both the channels MSS 5 (0.6-0.7
microns wavelength) and MSS 7 (0.8-1.1 microns wavelength) separately,
for the Landsat coverage of March 3, 1973. The plot (Fig. 3) shows the
relationship between the apparent and actual snowcover values to be
rectilinear and naturally the actual snowcover value is always greater
than the corresponding apparent snowcover value. The above example
represents a rather simplified case where the ground slope, its
direction and morphological features are regular and almost similar in
geographical orientation. With the variation in above conditions, the
relationship of actual and apparent snowcovers could be fairly complex.

Comparison of Snowcover Areas on MSS 5 and MSS 7

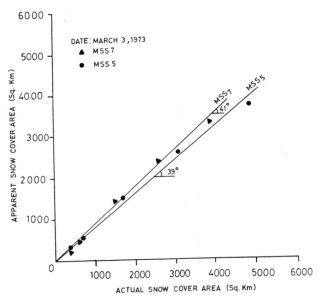

3. Relationship between actual snowcover area and apparent snowcover area.

Studies have also been carried out to find the extent of difference in snowcover area on MSS 5 and MSS 7 of the same Landsat coverage (March 3, 1973). Fig. 4 shows that the relationship between the

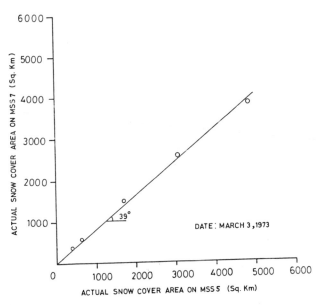

4. Relationship between snowcover areas on MSS 5 (visible) and MSS 7 (infrared) wavelength bands.

snowcover area on MSS 5 and the snowcover area on MSS 7 in various
subcatchments is linear. The infrared band (MSS 7) invariably gives
less snowcover area than the visible band (MSS 5). This appears to be
due to the presence of moisture on snow-surface in lower altitude
fringes of the snowcovered areas (Thomas and Lewis, 1978; O'Brien and
Munis, 1975). The actual snowcover area on MSS 5 (visible) has been
used as the snowcover area in the following analysis.

SNOWCOVER AND SNOWMELT RUNOFF RELATION

The study of hydrographs of various discharge sites and the
meteorological data from the Pandoh catchment reveals that the arrival
time of the monsoon in the area is after 15th of June. The daily
discharge during months December-February is quite steady and touches a
minimum which can be used as the base flow. The excess daily discharge
over the base flow, in the days of no concurrent precipitation, can be
attributed to snowmelt. Cumulative runoff, excess over the base flow,
in the period between the date of Landsat coverage and 15th June (15th
June being the date up to which no significant concurrent rain
precipitation is observed) has been computed for various sites and for
various years. The values of snowcover area and subsequent snowmelt
runoff were plotted on semi-log paper. It has been found that a
separate rectilinear relationship exists for each of the subcatchments
(Fig. 5). It would be undesirable to try to fit a single generalized
common curve for all of the above points, as has been proposed by other
workers in other areas, like, by Odegaard and Ostrem (1977) for
Norwegian catchments, and Rango et al. (1977) for Indus and Kabul river
catchments in the Himalayas.

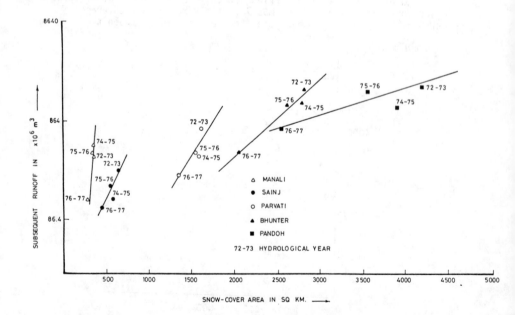

5.   Plot showing the relationship between snowcover area and
     subsequent snowmelt runoff up to 15th June - this being a date
     up to which no significant rain contribution occurs.

Effect of Geological-Geomorphological Factors

It is clear from Fig. 5, that the regression curves for smaller subcatchments are steeper than for larger subcatchments, i.e., for the same change in snowcover area, the snowmelt runoff is more in smaller subcatchments than in larger subcatchments. This seems to be because of the following reasons: (i) In smaller subcatchments, the slopes are generally steeper (owing to lower stream order) and hence snowmelt is more likely to reach the stream. In contrast to this, in larger subcatchments, slopes are generally gentler and percolation of snowmelt into ground may occur more readily. (ii) The smaller subcatchment being of lower stream order will have higher average altitude and therefore is likely to have greater snow thickness, in contrast to the larger subcatchment which will have lower average altitude and consequently thinner snow column.

Further, morphology of a subcatchment is controlled by the geology of the subcatchment and hence, an appraisal of the geology of the area seems necessary. The Pandoh catchment has been fairly well geologically mapped in recent years by different workers like Fuchs and Gupta (1971), Chaku (1972), and Frank et al. (1973). A simplified map after Frank et al. (1973) has been used in the following discussions. The distribution of various geological units in various subcatchments is shown in Fig. 6. Lithologically the various rock types exposed are: quartzites, gneisses, granites, schists, slates, marbles and metasedimentaries.

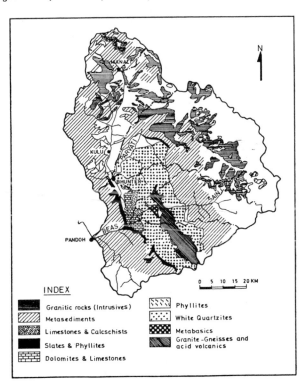

6. Geological map of the Pandoh catchment (simplified after Frank et al., 1973).

A look at the generalized geological map of the area (Fig. 6) shows that large parts of the catchments corresponding to Manali, Parvati and Sainj are occupied by competent rocks like quartzites, granites and gneisses. On the contrary the catchments of Bhunter and Pandoh comprise in addition less competent rocks like slates, phyllites and shales. It is well known that the competent rocks have a tendency to form steeper slopes, i.e., smaller subcatchments at higher elevations. On the other hand the less competent rocks will form gentler slopes resulting in larger subcatchments at lower elevations. This seems to be an important factor in controlling the snowcover and snowmelt runoff relation.

Hence, it can be generalized that morphological factors as controlled by geo-environmental conditions play an important role in shaping the slope of the regression curves for snowcover and snowmelt runoff relation, and that for each type of geologic subcatchment, a different empirical curve may be operative.

CONCLUSIONS

Based on the above work, the following broad conclusions may be drawn:

(i)   The snowcover area mapped from the Landsat images can be used for estimating snowmelt runoff expected in the ensuing snowmelt season.

(ii)   The relationship between snowcover area on Landsat images and subsequent snowmelt runoff is affected by topographical and geological factors and for each type of geo-environmental subcatchment, a separate empirical curve may exist.

ACKNOWLEDGEMENTS

Grateful thanks are due to Dr. G. Ostrem, Norwegian Water Research and Electricity Board, Oslo, for inspiration. Dr. V. K. S. Dave, Professor of Geology, Roorkee offered fruitful comments. Mr. S. N. Rao, Research Scholar, Department of Earth Sciences, helped in computations. The meteorological and river discharge data was made available by the Bhakra-Beas Management Board.

REFERENCES

Chaku, S. K., 1972, Geology of Chauri Tehsil and Adjacent area, Chamba Dist., H. P. Himalayan Geology, No. 2, p. 404-414.

Frank, W., Hoinkes, G., Christcine Miller, Purtscheller, Richter, W., and Thoni, M., 1973, Relation between metamorphism and orogeny in a typical section of the Indian Himalayas. Tschermaks Min. Petr. Mitt., Springer-Verlag, p. 303-312.

Fuchs, G., and Gupta, V. J., 1971, Palaeozoic stratigraphy of Kashmir, Kishtwar and Chamba (Punjab Himalayas). Verh. Geol. Bundesanst, Vienna, p. 68-97.

O'Brien, H. W., and Munis, R. H., 1975, Red and near infrared spectral reflectance of snow. In: A Rango (ed), Operational Applications of Satellite Snowcover Observations (NASA SP-391), Scientific and Technical Information Office, NASA, Washington, D.C., pp. 345-360.

Odegaard, H. A., Ostrem, G., 1977, Application of LANDSAT imagery for snow mapping in Norway. Norwegian Water Resources and Electricity Board, Publication, pp. 1-61.

Rango, A., Salomonson, V. V., Foster, J. L., 1977, Seasonal streamflow estimation in the Himalayan region employing meteorological satellites. Water Resources Research, Vol. 13, No. 1.

Singh, R., and Mathur, B. S., 1976, Snowmelt estimation of the Beas catchment using meteorologic parameters. Proc. IITM Symposium on Tropical Monsoons. Pune, 8-10 Sept. 1976.

Thomas, I. L. and Lewis, A. J., 1978, Snow field assessment from Landsat. Photogrammetric Engineering and Remote Sensing 44(4), pp. 493-502.

Y

w

)

6

Y

w

6